THE 125 BEST BIRDWATCHING SITES IN SOUTHEAST ASIA

Wang Bin

Azure-breasted Pitta, seen here in the lowland forests of Mindanao, the Philippines.

THE 125 BEST BIRDWATCHING SITES IN SOUTHEAST ASIA

Edited by
Yong Ding Li & Low Bing Wen

Contributing authors

Abdelhamid Bizid
Agus Nurza Zulkarnain
Alpian Maleso
Anuar McAfee
Ayuwat Jearwattanakanok
Bim Quemado
Carlos N. G. Bocos
Carmela P. Española
Chairunas Adha Putra
Cheong Weng Chun
Chi'en C. Lee
Choy Wai Mun
Colin R. Trainor
Con Foley
David Blair
Frédéric Goes
Hanom Bashari
Heru Cahyono
Ingkayut Sa-ar
Irene Dy
Janina Bikova
Khaleb Yordan
Le Manh Hung
Lim Kim Seng
Lisa M. Pangutulan
Low Bing Wen
Marc Thibault
Mark J. Villa
Mikael Bauer
Pete Simpson
Philippe Verbelen
Pyae Phyo Aung
Robert O. Hutchinson
Shita Prativi
Sofian Zack
Souvanhpheng Phammasane
Thet Zaw Naing
Wich'yanan Limparungpatthanakij
Wong Tsu Shi
Yann Muzika
Yeo Siew Teck
Yong Ding Li

JOHN BEAUFOY PUBLISHING

Dedicated to all the conservationists working tirelessly on the ground in Southeast Asia.

This edition published in 2018 by John Beaufoy Publishing
11 Blenheim Court, 316 Woodstock Road, Oxford OX2 7NS, England
www.johnbeaufoy.com

10 9 8 7 6 5 4 3 2 1

Photo Credits
Front cover: Bornean Bristlehead © Bjorn Olesen.
Back cover: clockwise from top left: Ward's Trogon © Yann Muzika; Vietnamese Greenfinch © Abdelhamid Bizid; Indigo-banded Kingfisher © Wang Bin; Chattering Lory © James Eaton; Red-eared Fruit Dove © James Eaton; Red-throated Barbet © Cheng Heng Yee.
Spine: Banded Kingfisher © Con Foley

ISBN 978-1-912081-52-3

Edited by Krystyna Mayer
Cartography by William Smuts
Design by Gulmohur Press
Project Management by Rosemary Wilkinson

Printed and bound in Malaysia by Times Offset (M) Sdn. Bhd.

CONTENTS

Foreword

I visited Southeast Asia for the first time in 1995, when I travelled to Tangkoko Nature Reserve in northern Sulawesi to spend time with two friends who were studying the park's hornbills and macaques. Back then there was no field guide to Sulawesi's birds, so my friends had made their own by photographing bird specimens in the collection of the American Museum of Natural History. And, of course, we had no sound recordings of the birds. I had a wonderful time nonetheless—I will never forget seeing 75 giant Knobbed Hornbills flying out of an immense fig tree, the whooshing of their wings sounding like the engines of a jet plane—but I kept wondering what I was missing.

Twenty years later I returned to Sulawesi with a group of birdwatchers, this time armed with a field guide to the birds of Wallacea and an iPhone full of bird songs. Our leader knew the best places to go, and we saw an astounding number of species, including many of the ones that had eluded me on my previous visit.

Without question, our knowledge of Southeast Asian birds has grown by leaps and bounds. Birdwatchers keen to study the region's avifauna now have the benefit of fine field guides covering Indonesia, Malaysia, Singapore, Thailand, Borneo, Sumatra, and almost anywhere else they might care to go. A growing number of these books are available in languages other than English, allowing people from throughout the region to enjoy their local birds. The songs and calls of virtually every species are also now freely available online.

But you still have to know *where* to go to find the birds. This huge void has now been filled by this splendid book, edited by Yong Ding Li and Low Bing Wen, two of Asia's preeminent field ornithologists. By enlisting the participation of experts from across the region, Yong and Low have produced the definitive bird-finding guide to Southeast Asia, giving directions to all the key sites, information on what birds to expect and tips on how to find them.

With the publication of this book, there has never been a better time to go in search of birds in Southeast Asia. And, sadly, there may never be a better time. The region's growing economies and human population have placed unprecedented pressure on natural ecosystems, resulting in rampant habitat destruction and overexploitation of many plants and animals. The numbers are sobering. In recent years, Southeast Asia has had the highest rate of deforestation of any tropical region. Indonesia also has the second highest number of imperilled bird species in the world (eclipsed only by Brazil). The Bali Myna teeters at the brink of extinction, a victim of the cage-bird trade, while the surreal Helmeted Hornbill was recently declared Critically Endangered due to relentless hunting for its casque. One need only stand atop a hill in Borneo, seeing nothing but oil-palm plantations from horizon to horizon instead of the forests that once were there, to understand the gravity of the situation.

Yong and Low are keenly aware of these dangers. For each site, they provide a concise summary of the threats it faces, from illegal logging to poaching and pollution. Implicitly, then, the authors are encouraging birdwatchers to see not only the birds, but also the threats to them. The book is both a bird-finding guide and a call-to-arms for those concerned about the region's biodiversity.

This raises the inevitable question–what can birders do to help? One easy way is to archive your bird observations in a database like eBird, so that they can be used now and in the future to monitor population trends, identify important areas to protect and record changes in distributions of various species. Another essential step is to support local and international conservation organizations that are trying to protect the region's biodiversity. There are plenty of organizations staffed with dedicated people who are trying to make a difference; they deserve and need your support. Finally, think about your own actions as they may affect Southeast Asian birds and biodiversity in general. For example, whether you are buying lumber to build a house or a chocolate bar for a snack, try to find out where the raw materials came from and whether they were grown and harvested in a sustainable way. Urge your friends to do the same.

The 125 Best Birdwatching Sites in Southeast Asia is your ticket to adventure and to all the joys that come from seeking some of the world's most beautiful and interesting birds. But it is also a wake-up call, a reminder of all that must be done to save those birds—and so many other species—while there is still time.

David S. Wilcove
Professor of Ecology, Evolutionary Biology, and Public Affairs
Princeton University

Introduction

Southeast Asia is a region steeped in culture and a history dating back to ancient times. This diverse region straddles the Asian tropics between south China, eastern India and Australia, and is one of the world's great bastions of biodiversity. Southeast Asia overlaps with four of the world's most important biodiversity hotspots, and collectively is home to more than 20% of the world's flora and fauna. More than 2,500 bird species have been recorded in Southeast Asia's 11 countries, a diversity equalled by few other parts of the globe. Many Southeast Asian countries, such as Indonesia, Myanmar and Thailand, have national lists in excess of 1,000 species. In addition, more than 1,000 bird species are unique to the region, and many of these are downright bizarre or charismatic. Some of the best examples include the **Helmeted Hornbill**, **Bornean Bristlehead**, **Philippine Eagle** and **Standardwing**.

With so many birds to be found and such mind-boggling levels of biological and cultural diversity, Southeast Asia ought to be on the radar for every serious birdwatcher. However, many birdwatchers planning to visit Southeast Asia quickly encounter two major impediments. First, as the region's economies develop and its population soars well past 600 million, forests are rapidly being cleared, rivers dammed and swamps drained, and in their places spring factories, cities, plantations and roads. Second, Southeast Asia's birdlife is poorly studied and new species continue to be described yearly. Historically, there has been a very limited selection of good bird books that one takes along on a birdwatching trip to this part of the world, although this is rapidly changing as new bird guides to Southeast Asia appear every year. Rarer still are books that provide a broad overview of the region's distinctive birdlife or its key birdwatching sites.

The 125 Best Birdwatching Sites in Southeast Asia was conceived to fill the gaps in information in book form, and is the result of numerous conversations with our peers working in the region, whether as birdwatching guides, bird-tour leaders, researchers or conservationists. It is the collective effort of many authors, all with intimate knowledge of the places they write about. We are also grateful to the individuals who have shared their knowledge or images of the places featured here. We hope that the book will not only intrigue armchair naturalists with the remarkable birdlife of Southeast Asia, but also encourage birdwatchers and general ecotourists to visit the region, and in so doing create the economic incentives needed to protect its incredible biodiversity. While the book is not intended to be a photographic guide to the birdlife of Southeast Asia, it includes images of more than 550 bird species, including many of the region's rarest and most elusive.

Organization of the Book

This book is divided into three sections, namely the introduction, the systematic section and the appendices, which contain useful practical information.

The introductory section provides a broad overview of Southeast Asia's geography, climate and habitats, and a synopsis of the key bird families that occur in the region. The systematic section provides details on each of the 125 sites across the region. Each profile describes the nearest major towns to the birdwatching site, key bird species and other wildlife, and the best time of year to visit. This information is by no means exhaustive, and some details may have changed by the time of publication. A list of the key bird species for each site is included, although these lists do not capture every species that occurs at each site. Birds that are endemic to a particular country have an asterisk next to their names. For consistency in names, the nomenclature and taxonomy recommended by the International Ornithological Union's checklist (v.7.3) has been adopted, which we consider to be the most authoritative source of bird names.

The appendices contain useful information for birdwatchers and naturalists planning visits to the region, including contacts of local birding groups and tour operators.

Chi'en C. Lee

The bizarre ***Standardwing*** *bird-of-paradise of the northern Moluccan islands captured the imagination of the naturalist Alfred Russel Wallace.*

Southeast Asia's Climate

Much of Southeast Asia experiences a subtropical to tropical climate characterized by high mean annual temperatures and rainfall throughout the year. However, depending on the time of year and the part of the region visited, there are bound to be distinct differences in the local climate. As you travel further north from the equator and further east in the region, rainfall patterns become increasingly seasonal. Torrential downpours, the bane of any birdwatcher, are strongly influenced by the monsoons (north-east and south-west), which vary depending on the time of year and local geography. In many mountainous regions, cloudy and misty conditions usually start to develop from late morning, eventually leading to bouts of showers.

Here, Southeast Asia has been divided into five regions based on the broadly similar climatic conditions encountered in each region.

Continental Southeast Asia

This region, comprising the countries of Myanmar, Laos, Cambodia, Vietnam and much of Thailand, experiences a tropical monsoonal climate. Residents in this part of the world tend to highlight three distinct seasons:

Cool and dry During the north-east monsoon between November–March, cold, dry winds blow from the Himalayas and China towards the South China Sea, reducing temperatures, humidity and rainfall across much of the region. This is one of the best times to visit the region for birdwatching.

Hot and dry The inter-monsoon period in March–May is characterized by light winds and high daytime temperatures. This is the hottest time of the year in many parts of Thailand, Myanmar, Cambodia and Vietnam.

Hot and rainy The rainy season coincides with the south-west monsoon between June–September, as winds blowing in from the coast carry with them moisture-laden clouds. Thunderstorms and extended periods of torrential rain are common, resulting in frequent flooding and landslides. Some national parks in the region may be closed during this period.

Thai-Malay Peninsula & the Greater Sundas

This region, comprising peninsular Thailand as well as Malaysia, Singapore, Brunei and western Indonesia, experiences an equatorial climate with high temperatures, humidity and rainfall throughout the year. However, rainfall is usually most pronounced during the north-east monsoon between November–March, when winds blowing down from continental Asia pick up moisture over the South China Sea and lead to long periods of torrential rain. Most birdwatchers planning trips to Malaysia and western Indonesia try to avoid the monsoon months at the end of the year. Conversely, the change in wind direction towards mainland Southeast Asia between June–September coincides with comparatively drier periods with shorter spells of torrential rain.

Philippine Islands

The islands of the Philippines experience considerable variation in climate depending on their position in the archipelago. The Philippines, while broadly described as having a tropical climate, experience increased heavy rainfall at different times of year throughout the numerous islands. Simply put, the northern half of the country is wetter during the south-west monsoon, while the southern half is either wetter during the north-east monsoon, or experiences nearly consistent rainfall through the year. Due to its position on the Pacific seaboard, much of the Philippines is exposed to the damaging tropical storms and typhoons that blow west towards China and Vietnam from the Pacific between June–November.

Wallacea

Climatic patterns are variable across Wallacea. Northern Sulawesi and the north Moluccas experience an equatorial climate, with no major fluctuations of temperature or rainfall throughout the year. Southern and central Sulawesi, as well as the southern Moluccas, tend to have a more seasonal climate, with distinct wet and dry seasons. The period between November–April is particularly wet and usually avoided by birdwatchers.

The Lesser Sundas are one of the most arid parts of Southeast Asia, with many areas receiving less than 1,200mm of rainfall. Most of the islands experience very distinct wet and dry seasons. In Flores and Timor the months of April–October are usually the driest time of the year, and they are visited by many birdwatchers during this period.

New Guinea

New Guinea's proximity to the Equator and the tropical waters of the Western Pacific ensures an equatorial climate with high rainfall throughout the year. Much of Indonesian New Guinea receives well over 2,500–4,500mm of rainfall annually, with most rain falling between January and April, and less in other months. Despite this, many locations across the island show little to no seasonal variation in precipitation and rain, especially in the form of thunderstorms, all year round. There are a number of areas of rain shadow across the island, usually adjacent to mountains (like the Cyclops Range, the lowlands south of the Central Cordilleras) where rainfall is markedly lower. Much of New Guinea is not on the path of the many destructive cyclones that affect northern Australia.

Southeast Asia's Geography

Southeast Asia's birdlife is the result of the region's complex biogeography, arising from an interplay between tectonic activity, climatic patterns and changing sea levels over time. Mountains and islands are particularly important in shaping the region's birdlife by creating physical barriers to species movements and driving speciation, while providing new niches for species to colonize. At various periods during the Pleistocene, sea levels were lower and many of the major islands of present-day western Indonesia on the Sunda Shelf and the Thai-Malay Peninsula were united as a larger land mass through various exposed land connections – Sundaland. This allowed bird species to disperse widely, and explains why the avifauna of Borneo, Sumatra and the Thai-Malay Peninsula is so similar. Similarly, New Guinea and its satellite islands were linked to Australia on the Sahul Shelf when sea levels were lower, and thus share a broadly similar avifauna.

Where the seas are deep, especially off the continental shelves, many islands remained as islands even when sea levels were lower during the ice ages, hampering the movement of plants and animals. Situated between Borneo and New Guinea is Wallacea, a well-known biogeographic region named after the eminent naturalist Alfred R. Wallace. Wallacea's island groups are spread over deep seas sandwiched between two continental shelves. Therefore, even when sea levels were lower in the past, Sulawesi, the islands of Maluku and the Lesser Sundas were never bridged to larger land masses to the east or west. Over many millennia of isolation, many endemic species have evolved on these remarkable islands.

Islands

The region from the Thai-Malay Peninsula east to the islands of Wallacea and New Guinea is often defined by geographers as 'insular Southeast Asia'. The vast island archipelagos of Indonesia and the Philippines that dominate insular Southeast Asia are collectively the largest clusters of islands on the planet, spanning an arc of nearly 5,000km between the Indian and Pacific Oceans. These islands are exceptionally important for birds and form multiple centres of bird endemism. Many islands larger than 500km^2 contain at least one endemic bird species, but even small islands like Boano, Tanahjampea, Damar and Calayan have their own endemics.

Southeast Asia's largest islands are New Guinea, Borneo, Sumatra, Sulawesi, Java, Luzon and Mindanao. Borneo, which contains the territories of three nations, is the third largest island in the world and is itself geologically complex, with tall mountains, long rivers and

Yong Ding Li

The Lesser Sundas, or Nusa Tenggara, feature thousands of islands between Lombok and the Tanimbars. Seen here is Pulau Komodo, with Pulau Padar and Rinca in the background.

Bjorn Olesen

The Raja Ampat islands off north-west New Guinea are characterized by extensive limestone formations, including cliffs that plunge steeply into the sea.

vast rainforests (until recent history). New Guinea, on which the easternmost provinces of Indonesia lie (Papua and West Papua) is larger, more rugged and supports some of the largest contiguous rainforests left in the world. Java (Indonesia) and Luzon (the Philippines) are among the most populous islands on the planet, supporting densities of 1,100 people/km^2 and 440 people/km^2 respectively. Further east, the convoluted 'K'-shaped island of Sulawesi is far less populated than Java, but is one of the most mountainous islands in the world, with chains of mountains running through all its main peninsulas.

Beyond the large islands there are many thousands of smaller islands scattered from the northern tip of Sumatra (like Weh and Nasi), to the Aru, Kai and Tanimbar clusters further east, and the Raja Ampat Islands off north-western New Guinea. The smaller Philippine islands start from the Batanes cluster in the far north, through the Visayas, and end in the Sulu group near the tripoint of Malaysia, Indonesia and the Philippines. Many of Southeast Asia's smaller islands are seldom visited and are likely to be the scenes of future species discoveries.

Mountains

Southeast Asia is characterized by mountainous landscapes, with mountains forming a constant feature of most countries on continental Southeast Asia. The tallest peaks, including Hkakabo Razi at 5,881m asl, together with a number of other surrounding peaks (like Gamlang Razi) exceeding 5,000m asl, are concentrated in northernmost Myanmar along a sub-range of the Eastern Himalayas that sweeps east into China's Yunnan province. The tectonic forces driving the rise of the Himalayas are also responsible for the nearly contiguous chains of low mountains on Myanmar's western frontier with India (like the Patkai-Naga Hills), peaking at Saramati (3,826m asl).

From the Eastern Himalayas a number of lofty sub-ranges sweep east into northern Vietnam, reaching 3,143m asl at Fan Si Pan on the Hoàng Liên Son Range. A series of low mountains forms a rugged bulwark along the long Thai–Burmese border (Tenasserim-Dawna or Tanaosri Hills), and eventually terminates at Peninsular Malaysia's Titiwangsa Range. Another major chain, the rugged Tru'ò'ng So'n (Annamite) Mountains, runs along nearly the entire length of Vietnam, forming a natural barrier between Laos and Vietnam. The Tru'ò'ng So'n reaches its highest point at Phou Bia (2,819m asl) in the north in Laos, and Ngọc Linh (2,598m asl) in the south in Vietnam,

Thet Zaw Naing

The snow-capped Burmese Himalayas near Hponkan Razi.

The Trường Sơn (Annamite) Mountains as seen from east-central Laos.

before slowly giving way to the Đà Lạt Plateau and the lowlands of the Mekong Delta.

All of Southeast Asia's largest islands have mountainous interiors. New Guinea is bisected by an imposing central cordillera consisting of the Jayawijaya and Sudirman ranges, which soars to 4,884m on Puncak Jaya. Borneo is also very rugged, and comprises a number of mountain ranges in its interior, the highest being the Crocker Range, which rises to 4,098m asl at the summit of Gunung Kinabalu. Sumatra's western backbone is framed by the Barisan Mountains, a long range that extends for nearly 2,000km and peaks at Gunung Kerinci (3,805m asl). The islands of Sulawesi, Java, Luzon and Mindanao are very mountainous, and all contain peaks exceeding 2,500m asl. Sulawesi in particular is geologically complex, having been formed from the collision of land masses originating from the Asian and Australian continents. It features many high peaks of more than 3,000m asl (like Gunung Rantemario) and deep lakes over -500m bsl (as in the Malili Lake system). For birdwatchers, many of Southeast Asia's mountains are associated with a high degree of avian endemism and are popular sites to visit.

Rivers & Lakes

A number of major rivers, with their sources in either the Himalayas or the eastern Tibetan Plateau, flow through Southeast Asia. The longest and best known is the Mekong River, which originates in the Tibetan Plateau in China and flows through all five countries of mainland Southeast Asia, before entering the South China Sea in Vietnam. While its significance to freshwater biodiversity is well known, the Mekong also plays a major role in shaping the avifauna of mainland Southeast Asia, notably the region's waterbirds. The Mekong is the main source of water for the great lake of Tonlé Sap in Cambodia, and the extensive wetland habitats that fringe the lake contain some of the largest congregations of breeding waterbirds left in the region.

After the Mekong, the next two longest rivers in Southeast Asia flow through Myanmar. Like the Mekong, the Salween River originates in the eastern Tibetan Plateau and snakes its way through the mountains of eastern Myanmar along the Shan and Karen Hills, and western Thailand. Flowing nearly parallel to the Salween in the west is the Irrawaddy River. Although the middle and lower reaches of the Irrawaddy have been degraded by human activities, its upper tributaries contain large areas of wetland

The Mekong River snakes through impressive karst landscapes in Laos.

and riparian forests that are important to many threatened species, including the **Masked Finfoot**, **Green Peafowl** and **White-winged Duck**. Some of this important wetland is protected in the Hukawng Valley Tiger Reserve, the largest protected area in continental Southeast Asia.

Numerous rivers and streams drain the islands of Southeast Asia. Many of the larger rivers in Borneo (like the Kapuas, Mahakam, Rajang and Kinabatangan), Sumatra (Batanghari), Sulawesi (Walanae), Luzon (Cagayan) and Mindanao (Agusan) are connected to major areas of shallow lakes, wetland and riparian vegetation that are of great importance to waterbirds. For example, the Kapuas River system in Indonesian Borneo is linked to the extensive freshwater wetlands of the Danau Sentarum lake system, which support large numbers of waterbirds. The Mamberamo River (670km) in New Guinea is Indonesia's longest river. It is one of the few large, undammed rivers left in the world, and flows through a watershed that remains mostly covered in pristine rainforest.

Bird Habitats

Spanning a wide latitudinal range of 28° N to 10° S, Southeast Asia features an impressive diversity of tropical and subtropical biomes, packed into an area half the size of Brazil. The region is located almost entirely within the tropics. Consequently, the dominant natural vegetation type is broadleaved evergreen forests, particularly lowland rainforest and montane forest. These habitats are of great interest to birdwatchers and naturalists, given their high levels of species diversity and endemism. While Southeast Asia's forested landscapes appear superficially similar to the untrained eye, there is considerable variation in plant communities and forest structure across the region as a result of regional variations in climate, soil substrate and biogeography. Described here are the most important vegetation types and habitats in the region, and some of their representative bird species.

Lowland & Hill Rainforests

Among the various forest types in Southeast Asia, lowland and hill rainforests support the highest plant and animal diversity in the Old World hectare for hectare. Rainforests are characterized by high and relatively aseasonal distribution of rainfall, although those at higher latitudes on mainland Southeast Asia are subject to more seasonal rainfall patterns, and thus contain a more prominent representation of deciduous tree species. Southeast Asian rainforests, especially those in the western half of the region, are dominated by trees from the dipterocarp family and are thus widely referred to by botanists as 'dipterocarp forests'. These lowland and hilly rainforests comprise the dominant vegetation on the Thai-Malay Peninsula, Borneo, Sumatra, the Philippine Islands and New Guinea. Due to their accessibility and the high commercial value of many timber trees, much of these lowland forests have been cleared or are degraded by logging.

Tropical Asian bird communities reach their highest diversity in lowland rainforests, and more than 200 species may reside in a square kilometre of virgin forest on the Thai-Malay Peninsula, Sumatra or Borneo, including trogons, pittas, cuckoos, babblers, bulbuls and sunbirds. Some lowland rainforest sites, like Peninsular Malaysia's Belum-Temenggor forest complex and adjacent parts of southern Thailand, support 15 woodpecker and 10

Yong Ding Li

Lowland dipterocarp forest in the Malaysian state of Terengganu.

hornbill species, the highest diversity for both families within a single area anywhere in the world. Towards the Philippines and Wallacea, groups like babblers and woodpeckers are poorly represented, but pigeons, kingfishers, parrots and whistlers become more prominent. A lowland rainforest on the southern Philippine island of Mindanao, for example, can be expected to host more than 100 resident species, with pigeons, parrots and kingfishers being particularly well represented. These include endemics such as the **Pink-bellied Imperial Pigeon**, **Hombron's** and **Winchell's Kingfishers** and **Blue-crowned Racket-tail**. Likewise, rainforests on Sulawesi and its outlying islands support diverse assemblages of parrots, pigeons, kingfishers and cuckooshrikes.

Further east on the island of New Guinea, lowland rainforests are extremely species-rich, and in many cases, support bird assemblages in excess of 200 species. New Guinea's lowland bird communities share many groups with Wallacea (such as parrots, pigeons and kingfishers), but also support many Australasian groups not found further west such as pitohuis, jewel-babblers and many more birds-of-paradise.

Peat & Freshwater Swamp Forests

Swamp forests occur primarily on low-lying riparian and coastal areas, and are subjected to regular inundation. Peat-swamp forests are most extensive in low-lying parts of coastal Borneo and eastern Sumatra, with smaller patches in north-west Sumatra, the Thai-Malay Peninsula (south-east Pahang) and Sulawesi (like Rawa Aopa).These forests are distinguished by their deep, highly combustible peat substrates formed from the accumulation

The Kluet peat swamp in Aceh is one of the last remaining peat swamps on the west coast of Sumatra.

of plant matter over thousands of years. Peat swamps are drained by numerous blackwater streams and rivers. A handful of Sundaic species are strongly associated with peat swamps, particularly the **Hook-billed Bulbul** and **Black Partridge**.

Freshwater swamp forests, on the other hand, have a diversity of flora similar to that of adjoining lowland rainforests. In some parts of mainland Southeast Asia a distinct type of swamp-forest tree, the Paperbark tree (*Melaleuca cajuputi*), is a major species, while in parts of the Moluccas and New Guinea these forests are dominated by pandans and palms such as the Sago Palm (*Metroxylon sagu*) and is home to various endemic rail species. Some of the best examples of freshwater swamp forests in the region are found around the Tonlé Sap Lake in Cambodia, and coastal low-lying areas of the Malay Peninsula, Sumatra and Borneo.

Heath Forests

Heath forests, or keranggas, as they are known in parts of the region like Malaysia, develop on sandy soils with low nutrient content. Here, plant diversity is lower, and trees are thinner and of lower stature. Heath forests are best represented on Borneo, and to a lesser extent in Sumatra and Peninsular Malaysia. No birds are restricted to heath forests, but a few species are more regularly encountered here than in other forest types, notably the **Scarlet-breasted Flowerpecker** and **Grey-breasted Babbler**.

Montane Evergreen Forests

*Southern conifers (*Podocarpus *spp.) are characteristic trees in Timor's beautiful montane forests.*

The term 'montane evergreen forest' is broadly used here to refer to broadleaved forests on uplands, usually above 800m asl, and can be further subdivided into lower montane, upper montane and subalpine forests. Montane

Montane rainforest on Sumatra's Gayo Highlands in Aceh.

evergreen forests remain fairly extensive on the mountain ranges of mainland Southeast Asia, growing to an altitude of over 3,500m asl on the Burmese Himalayas. On various large islands in the region, montane forest occurs to beyond 2,500m asl. Examples include Kinabalu (Borneo), Kerinci (Sumatra), Apo (Mindanao), Pulag (Luzon) and Rantemario (Sulawesi). Further east, montane forest soars to nearly 3,800m in the Central Cordilleras of Indonesian New Guinea. Montane forests are dominated by different assemblages of plants from those of hill and lowland forests, with notable representation of *Pandanus* spp., Fagaceae (oaks and chestnuts), Lauraceae (laurels), Myrtaceae and Podocarpaceae (like *Dacrydium* spp.). On Timor and New Guinea, southern conifers (like *Podocarpus and Nothofagus* spp.) are a prominent component of montane evergreen forests.

Depending on the degree of rainfall seasonality, altitude and disturbance, Southeast Asia's montane forests may support either mostly evergreen vegetation, or a mixture of evergreen and deciduous elements. With greater cloud cover and elevation, montane forests often become increasingly stunted, and individual trees support a multitude of epiphytes and bryophytes, thus the terms 'mossy forest' and 'cloud forest'.

The bird communities of much of Southeast Asia's montane forests show strong Sino-Himalayan influences (nearly all genera are shared with the Himalayas), with many species of partridge and pheasant, nuthatch, leaf warbler and laughingthrush. In Wallacea, montane bird communities lack laughingthrushes, nuthatches and partridges, while groups with Australasian affinities like whistlers, parrots, honeyeaters and white-eyes become more prominent, together with a handful of leaf warblers, flycatchers (*Ficedula* spp.) and one species of wren-babbler (**Pygmy Wren-babbler**). Bird endemism is more pronounced in montane forests compared with the lowlands, and many of the region's mountain ranges harbour restricted range species. Mountains on the Đà Lạt and Di Linh Plateaus in southern Vietnam, for example, host at least six threatened species with very small ranges, like the **Grey-crowned Crocias** and **Collared Laughingthrush**. Peninsular Malaysia's mountains are home to four restricted-range species, including the **Malayan Whistling Thrush** and **Mountain Peacock-Pheasant.** Similarly, further east in Indonesian New Guinea, the majority of regional endemics are confined to montane forests, including charismatic species such as the **Western Parotia** and **Arfak Astrapia**.

Coniferous Forests

In some upland parts of Southeast Asia characterized by strongly seasonal rainfall and regular bushfires, extensive stands of coniferous forests can be found. These are often dominated by nearly pure stands of *Pinus kesiya* (Benguet Pine, as it is known on Luzon) or *P. merkusii*. Some of the best examples can be found in Luzon's Central Cordillera, in the Shan Hills on the Thai-Burmese border (like Doi Chiang Dao), and on Laos' Nakai Plateau. Although the avifauna of these forests tends to be species

Yong Ding Li

The coniferous forests on Vietnam's Đà Lạt Plateau.

poor, they support distinctive communities of nuthatches, woodpeckers, tits, bushtits and many finches, notably the **Red Crossbill** and various greenfinches (*Carduelis* spp.).

In the Burmese Himalayas, especially on Hponkan and Hkakabo Razi, coniferous forests dominated by firs (*Abies* spp.) and hemlocks (*Tsuga* spp.) extend well above 3,000m asl. The bird communities of these forests are rich in leaf warblers, tits, woodpeckers and finches, and not unlike those in neighbouring parts of India and China.

Mixed & Dry Deciduous Forests

As rainfall becomes more seasonally distributed and the dry season lengthens, deciduous tree species, including the ubiquitous Teak (*Tectona grandis*), become an increasingly prominent component of broadleaved forests. Human disturbance and natural fires also favour deciduous species. Mainland Southeast Asia, Java and the Lesser Sundas are most strongly affected by seasonal variations in rainfall due to the monsoons, and are where such deciduous forests are most extensive.

Parts of Myanmar, Thailand, Cambodia and Laos are covered in a distinctive type of deciduous forest known as 'dry dipterocarp forest'. They are subject to a particularly long dry season and are dominated by a few species of deciduous dipterocarps. Given the regular incidence of forest fires, the understorey tends to be grassy and stands of bamboos are common. Bird communities in these deciduous forests are characterized by woodpeckers (like the **Black-headed Woodpecker**), starlings, and corvids like jays, treepies and magpies. In northern Cambodia dry dipterocarp forests are important habitats for Critically Endangered waterbirds like **Giant** and **White-shouldered Ibises**, which primarily breed and forage within them. In dryer parts of the Philippines these seasonally dry forests are characterized by the Molave tree (*Vitex* sp.), and are thus known as Molave forests.

Where rainfall is higher, moist deciduous forests occur and are characterized by many more evergreen tree species. The lowland forests of Timor, Flores and the Tanimbars are classic examples of moist deciduous forests and support a relatively rich birdlife, including many birds with Australasian affinities such as whistlers, honeyeaters, parrots and pigeons.

Frederic Goes

Dry dipterocarp forest in Cambodia's Preah Vihear province.

Karst (Limestone) Forests

Karst forests are most extensively distributed in areas on hilly limestone substrates across mainland Southeast Asia, and to a lesser extent throughout the rest of the region. Forests on limestone substrates tend to be of a shorter stature and trees are generally thinner. These forests support a rich herbaceous understorey. Some of the best and most extensive examples of limestone forests in the world can be found in central Vietnam and adjacent parts of Laos, especially in the Phou Hin Poun and Phong Nga–Kẻ Bàng protected landscapes. Besides many bird species shared with rainforests, species restricted in distribution to karst forests include the **Sooty Babbler**, **Limestone Wren-Babbler**, and the recently described **Limestone Leaf Warbler** and **Bare-faced Bulbul**, which underscores how poorly studied the fauna of limestone forests is.

Mangrove Forests and Mudflats

Mangroves are widespread across Southeast Asia's long coastlines, but are most extensive in low-lying areas of Sumatra, Borneo and New Guinea drained by large, slow-flowing rivers.

Low Bing Wen

The Phou Hin Poun National Biodiversity Conservation Area in Laos contains some of the largest areas of karst forest in the region.

Yong Ding Li

Monsoon forest on limestone in Yamdena, Tanimbar Islands.

On parts of Sumatra, for instance, mangrove forests form an almost contiguous belt and can extend more than 35km upriver from the coast on the Banyuasin Peninsula. While mangrove forests generally support low plant diversity, Southeast Asia's mangrove flora is recognized by botanists to be the most species rich among the world's mangrove forests. Mangrove bird communities are low in diversity, but a number of species are restricted to this habitat, notably the Copper-throated Sunbird, Brown-winged Kingfisher and Mangrove Pitta to name a few. Additionally, the large expanses of tidal mudflats that often abut mangrove forests are particularly important to large wintering flocks of shorebirds and terns, as well as uncommon resident waterbirds like the Great-billed Heron, Milky Stork and Lesser Adjutant.

Low Bing Wen

Mangrove creeks at Tangkoko Nature Reserve, Sulawesi.

Coastal Beach Forests

Beach forests occur mostly 5–50m from the coast and can be found throughout Southeast Asia's coastline in the absence of mangroves. On small islands beach forests are usually the dominant form of vegetation. Most of the region's beach forests have been cleared and replaced with stands of coconut or coastal resorts. On the Thai-Malay Peninsula and the Greater Sundas, bird communities of beach forests comprise a few species of bulbul, sunbirds, tailorbirds, babblers, and pigeons such as the Nicobar Pigeon and Pied Imperial Pigeon. Some of the best examples can be found on the island groups fringing western Peninsular Thailand and Malaysia (like Langkawi, Tarutao and Ko Phi Phi), the islands off northern Borneo and the Philippines. On small islands in the Moluccas, several small island specialists can be found, including the Beach Kingfisher, Grey Whistler and Olive Honeyeater.

Dry Scrub Forests & Xerophytic Scrub

Low-stature scrub forest and xerophytic (semi-desert) scrub occur in the driest areas of central Myanmar, and are best exemplified by the vegetation around Bagan. These scrubby habitats are dominated by short trees and shrubs, including many deciduous species, as well as dry-adapted plants with reduced leaves (like *Acacia*), or are succulent. A handful of species are endemic to this habitat, including the Hooded Treepie, White-throated Babbler and Jerdon's Minivet.

Natural Grasslands & Savannah

These habitats occur in parts of Southeast Asia that experience pronounced dry seasons and are subject to regular fires, flooding or a combination of both. Grasslands in the region often support sparse tree cover, represented by *Acacia*, *Eucalyptus* (Timor and surrounding islands) and various palms like *Borassus*. Some of the best examples of natural grasslands can be found in eastern Java, the Lesser Sundas (as in Timor and Komodo), parts of Cambodia where it intergrades into dry dipterocarp forests, and the upper tributaries of the Irrawaddy River. For example, riparian strips along Myanmar's Chindwin River are extensively covered with floodplain grassland, broken by patches of forests. These grasslands are particularly important for species like Jerdon's Bush Chat, Green Peafowl, finches, and various large waterbirds and vultures. The grasslands that fringe Cambodia's Tonlé Sap Lake are the last stronghold of

Thet Zaw Naing

Indawgyi Lake is one of the most important wetlands for wintering ducks in Myanmar.

the Critically Endangered Bengal Florican. Natural grasslands are also important wintering habitats for a variety of raptors and passerines, such as the globally threatened Greater Spotted Eagle and Manchurian Reed Warbler.

Alpine Shrubland & Grassland

Alpine shrubland and grassland occurs above the treeline of the highest mountains in the region, and usually above 3,800m elevation. Alpine shrubland is characterized by low-stature vegetation dominated by rhododendrons, *Impatiens*, *Vaccinium* and various grasses. In the Burmese Himalayas, these high-elevation habitats support the Alpine Accentor, tits, pipits and various finches. In Indonesian New Guinea, species such as the Snow Mountain Quail and Western Alpine Mannikin are dependent on such habitat.

Freshwater Wetlands

These are widely distributed across low-lying parts of Southeast Asia, especially on the fringes of waterbodies like floodplain lakes and rivers. Depending on the depth and frequency of inundation, these lakes or rivers may be fringed by marshland, inundated grassland or swamp forests (described above). Naturally occurring freshwater wetlands in Southeast Asia are best exemplified by Cambodia's vast Tonlé Sap Lake and a number of smaller systems in Myanmar (like Inle and Indawgyi Lakes), Thailand (Chiang Saen and Bueng Borapet Lakes) and Vietnam (Tràm Chim National Park). Examples of such habitats can also be found across the Greater Sundas (as at Danau Sentarum), the Philippines (Agusan Marsh), Sulawesi (Danau Tempe) and Indonesian New Guinea on the riverine floodplains of the Trans-Fly region. The wetlands are particularly important for many waterbirds, as well as wintering shorebirds and passerines. Tonlé Sap Lake, for example, hosts globally significant colonies of the threatened Milky Stork, Greater Adjutant, Spot-billed Pelican, Sarus Crane and many other large waterbirds now rare across Southeast Asia.

Riverine Sand Bars

Extensive areas of sand bars can be found along some of the large rivers in mainland Southeast Asia. These unique habitats are important for waterbirds and swallows like the Great Stone-curlew, Small Pratincole, Black-bellied Tern and Grey-throated Martin. One passerine, the Mekong Wagtail, is nearly restricted in its distribution to riverine sand bars and adjoining patches of scrub. Such habitats used to be more extensive in the region, but are now limited to small stretches along the Mekong and Irrawaddy Rivers due to indiscriminate dredging of rivers, damming and agriculture.

Bird Migration in Southeast Asia

The entire region of Southeast Asia lies within the East Asian-Australasian Flyway, one of the world's great migratory corridors. The flyway spans the eastern third of the Asian continent, parts of Alaska, and extends southwards in a great arc into Australia, New Zealand and the islands of the Western Pacific. More than 550 species of migratory landbird and waterbird (or a quarter of Southeast Asia's birds) are known to use the flyway annually. Many species, like the shorebirds that breed in arctic wetlands and the warblers that dwell in the Siberian taiga, fly more than 5,000km to spend the northern winter in Southeast Asia. One of the best known, and perhaps most abundant wintering bird in Southeast Asia is the humble **Arctic Warbler**. This drab passerine bird is found across the taiga forests of Eurasia and Alaska. From September onwards, millions of the birds arrive in the region and can be seen anywhere from city parks to dense rainforests.

Depending on the latitude and various environmental and weather conditions, migratory birds arrive in Southeast Asia mostly from late July onwards. Shorebirds are often among the first migratory species to arrive in the region, showing up in good numbers at various wetlands from July. From September onwards, many waves of landbird migrants make their journeys across Southeast Asia. Described by birdwatchers as the 'autumn passage' or the 'migration' period, this is the best time to observe a good diversity of flycatchers, warblers and robins anywhere from northern Vietnam to Singapore, the islands of the Philippines, Malaysia and Indonesia. During poor weather, migratory landbirds often concentrate to rest in small patches of woodland and forests (or even ships in the middle of the sea), with the result that some of the best sites to see a good variety of species are small islands such as Ko Man Nai (Thailand), Mantanani (Malaysia), Batanes (the Philippines) and Con Lu (Vietnam). Many birdwatchers also make trips to city parks and remnant stands of vegetation in urban areas to observe migratory landbirds, particularly in Thailand (for example Rot Fai Park, Bangkok) and Vietnam (for example Hanoi Botanical Gardens).

For birdwatchers with an interest in birds-of-prey, late October to November is undoubtedly the best period to observe large numbers of migrating hawks, kites and buzzards. Perhaps the best site to see large numbers of migratory birds-of-prey in the region is Khao Dinsor, Chumphon in southern Thailand, Cape San Agusdin in Mindanao and Sangihe Island off northern Sulawesi. As many as half a million raptors may migrate through Khao Dinsor in autumn, including large congregations of **Crested Honey Buzzard**, **Black Baza**, **Grey-faced Buzzard** and **Chinese Sparrowhawk**, and many other landbird migrants. Some raptor species, such as the **Chinese Sparrowhawk**, will migrate as far as Indonesian New Guinea. Recent surveys have found wintering **Chinese Sparrowhawks** on Dolak Island on the southern coast of New Guinea.

Shorebirds are the best-known group of migratory birds, and also among the easiest to observe (with the aid of a telescope), since large, conspicuous groups of sandpipers and plovers can be seen in open coastal mudflats, sandy beaches, rocky shorelines, freshwater marshes and even man-made fish ponds and salt pans. The vast tidal mudflats (and their adjacent scrub and working wetlands) that fringe the Melaka Straits on the coasts of Peninsular Malaysia and Sumatra (Indonesia), the Inner Gulf of Thailand, and the Gulf of Mottama (on the Andaman sea coast of Myanmar) are the most important wintering areas for large congregations of wintering shorebirds in the region, including the highly endangered **Spoon-billed Sandpiper**, **Nordmann's Greenshank** and **Great Knot**.

Abdelhamid Bizid

*Shorebirds like the **Black-tailed Godwit** migrate vast distances from their breeding grounds in northern Asia to winter on the coasts of Southeast Asia.*

Bird Conservation Issues

Southeast Asia's bird species are imperilled by a multitude of threats. As the human population in the region soars, demand for forest products, timber and agricultural land continues to rise at the expense of the wilderness. In the early 20th century, more than four-fifths of the region's area was covered by forests, swamps and natural grasslands. However, by the late 2000s, worrying estimates made from satellite imagery show that less than half of Southeast Asia's original forests still stand. As old-growth forests continue to fall to loggers, biodiversity-poor monocultures of oil palm, acacia and rubber trees take their place. In addition, the harvesting of wild bird populations for the flourishing pet-bird trade and human consumption places enormous pressures on the populations of some species. Those of species highly sought after by the bird trade, like the **Black-winged Starling**, **Straw-headed Bulbul**, **Java Sparrow** and many parrots, have crashed as a result of unsustainable harvesting, and the species concerned are now extremely rare in their original ranges.

At present, no Southeast Asian bird species that is endemic to the region has been declared extinct. However, a growing number of species are listed as Critically Endangered, and a handful of these have not been recorded for many decades. These 'missing birds' include the **White-eyed River Martin** (Thailand), **Javan Lapwing** (Java, Indonesia), **Black-browed Babbler** (Kalimantan, Indonesia), **Rück's Blue Flycatcher** (Sumatra, Indonesia), **Pink-headed Duck** (Myanmar) and **Sulu Bleeding-heart** (Philippines). Many more have been red listed by the International Union for the Conservation of Nature (IUCN), and Indonesia and the Philippines currently possess some of the highest numbers of threatened species among countries in the world.

Ayuwat Jearwattanakanok

*The Critically Endangered **White-eyed River Martin** was first discovered in the wetlands surrounding Bueng Borapet among flocks of Barn Swallows. (Original painting by Thai artist Ayuwat Jearwattanakanok).*

Cheng Heng Yee

*The **Helmeted Hornbill** has been uplisted to Critically Endangered due to excessive hunting for its prized casque.*

Habitat Loss & Degradation

Habitat loss is undoubtedly the main threat to Southeast Asia's birdlife, including migratory species. The region currently experiences the highest rates of forest loss in the tropics, and even a better understanding of deforestation and its impact on biodiversity has done little to slow the loss of tropical forests. Uncontrolled forest fires, many of which are associated with clearance for agriculture, have worsened the situation, and swaths of peat swamps continue to be lost annually to these fires. Thailand and Java have lost most of their forest cover in the past few centuries. Borneo, Sumatra and Indonesian New Guinea, have suffered significant deforestation in recent years as a result of agricultural expansion driven by the global demand for cash crops, especially oil palm, combined with weak forest governance. The best example of a bird threatened by deforestation is the endangered **Gurney's Pitta**. Rediscovered after a hiatus of 70 years in the 1980s, the site of its rediscovery in south Thailand has suffered significant degradation and deforestation, exacerbated by corruption and a lack of proper protection. Had it not been for the discovery of new populations in Myanmar, this species would definitely be teetering on the brink of extinction.

Besides forests, Southeast Asia's wetlands have suffered heavily from clearance, drainage, hydroelectric development,

James Eaton

*Back from the brink – the Critically Endangered **Silvery Pigeon** disappeared for 70 years before populations were rediscovered in the Mentawai islands in Indonesia.*

overharvesting and pollution. Consequently, many of the region's large waterbirds are on the verge of extinction. This decline in waterbird populations is particularly disconcerting, especially in the context of historical ornithological texts where colonial-era ornithologists reported large flocks of pelicans, cranes and storks in the region's wetlands, including cities like Bangkok. Likewise, the flocks of **Sarus Cranes** and **Spot-billed Pelicans** that once frequented the wetlands of the Philippines are now no more. In recent years, the massive dams planned along the channels of the Mekong, Irrawaddy and Salween Rivers threaten to destroy riparian wetlands and sandbars, imperilling the species that depend on these habitats like the **River Lapwing** and **Black-bellied Tern**.

The Pet-bird Trade

After habitat loss, the pet-bird trade is the next most significant threat to Southeast Asian bird populations. Investigations by conservation organizations like TRAFFIC and other researchers in recent years have revealed that the thriving trade in wild birds has reached unsustainable levels, threatening even species that were once abundant and widespread (like the **Red-whiskered Bulbul**, **Pied Myna** and **Orange-headed Thrush**). Other birds, like pittas, were formerly rare in the trade, but have become increasingly ubiquitous in recent years. Part of the problem with the pet-bird trade is the massive demand from Indonesian markets, where keeping pet birds is seen as integral to local culture and consequently weakly regulated.

Over 400 bird species have been reported from the bird trade based on recent surveys of major bird markets in Indonesia, including many songbirds and parrots. One of the best-documented declines of a songbird driven by the pet-bird trade is that of the **Straw-headed Bulbul**. This spectacular bulbul, the largest in the world, is all but extirpated from its range in Thailand and Indonesia. The only surviving viable populations persist today in a few protected areas in Malaysia, and in Singapore where poaching is under control.

Hunting of Birds for Food

Human consumption of wild birds remains rampant across Southeast Asia, especially in parts of Thailand, Vietnam, Laos and Indonesia. As well as being hunted for subsistence, birds are trapped for sale in food markets and for religious release at worrying levels. While large birds such as gamebirds and waterfowl are usually preferred, smaller songbirds are also taken for consumption when local populations of larger birds have collapsed. This problem has reached epidemic proportions in Laos, and it is commonplace to see bundles of flycatchers, bulbuls, leafbirds and other songbirds openly sold in markets for consumption.

Lee Tiah Khee

*The **Straw-headed Bulbul** has declined precipitously due to the unsustainable Indonesian pet bird trade.*

Andrew Chow

Hunting of birds for food is still a widespread practice in many parts of Southeast Asia such as Laos and Vietnam.

Southeast Asian Birdlife at a Glance

The Southeast Asian region is especially rich in birds, with an avifauna of more than 2,500 species representing over 70 families. A few families, including the Eupetidae (Rail-babbler), Pityriaseidae (Bornean Bristlehead) and Hylocitreidae (Hylocitrea), are entirely restricted to Southeast Asia. This section describes the major bird families that occur in Southeast Asia, with an indication of the number of species in each family .

New Species Discoveries in Southeast Asia

Southeast Asia has one of the least studied avifauna in the world, and new species are described here every year. There are two ways in which this can happen. First, subspecies of wide-ranging species may be elevated to full species as a result of newer genetic evidence, and of better knowledge of vocalizations and plumage patterns. Recognition of a new species in this way is often referred to as 'splitting' in the birdwatching literature. The rapid advancement of molecular techniques (such as next-generation sequencing) in recent years has played a major role in driving the current revolution in bird taxonomy. Some examples of species where distinct subspecies are now considered full species include the **Pied Fantail**, **Oriental Magpie-Robin**, **Crimson Sunbird** and **Purple Swamphen**. Birdwatchers in Southeast Asia can thus see their bird lists expanding without even having to go into the field, defined as 'armchair ticks' in birdwatcher jargon.

New species are occasionally discovered when field researchers stumble upon a bird previously unknown to science. This may even happen in museums, involving collected specimens that have been forgotten by researchers (like the **Cinnabar Boobook**). As more researchers explore the remote islands of Indonesia and forests of Indochina, undescribed forms that potentially represent new species are found. Some remain undescribed, sometimes for years, because of the difficulties involved in obtaining specimens. Two classic examples are the 'Spectacled Flowerpecker' from Borneo and a distinctive parrotfinch from the montane forests of Timor. Others have just been recently described, for instance the **Sulawesi Streaked Flycatcher** (Indonesia) and **Bare-faced Bulbul** (Laos).

Teo Nam Siang

*The **Sulawesi Streaked Flycatcher** was known to ornithologists for more than 20 years, but was not officially described until 2014 (original painting by Singaporean wildlife artist Teo Nam Siang).*

Synopsis of Key Bird Families

Cassowaries – 3 species

Cassowaries are primarily endemic to the island of New Guinea, where they inhabit both lowland and montane forests. Although they are known to attack humans when provoked, these shy, fruit-eating birds are glimpsed by only the most fortunate of observers. Often, the only sign of their presence is the dinosaur-like footprints imprinted along forest trails.

Philippe Verbelen

*The **Southern Cassowary** is the largest member of the family, and also the third largest bird in the world.*

Magpie Goose – 1 species

This large and unmistakable waterbird breeds on the floodplains of southern New Guinea, where it occurs in large, noisy flocks. These flocks are seasonally supplemented by migrants from Australia.

Ducks, Geese & Swans – 44 Species

A large family of familiar-looking aquatic birds with a worldwide distribution, ducks and geese have diagnostic webbed feet and flattened bills. Most species are herbivores, although some like mergansers specialize in a diet of fish. In Southeast Asia many species occur as winter migrants (like the **Garganey**, **Baer's Pochard** and **Northern Pintail**), arriving at various wetland sites across the region from their breeding grounds in temperate parts of Asia. A few species are resident in the region, including the endangered **White-winged Duck**. One species, the **Pink-headed Duck**, may be extinct.

Kin Yip Ho

*The Critically Endangered **Baer's Pochard** sometimes appears at Indawgyi (Myanmar) or Bueng Borapet (Thailand) Lakes in winter.*

Megapodes – 14 Species

This small family of ground-dwelling birds is best known for building large mounds with substrate and decaying matter, used to incubate eggs. Most species in Southeast Asia inhabit the islands of Wallacea in Indonesia, including the **Maleo**, while one species, the **Philippine Megapode**, occurs in the Philippines and islands off Borneo. Further east, Indonesian New Guinea is home to five species of brushturkey and the endemic **Biak** and **New Guinea Megapode**. Megapodes are characterized by small heads and large, strong legs with powerful claws for mound construction.

Bjorn Olesen

*The **Maleo** is one of Sulawesi's most charismatic large birds.*

Pheasants & Partridges – 59 Species

A large family best represented in the Old World, many pheasants and partridges are endemic to Southeast Asia and thus highly sought after by birdwatchers. Some of the best-known species include the stunning **Great** and **Crested Argus**, and the **Green Peafowl**, the latter the largest pheasant in the world on record. The **Crested Argus** also holds the world record for having the longest feathers of any bird. Most pheasants are sexually dimorphic, with males being noticeably larger than females and possessing striking plumage colouration. The majority are sedentary, and found in Southeast Asia's lowland and montane forests. Hunting and habitat loss have resulted in many of the region's species being globally threatened, and some, like Vietnam's **Edwards's Pheasant**, are now very rarely recorded.

Robert O. Hutchinson

*All but one species of peacock-pheasant occur in Southeast Asia, and the **Palawan Peacock-Pheasant** is undoubtedly the cream of the crop.*

Storm Petrels – 3 Species

Storm petrels are a group of small seabirds found throughout the world's oceans. They are strictly pelagic, returning to nesting colonies only to breed, and are characterized by their erratic, fluttering flight style and distinctive foraging method of picking food off the water's surface. The best-known representative in Southeast Asia is **Swinhoe's Storm Petrel**, which migrates through the region between its breeding grounds off the Korean Peninsula and the Russian Far East, and the wintering grounds in the Indian Ocean.

Petrels & Shearwaters – 7 Species

A large and diverse family of seabirds, many petrels and shearwaters undertake poorly documented, long-distance migrations during the non-breeding season. Petrels and shearwaters are pelagic in habits and adept surface feeders. The waters of Southeast Asia serve as wintering grounds for a few species like the mysterious **Heinroth's Shearwater**, while many species, such as the **Streaked Shearwater**, pass through the region as passage migrants.

Grebes – 6 Species

This small family of aquatic birds is well known for its elaborate courtship rituals. Grebes are excellent divers with powerful legs positioned at the rear of the body for propulsion. They also develop striking plumage colouration during the breeding season. Besides the resident **Little**, **Tricolored** and **Australasian Grebes**, three migratory species may turn up in Southeast Asia during the northern winter.

Tropicbirds – 3 Species

The tropicbirds are a family of elegant pelagic seabirds characterized by elongated tail feathers and brightly coloured bills that contrast sharply with their white plumage. All are solitary breeders and disperse widely throughout tropical waters outside the breeding season. Two species have been recorded breeding on several Indonesian islands.

Storks – 10 Species

Storks are large waterbirds with characteristic long legs and necks. Southeast Asia features more than half the world's stork species, although many have declined greatly due to habitat loss and hunting. **Storm's Stork** is restricted to the tropical forests of the Sundaic region and is the world's most threatened stork. Other species, like the threatened **Woolly-necked Stork** and **Greater Adjutant**, have small but significant populations in Southeast Asia.

Chairunas Adha Putra

*The **Milky Stork** and **Lesser Adjutant** share the same mudflats on the eastern coast of Sumatra.*

Ibises & Spoonbills – 12 Species

These large waterbirds are characterized by their distinctive bill shapes – the bills are long and decurved in ibises, and straight and spatulate in spoonbills. Southeast Asia is home to two of the rarest and most localized ibises in the world, namely the **Giant** and **White-shouldered Ibises**. The globally threatened **Black-faced Spoonbill** is a rare but regular winter visitor to northern Vietnam.

Herons & Bitterns – 30 Species

These familiar waterbirds have a cosmopolitan distribution and diverse ecological habits. Some, like the *Nycticorax* night herons, are nocturnal, others develop ornamental plumes during the breeding season, and still others undertake long-distance migrations to winter in the tropics. Bitterns tend to be secretive, keeping to dense vegetation when they hunt. Some of the most threatened herons occur in Southeast Asia, such as the Critically Endangered **White-bellied Heron**, now very rare in northern Myanmar's mountain rivers, and the crepuscular **White-eared Night Heron**, which breeds in northern Vietnam.

Le Manh Hung

*The **White-eared Night Heron** is one of the least-known members of the heron family. Ba Be National Park in northern Vietnam is the best place to see it in the region.*

Pelicans – 4 Species

Pelicans are very large waterbirds easily recognized by their distinctive pouched bills. They are gregarious, nesting in large colonies and hunting cooperatively by encircling schools of fish. In Southeast Asia a combination of water pollution, overfishing and habitat loss has resulted in the formerly widespread **Spot-billed Pelican** being extirpated from much of its original range, with the last breeding populations left in Cambodia.

Frigatebirds – 3 Species

Frigatebirds are best known for their unique courtship display, in which males inflate their red throat pouches to attract prospective mates. They are also kleptoparasites, regularly harassing other seabirds into regurgitating their catch as they return to their nesting colonies. Three species, including the Critically Endangered **Christmas Frigatebird**, are recorded in Southeast Asia, and large numbers can often be seen returning to island roost sites (like Mantanani) or in Jakarta Bay.

Boobies – 4 Species

Boobies and gannets are a small family of seabirds found throughout the world's oceans. Unlike many seabirds, boobies hunt for fish and cephalopods by plunge diving from a height. In Southeast Asia all four recorded species, except **Abbott's Booby** which breeds only on Christmas Island, breed widely on tropical islands and coral atolls throughout the Indo-Pacific.

Cormorants & Shags – 5 Species

These generally dark-coloured, distinctive-looking waterbirds are capable of diving underwater in pursuit of fish. They are gregarious and groups are regularly seen perched on trees and along the banks of large wetlands, drying their outstretched wings in between fishing forays.

Darters – 2 Species

Large waterbirds that look superficially similar to cormorants, darters can be distinguished by their slender, unhooked bills and long, serpentine necks. The **Oriental Darter** occurs throughout the region, but is largely confined to large, fairly undisturbed wetlands, although good numbers occur in the small wetlands in and around the Indonesian capital of Jakarta.

Ospreys – 2 Species

Ospreys are large, fish-hunting specialists of freshwater and coastal wetlands. While generally regarded as a monotypic genus with a cosmopolitan distribution, several authorities, such as the IOC, have recognized the **Eastern Osprey**, which breeds on the eastern islands of Indonesia, as a separate species.

Kites, Hawks & Eagles – 82 Species

Often known as raptors or birds of prey, kites, hawks and eagles comprise a large and diverse family of predatory birds with a cosmopolitan distribution. While some, like *Accipiter* hawks, are specialist predators of smaller birds others, such as vultures, feed mostly on carrion, while serpent eagles tackle mostly reptiles. Some of the rarest and most sought-after raptors are endemic to Southeast Asia, including the impressive **Philippine Eagle**, by many measures the largest eagle in the world. The island of Sulawesi is notable for its high diversity of endemic raptors, including four endemic species of sparrowhawk, a hawk-eagle and a honey buzzard. A similar situation is observed on New Guinea, which is home to a host of endemic raptors including the enigmatic **Papuan Eagle** and **Doria's Goshawk**.

Abdelhamid Bizid

*The **Spot-tailed Sparrowhawk** occurs in Sulawesi's lowland and montane forests.*

Bustards – 3 Species

This family of large, omnivorous birds with largely terrestrial habits features two resident species in the region. The Critically Endangered **Bengal Florican** is restricted to remnant grassland around Cambodia's Tonlé Sap Lake, and possibly Vietnam. Members of the family engage in captivating courtship displays that, in the case of floricans, involve males performing leaps above the grassland before floating back down to the ground repeatedly.

Markus Handschuh

*Less than 500 **Bengal Floricans** inhabit the grasslands that fringe Cambodia's Tonlé Sap Lake.*

Finfoots – 1 Species

The sole representative of this family in Southeast Asia is the globally threatened **Masked Finfoot**, a rarely encountered migratory species whose movements remain poorly understood. Breeding in Bangladesh's Sundarbans, northern Myanmar and Cambodia, it is suspected to migrate to parts of western Southeast Asia in winter. This species is usually seen foraging along the densely forested banks of rivers and lakes.

Yann Muzika

*Within Southeast Asia, the endangered **Masked Finfoot** breeds mostly in the wetlands of the upper Chindwin in the Tamanthi Wildlife Sanctuary.*

Rails, Crakes & Coots – 41 Species

Rails form a large, cosmopolitan family of terrestrial birds that are generally associated with wetlands. Most species are shy and extremely elusive, including the aptly named **Invisible Rail** of the Moluccas, often preferring to run when threatened. Although many species are resident in the region some, like **Band-bellied** and **Slaty-legged Crakes**, undertake lengthy migrations to spend the northern winter in Southeast Asia's wetlands and forests. Indonesian New Guinea is home to a variety of secretive forest rails and the **New Guinea Flightless Rail**.

Robert O. Hutchinson

*The shy **Invisible Rail** dwells in the sago swamps of Halmahera.*

Cranes – 5 Species

Cranes constitute a family of very large waterbirds characterized by predominantly grey or white plumage, and long legs and necks. The best-known example in Southeast Asia is the **Sarus Crane**, the tallest member of the family and a resident of the wetlands and plains of northern and central Myanmar, Cambodia and Vietnam.

Buttonquails – 7 Species

This family of small, terrestrial birds is characterized by its polyandrous mating habits, with females generally sporting more attractive plumage and males being heavily involved with the rearing of offspring. In the field, buttonquails are extremely retiring and difficult to see, often darting mouse-like into dense cover and only undertaking short, direct flights as a last resort to escape danger. Luzon's **Worcester's Buttonquail**, a species known in modern times solely from trapped specimens, is among the most poorly known representatives of this family.

Cheng Heng Yee

__Barred Buttonquail__ occurs widely across Southeast Asia.

Stone-curlews & Thick-knees – 4 Species

These are long-legged, largely nocturnal shorebirds with large eyes and powerful bills. The largest member of the group, the **Beach Stone-curlew**, occurs on remote sandy beaches, reefs and coral atolls across much of insular Southeast Asia, while the similar-looking **Great Stone-curlew** is dependent on sandy bank habitat of large, unpolluted rivers.

Crab-plover – 1 Species

An unmistakable shorebird with predominantly white plumage and a large black bill, the **Crab-plover** breeds in the Middle East and disperses widely during the non-breeding season. A few individuals sometimes reach the coasts of peninsular Thailand, especially some of the more secluded beaches of Phang Nga province.

Stilts & Avocets – 3 Species

These lanky waterbirds have generally pied plumages. While stilts have straight bills, those of avocets are upturned and are swept from side to side while foraging. Members of this family are migratory and disperse widely during the non-breeding season.

Plovers – 17 Species

The plovers are mainly short-billed shorebirds that hunt by making short sprints in pursuit of prey like small crabs. Many familiar members, like **Pacific Golden** and **Grey Plovers**, are accomplished fliers and migrate long distances from the Arctic tundra to the coastal wetlands of Southeast Asia where they spend the winter. A few species, like **Javan** and **Malaysian Plovers**, are sedentary and occur only in the region. The Critically Endangered **Javan Lapwing** used to dwell in the low-lying swamps in coastal Java, but there are no recent records and it may already be extinct.

Jacanas – 3 Species

Jacanas are characterized by their particularly long toes, which enable them to live on mats of floating vegetation where they forage and breed. Many members are polyandrous and some, like the **Pheasant-tailed Jacana**, are migratory.

Kin Yip Ho

*The elegant **Pheasant-tailed Jacana** is the most widespread member of this unusual family in Southeast Asia.*

Sandpipers & Snipes – 49 Species

This very diverse family of waterbirds of varying sizes and bill shapes is generally associated with freshwater and coastal wetlands. Many members, like the **Black-tailed Godwit** and **Curlew Sandpiper**, breed in the temperate and sub-arctic wetlands, and moult from their striking summer plumages to drab winter plumages when they arrive in the tropics. Many sandpipers are gregarious and form large flocks

Wich'yanan Limparungpatthanakij

*One of the best places to see the **Spoon-billed Sandpiper** are the coastal salt pans of central Thailand.*

on migration and at their wintering grounds. Two of the most threatened members of this family, **Nordmann's Greenshank** and the **Spoon-billed Sandpiper**, spend the winter in the coastal mudflats of Southeast Asia and are highly sought-after by birdwatchers around the world.

Pratincoles – 4 Species

This family of small shorebirds is characterized by their short legs and bill, forked tail and long, pointed wings – recalling a large swallow. These adaptations allow the birds to hawk for insects while in flight, a hunting technique unusual among waterbirds. Both **Oriental** and **Australian Pratincoles** exhibit migratory movements.

Gulls, Terns & Skimmers – 30 Species

A big family of seabirds with distinctive white-and-grey adult plumages, gulls are large with broad wings and rounded tails, and moult through a range of intermediate plumages as they mature. The migratory **Brown-headed** and **Black-headed Gulls** are the best known species in Southeast Asia. Terns are smaller and have forked tails, and most species hunt by plunge diving. Skimmers are distinctive, tern-like birds with a unique bill shape, sporting a significantly longer lower mandible that is used to detect fish by touch when hunting. The only skimmer in Southeast Asia, the **Indian**

Francis Yap

*The **Aleutian Tern**, an enigmatic breeder of the North Pacific, is now known to winter primarily in the waters of Southeast Asia.*

Skimmer, is now on the brink of extinction in the region due to extensive habitat loss. The Critically Endangered **Chinese Crested Tern** has been recently discovered wintering in the waters off the islands of Mindanao (Philippines) and Seram (Indonesia).

Skuas – 4 Species

These robust, gull-like seabirds are infamous for their kleptoparasitic behaviour. Most species are migratory, and wintering birds regularly harass congregations of smaller seabirds in Southeast Asia's seas. Skuas moult through a variety of immature plumages as they mature, making the identification of immature skuas difficult.

Pigeons & Doves – 135 Species

This familiar yet diverse group of birds exhibits considerable variation in size, plumage patterns and habits. They are characterized by short bills and necks, and a predominantly frugivorous diet. While many, like fruit doves and green pigeons, are arboreal, the elusive and aptly named ground doves, which include the bleeding-hearts of the Philippines and the mysterious **Wetar Ground Dove** of the Lesser Sundas, are shy, ground-dwelling species.

Philippe Verbelen

*The spectacular **Southern Crowned Pigeon** is localised to the lowlands of the southern half of New Guinea.*

James Eaton

*Seeing the enigmatic **Wetar Ground Dove** requires a mini-expedition to its namesake island of Wetar, located north of Timor.*

Indonesian New Guinea is home to a myriad of exquisite fruit doves as well as all four species of crowned pigeons. Due to hunting pressure and habitat loss, many pigeons, such as the **Silvery Wood Pigeon**, **Pale-capped Pigeon** and **Flores Green Pigeon**, have declined greatly throughout their range.

Cuckoos – 66 Species

Cuckoos are insectivorous land birds best known for brood parasitism, with numerous species laying their eggs in the nests of a range of host species. However, many species, including malkohas and most coucals, are non-parasitic and raise their own young. Some species, like the **Oriental Cuckoo** and **Rufous Hawk-Cuckoo**, undertake long migrations to winter in the region.

Francis Yap

*The **Violet Cuckoo** is widespread across Southeast Asia, but difficult to see due to its arboreal habits.*

Barn Owls – 10 Species

These owls are characterized by their heart-shaped facial discs and long legs. Many species

Robert O. Hutchinson

*The **Minahassa Masked Owl** is one of two* Tyto *owls that occur in Sulawesi's forests.*

utter a variety of unearthly screeches and hisses. Within Southeast Asia several species are endemic to islands of Wallacea (for example **Taliabu** and **Moluccan Masked Owls**), and remain poorly studied. The **Oriental Bay Owl**, the only member of its genus in Southeast Asia, is smaller and characterized by its mottled brown plumage, the presence of ear-tufts and a horseshoe-shaped facial disc.

Owls – 72 Species

This is a large family of nocturnal birds characterized by their cryptic brown plumages, large eyes and broad facial discs. Some owls, like the **Buffy Fish Owl**, hunt primarily aquatic prey, while the smaller scops owls consume mostly insects and reptiles. Two of the best represented groups in the region are the *Ninox* hawk owls and *Otus* scops owls, both featuring many endemic or restricted-range species.

Chi'en C. Lee

*The **Halmahera Boobook** is an inhabitant of lowland rainforest on islands in the northern Moluccas.*

Frogmouths – 13 Species

This is a family of nocturnal birds characterized by their thick, hooked bills and large gapes. They are cryptically plumaged and highly arboreal in habits, preying on insects captured by foliage gleaning or ambush predation. The majority of the world's frogmouths are confined to Southeast Asia, and more species occur in the forests of Borneo than anywhere else in the region. The **Dulit Frogmouth**, a poorly known member of the family, is now known to be locally common in Sarawak's central highlands.

Con Foley

*The nest of the **Malaysian Eared Nightjar** was first documented in 2015.*

Nightjars – 14 Species

Nightjars are nocturnal birds characterized by long, pointed wings, short bills and tiny legs. All nightjars hawk for insects repeatedly from favoured perches, and roost on the ground or on a low perch during the day. Eared nightjars are so named for their elongated crown feathers; they also lack the white wing and tail patches found on other nightjars. Due to their nocturnal habits the distributions of a number of species, like the **Satanic Nightjar** and **Bonaparte's Nightjar**, remain poorly known.

Owlet-nightjars – 7 Species

A small Australasian family of nocturnal birds that bear a superficial resemblance to frogmouths, owlet-nightjars are generally seen perched upright in the middle storey of rainforests. Like nightjars, they feed by hawking for insects from favoured perches. During the day they roost in tree cavities and foliage tangles. One species, the **Moluccan Owlet-nightjar**, occurs in Wallacea (north Moluccas). The island of New Guinea is the centre of owlet-nightjar diversity and at least six species can be seen in Indonesia's Papua and West Papua provinces, including both **Feline** and **Mountain Owlet-nightjars**.

Treeswifts – 4 Species

This small family, closely related to typical swifts, is restricted to tropical Asia and Australasia. Treeswifts differ from typical swifts by having more striking plumage with prominent crests or facial stripes, and also by being distinctly less aerial in habits. Individuals often make foraging sorties from a favoured perch in the canopy, and their tiny, cup-shaped nests are also constructed on tree branches.

Swifts – 38 Species

The swifts comprise a large and diverse family of birds that spend the vast majority of their lives in the air. They include needletails, arguably some of the fastest fliers among all birds, as well as the *Aerodramus* swiftlets, of which some species are known to use echolocation to navigate. Some species, notably the **Edible-nest Swiftlet**, is known to construct its nests from saliva, a sought-after delicacy in Chinese communities.

Francis Yap

***Black-nest Swiftlets** often occupy abandoned buildings and bunkers, as shown here on the island of Sentosa in Singapore.*

Trogons – 11 Species

A family of attractive, forest-dwelling birds distributed across the tropics, trogons are some of the most sought-after birds in Southeast Asia. The family is sexually dimorphic, with males usually being more attractive than females. Trogons are mostly insectivores, regularly observed sallying for insects in the understorey of the rainforest. They are otherwise unobtrusive, occasionally participating in feeding flocks moving through their territory. The lowland forests of Peninsular Malaysia, Sumatra and Borneo are where trogons are best represented, with four co-occurring species inhabiting many sites.

Con Foley

***Diard's Trogon** occur in the Thai-Malay Peninsula, Sumatra and Borneo.*

Rollers – 4 Species

So named because of their acrobatic display flights, rollers are a small family of colourful birds characterized by their large bills and robust build. Insects are captured on the wing during aerial sorties from favoured perches, usually on the tops of dead trees or other tall structures. Rollers are vocal and often give away their presence with harsh croaks and cackles. The best-known species in the region is the **Oriental Dollarbird**.

Abdelhamid Bizid

***Azure Dollarbird** is one of two rollers endemic to Wallacea.*

Kingfishers – 60 Species

These brightly coloured, carnivorous birds are characterized by their upright posture when perched and long, dagger-shaped bills. Well over half the world's species occur in Southeast Asia's forests and wetlands. The forests of Indonesian New Guinea and its surrounding islands are home to various endemic kingfishers including the **Shovel-billed Kookaburra** in the lowland rainforests as well as the **Biak** and **Numfor Paradise Kingfishers** that are readily encountered on their namesake islands. There is significant variation in ecology and feeding behaviour among kingfishers. While some species, like the **Blue-eared Kingfisher**, are associated with freshwater and coastal wetlands, where they plunge-dive for aquatic prey,

Dubi Shapiro

***Winchell's Kingfisher** is endemic to the islands of the southern Philippines.*

others like the *Actenoides* kingfishers inhabit rainforests often far from water, and take a variety of terrestrial prey.

Bee-eaters – 8 Species

This distinctive family of insectivorous birds is characterized by largely green plumage, long, down-curved bills and long wings. Bee-eaters specialize in hunting bees and other aerial insects, which are captured on the wing and beaten against favoured perches to remove the sting before consumption. Many species, like the **Blue-throated Bee-eater**, are short-distance migrants and colonial nesters, excavating their burrows on exposed riverbanks and cliffs. Other species, like the **Red-bearded Bee-eater**, are sedentary residents in forested areas and do not form colonies.

Hornbills – 27 Species

Hornbills are large, predominantly forest-dwelling birds readily identified by their colourful bills, and are among Southeast Asia's most recognizable birds. They are omnivorous, feeding primarily on fruits but also taking live prey on occasion, particularly when raising young. Hornbills have some of the most unusual breeding habits among birds, with the female sealing herself inside a tree cavity during the incubation and chick-rearing period, and depending entirely on the male to provide food. Many species in Southeast Asia are endemic to the region and some, like **Walden's** and **Sulu Hornbills** of the Philippines, are now highly threatened due to habitat loss and hunting.

Lorenzo Vinciguerra

Walden's Hornbills *now occur only in the mountains of Panay, where populations are closely guarded.*

Asian Barbets – 27 Species

The Asian barbets are a family of largely forest-dwelling frugivores, with most species occuring in Southeast Asia. They are characterized by their large heads and large, bristle-fringed bills. Many species have distinctive facial patterns and mostly green plumage, with **Brown** and **Sooty Barbet** notable exceptions. Highly arboreal in habits, some species like the **Red-throated Barbet** spend much of their lives in the canopy, where they can be very difficult to see. A recent taxonomic revision has transferred all the barbets into the genus *Psilopogon*, which formerly held only the **Fire-tufted Barbet**.

Honeyguides – 2 Species

A small family of predominantly African birds that superficially resemble bulbuls in shape and structure, the two Asian representatives have a disjointed distribution within the region. Honeyguides are partial to beeswax and, like cuckoos, are brood parasites that lay their eggs in the nests of various host species. Although the ecology of the Asian species remains poorly known, the **Malaysian Honeyguide** is thought to be a brood parasite of the Brown Barbet.

James Eaton

Malaysian Honeyguides *are unobtrusive and easily overlooked.*

Woodpeckers – 53 Species

This iconic family of arboreal birds is famed for its ability to drill into trees to excavate invertebrates. This ability is also used to excavate nesting holes for breeding. Woodpeckers are often sexually dimorphic, with males usually having more attractive facial patterns and colouration than females. Southeast Asia's forests support almost a quarter of all known species, many of which are endemic to islands in the region (like the **Red-headed Flameback**). Woodpeckers range in size from the diminutive **Rufous Piculet** to the **Great Slaty Woodpecker**, the largest extant woodpecker in the world.

Falcons – 18 Species

A family of small to medium-sized raptors ranging from the tiny, mostly insectivorous falconets to the familiar **Peregrine Falcon**, the fastest flier in the avian world. Formerly thought to be related to hawks and eagles, recent taxonomic work has found falcons to in fact be close relatives of parrots and pigeons. Most falcons have long, pointed wings, affording them excellent aerial manoeuvrability and enabling them to capture prey in mid-air. Several species, like the **Amur Falcon**, are rare winter vagrants to Southeast Asia, while resident species including the **White-rumped Falcon** are generally restricted to the region's forests.

Cockatoos – 8 Species

An easily recognizable family of large parrots with predominantly white plumage, cockatoos are gregarious frugivores that occur in the Philippines and throughout much of the Wallacean islands. Feral populations of some species, like the **Tanimbar Corella**, have established themselves in several cities. Renowned for their attractive appearance and intelligence, many cockatoos are threatened by overexploitation for the pet trade, and only a few remaining populations of some species, such as the **Red-vented Cockatoo**, survive in the wild.

Old World Parrots – 95 Species

One of the world's most recognizable bird families, parrots are characterized by their vividly coloured plumage and powerful, hooked beaks. The island of New Guinea is a global hotspot for parrot diversity, and the Indonesian half of the island supports a multitude of endemic species ranging from the prehistoric-looking **Pesquet's Parrot** to the tiny **Geelvink Pygmy Parrot**. Many species, especially lorikeets, racket-tails and the *Psittacula* parakeets, are gregarious, and large flocks can often be seen at favoured feeding and roosting trees. Others, like the *Loriculus* hanging parrots and **Blue-rumped Parrot**, usually occur in pairs. A small number of species, including Buru's **Black-lored Parrot**, have nocturnal habits. Like cockatoos, many parrots are highly sought after in the pet trade and consequently threatened by overexploitation throughout Southeast Asia's forests.

James Eaton

*The exquisite **Great-billed Parrot** is a Wallacean speciality.*

Broadbills – 11 Species

This is a predominantly Southeast Asian group of birds, with only a handful of representatives in Africa. Broadbills are noted for their large heads and aptly named bills, which are adaptations to catch large insects in the forest canopy. Most species are vividly coloured, *Calyptomena* broadbills are predominantly green, and other species like **Visayan** and **Banded Broadbills** sport hues of red and pink. All members are forest specialists.

Michelle & Peter Wong

*All three species of Calyptomena broadbill, including the stunning **Hose's Broadbill**, occur on Borneo.*

Pittas – 35 Species

One of the most vividly coloured groups of songbirds and a favourite among birdwatchers, pittas reach their highest diversity globally in Southeast Asia. The number of pitta species in the region has increased by nearly a third due to the recent recognition of eight subspecies of the 'Red-bellied Pitta' as distinct species. Pittas are characteristically short tailed and long legged, and spend much of their lives on the forest floor, where they hunt for soft-bodied invertebrates like worms and snails. Many species have white wing-patches that are visible in flight. Some of the best-known species include the striking **Gurney's Pitta** of Thailand and Myanmar.

Cheng Heng Yee

*Seeing the near-mythical **Giant Pitta** remains a dream for most birdwatchers visiting Indonesia or Malaysia.*

Bowerbirds – 14 Species
One of the most iconic families of the Australasian region, bowerbirds are renowned for the bowers built by males and used as a stage to attract mates. The bowers are elaborate structures constructed out of sticks and decorated with a wide variety of natural and man-made objects. This family also includes the catbirds, so named because of their cat-like vocalizations. All members of this family are primarily frugivorous.

Australasian Treecreepers – 1 Species
This small family is largely endemic to Australia with a single representative found in the mountains of New Guinea, the **Papuan Treecreeper**. Although they share similar habits to Asia's treecreepers, their closest relatives are the bowerbirds of Australasia. These insectivorous birds are often observed in foraging parties, climbing up the trunks and larger branches of trees in search of insects.

Daniel López Velasco

*The **Emperor Fairywren** occurs widely across the lowlands of New Guinea.*

Australasian Wrens – 5 Species
The five species of fairywren found in Indonesian New Guinea are handsome but secretive inhabitants of thickets and the understorey of lowland and hill forests. These insectivores live in extended family groups, and often show striking sexual dimorphism in plumage colouration.

Honeyeaters – 83 Species
Honeyeaters are a large and diverse group of songbirds that feed primarily on nectar, and are thus important for the pollination of many plant species. A number have evolved long, decurved beaks and brush-tipped tongues to help them access nectar deep within flowers. On the island of New Guinea, honeyeaters are among the main songbird families with an incredible diversity of species present in all of the island's ecosystems. One of the most desired is the **MacGregor's Honeyeater**, a species formerly considered a bird-of-paradise. Only one species, the **Indonesian Honeyeater**, occurs west of Wallace's Line.

Chi'en C. Lee

*The gaudy **Red-collared Myzomela** is one of many species of honeyeaters that dwell in the montane forests of New Guinea.*

Australasian Warblers – 22 Species
As their family name suggests, the Australasian warblers are a group primarily confined to New Guinea and Australia. This family is well-represented on New Guinea, with a variety of mouse-warblers and scrubwrens inhabiting the island's rainforest as well as numerous gerygones. However, its most unusual member is the handsome **Goldenface**, formerly considered a whistler. In Southeast Asia the **Golden-bellied Gerygone** is the most widespread representative of the group, occuring in diverse habitats from city parks to virgin forests. Two other gerygones are endemic to the Lesser Sundas islands. Gerygones are petite, arboreal birds, and are usually seen gleaning leaves in the canopy for insect prey. In contrast, mouse-warblers and scrubwrens are secretive denizens of the understorey and forest floor where they creep through the vegetation in search of insects.

Australasian Babblers – 2 Species
Formerly lumped with the babblers of the Old World, recent research has found this small family to be more closely related to logrunners. The two species found in Indonesian New Guinea inhabit very different habitats, with **Papuan Babblers** being found in the rainforests and **Grey-crowned Babblers** in the savannah regions. Both species live in large family groups and forage for insects among the vegetation with the aid of their long, decurved bills.

Logrunners – 1 Species

Logrunners are medium-sized, ground-dwelling birds found in the rainforest of Australia and New Guinea. One species is endemic to New Guinea, and is uncommonly observed in the island's montane forest. Logrunners are carnivorous and reveal their prey by scratching the leaf litter with powerful legs, supported by their unusual spiny tails.

Daniel López Velasco

*The skulking **Papuan Logrunner** is regularly observed in the upper montane forests of the Snow Mountains.*

Satinbirds – 3 Species

Satinbirds are enigmatic, fruit-eating birds endemic to the mountains of New Guinea. Once thought to be birds-of-paradise, recent taxonomic work has found that they are most closely related to berrypeckers and longbills. In all three species, males are significantly more colourful than females, and do not assist their partners in the raising of chicks.

Berrypeckers and Longbills – 10 Species

This family is endemic to New Guinea and comprises 10 species (six berrypeckers and four longbills) of forest-dwelling birds that are predominantly frugivorous. Berrypeckers are generally found in the highlands, with the notable exception of the **Black Berrypecker**. Longbills, on the other hand, are found at lower elevations and bear a superficial resemblance to female sunbirds.

Painted Berrypeckers – 2 Species

A small family of colourful frugivorous birds endemic to New Guinea's montane forests, their taxonomic relationship with other passerines is the subject of further study. In the field, both species are gregarious and can often be seen in small flocks on fruiting trees.

Daniel López Velasco

*The arboreal **Tit Berrypecker** is usually observed in small parties.*

Whipbirds, Jewel-babblers and Quail-thrushes – 5 Species

This family of medium-sized, terrestrial birds is often classified together. Indonesia's Papua and West Papua provinces are home to three species of jewel-babbler and a single whipbird and quail-thrush species, all of which are shy, retiring inhabitants of the forest floor. Many species are also accomplished singers with a range of vocalizations.

Woodshrikes & Allies – 6 Species

This group of large-headed, chunky birds is characterized by mostly grey, white and black plumage. Genetically, they are most closely related to the vangas of Madagascar. In Southeast Asia woodshrikes and flycatcher-shrikes are familiar participants of mixed feeding flocks in the forest canopy, where they occur in pairs or small groups.

Boatbills – 2 Species

A small family of insectivorous birds found in Australia and New Guinea, boatbills are characterized by their broad, flat bills. Regular participants in foraging parties, they are often seen gleaning and sallying for insects in the forest canopy. The two species found in Indonesian New Guinea are separated by elevation, with the endemic **Black-breasted Boatbill** replacing the more widespread **Yellow-breasted Boatbill** in the highlands.

Daniel Lopez Velasco

*Mixed flocks in the montane forests of New Guinea are regularly joined by the attractive **Black-breasted Boatbill**.*

Bristlehead – 1 Species

The **Bornean Bristlehead**, a chunky, red-and-black bird with a powerful bill hooked at the tip is one of the most recognizable birds in Southeast Asia's tropical forests. Bristleheads are so named for the unusual yellow patch of bristles on the crown. The only representative of this family is endemic to Borneo, and is undoubtedly one of the island's most sought-after birds by visiting birdwatchers.

Woodswallows & Allies – 11 Species

Bjorn Olesen

*The well-marked **Hooded Butcherbird** is widespread in the lowlands of New Guinea.*

This family is primarily an Australian group with only a handful of representatives (woodswallows) occurring west of New Guinea. Woodswallows are noted for their superficial resemblance to true swallows, but are chunkier in appearance and have distinctive semi-triangular wings. They are gregarious and usually seen in small groups to flocks with as many as 100 individuals. This family also includes the two peltops species and several butcherbirds that inhabit the island of New Guinea. Peltops are arboreal forest birds that sally for insects from exposed perches, while butcherbirds are active arboreal carnivorous birds that move through the canopy in search of small animals.

Mottled Berryhunter – 1 Species

Another taxonomic oddity, the **Mottled Berryhunter** used to be classified with the whistlers, until recent taxonomic work found it was most closely related to woodswallows. It is endemic to New Guinea, where it inhabits the island's mid-montane rainforest, feeding on small fruiting trees in the understorey.

Ioras – 3 Species

Ioras are a small group of mostly yellow or green birds confined to Southeast Asia and the Indian subcontinent. All members exhibit sexual dimorphism. Ioras are primarily arboreal birds that occur in pairs or small family groups. The widespread **Common Iora** is the best-known member of the group and is a familiar bird of city parks across the region.

Cuckooshrikes & Minivets – 59 Species

This is a diverse group of insectivorous birds that inhabits the forest canopies across Southeast Asia. While minivets are easy to recognize with their striking hues of black, yellow, orange and red, the *Coracina* cuckooshrikes are mostly chunky and nondescript birds with greyish plumages. Minivets and cuckooshrikes are regular participants of mixed feeding flocks. The island of New Guinea supports numerous endemic cuckooshrikes, trillers and cicadabirds but has no minivets.

Con Foley

*The **Scarlet Minivet** is a widespread forest canopy species across the Oriental region.*

Sittellas – 2 Species

This small, distinctive family of insectivorous birds inhabits the forest canopies of New Guinea and Australia. Sittellas have similar habits to nuthatches, moving through the canopy in vocal flocks and creeping along tree trunks and branches in search of insects. The two species endemic to New Guinea inhabit the island's montane rainforest.

Daniel López Velasco

*The sittellas, such as this **Black Sittella**, are the Australasian equivalents of the nuthatches, creeping along branches and trunks as they forage.*

Ploughbill – 1 Species

The **Wattled Ploughbill** is an unmistakable inhabitant of New Guinea's montane rainforest. It uses its peculiar, wedge-shaped bill to extract insects from dead wood and bamboo stems and can sometimes be found in mixed foraging flocks together with other insectivores. Recent taxonomic work found its closest relatives to be the sittellas.

Australo-Papuan Bellbirds – 2 Species

Formerly classified with the whistlers, this small family of stocky, crested birds is endemic to Australia and New Guinea. Of the two species found in Indonesia, the **Rufous-naped Bellbird** is a secretive inhabitant of the montane forests, while the **Piping Bellbird** is more often heard than seen in the understorey of lowland rainforests.

Daniel López Velasco

*The **Rufous-naped Bellbird** was described for science by the noted English ornithologist, Philip Sclater.*

Whistlers – 34 Species

Occupying a broad range of habitats from coastal mangroves to montane forests, whistlers are so named for their distinct and rich songs. Most are large-headed, sluggish birds with mostly brown-and-grey plumage, while a few, like **Black-chinned** and **Bare-throated Whistlers**, sport bright hues of yellow. While continental Southeast Asia is only represented by the **Mangrove Whistler**, a coastal species, whistlers are a distinctive part of the avifauna of the Philippine and Wallacean regions, where many species occur. Many species are also present in Indonesian New Guinea, including several shrikethrushes and pitohuis such as the **Black** and **White-bellied Pitohui**.

James Eaton

*The **Sangihe Shrikethrush** is now considered to be a whistler, closely related to Sulawesi's Maroon-backed Whistler.*

Shrikes – 8 Species

Shrikes are highly recognizable birds of open-country habitats throughout Southeast Asia. All species are large headed with distinctively hooked bills built for their carnivorous diet. Many have a habit of impaling their prey on sharp thorns as a means of temporary food storage. Although most shrikes in Southeast Asia are widely distributed, the **Mountain Shrike** is restricted to the montane forests of the Philippines.

Vireos & Greenlets – 9 Species

For many years, vireos and greenlets formed a purely American family of birds, with most of its members in South America. Taxonomic work based on molecular techniques in recent years has since found the striking **White-bellied Erpornis** and the shrike-babblers to in fact be members of this group. All members are regular participants of mixed foraging flocks in the forest canopy.

Figbirds & Orioles – 25 Species

This is a small group of medium-sized, stout-looking songbirds that feed mostly on fruits and insects. Many species, like **Dark-throated** and **Black-naped Orioles**, exhibit bright hues of yellow that contrast with black plumage patches, or in the case of figbirds a predominantly green plumage. While many orioles are familiar birds of gardens, the **Black Oriole** of Borneo's mountains is one of the region's least-known birds and was only rediscovered in recent years. This family includes four species of pitohui that can be found in Indonesian New Guinea. One of them, the handsome **Hooded Pitohui**, is well-known for being one of the few poisonous birds in the world.

Daniel López Velasco

*The **Raja Ampat Pitohui** is endemic to the lowland forests of its namesake island archipelago.*

Drongos – 13 Species

Some of the most familiar birds in Southeast Asia, drongos are distinctive insectivorous birds with glossy black or grey plumage, forked tails and strong bills. Many species are regular participants of mixed feeding flocks. The **Greater Racket-tailed Drongo** occurs across much of the western half of Southeast Asia and is probably the most widespread member of the group. **Wallacean** and **Hair-crested Drongos** are widely distributed across the islands of the Philippines and Wallacea, and some of the island races may in fact be distinct species.

Fantails – 34 Species

These long-tailed, short-billed insectivorous birds are named for their frenetic habit of opening and closing their tail feathers as they move through the forest, usually with mixed flocks. In Southeast Asia they reach their greatest diversity on the islands of Wallacea where many species are endemic, and multiple species may co-occur in a particular habitat. Fantails are also well-represented in New Guinea where endemic members include three species of thicket fantail and the **Drongo Fantail**, with the latter now found to be related to the silktails of Fiji. The latest addition to this family is Sangihe's **Cerulean Paradise Flycatcher**, as revealed by recent genetic studies.

Monarchs – 36 Species

Monarchs and paradise flycatchers are a distinctive group of insectivorous birds formerly lumped with the Old World flycatchers. Most monarchs are arboreal, forest-dwelling birds with broad bills and conspicuous rictal bristles. While some of the paradise flycatchers, like the **Amur Paradise Flycatcher**, are long-distance migrants from north-east Asia, many others, like the **Maroon-breasted Philentoma**, are sedentary forest birds. A variety of monarchs are also present in Indonesian New Guinea, including the **Torrent-lark**, an inhabitant of forested streams that builds nests out of mud.

Con Foley

*Both species of philentoma, including the **Maroon-breasted Philentoma**, occur in the lowland forests of the Malay Peninsula and the Greater Sundas.*

Crows, Jays & Magpies – 33 Species

Collectively known as corvids, this large group comprises large-bodied, omnivorous songbirds that occupy a great range of habitats from urban areas to tropical forest. While many crow species are primarily black, a number of magpies, like the **Red-billed Blue Magpie**, have brightly coloured plumages with vivid hues of green and blue. Some of the most threatened species in Southeast Asia come from this group, including the **Banggai Crow** of Peleng in Wallacea, and the **Javan Green Magpie**.

Melampittas – 2 Species

This small family of all-black, ground-dwelling songbirds is found only in New Guinea's montane rainforest. Although related to birds-of-paradise, melampittas are primarily terrestrial insectivores, preferring to hop or run along the forest floor in search of prey. The **Greater Melampitta** is never found far from limestone sinkholes, which are used as roosting and nesting sites.

Daniel López Velasco

*The elusive **Lesser Melampitta** is best detected by its distinctive clicking call.*

Ifrit – 1 Species

Recent taxonomic work has assigned the **Blue-capped Ifrit** to its own family and found it to be distantly related to crows and Birds-of-paradise. The species is endemic to New Guinea's montane rainforest where it can often be encountered in vocal foraging parties, creeping along mossy trunks and branches in search of insects. It is also one of the few birds worldwide that are known to be poisonous.

Daniel López Velasco

*The skin and feathers of the **Blue-capped Ifrit** is known to contain toxins, likely derived from its insectivorous diet.*

Birds-of-paradise – 27 Species

The birds-of-paradise are a mainly Australasian group, with the vast majority of the species occuring on New Guinea. Many species are sexually dimorphic, the male plumage being particularly exquisite, with highly elaborate and extended feathers. Indonesian New Guinea is home to more than 20 species in this family, including a suite of highly sought-after regional endemics such as **Western Parotia** and **Wilson's Bird-of-paradise**.

Australasian Robins – 26 Species

This large family of insectivorous birds is endemic to Australasia and comprises a variety of species including the scrub robins and ground robins. Most of the 26 species found in New Guinea are endemic to the island, with a handful restricted to the Indonesian part of the island. Many of New Guinea's robins inhabit the understorey of rainforests where they are best detected by their distinctive, musical songs.

Daniel López Velasco

*The enigmatic **Greater Ground Robin** has been more frequently observed in recent years in the Snow Mountains.*

Rail-babbler – 1 Species

Looking somewhat like a cross between a rail and a small pheasant, the monotypic **Rail-babbler**, the sole representative of the family Eupetidae, is one of the most sought-after Southeast Asian birds. This taxonomic oddity dwells in the tropical forests of the Thai-Malay Peninsula, Sumatra and Borneo, and is characterized by its long-necked appearance, rich orange-brown plumage and rail-like gait.

Cheng Heng Yee

*The **Rail-babbler** occurs in the lowland forests of the Malay Peninsula and Sumatra, but appears to be found only in submontane forests on Borneo.*

Hylocitrea – 1 Species

Yet another of Southeast Asia's many taxonomic puzzles and now put into a single-species family, the unusual **Hylocitrea** was for a long time classified with the whistlers, until recent taxonomic work found its closest relatives to be the waxwings. The **Hylocitrea** is a chunky, brownish-olive bird that dwells in the cloud forests of Sulawesi's high mountains.

Robert O. Hutchinson

Lore Lindu National Park is the best place to see the Hylocitrea.

Fairy Flycatchers – 3 Species

The fairy flycatchers are a newly established group of insectivorous birds that were once classified with the Old World flycatchers. All the Southeast Asian members of this family are small, forest-dwelling birds of the canopy, including the familiar **Grey-headed Canary-flycatcher**.

Tits & Chickadees – 14 Species

This is a big group of large-headed, small-billed insectivorous birds that reach their highest diversity in temperate Eurasia and North America. Many species are active feeders that participate in mixed foraging flocks. In Southeast Asia the number of tit species declines along the west–east axis, with most species found in the high mountains of Myanmar, and only one in Wallacea. One of the most distinctive members of the group in the region is the yellow-and-black **Sultan Tit**, a montane forest bird across much of continental Southeast Asia.

Wang Bin

*The attractive **Sultan Tit** is a regular participant of mixed foraging flocks.*

Larks – 6 Species

This family comprises mostly brownish songbirds best known for their varied songs, which are uttered when birds are hovering in the air during a display flight. Most larks inhabit open habitats like grasslands and open scrub. The best-known member of the family in the region is **Horsfield's Bush Lark**, a widespread species associated with open country and occuring from mainland Southeast Asia to the Philippines and Lesser Sundas.

Bulbuls – 65 Species

Bulbuls are undoubtedly some of the most familiar birds in Southeast Asia, with species occupying every habitat from urban parkland to deep forest. They are characterized by their small bills, relatively short wings and upright stance. Many species, like the **Straw-headed Bulbul**, sing rich songs and are widely persecuted for the pet-bird industry. Bulbuls reach their highest diversity in the tropical rainforest of the Malay Peninsula and Sumatra,

Agus Nurza Zulkarnain

*The **Orange-spotted Bulbul** has been extensively trapped for the pet-bird trade, and is now scarce across Sumatra and Java.*

where more than 10 species can co-exist. Some bulbuls, like the **Asian Red-eyed Bulbul**, are predominantly brown while others, such as the **Scaly-breasted Bulbul**, have attractive plumage colouration comprising black, yellow and white.

Swallows & Martins – 17 Species

This is a widely distributed family of aerial insectivorous birds distinguished by their long, pointed wings, slender bodies and small bills. Many species build their nests out of mud in natural or man-made structures, while some, like the *Riparia* sand martins, excavate their burrows on sand banks along rivers. The **White-eyed River Martin**, a species first discovered in central Thailand, is the most endangered member of the group, with no confirmed sightings reported since the late 1970s.

Wren-Babblers – 2 Species

The wren-babblers or cupwings are a small but highly recognizable group of diminutive, short-tailed songbirds that inhabit the forest floor of Southeast Asia's montane forest. They were formerly classified with other ground-dwelling babblers due to superficially similar habits. The most widespread member of the family in Southeast Asia, the **Pygmy Wren-babbler**, occurs from Myanmar to Timor and spends much of its life scurrying on the forest floor in search of invertebrates.

Yann Muzika

*The forest floor in the misty montane forests of northern Myanmar is the home of the **Scaly-breasted Wren-babbler**.*

Cettia Bush Warblers & Allies - 25 Species

This diverse group of small insectivorous birds has its global stronghold in Southeast Asia, and includes the mostly ground-dwelling and solitary tesias and stubtails, the colourful *Abroscopus* warblers and the namesake bush warblers. Many species are difficult to identify because of their drab colouration and very skulking habits. While many of the *Horornis* bush warblers are long-distance migrants, other like the charismatic **Chestnut-headed Tesia** are sedentary species that dwell in montane forests.

Con Foley

***Chestnut-headed Tesias** may be observed in the montane forests of Thailand and north Vietnam.*

Bushtits - 4 Species

This is a small family of long-tailed, small insectivorous birds with short, stubby bills. Bushtits are highly active and often form small parties as they forage, keeping in contact with each other in dense vegetation with twitters and churrs. Bushtits in the region mostly occur in the montane forests of mainland Southeast Asia, the sole exception being the **Pygmy Bushtit**, which is endemic to the Indonesian island of Java.

James Eaton

*Noisy parties of the **Pygmy Bushtit**, one of the smallest passerines in the world, forage in the montane forests of Java.*

Leaf Warblers & Allies - 45 Species

Phylloscopus and *Seicercus* warblers, the two genera in this family, are infamous among birdwatchers for being notoriously difficult to identify because of the often subtle differences in call and plumage between species. Leaf warblers (*Phylloscopus* spp.) are small, arboreal insectivorous birds with mostly dull green, brown and grey plumage, while *Seicercus* warblers tend to be more colourful with bright yellow and green hues. In numerous cases species are best identified by their distinctive songs. In Southeast Asia many leaf warblers are latitudinal or altitudinal migrants, while species inhabiting the islands of the Greater Sundas, Philippines and eastern Indonesia (like **Island** and **Negros Leaf Warblers**) tend to be sedentary. Two leaf warbler forms on the Banggai and Sula islands probably represent undescribed species.

Con Foley

*The **Pale-legged Leaf Warbler** is a widespread winter visitor to the forests of mainland Southeast Asia.*

Reed Warblers & Allies - 9 Species

Reed warblers and their allies are a large group of mostly drab brown birds with fairly long bills. Most reed warblers, as their name implies, skulk in dense areas of waterlogged vegetation by the edges of wetlands. The best-known member of the family in Southeast Asia is the **Oriental Reed Warbler**, a winter visitor to wetlands across much of the region. Two of the rarest and most restricted range members of this family, **Large-billed** and **Manchurian Reed Warblers**, winter in Southeast Asia's wetlands.

Abdelhamid Bizid

*The **Chestnut-backed Bush Warbler** occurs in the montane forests of Sulawesi, with subspecies in Buru and Seram probably being distinct.*

Grassbirds & Allies – 26 Species

Grassbirds and their allies are a diverse group of small insectivorous birds particularly well-represented in Southeast Asia. The best-known representatives in the region are the grasshopper warblers and *Bradypterus* bush warblers, all highly-skulking birds of dense vegetation. Grasshopper warblers are so named for their complex, high-pitched songs that have been likened to the sounds made by grasshoppers. While many species exhibit migratory behaviour, the *Robsonius* ground warblers of Luzon and Timor's **Buff-banded Thicketbird** are largely sedentary birds of tropical forests and forest edges. An undescribed bush warbler occurs on the island of Taliabu.

Cisticolas & Allies – 26 Species

Cisticolas, prinias and tailorbirds are another group of birds that was previously classified under Old World warblers, a 'waste-bin' group. This diverse group of small insectivorous birds is characterized by generally long tails and fairly long, thin bills. Many species, like the **Zitting Cisticola** and **Yellow-bellied Prinia**, inhabit open country, but most tailorbirds (*Orthotomus* spp.) are forest dwellers. The Philippine islands are particularly rich in endemic tailorbirds, including several striking species. The **Cambodian Tailorbird** of the Mekong floodplains was described as recently as 2013.

Mohamad Zahidi

*The **Ashy Tailorbird** occurs mostly in the mangrove and riparian forests of the Malay Peninsula, Sumatra and Borneo.*

Babblers – 38 Species

A former 'waste-bin' group containing more than 300 species, the use of molecular systematics has helped trim the babbler family down to its present size of about 55 species spread across Asia. Typically, babblers comprise a varied group of small to medium-sized birds with fairly short, rounded wings and proportionally short tails. Most babblers live in small family groups and are not migratory. Within the family, babblers range from the scimitar babblers with long, decurved bills, to the mainly brown *Stachyris* and *Macronous* babblers, and the extremely skulking *Spelaeornis* babblers. Many members of this family utter far-carrying, melodious songs.

Fulvettas & Ground Babblers – 53 Species

The Pellorneidae is a diverse group of Asian and African 'babblers' carved out of the Timaliidae in recent years thanks to insights from molecular systematics. The *Alcippe* fulvettas (like the **Mountain Fulvetta**) are a group of small, skulking insectivorous birds, usually occuring in family groups that forage in thick vegetation cover by the forest edge. A number of mostly ground-dwelling 'wren-babblers' of the genera *Kenopia*, *Ptilocichla* and *Napothera* are also placed in this group, and include sought-after species like **Large** and **Falcated Wren-Babblers**. While the *Graminicola* grassbirds are mainly species of grassland and swamps, the remaining genera like the *Malacocinla* and *Pellorneum* babblers are predominantly shy birds of the forest floor and understorey.

Bjorn Olesen

***Striped Wren-Babblers** occur in the lowland forests of western Indonesia and Malaysia.*

Laughingthrushes – 73 Species

Yet another group of former 'babblers', the laughingthrushes are a large and diverse group of medium-sized birds that reach their greatest diversity in the tropical and subtropical forests of mainland Southeast Asia and the eastern Himalayas. Most laughingthrushes live in small, noisy parties in the forest understorey, and while no species are truly migratory, those occuring in montane forests may disperse to lower elevations in the winter months. Besides the true 'laughingthrushes' (like **White-necked** and **Orange-breasted Laughingthrushes**), this diverse group includes other former 'babblers' like the cutias, mesias, liocichlas and sibias.

Cheng Heng Yee

*Most laughingthrushes, including the **Spot-breasted Laughingthrush**, skulk in dense thickets.*

Sylviid Babblers – 20 Species

As a result of the use of molecular approaches in taxonomy, many 'babbler' species formerly grouped with 'Old World warblers' have now been reclassified. Sylviid babblers are small to medium-sized birds that occur in noisy parties in Southeast Asia's forests and scrub habitats. Most sylviid babblers in the region are parrotbills, a group of small forest birds with distinctive short, stubby bills, and are spread over seven genera. Other distinctive groups include the fulvettas (*Fulvetta* spp., distinct from the *Alcippe* fulvettas) and the *Chrysomma* babblers, most notably Myanmar's recently rediscovered **Jerdon's Babbler**.

White-eyes – 60 Species

The white-eyes are a large family of Old World songbirds with a rather complex taxonomic history. As the taxonomic relationships of the 'babblers' became better understood, many are now grouped under this family, including most of the Philippine 'Stachyrine babblers' (like **Chestnut-faced** and **Flame-templed Babblers**). At the same time, taxonomists removed some former 'white-eyes' like the Cinnamon Ibon (Mindanao) and Madanga (Buru). In general, white-eyes and their allies are small, gregarious forest birds with green, olive and brown in their plumage. Many typical white-eyes, like the **Oriental White-eye**, also sport a ring of fine white feathers around the eye.

Dubi Shapiro

***The Cream-browed White-eye** is endemic to the Lesser Sundas.*

Fairy-bluebirds – 2 Species

Fairy-bluebirds are a small family of mainly frugivorous songbirds that are unique to tropical Asia. Both members of the family have mostly iridescent blue plumage, and exhibit significant sexual dimorphism, with females being duller than males. The widespread **Asian Fairy-bluebird** occurs across most of western Southeast Asia, extending into Palawan, while the **Philippine Fairy-bluebird** is endemic to the Philippine islands (excluding Palawan).

Elachura – 1 Species

Previously considered a 'wren-babbler' based on its morphology, the **Spotted Elachura** of subtropical and tropical Asia was found to be so distinctive from other 'babblers' based on DNA studies that it was grouped into a newly established family. Superficially, it resembles the wren-babblers, with its short tail and understorey-dwelling habits. In Southeast Asia it occurs mainly in the mountains of Myanmar, Laos and Vietnam, with recent records from northern Thailand.

Tong Menxiu

*Research by Per Alström and colleagues allocated the **Spotted Elachura** to a family of its own.*

Nuthatches – 11 Species

These small, distinctive, mostly arboreal songbirds are characterized by their large heads, short tails, largely bluish-grey upperparts and creeping posture. Most species participate in mixed flocks and forage by creeping along tree trunks and branches. Southeast Asia is not just home to the greatest diversity of nuthatches in the world (more than four species co-occur in some sites in the Burmese Himalayas and northern Thailand), but also the largest (**Giant Nuthatch**) and the most charismatic (**Beautiful Nuthatch**) members of the family.

Treecreepers – 4 Species

Treecreepers are closely related to the nuthatches, and not surprisingly have similar foraging habits, creeping up and down branches in search of insects. All species are characterized by their slightly decurved bills and largely brown upperparts that are spotted with pale patches. The best-known species in the region is **Hume's Treecreeper**, a regular mixed flocked participant in the upland forests of northern Thailand.

Starlings & Rhabdornises – 47 Species

Bim Quemado

*Rhabdornises, such as this **Stripe-headed Rhabdornis**, are now widely considered to be aberrant starlings.*

The starlings are a large and diverse family of songbirds familiar to most people due to their vocal habits and ability to adapt to human-modified environments. Starlings are mostly medium-sized birds with a fairly upright posture, long legs and sharp bills; many species have dark, glossy plumages (like *Aplonis* and *Basilornis* starlings). The rhabdornises comprise three similar-looking species in a single genus, all with long, decurved bills. They forage for insects by creeping and hopping up branches. Rhabdornises were found to be most closely related to the starlings, although

Ayuwat Jearwattanakanok

*The **Chestnut-tailed Starling** is regularly seen in the dry forests of Thailand and Cambodia.*

their placement within the Sturnidae may be temporary. Many starlings, like the **Bali Myna** and **Black-winged Starling** of Java and Bali, are highly threatened due to uncontrolled poaching for the pet trade.

Thrushes – 42 Species

The thrushes are a large group of mostly medium-sized, plump-looking songbirds; many are ground dwellers and are broadly omnivorous, taking invertebrates and fruits. Some species, like the *Turdus* thrushes (**Grey-sided Thrush**), and a handful of *Geokichla* species (**Siberian Thrush**) are migratory, occuring in Southeast Asia as winter visitors. A number of species with complex taxonomic histories are now considered to be thrushes, based on recent DNA studies – the best examples include Sulawesi's **Geomalia** and Borneo's **Fruithunter**.

Khaleb Yordan

*Like other cochoas, the shy **Javan Cochoa** is usually first detected by its shrill whistle.*

Chats & Old World Flycatchers – 143 Species

The flycatchers and chats form a very large and diverse group of small-bodied songbirds. Many species, especially the flycatchers, are arboreal insectivores, sallying from a perch to capture insects. The chats and robins, which also include the rock thrushes and whistling thrushes, are mostly ground-dwelling birds and are characterized by their generally long tails, large heads and long legs. A number of them, like the stonechats, prefer open habitats, while the *Phoenicurus* redstarts including the

James Eaton

*The Tanimbar subspecies of the Rufous-chested Flycatcher is now considered a distinct species, the **Tanimbar Flycatcher**.*

distinctive **White-capped Redstart** inhabit fast-flowing streams. Across Southeast Asia flycatchers and chats are some of the most familiar examples of the region's birdlife, and occupy a wide range of habitats from urban parks to pristine montane forests. Many species, including the mysterious **Rufous-headed Robin** and **Blackthroat** of central China, are migrants to the region. Recent studies have found Mindanao's **Bagobo Babbler** to be an aberrant chat, closely related to the shortwings.

Leafbirds – 10 Species

Leafbirds are a family of medium-sized, arboreal birds found across the forests of mainland Southeast Asia, the Greater Sundas and the Philippines. All except one species are found in this region. Leafbirds sport a predominantly green plumage and exhibit sexual dimorphism. Some species are popular as pet birds, and wild populations have declined as a result of trapping.

Francis Yap

Prized for its varied song, the ***Greater Green Leafbird*** *is now difficult to find in the forests of Sumatra.*

Flowerpeckers – 39 Species

These are small, arboreal birds that are mainly nectar feeders and reach their highest diversity in the Philippines and Greater Sundas; many species also feed on sticky mistletoe berries. As a group, flowerpeckers are characterized by their short tails, short and slightly curved bills, and feathery tongue-tip. Many species, like the **Scarlet-breasted Flowerpecker** of the Greater Sundas, are brightly coloured and exhibit strong sexual dimorphism.

Sunbirds – 48 Species

Like the flowerpeckers, most sunbird species are nectar feeders, although many take insects opportunistically. Sunbirds are mostly small-bodied birds with fairly long, decurved bills. A number of species, especially the *Aethopyga* sunbirds, have very colourful plumages. By contrast, the larger *Arachnothera* spiderhunters and the **Purple-naped Sunbird** are mostly

Abdelhamid Bizid

The ***Pale Spiderhunter****, endemic to the island of Palawan in the Philippines, is a recent split from the widespread Little Spiderhunter.*

olive-green and yellow. Recent studies of the taxonomic relationships among Philippine sunbirds have elevated numerous distinct subspecies to full species, including sunbirds in the **Metallic-winged** and **Apo Sunbird** complex.

Old World Sparrows – 5 Species

Sparrows, particularly the **Eurasian Tree Sparrow**, are some of the most easily recognized birds in the region, due to their ability to co-exist with humans. They are characterized by their short tails, rather large heads and short, stubby bills adapted for seed-eating. The **Cinnamon Ibon** of Mindanao was formerly thought to be a white-eye, but recent studies have found it to be an aberrant forest sparrow.

Yann Muzika

Looks can be misleading – the ***Cinnamon Ibon*** *of Mindanao is in fact a sparrow.*

Weavers – 3 Species

Weavers are superficially similar to the sparrows in terms of proportions and their thick, short bills. Breeding males tend to have bright yellow colouration in their plumage, in contrast to that of the concolourous females. All species build elaborate hanging nests out of plant material – a procedure that is also part of the courtship process.

Waxbills, Munias & Allies – 35 Species

Munias and waxbills form the largest group of seed-eating birds in Southeast Asia. Many species are familiar to people, being popularly traded in pet shops and persecuted by rice farmers across the region. Like the weavers, munias and waxbills are characterized by their small sizes, stout appearance and thick bills. Most munias and waxbills are birds of open habitats, but the parrotfinches occur in forests and forest edges where many species disperse nomadically to feed on flowering bamboo.

Daniel López Velasco

The unobtrusive ***Mountain Firetail*** *occurs mainly above 2,700 m in New Guinea's Central Cordillera.*

Wagtails & Pipits – 19 Species

Most wagtails and pipits occur as winter visitors to Southeast Asia. They are generally long-legged, ground-dwelling birds. A handful of species are residents in the region, most notably the **Paddyfield Pipit**, a familiar bird of open grassland. The unusual **Madanga** of Buru dwells in montane forests – for many years it was believed to be an aberrant white-eye, until DNA studies revealed it to be closely related to the pipits. While the wagtails generally have colourful breeding plumages and have the distinctive habit of bobbing the tail, pipits are mostly cryptically coloured brown.

Francis Yap

The unusual ***Forest Wagtail*** *is a winter visitor to much of western Southeast Asia.*

Finches – 26 Species

This is a group of large-headed, seed-eating birds with thick, stubby bills. Many species have long, clearly notched tails. Superficially, finches resemble munias and sparrows. Most form small flocks after the breeding season. In Southeast Asia finches reach their highest diversity in the mountains of northern Myanmar, where various species co-occur. Elsewhere in the region a handful of bullfinch species, the **Red Crossbill** and the **Mountain Serin** occur in pine or mixed forests in high mountains.

Thang Nguyen

The south Vietnamese race of the ***Red Crossbill*** *likely represents a distinct species.*

Buntings – 14 Species

Buntings form a large family of small-bodied, long-tailed, seed-eating birds that are best represented in temperate parts of eastern Asia. Most species are winter visitors to the region from further north, except for **Godlewski's** and **Crested Buntings**. The best-known bunting in the region is probably the **Yellow-breasted Bunting**, a species that has declined drastically due to overexploitation for food and the pet trade. Other wintering buntings like the **Little Bunting** are also in decline.

Birdwatching Conditions in Southeast Asia

Most birdwatching in Southeast Asia revolves around visiting tropical forests, and to a lesser extent wetlands and open grasslands. Birdwatching in the tropical forests of the region is usually very safe and the majority of trips pass without incident. However, there are various annoyances that visitors should be aware of and be prepared for. Biting insects (midges, mosquitoes and horseflies), ticks and leeches can be found in many forested areas and can make birdwatching at these places unpleasant. Visitors should consider protecting against these insects with appropriate attire (for example long-sleeved shirts) and repellents. Leeches are most prevalent in rainforests, especially areas where large mammals are present, and can be difficult to avoid although they are not known to spread diseases. Some birdwatchers use special 'leech socks' to minimize instances of leech bites. Venomous snakes occur across Southeast Asia, and although they are rarely encountered it is good practice to stay away from any snake that cannot be identified conclusively. For example, the highly venomous King Cobra occurs throughout Southeast Asia and can be aggressive if confronted.

Due to widespread hunting pressure it is uncommon to encounter large mammals in the field, including predatory large cats like Leopards and Tigers. However, in the big national parks of mainland Southeast Asia, Sumatra, Borneo and Peninsular Malaysia, where large mammals occur, it is important to take precautions. There have been isolated incidents of tourists being gored by wild elephants in Thailand and Malaysia, and visitors should stay clear of areas where elephants occur on a regular basis.

In well-developed national parks and reserves such as those in Thailand, Peninsular Malaysia and Singapore, facilities for visitors such as marked trails and boardwalks are often available and easy to navigate, and can thus be visited independently. At some sites where birdwatching is done primarily along roadsides, one needs to be wary of speeding vehicles. Elsewhere in Southeast Asia, facilities for visitors tend to be very basic to non-existent. Some remote sites may involve tricky stream crossings, climbing steep slopes or walking through poorly marked trails. In these cases it is important, at least for safety reasons, to be accompanied by local guides and bird-tour leaders who know the local conditions best, and can arrange for help when needed.

Glossary of Terms

arboreal Living in trees
ban Lao term for village
biodiversity hotspot Biogeographic region known for significant biodiversity that is highly threatened by human activity
biogeography Study of the distribution of plants and animals in relation to geography
boeng Khmer word for lake
bukit Indonesian and Malay word for hill
chaung Burmese word for stream
crepuscular Active during twilight hours
danau Indonesian word for lake
deciduous (of plants) Tendency to shed leaves
deo Vietnamese word for mountain pass
dimorphic Occurring in two unique (colour) forms
dipterocarp Family of tall forest trees well-represented in Southeast Asia
doi Lanna (northern Thai) word for mountain
endemic Restricted to a certain place (in describing distributions of species)
epiphyte Plant that grows on a tree, but does not parasitize it
evergreen Plant that retains its leaves throughout the year
gunung Indonesian and Malay word for mountain
karst Limestone
keranggas Malay word for forest on nutrient poor soil
khao Thai word for mountain or hill
molave *Vitex pauciflora*, tree widespread in Southeast Asia
monotypic Containing only one member, in a family or genus
next generation sequencing Group of molecular technologies used to describe DNA sequences
Nusantara Indonesian term for the Indonesian archipelago
Nusa Tenggara Indonesian term for the Lesser Sunda Islands
nocturnal Active at night
Oriental region a biogeographic region consisting of Southeast Asia, the Indian subcontinent and southern China
pantai Indonesian and Malay word for coast or beach
peat-swamp forest A type of tropical forest that grows on peat substrates
Pleistocene First epoch of the Quaternary period (2.5 million to 11,700 years ago)
riparian Located near the banks of a river
poco Manggarai (west Flores) term for volcano
rawa Malay and Indonesian term for a swampy area
sedentary Not showing migratory behaviour
skulk Habit of hiding in dense cover
speciation Evolutionary process that leads to the formation of new species
sungai/sungei Malay and Indonesian word for river and stream
Sunda region Region that includes the Malay Peninsula, Borneo, Sumatra and Java
Sundaland Large exposed continental shelf that contained the land masses of the Sunda region
tectonic Relating to the Earth's crust and the geological processes taking place involving it
terrestrial Living on land
tropeang Khmer word for small pond
Wallacea Region of islands between Borneo and New Guinea

BRUNEI DARUSSALAM

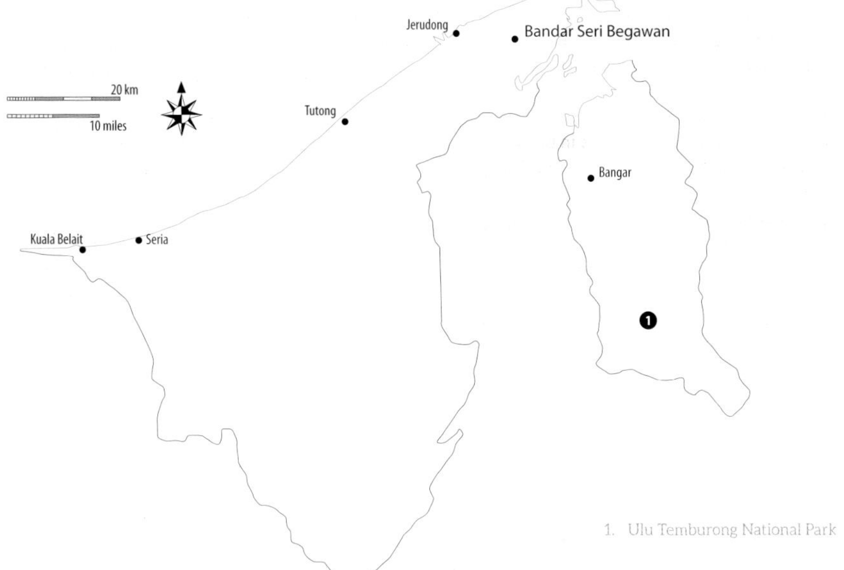

1. Ulu Temburong National Park

The Sultanate of Brunei Darussalam sits in the north-western corner of Borneo, completely surrounded by the East Malaysian state of Sarawak. The four western districts of Brunei are separated by Brunei Bay from the hilly Temburong district, where the country's highest peak of Bukit Pagon (1,850m asl) rises. The western districts of Brunei occupy more than three-quarters of the country's land area, and are relatively flat with extensive lowland dipterocarp and pristine peat-swamp forests.

Oil-rich Brunei is a conservation anomaly in Southeast Asia. As the rest of Borneo's rainforests continue to be converted to oil-palm monoculture at a rapid pace, more than 70% of Brunei's total land area remains covered by rainforest and is protected as reserves or national parks. Even so, it is still one of the most under-birded countries in Southeast Asia, with most birdwatchers heading to Borneo opting to visit the popular rainforest sites in Sabah instead. The few birdwatching reports from this country suggest a rich diversity of birdlife, in particular birds associated with swamp forests such as Storm's Stork and the Large Green Pigeon. In recent years the launch of the international Borneo Bird Race, which takes place over sites in Sabah, Brunei and Sarawak, has exposed more visiting birdwatchers to the key birdwatching sites in Brunei, especially Ulu Temburong National Park.

KEY FACTS

No. of Endemics
0

Country List
465 species

Top 5 Birds
1. Bornean Bristlehead
2. Grey-breasted Babbler
3. Cinnamon-headed Green Pigeon
4. Hook-billed Bulbul
5. Scarlet-breasted Flowerpecker

Climate

Brunei experiences an equatorial climate, so that it is hot and humid throughout the year with plenty of rain. Rain can occur at any time, but is most intense during the north-east monsoon in November–January. The inter-monsoonal months of February–March constitute the driest period of the year.

Access, Transportation & Logistics

The majority of foreign visitors enter Brunei via the international airport near its capital Bandar Seri Begawan, although it is possible to enter the country overland from the East Malaysian state of Sarawak. Overseas visitors from many countries do not require a tourist visa to visit, and can stay in the country for between 14 days and three months, depending on their nationality. As a reflection of the country's wealth, transport infrastructure is excellent, but public transport options are limited due to high car ownership. Consequently, hiring a vehicle provides the best way to explore the country. English is widely spoken and understood, but less so in rural areas, so knowledge of some basic Malay phrases is useful.

Health & Safety

There are no major health concerns in Brunei, although vaccinations for Hepatitis A and B and typhoid are recommended for visiting birdwatchers. Mosquito-borne viruses like dengue fever and Japanese encephalitis are also present, and appropriate precautions should be taken.

Due to a high standard of living and strict laws, Brunei is one of the safest countries in the region. Visitors should be aware that since 2014, Brunei's government has implemented an Islamic Sharia penal code according to which punishments for various offences may be considered harsh by Western standards. Visitors should be aware of these offences to avoid complications during their visits.

Birdwatching Highlights

Species lists for Brunei are few and far between, but local and visiting birdwatchers have recorded close to 300 species across several accessible forest sites throughout the country, such as Ulu Temburong National Park and the swampy forests at Belait. Birdwatchers planning short visits to Brunei should consider spending a few days in the wonderfully pristine lowland forests of Ulu Temburong. The alternative is to spend time at the forest sites and wetlands in western Brunei, focusing on sites like the forests along the Mumong-Balai road (Belait), Badas, Lamunin and Labi, and the wetlands at Seria and Wasan, where various waterbirds, including terns and migratory waders, can be seen.

Species of particular interest to birdwatchers in Brunei include several that are associated with undisturbed swamp forests of Belait, in particular the Grey-breasted Babbler, Wrinkled Hornbill, Hook-billed Bulbul, Scarlet-breasted Flowerpecker and Cinnamon-headed Green Pigeon. The Mumong-Balai road passes through some disturbed areas of swamp forest and is a good site in which to see some of these specialities. A number of sought-after species associated with pristine lowland dipterocarp forests have also been reported, including the sought-after Storm's Stork, Wallace's Hawk-Eagle, Reddish Scops Owl, Large Green Pigeon, Bornean Bristlehead and the undescribed 'Spectacled Flowerpecker'.

Lim Kim Chuah

*A speciality of Brunei's peat swamp and kerangas forests is the breathtaking **Scarlet-breasted Flowerpecker.***

Jimmy Chew

*The **Black-thighed Falconet** appears to replace the White-fronted Falconet in Brunei, and may be seen along the forest edge in Badas and Belait.*

Ulu Temburong National Park

Lim Kim Seng

Lim Kim Keang

Extensive areas of undisturbed lowland dipterocarp forests cloak more than half of Brunei, in contrast to the logging, forest fires and oil-palm monoculture that has overrun the rest of Borneo.

KEY FACTS

Nearest Major Towns
Batang Duri, Bangar

Habitats
Primary lowland and hill rainforests, submontane forest

Key Species
Lowland White-crowned and Helmeted Hornbill, Brown Barbet, Malaysian Honeyguide, Blue-banded Pitta, Rail-babbler, Bold-striped Tit-Babbler, Bornean Wren-Babbler, Grey-chested Jungle Flycatcher, Malaysian Blue Flycatcher, Yellow-rumped and Yellow-vented Flowerpeckers
Submontane Bornean Bulbul
Winter Red-rumped Swallow, Asian House Martin, Grey Wagtail

Other Specialities
Sunda Colugo, Sunda Pangolin, Horsfield's Tarsier, Hose's Langur*, Bornean Gibbon*, Yellow-throated Marten, Plain Pygmy Squirrel*, Tufted Ground Squirrel*, Binturong, Bearded Pig

Best Time to Visit
February–October

This park covers 550km^2 of primary, hill and submontane forests in Temburong District, eastern Brunei. Its topography is hilly, with steep slopes, narrow ridges and impressive waterfalls. The highest points in the park are at Bukit Belalong (913m asl) and Bukit Pagon (1,850m asl) on the border with Sarawak. There are no roads in this remote region and access is by river longboats plying the Sungai Temburong. Ulu Temburong is also famous for its Canopy Walkway, which spirals 50m above the forest floor, enabling scenic views of the rainforest and close-up looks at the forest canopy. For the birdwatcher, the main draw is the majority of Borneo's lowland and submontane endemics that occur in this park, alongside species that are absent in northern Borneo (Sabah), such as the **Black-thighed Falconet**, **White-rumped Shama** and **Grey-breasted Spiderhunter** as well as the undescribed **'Spectacled Flowerpecker'**. Most visitors concentrate on the Canopy Walkway and the resort grounds, and also undertake river trips along the Temburong River to see the **Oriental Darter**.

Birdwatching Sites

Headquarters & Ulu Ulu Resort Grounds

Situated south of the confluence of two major rivers, the Sungei Temburong and Sungei Belalong, the extensive headquarters and resort complex offer a first opportunity to get to grips with the birds of the riverine forest and lowland rainforest edge. The fruiting figs by the resort grounds attract a variety of barbets, bulbuls and leafbirds. Widespread Sundaic species commonly found around the headquarters include the **Black-and-red Broadbill, Grey-chested Jungle Flycatcher** and **Malaysian Blue Flycatcher**. The endemic **Brown Barbet** and **Yellow-rumped Flowerpecker** are common in the resort grounds, as is the shy

Con Foley

Finsch's Bulbul *can be seen from Ulu Temburong's canopy walkway.*

Bold-striped Tit-Babbler. Also interesting are a few species that are absent in North Borneo but found in the resort grounds, such as the White-rumped Shama and Grey-breasted Spiderhunter.

Sungei Temburong

Birds that frequent the river, which can be explored by hiring boats from the park headquarters, include the Oriental Darter, Grey-headed Fish Eagle, Blue-eared Kingfisher and White-chested Babbler.

Con Foley

*The **Scarlet-rumped Trogon** is common in Ulu Temburong's lowland forest.*

Canopy Walkway

There is an excellent network of boardwalks, suspension bridges and steel towers than span a total of 7km, with the best for birdwatching being the 50m-tall Canopy Walkway. The forest below the Canopy Walkway itself is excellent for babblers, including the endemic Bornean Wren-Babbler and the scarce and monotypic Rail-babbler. At the lofty towers, it takes some effort to climb to the top, and it is worth spending a whole morning up in the canopy to watch for hornbills, as seven species occur here, as well as a diversity of barbets, broadbills and bulbuls. The scarce Malaysian Honeyguide has been recorded here, as well as the localized Black-thighed Falconet and Finsch's Bulbul.

Access & Accommodation

The journey to the national park is an adventure in itself. It starts in the capital city of Bandar Seri Begawan, where you can take a 'water taxi' to Bangar town. The journey takes about an hour and allows you to see extensive mangrove forests and the resident Proboscis Monkeys, a Bornean mammal endemic. Next, you need to travel by road for half an hour to the next town, Batang Duri, where a longboat can be hired to take you upstream to the park headquarters and resort. This last journey usually takes an hour, depending on season. Accommodation is situated beside the river in the form of the eco-friendly Ulu Ulu Resort, a joint venture between a local tour operator and the Brunei government. Visits can be arranged through the many tour agents based in Bandar Seri Begawan.

Conservation

This park was designated as Brunei's first national park in 1991 and is managed by the Forestry Department of the Ministry of Industry and Primary Resources. There is hope for the future, as the park is located near the border with Sarawak and Kalimantan, where the 'Heart of Borneo' initiative is being planned to protect some 220,000km^2 of Borneo's rainforests.

Bjorn Olesen

*The **Yellow-rumped Flowerpecker**, a Bornean endemic, is commonly seen around the Ulu Ulu resort grounds.*

CAMBODIA

1. Kulen Promtep
2. Ang Tropeang Thmor
3. Kampi Pools
4. Prek Toal
5. Stoung-Chikreng
6. Seima
7. Bokor
8. Phnom Aural

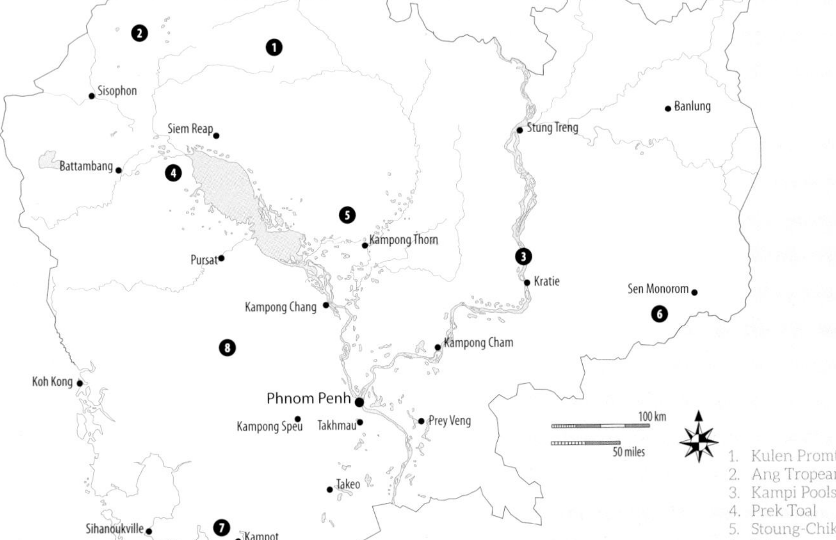

Sprawling across the low-lying floodplain of the lower Mekong River, the Kingdom of Cambodia is relatively flat compared with its neighbours. Southeast Asia's largest lake, the Tonlé Sap, is Cambodia's best-known geographical feature, and is fed by the mighty Mekong via the Tonlé Sap River. With the rains of the south-west monsoon, the lake expands to nearly 16,000km^2. Cambodia is bounded to its north-east by the tail end of the Annamite Mountains, and to the south by the Cardamom Mountains, which rise to the country's highest point at Phnom Aural (1,810m asl).

Marred by a recent history of violence and genocide during the turbulent Khmer Rouge regime, Cambodia was deemed too dangerous for the visiting birdwatcher and was largely overlooked for many years as a birdwatching destination. However, due to significant improvements in tourist infrastructure and increased foreign interest in local conservation projects, the country now offers visiting birdwatchers the opportunity to connect with some of the rarest waterbirds in the world, complemented by a supporting cast of country endemics and regional specialities. The opportunity to visit some of the last great wilderness areas in Indochina, coupled with cultural highlights such as the renowned temples of Angkor Wat and the sobering Tuol Sleng Museum in Phnom Penh, make the country an increasingly popular birdwatching destination.

KEY FACTS

No. of Endemics
3 (Cambodian Tailorbird, Chestnut-headed Partridge, Cambodian Laughingthrush)

Country List
613

Top 5 Birds
1. Giant Ibis
2. White-rumped Falcon
3. Bengal Florican
4. Chestnut-headed Partridge
5. White-winged Duck

Climate

Cambodia experiences a tropical monsoon climate with distinct wet and dry seasons. The warmest and driest time of the year occurs between November–April, during which diurnal temperatures can soar to more than 40° C. During the rainy season in May–October, torrential rain floods a large expanse of the Tonlé Sap Lake and renders many of the unpaved roads unpassable, severely limiting access to birdwatching sites.

Access, Transportation & Logistics

The vast majority of foreign visitors enter the country via the tourist hub of Siem Reap in the country's north-western region, located on the

Bjorn Olesen

The Critically Endangered ***Red-headed Vulture*** *is now best seen in Southeast Asia in northern Cambodia.*

northern banks of the Tonlé Sap Lake and close to the world-renowned temples of Angkor Wat. An alternative entry point is via the country's capital Phnom Penh, located in the south of the country. Most foreign visitors require a visa to enter Cambodia, which can be obtained from both airports on arrival. Transport infrastructure is generally neglected, and roads are unpaved and poorly maintained away from the major towns and cities. To further complicate matters, English is not spoken or understood in rural areas. Nevertheless, birdwatching sites around the Tonlé Sap Lake and Mekong River are still fairly accessible via day trips from Siem Reap or Phnom Penh.

Health & Safety

Vaccinations for Hepatitis A and B and typhoid are recommended for visiting birdwatchers. Malaria is prevalent in remote areas and prophylaxis should be taken, although this is less of a problem in the dry season when most birdwatchers visit.

Unexploded land mines are an unwelcome legacy of Cambodia's violent past and a serious safety concern in remote areas. Although prominent warning signs are erected wherever minefields are present, visiting birdwatchers are strongly recommended to keep to walking paths and follow the directions of local guides to avoid the possibility of serious injury.

Birdwatching Highlights

A comprehensive two-week trip can record upwards of 250 species. Day trips around the Tonlé Sap Lake yield globally threatened waterfowl, including the Greater Adjutant, Milky Stork, Sarus Crane and Critically Endangered Bengal Florican, while wintering passerines such as the Yellow-breasted Bunting and Manchurian Reed Warbler skulk in the reedbeds.

The remote Preah Vihear and Stung Treng provinces in the country's north is the focal point of any birdwatching trip to the country, where you can expect to encounter the Critically Endangered Giant and White-shouldered Ibises, up to four species of vulture at Southeast Asia's most accessible vulture restaurant, the enigmatic White-winged Duck, and a multitude of woodpeckers and other regional specialities including the White-rumped Falcon and Pale-capped Pigeon.

To the east, the picturesque town of Kratie offers the best opportunity to see the near-endemic Mekong Wagtail, and all the three country endemics can be seen in the southern province of Kampong Speu, where Cambodian Tailorbirds lurk in riverine scrub around Phnom Penh, while both the Cambodian Laughingthrush and Chestnut-headed Partridge require a strenuous trek up Phnom Aural, the country's tallest mountain, for viewing opportunities.

Con Foley

The ***Giant Ibis*** *is Cambodia's national bird.*

Kulen Promtep Wildlife Sanctuary & Preah Vihear Protected Forest

Frederic Goes

Bjorn Olesen

Thmat Baeuy is the best site in the world to observe the mythical ***Giant Ibis****, Cambodia's national bird.*

Kulen Promtep Wildlife Sanctuary (4,100km²) and Preah Vihear Protected Forest (1,900km²) form two large adjacent protected areas in Cambodia's northern plains. Together, this extensive landscape of lowlands and gentle hills represents the largest remaining tract of deciduous dipterocarp forest in Southeast Asia. The savannah-like habitat, with an open canopy and grassy understorey, has been created and is maintained by regular fires. Extending further east of the Mekong, these forests form the last global or regional refuge for five Critically Endangered species, namely **Giant** and **White-shouldered Ibises**, as well as three vulture species. The seasonal pools and Angkor-era reservoirs peppering the landscape also make it a stronghold for a suite of other rare waterfowl, such as the **Sarus Crane**. No less than 17 species of woodpecker are found here, as well as the **Indian Spotted Eagle**, **Spotted Wood Owl** and **White-rumped Falcon**. Additionally, tracts of semi-evergreen and gallery forests support the **White-winged Duck**, **Green Peafowl** and **Pale-capped Pigeon**. In all, 360 bird species, representing 60% of Cambodia's avifauna, have been recorded at these two important sites.

KEY FACTS

Nearest Major Town
Tbeng Meanchey (Preah Vihear Province)

Habitats
Dry deciduous and mixed evergreen forests, wetlands, open grasslands

Key Species
White-winged Duck, Green Peafowl, Siamese Fireback, Woolly-necked Stork, Lesser Adjutant, Giant and White-shouldered Ibises, White-rumped, Slender-billed, Himalayan and Red-headed Vultures, Indian Spotted Eagle, Masked Finfoot, Pale-capped Pigeon, Spotted and Brown Wood Owls, Brown Fish Owl, Oriental Scops Owl, Black-headed, Great Slaty and White-bellied Woodpeckers, White-rumped Falcon, Collared Falconet, White-browed Fantail, Brown Prinia, Scaly-crowned Babbler
Winter Himalayan Vulture, Greater Spotted Eagle, Hainan Blue Flycatcher

Other Specialities
Asiatic Jackal, Banteng, Eld's Deer

Best Time to Visit
December–April

Con Foley

The area around Boeng Toal supports the healthiest population of vultures in Southeast Asia, and at least three species, including this ***Slender-billed Vulture****, can be seen at leisure around specially appointed hides.*

Birdwatching Sites

KULEN PROMTEP WILDLIFE SANCTUARY

Thmat Baeuy

Arguably the top birdwatching site in the area, most of the key species can be seen within three days. Local guides know the nesting or roosting (depending on season) sites of the **White-shouldered Ibis**, which makes locating it straightforward. To encounter the near-mythical **Giant Ibis**, walking from *tropeang* (forest pool) to *tropeang* is the usual practice. Listening for the loud, bugling call/duet uttered at dawn helps direct you to the right area. Approaching pools silently and with great care is required to allow observation of feeding ibises. The pools also host a variety of storks, including foraging **Lesser Adjutants**, and **Woolly-necked** and **Black-necked Storks** (rare). Vegetation along the banks is home to the handsome **Asian Golden Weaver**, as well as the rare wintering **Manchurian Reed Warbler**.

Bjorn Olesen

*The largest population of the Critically Endangered **White-shoulded Ibis** occurs in Cambodia's northern plains.*

In the open forest **White-rumped Falcons** sit unobtrusively in the canopy, while the **Collared Falconet** favours dead/burned trees in clearings. An impressive diversity of woodpeckers generally enlivens the proceedings, and features the specialized **Black-headed Woodpecker** (ubiquitous), **Streak-throated Woodpecker** (fairly common), and **Rufous-bellied** and **Yellow-crowned Woodpeckers** (uncommon). The large and vocal **Great Slaty Woodpecker** is regularly seen. Other species regularly noted in this habitat include the **Chinese Francolin**, **Rufous-winged Buzzard**, **Yellow-footed Green Pigeon**, **Blossom-headed** and **Alexandrine Parakeets**, **Indochinese Cuckooshrike**, **Red-billed Blue Magpie**, **Indochinese Bush Lark** and **White-browed Fantail**.

The grassy bamboo understorey teems with prinias, including the localized **Brown Prinia** and skulking **Chestnut-capped Babbler**. In the skies above, the uncommon **Indian Spotted Eagle** may also be seen. The local guides have become very good at locating **Spotted** and **Brown Wood Owls** at their day roosts, while nocturnal forays may yield the **Oriental Scops Owl** and **Brown Fish Owl**.

Last but not least, a riparian forest with stands of bamboo to the east of the village (about an hour on a bad track) supports the **Green Peafowl** and unpredictable **Pale-capped Pigeon**, in addition to other species that favour the wetter forest.

Prey-Veng Reservoir

This Angkor-era reservoir located two and a half hours north-west of Thmat Baeuy is mostly visited for the **White-winged Duck**, which may be seen fairly reliably from a hide at dawn or dusk. The **Black-necked Stork** and **Grey-headed Fish Eagle** can often be seen near the reservoir. The **Giant Ibis** also occurs in the surrounding deciduous dipterocarp forest alongside the usual suite of species.

Lim Kim Chuah

Kulen Promtep and Preah Vihear collectively protect the largest remaining tract of deciduous dipterocarp forest in Southeast Asia.

Stung Memay at Antiel

This remote site is not on Cambodia's birdwatching map yet, but is arguably the best site in the country for the enigmatic **Masked Finfoot**. The species has a small breeding population there and has been recorded annually since 2009 between July–February.

PREAH VIHEAR PROTECTED FOREST

Boeng Toal

A three-hour drive from Thmat Baeuy, the 'vulture restaurant' in Chhep district has recently moved 2km east of Veal Krous, where it was previously conducted. A cow is now slaughtered monthly and left for the vultures as part of a monitoring programme. **Red-headed**, **Slender-billed** and **White-rumped Vultures** can be observed at leisure from a hide near the feeding station. Rare winter treats include immature **Himalayan** and **Cinereous Vultures**, which occasionally join in the feast. The **Indian Spotted Eagle** may also be seen alongside many of the dry dipterocarp specialists mentioned above.

O'Koki

Situated near the border with Laos and four hours from Thmat Baeuy, O'Koki is another site where the endangered **White-winged Duck** can be regularly seen from a hide overlooking a feeding pool. Other birds found in the semi-evergreen forest here include the **Siamese Fireback**, **Oriental Bay Owl**, **Blyth's Frogmouth**, **Black-and-red Broadbill** and **Bar-bellied Pitta**.

Access & Accommodation

Preah Vihear's provincial capital of Tbeng Meanchey is easily accessed by regular buses running from Phnom Penh or any provincial town. However, to reach the birdwatching sites, private transport must be organized, preferably involving a four-wheel drive vehicle. Thmat Baeuy is a 45-minute drive north of Tbeng Meanchey, but birdwatchers usually travel straight from Siem Reap (three hours). Visitor facilities are available here, and include an eco-lodge and experienced (though Khmer-speaking only) local guides. Other sites are more remote and less travelled, with very basic accommodation (safari-style tents or village houses), although an eco-lodge has also recently been built at Prey-veng. No official permit or park entrance fee is required, but birdwatching tourism is exclusively organized through the Sam Veasna Center. Its services include transport, food, an English-speaking birdwatching guide and a set donation to a village conservation fund.

Benjamin Schweinhart

The ***White-rumped Falcon*** *is a speciality of Southeast Asia's dry deciduous forests.*

Conservation

Since wide-ranging surveys, including camera trapping conducted in 2001, unveiled their irreplaceable global biodiversity significance, the northern plains of Cambodia are a top priority for conservation in the region. Kulen Promtep WS (designated in 1993) is managed by the Ministry of Environment, while Preah Vihear PF (designated in 2002) is under the management of the Forestry Administration. Both benefit from strong financial and technical support provided by the Wildlife Conservation Society. Ranger support and capacity building, combined with community benefits through nest-protection programmes and birdwatching tourism, have significantly reduced hunting and poaching. However, in recent years the rate and scale of logging operations have surged in the wake of large Economic Land Concessions granted by the central government. Disturbingly, these are even overlapping protected areas and posing an insurmountable challenge for conservation to date, with significant habitat loss and fragmentation expected in the near future.

Ang Tropeang Thmor Sarus Crane Conservation Area

Frederic Goes

Bjorn Olesen

Ang Tropeang Thmor is the most accessible place in Southeast Asia to see large congregations of the regal ***Sarus Crane****.*

Ang Tropeang Thmor is a vast reservoir built during the reign of the Khmer Rouge. It covers 10km² of the 129km² that make up the Sarus Crane Conservation Area, but expands up to five times this area during the wet season. Elevated dikes demarcate the southern and eastern boundaries of the reservoir. In the north the wetland transitions into grasslands, ricefields and a peculiar savannah forest. During the dry season this site hosts the largest congregation of the eastern race of the **Sarus Crane** (ssp. *sharpii*) and **Knob-billed Ducks** in Southeast Asia. It also supports a large waterbird colony comprising **Painted** and **Milky Storks**, as well as **Spot-billed Pelicans**. The reserve is the only site in the region where the endangered Eld's Deer can be easily seen. The wetlands of Ang Tropeang Thmor are the most important site for wintering ducks in Cambodia, and act as an important dry-season refuge and staging area between the dry deciduous forest of the north and the Tonlé Sap Lake. Nearly 300 bird species have been recorded at this site, including many firsts for Cambodia.

KEY FACTS

Nearest Major Town
Siem Reap city (Siem Reap Province)

Habitats
Freshwater wetlands, grasslands, paddy fields, dry deciduous forest

Key Species
Knob-billed Duck, Milky, Painted and Black-necked Storks (April–May), Spot-billed Pelican, Bengal Florican, Sarus Crane
Passage Oriental Plover
Winter Greater Spotted Eagle, Pied Harrier

Other Specialities
Lyle's Flying Fox, Eld's Deer

Best Time to Visit
January–May; rainy season June–October

Birdwatching Sites

Prey Mwan to Phnom Srok District

From the National Route 6 turn off at the silk weaver statue; the laterite road passes through dry seasonal ricefields and offers excellent birdwatching. It is possible to see flocks of **Sarus Cranes**, storks, waders and shorebirds, as well as both **Eastern Marsh** and **Pied Harriers**.

Michelle & Peter Wong

*The vulnerable **Greater Spotted Eagle** is a regular winter visitor.*

Thnol Toteang (Waterbird Colony)

This site is located at the south-western end of the reserve, near the third water gate along the southern dike. A dirt track heading north passes through several rickety bridges and leads to a good observation post. Among the nesting **Painted Storks** and **Spot-billed Pelicans**, a couple of **Milky Storks** are usually present, although close scrutiny often reveals hybrids. In 2014, a lone **Black-faced Spoonbill** visited for a few days. **Knob-billed Ducks** are often seen in the wet fields in this area. The **Cotton Pygmy Goose**, **White-browed Crake** and various waterfowl thrive in the aquatic vegetation. The **Greater Spotted Eagle** is often seen soaring in the distance, along the forest edge towards the north.

Reservoir at Pong Ro

From the village of Pong Ro on the eastern dike, vast sections of the reservoir can be scanned at leisure for Palearctic ducks like the **Northern Shoveler** and **Garganey**, as well as jacanas and other waterfowl.

Kob Leav & Chork Thom

Access to the grassland belt at the centre of the site allows you to approach parties of **Sarus Cranes** feeding on the roots of the Water Chestnut (*Eleocharis dulcis*). However, these birds are highly mobile and sensitive to disturbance, frequently changing roosting and feeding areas throughout the season, so a local guide proves invaluable here. Late into the dry season, **Black-necked Storks** may be seen here. Other notable grassland species include the **Greater Painted-snipe**, **Lanceolated Warbler**

Bjorn Olesen

__Lesser Whistling Ducks__ are common in the area.

Bjorn Olesen

*Ang Tropeang Thmor is one of few regular sites for the **Black-necked Stork** in Cambodia.*

Ang Tropeang Thmor is a vast reservoir that is seasonally flooded and supports large congregations of waterfowl and grassland avifauna.

and various quails, buttonquails, bush larks and weavers. The sighting of a Bengal Florican (March–May) is a big bonus, since only 1–2 pairs are known to subsist at this site.

Kok Samrong

The savannah forest is worth exploring for those wishing to see resident Black Kites, as well as herds of the endangered Eld's Deer. Dry fields may hold Oriental Plovers on passage in March. Regional rarities such as the Lesser Kestrel and Short-toed Snake Eagle have also been recorded. To get a taste of birds of the deciduous forest, a drive further north towards Kon Khleang settlement is needed. Highlights here include the Rufous-winged Buzzard, Black-headed Woodpecker and occasional Spotted Wood Owl.

Access & Accommodation

Ang Tropeang Thmor is usually accessed from Siem Reap town. It takes one and a half hours to drive to the site, taking National Route 6 towards the Thai-Cambodian border, then heading north on a laterite dirt track. The turn-off is either through Phnom Srok district (Prey Mwan turn-off, at the silk weaver statue) or further west at Preah Net Preah district. For the birdwatcher travelling overland from the Poipet border post, the turn-offs are a 30-minute drive from Sisophon provincial town. Accommodation is available at the local conservation office. Meals can also be prepared on demand. The entrance fee includes a local ranger as guide, who is usually knowledgeable and helpful, but only speaks Khmer. No permit is required, although birdwatching tourism is exclusively organized through the Sam Veasna Center, which offers all-inclusive packages including English-speaking birdwatching guides.

Conservation

In 1997, the discovery of a large population of Sarus Cranes by the late Sam Veasna, Cambodia's best known conservationist, prompted a Royal Decree designating the site as a Sarus Crane Conservation Area in 2000. A local conservation team, supported by the Wildlife Conservation Society, has been established to monitor human activities and protect the site's wildlife and habitats. Poaching and the trade in wildlife have been significantly reduced, but still occur. However, the main threats are the conversion of grasslands for rice farming, and hydrological changes due to irrigation.

Mekong River at Kampi & Four-Arms Plain

Frederic Goes

Frederic Goes

An area of fast-flowing rapids near the town of Kratie is a regular haunt of the recently described Mekong Wagtail.

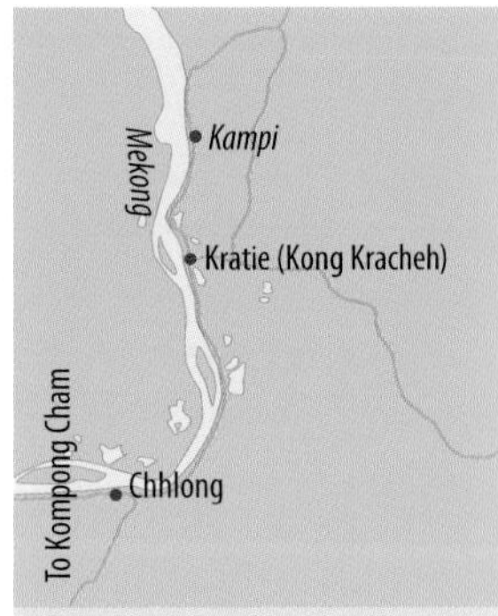

KEY FACTS

Nearest Major Towns
Kratie, Phnom Penh

Habitats
Rocky rapids, sandbars, marshes and riverine scrub

Key Species
Oriental Darter, Small Pratincole, Grey-throated Martin, Cambodian Tailorbird*, Asian Golden Weaver, Red Avadavat, Mekong Wagtail
Winter Pallas's Grasshopper Warbler, Lanceolated Warbler, Black-browed Reed Warbler, Yellow-breasted Bunting

Other Specialities
Irrawaddy Dolphin

Best Time to Visit
November–May

James Eaton

*The **Cambodian Tailorbird** is one of the few species dependent on scrub habitats on the Mekong floodplains.*

The stretch of the Mekong River running through Cambodia hosts two recently discovered species: the **Mekong Wagtail** (described in 2001) along the rapids upstream down to Kratie provincial town, and the endemic **Cambodian Tailorbird** (described in 2013) in the lower Four-Arms floodplain formed with the Tonlé Sap and Bassac Rivers around Phnom Penh. Kampi, a small village just north of Kratie, offers the opportunity to watch the Mekong subpopulation of the globally threatened Irrawaddy Dolphin, which uses the deep pools as a dry-season refuge, as well as the **Mekong Wagtail** along the rapids bordering the dolphin pools. An added bonus is an area of marshland at the edge of Kratie, where the **Asian Golden Weaver** is most reliably seen in Cambodia. Downstream from Kompong Cham, the wide, seasonally flooded shrublands are the home of the **Cambodian Tailorbird** as well as a range of wetland species.

Birdwatching Sites

Kampi Pools

Wooden boats docked at the riverbank of the 'Mekong Dolphin' watching site are easily rented and allow close approach to the **Mekong Wagtail**'s habitat. Note that some wintering **White Wagtails** also occur. The **Small Pratincole** and localized **Grey-throated Martin** are also present. Other riverine species likely to be encountered include the **Indian Spot-billed Duck**, **Indian Cormorant** and **Pied Kingfisher**.

Kratie Marsh

A short walk south-east of Kratie town brings you to an area of marshland that should yield views of the **Asian Golden Weaver**

Francis Yap

***Small Pratincoles** rest on open, sandy areas by the river.*

Yann Muzika

*A **Mekong Wagtail** forages in the scrub and open areas along the riverbank.*

(in breeding plumage starting from February), as well as **Baya** and **Streaked Weavers**. The **Red Avadavat** is occasionally seen here. Various rails, bitterns and wetland passerines are present, with **Pallas's Grasshopper Warbler** being common and fairly easy to spot.

Kompong Cham Shrublands

Any area of dense evergreen scrub accessible from Highway 6 or 7 to Phnom Penh may yield the **Cambodian Tailorbird**. The usual spot is called Prek Ksach, a reclaimed area east of the Prek Dam Bridge on Highway 7. Weavers, including the **Asian Golden Weaver**, can also be found. In addition to the species occurring in Kampi, more widespread waterfowl such as the **Cotton Pygmy Goose**, **Black** and **Cinnamon Bitterns**, **Black-backed Swamphen** and terrestrial birds such as the **Green-billed Malkoha**, **Racket-tailed Treepie** and **Plain-backed Sparrow** are also present. In the winter months, scattered shorebirds including Cambodia's only records of the white-faced form (ssp. *dealbatus*) of the **Kentish Plover** to date are an added attraction.

Access & Accommodation

Kratie town is served by most bus companies from the capital and all provincial towns. A passenger or merchant boat trip up the Mekong from Phnom Penh or Kompong Cham is also possible but slower. Kratie has an array of river-front hotels. To reach Kampi (12km north) and the marsh, it is best to rent a car or moto-taxi from Kratie town. Prek Ksach is about 30–45 minutes' drive from Phnom Penh, and an hour and a half from Kompong Cham, where a range of accommodation is available.

Conservation

Neither site is formally protected or under any active conservation management. Despite legal protection of the dolphins, they are still in decline due to entanglement in fishing nets. The Kampi site constitutes the southernmost limit of the Mekong Wagtail's range. Larger populations associated with a broader community of sensitive riverine-nesting birds (River Tern, River Lapwing, Great Stone-curlew, White-shouldered Ibis, Green Peafowl) are found in the braided Mekong channel further north, where field conservation activities are undertaken. No dedicated protected area currently incorporates the Cambodian Tailorbird's core range. Reclamation of large areas of floodplain around the capital for land development is the main threat faced by the species.

Francis Yap

*Wintering **Pallas's Grasshopper Warblers** are present in the scrub by the Mekong.*

Prek Toal 'Bird Sanctuary'

Frederic Goes

Frederic Goes

The flooded forest of the Tonlé Sap Biosphere Reserve is the most important waterbird nesting site in Southeast Asia.

KEY FACTS

Nearest Major Town
Siem Reap city (Siem Reap Province)

Habitats
Freshwater swamp forest, freshwater wetlands (e.g. marshes), streams

Key Species
Greater and Lesser Adjutants, Milky and Painted Storks, Asian Openbill, Black-headed Ibis, Spot-billed Pelican, Oriental Darter, Grey-headed Fish Eagle

Other Specialities
Hairy-nosed and Smooth-coated Otters, various watersnakes

Best Time to Visit
November–March

Wang Bin

Grey-headed Fish Eagles *are often seen the flooded forests around the lake.*

Featured in all travel guides to Cambodia as the 'bird sanctuary', Prek Toal is Southeast Asia's last breeding stronghold for a suite of globally threatened large waterbirds. Located in the north-western belt of the Tonlé Sap, its 250km² of relatively undisturbed flooded forest with a maze of creeks and pools provides nesting trees and an abundance of fish, which sustain more than 30,000 breeding pairs of large colonial waterbirds. Between September–January these birds converge from across the region to nest at this site, forming vast mixed species colonies. They include the **Oriental Darter**, **Spot-billed Pelican**, and globally threatened storks such as the **Milky Stork** and **Greater Adjutant**. Besides a supporting cast of cormorants, terns, waders, rails and other waterfowl, piscivorous raptors like the **Grey-headed Fish Eagle** are ubiquitous. The sheer density of waterfowl makes a visit to Prek Toal a memorable experience.

Birdwatching Sites

Visiting birdwatchers to Prek Toal should bear in mind that the lower the water levels in the lake, the more birds there will be, but accessibility will be compromised as well.

Breeding Waterbird Colonies

Viewing at least one of the breeding colonies from a watchtower is a necessity for any visiting birdwatcher. Colony composition and location vary from year to year, so no single spot can be recommended. In addition, the time of year dictates which colonies are the most attractive and accessible. The most often visited colony is accessed through the Prek Spot stream. Alternatively, further south of this there is yet another complex

James Eaton

*The **Greater Adjutant** occurs in small numbers at Prek Toal.*

of colonies accessed through the Prek Preah Daem Chheu and the narrow Prek Day Kray Kreng.

Reaching a waterbird colony may take several hours, and require transfer to small wooden canoes when the water level is low. Rangers have established watch platforms in the canopies of tall trees, from where scope views allow detailed study and scanning for target species. Climbing the artisanal wooden ladder as well as the softly moving bamboo-floored top platform may discourage some, but they are sturdy and the rewards are worth the effort.

In November and December the forest is still submerged and only swimming and diving waterfowl, including the Spot-billed Pelican and Oriental Darter, are actively nesting on the partially exposed trees. However, the sought-after Greater Adjutant generally settles in by November, and the high water level allows close approach to any nesting tree. By January, the full set of species is present, ensuring *ad libitum* observation of the Painted Stork, Asian Openbill, Lesser Adjutant and Spot-billed Pelican. A large stork-pelican colony usually holds a decent number of Greater Adjutant nests, as well as a few pairs of Milky Storks, although gaining a satisfying view of both species from a single platform is another matter entirely.

Flooded Forest & Streams

When water levels permit, a boat can navigate through the flooded forest and streams. This ensures an uninterrupted spectacle of a variety of waterbirds going about their daily routine, often in impressive flocks. Innumerable herons, egrets and cormorants are the key birds seen, along with the occasional bittern, rail or Stork-billed Kingfisher. The Racket-tailed Treepie and Green-billed Malkoha generally flush from the trees, while an opportunistic Peregrine Falcon may hover around a boat. In winter the Chestnut-winged Cuckoo is regularly reported from the forest, while the unobtrusive Masked Finfoot is very occasionally seen along the streams.

Tonlé Sap Lake Shore

As water levels drop, boats have to make a detour along the lake shore to reach the colonies. This also provides good birdwatching opportunities. Flocks of Spot-billed Pelicans forage there, while Whiskered Terns may number in the thousands. By April, falling water levels unveil mudflats that attract egrets and herons, Black-backed Swamphens and migratory shorebirds.

Access & Accommodation

Prek Toal is only accessible by boat and is a complicated and costly proposition for an independent birdwatcher. It is preferable to book a tour with the Cambodian Bird Guide Association (specialized birdwatching trip) or with Osmose (ecotourism visit); both have guides familiar with the birdlife. From Siem Reap, a car, tuk-tuk or moto-taxi (20–40 minutes) brings you to the Tonlé Sap shore at Chong Khneas or Mechrey harbour (less travelled). An hour-long crossing through the northern tip of the open lake brings you to Prek Toal village, the gateway to the reserve. The core area's office issues entrance tickets (priced according to group size) before assigning one ranger to each visitor's group and eventually organizing transfer to a local boat. The rangers know the sites very well and are usually good at spotting birds, but they may not be fluent in English. Basic concrete accommodation is available at the core area's station, but for a more comfortable stay book a floating homestay with local NGO Osmose.

Con Foley

*Large waterbirds such as the **Spot-billed Pelican**, which are otherwise difficult to see in Southeast Asia, are locally abundant at Prek Toal.*

Conservation

Aerial and boat-based surveys conducted in 1994 across the country uncovered the importance of Prek Toal's waterbird colonies and the principal threat to their survival: large-scale egg and chick collection by poor villagers for subsequent sale as food in the floating villages. In 1997, UNESCO designated the Tonlé Sap as a Biosphere Reserve, with three core areas strictly dedicated to biodiversity conservation. The Prek Toal core area stood out as the single most important site for large waterbird conservation in Southeast Asia. In 2001, a local conservation team was assembled, enrolling former bird collectors and villagers to ensure close protection of the colonies. The programme has been one of Asia's most inspiring conservation success stories: waterbird poaching was curtailed by 2005, and bird numbers have dramatically recovered. Prek Toal is currently the world's most important breeding site for the **Oriental Darter**, **Spot-billed Pelican**, and **Greater** and **Lesser Adjutants**, to name a few. However, the large fishing concession affording effective protection to the habitat was abolished in 2012, allowing unchecked access and leading to illegal and destructive fishing, associated habitat degradation, disturbance to colonies and a revival of poaching activities.

Lee Tiah Khee

__Asian Openbills__ are among the most common species of large waterbird at Prek Toal.

Frederic Goes

Suitable nesting trees are shared by multiple species of waterbirds in Prek Toal.

Stoung-Chikreng Bengal Florican Conservation Areas

Frederic Goes

Yong Ding Li

The extensive grasslands at Stoung-Chikreng support several globally threatened birds, but are themselves under threat from conversion to agricultural fields.

KEY FACTS

Nearest Major Towns
Kompong Thom, Siem Reap

Habitats
Freshwater wetlands (seasonally flooded grasslands), scrub, ricefields

Key Species
Bengal Florican, Sarus Crane, Common Buttonquail, Red Avadavat
Passage Oriental Plover (March–April)
Winter Great Bittern (rare), Pied and Eastern Marsh Harriers, Greater Spotted Eagle, Manchurian Reed Warbler, Bluethroat, Red Avadavat, Yellow-breasted Bunting

Other Specialities
Asiatic Jackal

Best Time to Visit
December–May

Markus Handschuh

*The Stoung-Chikreng grasslands are among the most important breeding sites in the world for the Critically Endangered **Bengal Florican**.*

The Tonlé Sap floodplain grasslands of the Stoung-Chikreng BFCA (74.5km^2) are the largest remaining tracts of natural floodplain grasslands in Southeast Asia. This highly threatened habitat comprises a mosaic of grasslands of varying heights interspersed with seasonal pools, marshes, scrub and some areas of deep-water rice cultivation. The area is the last stronghold of the Critically Endangered **Bengal Florican** (ssp. *blandini*) outside the Indian subcontinent, and hosts a suite of other grassland specialists such as the **Common Buttonquail**, threatened **Manchurian Reed Warbler**, **Lanceolated Warbler**, **Red Avadavat**, weavers, buntings and bush larks. Large waterbirds including the **Black-necked Stork** also visit the area. Raptors are ubiquitous, with harriers and the **Greater Spotted Eagle** being prominent wintering visitors. Rare passage migrants such as **Oriental Plover** are also regular in March–April, together with various shorebirds.

Birdwatching Sites

Prolay

About 18km after Kampong Kdey market on the road from Siem Reap towards Kompong Thom, the road curves left and a 'Cambodia Beer' banner is visible on the right just before a gas station. Driving through the banner, then along a dirt track, brings you to a tract of good-quality grassland. Oxcart tracks crisscross the vast grassland, and diligent scanning generally affords flight views or sightings of displaying male **Bengal Floricans**. Leaving the car usually flushes some **Lanceolated Warblers** underfoot and, if patient, a florican may be seen

Michelle & Peter Wong

The grasslands at Stoung-Chikreng provide important wintering habitat for the Critically Endangered ***Yellow-breasted Bunting****.*

crouched at a few metres. Finding the tricky **Manchurian Reed Warbler** requires close inspection of clumps of tall grass and scrub. Wintering **Pied** and **Eastern Marsh Harriers** are regularly seen quartering the landscape, and the **Greater Spotted Eagle** may also be present. Flocks of **Sarus Cranes**, **Painted Storks**, **Black-headed Ibises** and **Lesser Adjutants** are fairly regular, while a few **Black-necked Storks** visit annually. The **Common Buttonquai**l and **King Quail** are found in areas of short grass and fallow fields, and small flocks of **Red Avadavats** favour tall grassland, along with flocks of wintering **Yellow-breasted Buntings**.

The scrub-grassland-agricultural mosaic here supports **Pallas's Grasshopper Warblers**, large numbers of wintering **Bluethroats**, some **Siberian Rubythroats** and **Red-throated Pipits**. Rarities to keep an eye out for are the little-known **Rain Quail**, **Eastern Imperial Eagle**, **Eastern Grass Ow**l, **Short-eared Owl** and **Chestnut-eared Bunting**. Various waterbirds, rails and shorebirds can be seen at pools and wetlands. Between March–April small parties of migrating **Oriental Plovers** frequent dry fields with sparse vegetation.

Access & Accommodation

The Stoung-Chikreng grasslands are readily accessible from National Route Number 6 between Kompong Thom and Siem Reap. The turn-off is roughly equidistant from each provincial town, after an hour and a half drive. Many buses ply this route as well, with a bus from Phnom Penh taking at least four hours to reach the site. A more flexible option is to hire a private taxi. No formal accommodation is available at the site. Birdwatching visits are best organized through the Sam Veasna Center and Cambodian Bird Guide Association, which provide qualified English-speaking birdwatching guides and hire local villagers involved in the site's conservation.

Conservation

After Sam Veasna's breakthrough discovery in 1999, further surveys in 2000–2001 confirmed the importance of the Tonlé Sap grasslands for the Bengal Florican and a suite of threatened species. In 2006, key florican breeding areas were granted provincial protection, initially totalling 350km^2. By 2010, these were elevated to national protected areas as the 'Bengal Florican Conservation Areas', including the Stoung and Chikreng BFCA. Between these milestones major threats including widespread hunting and large-scale habitat loss were ongoing. Surveys conducted in 2005 and repeated in 2012 found that nearly half of the extant habitat had been lost and the florican population had declined by 44% in seven years. The conservation significance of the florican as well as the scale of ongoing threats makes this site a top priority for conservation in Cambodia.

A highlight at this site is the exquisite ***Pied Harrier****.*

Seima Protected Forest

Frederic Goes

Seima Protected Forest encompasses some of the largest remaining tracts of lowland forest in Indochina.

Located in southern Mondolkiri province and adjacent to south-central Vietnam, Seima Protected Forest (3,030km^2) is part of the Southern Annamite Lowlands ecoregion, and predominantly consists of evergreen and semi-evergreen (mixed) forest, a landscape very similar to Vietnam's Cat Tien. Seima supports populations of three regional endemics: the **Orange-necked Partridge**, **Germain's Peacock-Pheasant** and **Grey-faced Tit-Babbler**. It is also a global stronghold for the **Green Peafowl**, as well as for a number of threatened mammals such as the Black-shanked Douc, Buff-cheeked Gibbon and Banteng. The deciduous dipterocarp forest in the western sector, which extends into Kratie province, further supplements the site's conservation significance, but is not currently visited by birdwatchers. Nearly 350 species have been recorded to date at Seima.

KEY FACTS

Nearest Major Town
Sen Monorom (Mondulkiri Province)

Habitats
Semi-evergreen and evergreen forests, dry deciduous forest, bamboo groves

Key Species
Orange-necked and Green-legged Partridges, Germain's Peacock-Pheasant, Green Peafowl, Pin-tailed and Ashy-headed Green Pigeons, Annam and Red-vented Barbets, Pale-headed and Bay Woodpeckers, Blue-rumped and Bar-bellied Pittas, Grey-faced Tit-Babbler, Golden-crested Myna

Other Specialities
Black-shanked Douc, Buff-cheeked Gibbon, Asian Elephant, Banteng

Best Time to Visit
December–May

Birdwatching Sites

The park headquarters lie just north of Keo Seima district town, along the main road towards Sen Monorom, Mondolkiri's provincial town. The birdwatching trails leave from the road, bisecting the Protected Forest.

'Orange-necked Partridge' (ONP) Trail

The ONP Trail leaves the main road about 1km south of the headquarters, heading east through slash-and-burn fields. It quickly enters degraded evergreen forest with undergrowth comprising large bamboos, the favourite habitat of the **Orange-necked Partridge**. This trail is also good for observing some of the most sought-after species in Seima: **Germain's Peacock-Pheasant**, **Bar-bellied Pitta** and **Grey-faced Tit-Babbler**. **Pale-headed** and

Wong Lee Hong

*The **Bar-bellied Pitta** is a regional speciality that is regularly observed at Seima Protected Forest.*

Con Foley

*The **Racket-tailed Treepie** is not uncommon in areas of open forest.*

Bay Woodpeckers are fairly easy to find, and parties of **Great Slaty Woodpeckers** often vocalize on the larger trees. **White-browed Piculets** and **Yellow-bellied Warblers** favour the bamboo clumps and are fairly common here. The **Banded Kingfisher** is not hard to locate, while the **Blue-rumped Pitta** has been recorded annually but is decidedly more elusive.

O'Pam Trail

The O'Pam is a stream in the south-east corner of Seima PF. The trail starts just 300m south of the headquarters, and passes through similar lowland forest and bamboo habitats as the ONP Trail. The same suite of species can be found here, including the **Orange-necked Partridge**. Cambodian rarities that have been recorded along these two trails include the **Black-throated Laughingthrush**, **Collared Babbler** and **Pin-tailed Parrotfinch**.

Km 157

This trail passes a ranger outpost on the east side of the main road and enters taller, less disturbed semi-evergreen forest. Typically present here are **Pin-tailed Green Pigeon**, **Red-vented Barbet**, **Large Scimitar Babbler** and **Golden-crested Myna**. **Purple-naped Sunbird** and **Dusky Broadbill** are sometimes encountered. The **Ashy-headed Green Pigeon** is uncommon although annually reported.

Quarry (Km 159)

Further north a large clearing adjacent to the eastern side of the main road is a good spot to watch the **Green Peafowl**, which roosts here. Scanning for Black-shanked Doucs is often rewarded with sightings. Slowly driving along the main road in the late afternoon may also bring sightings of these primates, as they roost in large trees at the edge of the forest.

Andong Kraleung

This site lies west of the road, at the level of the O'Reang ranger station (the last one on the way to Sen Monorom). It consists of semi-evergreen forest with a waterfall. It has recently been assessed for birdwatching tourism, and should be opened soon. Updated information can be obtained at the park headquarters or via the Sam Veasna Center.

Access & Accommodation

Daily buses ply the Phnom Penh–Sen Monorom (Mondolkiri province) road. It takes a minimum of six hours to reach Keo Seima district town,

Cheng Heng Yee

James Eaton

Seima is exceptionally rich in woodpeckers with more than 16 species, including ***Great Slaty*** *(left) and* ***Black-and-Buff Woodpecker****.*

or five hours by private taxi. A few guesthouses offer basic accommodation there. The protected area headquarters is about 10km north. Sen Monorom (a drive of a further hour and a half) has a wider range of accommodation, but is further from the main birdwatching sites. No permit or entrance fees are required, but birdwatching trips are exclusively organized through the Sam Veasna Center, and include both English-speaking birdwatching guides and local guides.

Conservation

In 2000, wildlife surveys conducted in the then Samling logging concession in southern Mondolkiri province highlighted its conservation value for large mammals and birds. The site even yielded the first photograph of a wild Tiger in Cambodia. In 2002, the government declared a Biodiversity Conservation Area, later renamed as a Protected Forest, managed by the Forestry Administration with support from the Wildlife Conservation Society. Since then, active conservation management and enforcement support has greatly reduced the hunting and poaching of key species. Positive conservation outcomes have been confirmed with monitoring programmes showing increasing populations of primates, ungulates and Green Peafowl. Unfortunately, since 2010 the scale and intensity of logging and conversion for plantations has increased significantly, and ensuring the continued persistence of Seima PF is a now major conservation challenge.

Sam Thuong

Seima is the best place to see the range-restricted ***Germain's Peacock-Pheasant*** *in Cambodia.*

Bokor National Park

Frederic Goes

Chi'en C. Lee

The Bokor Plateau features many unusual rock formations.

KEY FACTS

Nearest Major Town
Kampot (Kampot Province)

Habitats
Hill evergreen forest, semi-evergreen forest

Key Species
Chestnut-headed Partridge*, Silver Pheasant, Black Eagle, Yellow-vented Green Pigeon, Spot-bellied Eagle-Owl, Red-headed Trogon, Great and Wreathed Hornbills, Moustached and Green-eared Barbets, Long-tailed Broadbill, Blue Pitta, Indochinese Green Magpie, Orange-headed Thrush, White-tailed Robin, Fire-breasted Flowerpecker
Winter Sulphur-breasted and Alström's Warblers, White-throated Rock Thrush

Other Specialities
Pileated Gibbon, Northern Pig-tailed Macaque

Best Time to Visit
December–March; rainy season May–November

White-throated Rock Thrush is a regular winter visitor to Bokor's forests.

Bokor National Park (1,500km²) encompasses the evocatively named Elephant Mountain Range, a southern outlier of the Cardamom (Krâvanh) Mountains. In the 1920s, the French colonists built a 30km road to access the spectacular plateau at 1,000m asl, where they developed a hill resort. To date, this remains the only accessible part of the park. Despite recent private developments, the plateau retains much of its natural charm. On clear days the view from the ridge is breathtaking and spans the whole Cambodian coastline. Stunted hill evergreen forest cloaks the plateau, while semi-evergreen forest covers the rest of the park. About 270 bird species have been recorded, and Bokor is known as the most accessible site to observe the localized **Chestnut-headed Partridge**, as well as a number of subspecies endemic to the Cardamom Mountains, such as the very distinctive subspecies of the **White-tailed Robin** (ssp. *cambodiana*). It is also an excellent place for seeing forest birds in general, including sought-after species such as the **Spot-bellied Eagle-Owl**, **Great** and **Wreathed Hornbills**, and **Indochinese Green Magpie**.

Birdwatching Sites

The Plateau

Aside from the spectacular view, the edge of the plateau offers good viewing opportunities for aerial birds like needletails and swifts, as well as **Great** and **Wreathed Hornbills**, and other large canopy species, with the rare advantage of being able to look at them from above. The **Wedge-tailed Green Pigeon**, a species rarely recorded elsewhere in Cambodia, is present here. Various places in the derelict French town are readily accessible and can be productive, featuring species such as the hyperactive **Black-throated Sunbird**.

James Eaton

Most birdwatchers visit Bokor for a chance to see the endemic ***Chestnut-headed Partridge****.*

Tea Farm

On the plateau a track to the left of the road (just after the ticket checkpoint) with an old Tea Farm signboard leads to a forested patch where the **Chestnut-headed Partridge** has been seen, as well as **Orange-headed** and **Eyebrowed Thrushes**. The trees above are home to the **Red-headed Trogon**, and foraging flocks contain **White-bellied Erpornis**, endemic subspecies of the **Lesser Racket-tailed Drongo**, **Kloss's Leaf Warbler**, **White-browed Scimitar Babbler**, **Streaked Wren-Babbler**, **White-tailed Robin** and **Fire-breasted Flowerpecker**. The **Black Eagle** is sometimes seen gliding along the forest edge.

Waterfall

Accessed via the right-hand road at the roundabout after the ticket checkpoint, the **Long-tailed Broadbill**, **Barred Cuckoo-Dove** and **Ochraceous Bulbul** may be sighted en route. The waterfall itself is rather devoid of birdlife, but a path starting on the opposite side quickly enters good forest, where the **Chestnut-headed Partridge** was rediscovered in 2000 after a gap of 64 years. Spending time on this track also allows you to see many of the endemic subspecies.

Access Road

The access road deserves some dedicated birdwatching time. Its upper stretches offer the best chance to surprise a **Silver Pheasant** or **Blue Pitta**. Flight views of **Great** and **Wreathed Hornbills** are also possible. The road offers good viewing opportunities for the localized **Yellow-vented Green Pigeon** (ssp. *modestus*), as well as the neat-looking **Blue-bearded Bee-eater**, **Moustached Barbet** and other species attracted to the fruiting fig trees. In particular, Km 14 features a huge strangler fig at a sharp bend where there is a hornbill nesting hole. Raptors are much in evidence, with soaring **Rufous-bellied Eagles** and **Himalayan Buzzards**, the latter often perching along forest clearings. The sought-after **Indochinese Green Magpie** is regularly detected here. During the winter months the **Silver Oriole**, which was last recorded in 2004, is a possibility. On the lower slopes there is a spot near Km 6 where the **Spot-bellied Eagle-Owl** has recently been seen.

Access & Accommodation

Regular bus services run daily between Phn Penh and Kampot with a travel time of 3–4 hours. From Kampot, taxis or moto-taxis may be rented for private tours, although a much cheaper option is to join a day trip in mini-vans, organized by most guesthouses. The park entrance is located 9km west of Kampot, along the coastal road towards Sihanoukville at the foot of the mountain. Entrance fees (20,000 Riels per vehicle) are paid there. The access road is marked with old Km milestones. Major hotels offer comfortable accommodation on the plateau itself. Local rangers can be hired at the park headquarters, but none can speak more than very basic English and they possess limited field knowledge. It is also best to avoid this site on weekends and public holidays.

Conservation

Bokor National Park was designated under the Royal Decree on protected areas in 1993. Major wildlife and vegetation surveys were conducted in 1998–2000 to document the conservation value of the park. When security allowed access to the vast Cardamom Mountains, the conservation significance of Bokor appeared rather relative, as it only supports a subset of the endemic biodiversity of the region. Furthermore, large-scale logging and plantations have degraded and continue to seriously degrade the habitat in the lower elevations. The montane sectors of the park remain less affected by logging, but poaching is rampant and widespread. In recent years large-scale tourist developments have resulted in further habitat degradation.

Con Foley

The ***Large Scimitar Babbler*** *forages in the forest understorey in small, noisy parties.*

Phnom Aural Wildlife Sanctuary

Frederic Goes

Frederic Goes

Large tracts of contiguous forest still exist on the slopes of Phnom Aural.

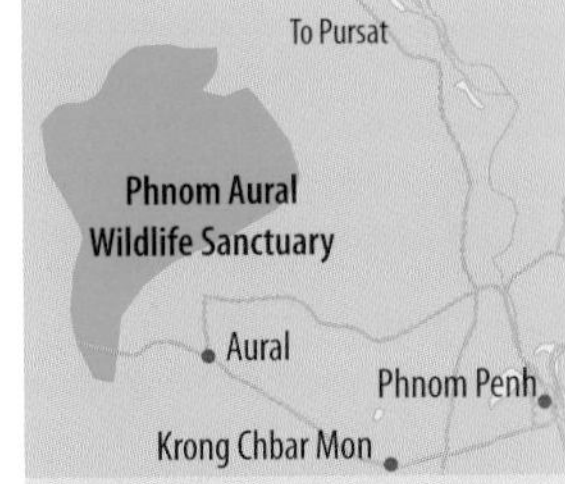

KEY FACTS

Nearest Major Towns
Kompong Speu, Kompong Chhnang, Pursat (Kompong Speu Province)

Habitats
Hill and montane evergreen and semi-evergreen forests, dry deciduous forests

Key Species
Silver Pheasant, Chestnut-headed Partridge*, Coral-billed Ground Cuckoo, Great and Wreathed Hornbills, Moustached Barbet, Silver-breasted Broadbill, Blue and Rusty-naped Pittas, Grey-chinned Minivet, Indochinese Green Magpie, Mountain Bulbul, Cambodian Laughingthrush*, Blue-winged Minla, Green Cochoa, White-tailed Robin
Winter White-throated Rock Thrush

Other Specialities
Pileated Gibbon, Cardamom Banded Gecko

Best Time to Visit
January–March; rainy season May–November

The endemic ***Cambodian Laughingthrush*** *is the primary objective of any trek up Phnom Aural.*

Situated on the eastern Cardamom Mountains, Phnom Aural Wildlife Sanctuary (2,500km^2) encompasses forested lowlands, rugged hills and mountains including Phnom Aural, Cambodia's highest peak at 1,813m asl. It is one of the few places left in Indochina where it is possible to walk from 90m to 1,771m asl through unbroken forest. Phnom Aural was virtually terra incognita until the early 21st century, when the first biological expeditions could be conducted safely. The higher elevations feature pristine montane evergreen forests and host the full range of the Cardamom's avian endemics, including the **Cambodian Laughingthrush** and **Chestnut-headed Partridge**, as well as endemic subspecies not found in Bokor such as the **Grey-chinned Minivet**, **Blue-winged Minla** and **Mountain Bulbul**. Large mammals like the Gaur are still present, and the endemic Cardamom Banded Gecko is common at 500–800m asl.

Birdwatching Sites

Altitude Camp

Rambles around the camp located at about 1,000m asl quickly provide encounters with noisy parties of **Cambodian Laughingthrushes**, while **Blue Pittas** may be seen in the surrounding forests. **Chestnut-headed Partridges** are present although unpredictable in movements. The understorey is inhabited by restless **Streaked Wren-Babblers**, as well as distinctive subspecies of the **White-tailed Robin** (ssp. *cambodiana*), **White-throated Fantail** (ssp. *cinerascens*) and diminutive **Chestnut-crowned Warbler** (ssp. *stresemanni*). In the forest canopy the **Mountain Bulbul** is conspicuous, while

Yann Muzika

***Blue Pittas** are relatively common in the forests of Phnom Aural.*

Yann Muzika

*Although widespread across Southeast Asia, **Great Hornbills** are always a delight to watch.*

the **Blue-winged Minla** and **Large Niltava** are comparatively unobtrusive. Other birds of interest include **Kloss's Leaf Warbler** and the **Green Cochoa**. Spotlighting forays may yield the **Mountain Scops Owl** and **Oriental Bay Owl**.

Phnom Aural Summit

The forests further up towards the summit hosts a similar avifauna, becoming poorer in diversity with altitude. Of particular note is the recent discovery of the **Rusty-naped Pitta** at higher elevations.

Access & Accommodation

Of all key birdwatching sites in Cambodia, Phnom Aural is by far the least accessible, and unsurprisingly, the least visited. It takes a five-hour drive from Phnom Penh to reach the foothill village. Along the National Route 4, take a turn-off north at Kompong Speu provincial town towards Aural district town. The dirt track is in relatively good condition up to the district centre, then heads further on to Sreken village at the foothills. There, local villagers familiar with the mountain should be hired as guides and cooks. Oxcarts or motorbikes then transport food and equipment to the base of Phnom Aural. Ascending the steep slope requires a strenuous climb over one long day, although this can be split into two days with an overnight stay at a mid-altitude camp. The mountain is uninhabited and there are no facilities whatsoever. The camps have a nearby stream, but any birdwatching trip there is nothing less than an expedition. The Sam Veasna Center for Wildlife Conservation organizes such trips on demand, including all the logistics and specialized English-speaking birdwatching guides. There is currently no official entrance fee or permit required.

Conservation

Despite its large size and symbolic appeal as home to Cambodia's tallest peak, Phnom Aural Wildlife Sanctuary remains one of Cambodia's many neglected protected areas. In the absence of strong support from an international conservation organization, park rangers, equipment and facilities are lacking. The largest part of the sanctuary lies in the lowlands and is devastated by logging and large agricultural concessions. Hunting is ubiquitous. Due to their relative inaccessibility, the higher mountains are still largely spared from these threats, although the harvesting of forest products and hunting still occur at these elevations.

INDONESIA

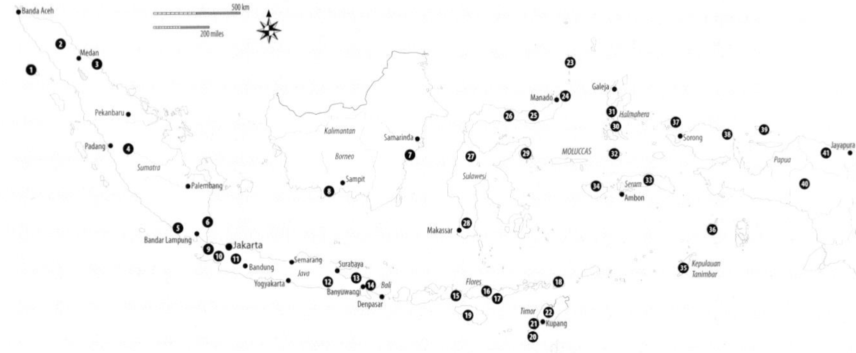

1. Simeulue Island
2. Gunung Leuser
3. Deli-Serdang
4. Kerinci-Seblat
5. Bukit Barisan Selatan
6. Way Kambas
7. Sungai Wain
8. Tanjung Puting
9. Carita
10. Gunung Halimun-Salak
11. Gunung Gede
12. Raden Soerjo
13. Baluran
14. Bali Barat
15. Komodo Island
16. Ruteng
17. Poco Ndeki
18. Alor and Pantar Islands
19. Lewa
20. Rote Island
21. Bipolo
22. Gunung Mutis
23. Sangihe Island
24. Tangkoko
25. Bogani Nani Wartabone
26. Nantu
27. Lore Lindu
28. Bantimurung
29. Peleng Island
30. Weda Bay
31. Aketajawe-Lolobata
32. Obi Island
33. Manusela
34. Buru
35. Yamdena
36. Kai Islands
37. Raja Ampat Islands
38. Arfak Mountains
39. Biak & Numfor
40. Maoke Mountains
41. Nimbokrang

Comprising more than 17,000 islands and home to an ever-increasing list of bird endemics (currently over 450 species – more than in any country on Earth), the sprawling Indonesian archipelago is the jewel in the crown of Southeast Asia from a birdwatching perspective. Global birdwatchers looking to make a dent in Indonesia's long list of country endemics have to make numerous trips to the largest archipelago in the world, while new discoveries abound in hidden nooks and crannies for the intrepid birdwatcher, with at least 10 potentially new species to science awaiting description.

A journey through the country's spectacular wilderness areas takes you from the lush equatorial rainforests of Sumatra and Kalimantan, home to familiar Asian bird families like trogons, hornbills, babblers and bulbuls, to the varied and more seasonal forests of the Lesser Sundas and Moluccas, where bird families reminiscent of Australia, such as parrots, whistlers and megapodes, take centre stage. The bizarrely shaped island of Sulawesi is where the biota of Australasia and Asia collide. In the far east of Indonesia, the rainforests of New Guinea are home to unforgettable birds-of-paradise, and complemented by a long supporting cast of parrots, pigeons, kingfishers and honeyeaters. In the past, digesting this vast avian diversity was challenging as few references were available but this has changed with a new field guide covering most of Indonesia by James Eaton and colleagues, and two new field guides focused on New Guinea.

KEY FACTS

No. of Endemics
c. 483 species (including Indonesian New Guinea)

Country List
c. 1,800 species (including Indonesian New Guinea)

Top 10 Birds
1. Bali Myna (Bali)
2. Sumatran Ground Cuckoo (Sumatra)
3. Maleo (Sulawesi)
4. Standardwing (Halmahera)
5. Bornean Peacock-Pheasant (Kalimantan)
6. Javan Hawk-Eagle (Java)
7. Ivory-breasted Pitta (Halmahera)
8. Wilson's Bird-of-paradise (Indonesian New Guinea)
9. Bare-throated Whistler (Flores)
10. Salmon-crested Cockatoo (Seram)

Climate

Indonesia has a tropical climate, associated with high humidity, temperature and rainfall. Rainfall is most intense throughout most of the Indonesian archipelago between November–March during the north-east monsoon, especially in the western islands of Sumatra, Borneo and Java. However, as you head further east, rainfall patterns become more seasonal, especially from eastern Java to the islands of the Lesser Sundas. For example, some islands receive significantly lower yearly rainfall compared with the national average, while others receive most of their rainfall during the south-west monsoon, which is associated with drier conditions throughout most of Indonesia. Indonesian New Guinea generally experiences very high rainfall all year round, with some seasonal variation depending on where you are.

Access, Transportation & Logistics

The majority of international visitors are likely to arrive in Indonesia via either of the two main international airports located in its capital, Jakarta, and in the city of Denpasar on the island of Bali. A visa on arrival is available for visitors from many countries and allows a stay of up to 30 days. Jakarta is well connected to many towns and cities on various islands throughout the country, but Bali is regarded as a more convenient point of entry for birdwatchers seeking to explore east Java, the Lesser Sundas and the Moluccas. Birdwatchers on a budget but with time to spare can consider travelling between islands by ferries operated by the national ferry company (PELNI).

Within individual islands transport options vary, ranging from rented cars to minibuses (angkut) and motorcycles (ojek). Many of the more remote birdwatching sites, particularly on outlying islands, are most easily reached by ojek (riding them can be an experience in itself). English is not widely spoken outside the main tourist sites, but Bahasa Indonesia is an easy language to pick up and learning a few common phrases greatly enhances the travelling experience, particularly on the outlying islands.

Visitors should be aware that the transport network in Indonesia can be very unreliable, with frequent flight delays, ferry cancellations due to bad weather or overloading of passenger vessels among other risks and inconveniences. Consequently, birdwatchers planning to undertake long trips around the country should plan some buffer time to account for such issues, particularly in the remote islands to the east of the country.

Health & Safety

Vaccinations for Hepatitis A and B and typhoid are recommended for visiting birdwatchers. Malaria is prevalent in remote areas and prophylaxis should be taken, although this is less of a problem on the islands of Java and Bali and during the dry season. Dengue fever is also present and appropriate precautions should be taken. Leeches, ticks and a variety of biting insects occur in Indonesia's forests. Of particular note is the presence of chiggers (biting mites) at several of the country's birdwatching sites, notably the renowned Tangkoko National Park and Gunung Ambang Nature Reserve on the island of Sulawesi.

Petty crime may be ubiquitous in some areas, especially in large cities, and visitors should keep a close eye on their valuables, particularly when using public transport. Additionally, Indonesia grapples with occasional insurgent activities in some parts of the archipelago, especially on Sulawesi and New Guinea. Travellers are advised to check the security situation with their local embassies when planning their birdwatching trips.

Birdwatching Highlights

A good number of the country endemics can be seen over the course of several dedicated trips focusing on the major islands of Indonesia, namely Sumatra and Java in western Indonesia, Sulawesi and Halmahera, and the Lesser Sundas islands of Sumba, Timor and Flores.

A three-week trip to Sumatra and Java is likely to yield about 300 species, including the majority of the avian endemics on both islands, which are concentrated in the mountains. The mountains of Sumatra harbour species such as Schneider's and Graceful Pittas, Sumatran Cochoa and Red-billed Partridge, while those on Java support the globally threatened Javan Hawk-Eagle, Spotted Crocias and White-bibbed Babbler, to name a few. The lowlands of Sumatra are full of forest birds, and when visiting the rainforests of Way Kambas, Bukit Barisan Selatan and Gunung Leuser one can expect to see many bulbuls, babblers, hornbills and flycatchers.

Heading east and across Wallace's Line, the islands of Sulawesi and Halmahera support a distinctly Wallacean avifauna, headlined by the bizarre Maleo on Sulawesi and the Standardwing on Halmahera. A three-week trip to Sulawesi and the north Moluccas is expected to yield upwards of 250 species, including a variety of endemic kingfishers in the Sulawesi lowlands, the monotypic Hylocitrea, the enigmatic Geomalia (now found to be an aberrant ground thrush) and the Satanic Nightjar in the mountains. Aside from the Standardwing, the island of Halmahera also supports the hulking Ivory-breasted Pitta, unobtrusive Scarlet-breasted Fruit Dove, bizarre Moluccan Owlet-nightjar and a variety of other regional endemics.

The other islands of the Moluccas are less accessible, but are home to some very special endemics. The forests of Seram are home to the incomparable Salmon-crested Cockatoo, Long-crested Myna, and a host of flycatchers, white-eyes, parrots and honeyeaters. The remote montane forests of Buru are equally remarkable, being home to the Black-lored Parrot, one of few nocturnal parrot species, and the Madanga, an aberrant pipit once thought by scientists to be a white-eye.

The Lesser Sundas islands of Sumba, Timor and Flores support more than 70 regional endemics, along with close to 200 other species, most of which can be encountered during a visit to these three islands over the course of three weeks. Avian highlights from this region include the Sumba Hornbill, two species of endemic boobook and the Red-naped Fruit Dove on Sumba, the Black-banded Flycatcher, Buff-banded Thicketbird and Jonquil Parrot on Timor, and the Flores Hawk-Eagle, Flores Crow and Flores Scops Owl on Flores.

Besides these three islands, exploring the other less visited islands in the Lesser Sundas can be very worthwhile, particularly Yamdena where extensive forests full of endemics remain.

A typical four-week trip around Indonesian New Guinea covers montane sites in the Arfak and Snow Mountains, the lowland rainforest around Nimbokrang and several islands including both Biak and Numfor as well as the island of Waigeo in the Raja Ampat island group. Birdwatchers can expect to encounter more than 230 species of New Guinea's avian endemics, including the vast majority of the 50 or so species found only in Papua and West Papua provinces. Birds-of-paradise unsurprisingly take centre stage here, with the peerless Wilson's Bird-of-paradise on Waigeo, Western Parotia in the Arfak Mountains and Twelve-wired Bird-of-paradise in the lowlands. The supporting cast is also impressive and includes the likes of Western Crowned Pigeon, Shovel-billed Kookabura, MacGregor's Honeyeater and several monotypic families such as the bizarre Wattled Ploughbill.

Sam Thuong

*The attractive **Salvadori's Pheasant** lurks in the cloud forests of Sumatra.*

James Eaton

*Endemic to the north Moluccas, the beautiful **Chattering Lory** has declined rapidly due to the pet trade.*

Chi'en C. Lee

*The remarkable **Black Sicklebill**, among the largest of the birds-of-paradise, is regularly seen in the Arfak Mountains of West Papua.*

Simeulue Island

Agus Nurza Zulkarnain and Yong Ding Li

Agus Nurza Zulkarnain

The Luan Balu area of Teluk Dalam is one of the best sites for birdwatching on Simeulue.

KEY FACTS

Nearest Major Town
Sinabang (Aceh)

Habitats
Lowland and hill rainforests, coastal scrub, mangrove forests

Key Species
Crested Serpent Eagle, Beach Stone-curlew, Silvery Pigeon, Barusan Cuckoo-Dove*, Nicobar Pigeon, Simeulue Scops Owl*, Great Eared Nightjar, White-bellied Woodpecker, Blue-rumped Parrot, Red-breasted Parakeet, Black-naped Monarch, Blyth's Paradise Flycatcher, Common Hill Myna
Winter Bar-tailed Godwit, Ruddy Turnstone, Forest Wagtail

Other Specialities
Long-tailed Macaque

Best Time to Visit
March–September

Located about 150km off the north-western coast of Sumatra (in Aceh Province), Simeulue is the northernmost of a chain of islands running along the western coast of Sumatra. Idyllic Simeulue remains largely covered with lowland and hill forests, while its coasts are fringed by extensive mangroves and sandy beaches (especially along the west coast) lapped by strong waves, making it popular with surfers. The island's hilly interior rises to about 576m asl, and there are two freshwater lakes, Danau Laut Tawar and Etutuk. Simuelue is of great interest to ornithologists, not just because of the endemic **Simeulue Scops Owl**; the island also sports a long list of species with endemic races, including taxa that may be eventually recognized as distinct at the species level (like the **Crested Serpent Eagle**, **Great Eared Nightjar**, **White-bellied Woodpecker** and **Blue-rumped Parrot**). Simuelue is also widely recognized by birdwatchers as the best place to see the recently rediscovered and Critically Endangered **Silvery Pigeon**, which appears to be regular at some parts of the island. More adventurous birdwatchers could consider a side excursion by boat to nearby Babi Island, where a robust and highly distinctive form of the **Black-naped Monarch** (ssp. *abbotti*) occurs.

Agus Nurza Zulkarnain

The Critically Endangered ***Silvery Pigeon*** *is one of the most threatened pigeon species in the world.*

Birdwatching Sites

Kahat

Located about 7km from the airport, Kahat (Teupah Tengah sub-district) on the south-eastern coast of Simeulue is a great site to see most of Simeuluc's forest birds. Here, the scrub and remaining lowland rainforest is one of the best sites to observe the **Crested Serpent Eagle** (ssp. *abbotti*) and **White-bellied Woodpecker** (ssp. *parvus*), among other forest species. At dusk, birdwatchers should look out for the **Simeulue Scops Owl** and

Great Eared Nightjar (ssp. *jacobsoni*) that can be found here. The taxonomic affinities of the Great Eared Nightjar on Simeulue remain uncertain and is believed to be an endemic species closer to Malaysian Eared Nightjar. The **Silvery Pigeon** may sometimes be seen in flight here, but care must be taken to separate it from the similar-looking **Pied Imperial Pigeon**, which is common on the island. Small troops of the highly distinct Simeulue form of the Long-tailed Macaque may be seen here as well.

Labuhan Bakti

This site is about 30km (50 minutes' drive) from the airport on the south-eastern coast of the island. The coastal scrub and mangrove place here is a good place to see shorebirds such as the resident **Beach Stone-curlew** as well as the migratory **Bar-tailed Godwit**, **Ruddy Turnstone** and **Curlew Sandpiper**. Other species possible from the roadside forest and along the number of tracks here include the **Crested Serpent Eagle**, **Great Eared Nightjar**, **White-bellied Woodpecker**, **Lesser Cuckooshrike** and **Black-headed Bulbul**, the latter being among the most common forest birds on the island.

Teluk Dalam Forest

The forest at Teluk Dalam sub-district north-

Yann Muzika

Simeulue Scops Owls are not difficult to find, being fairly common in the forests.

Agus Nurza Zulkarnain

*The endemic subspecies of the **Crested Serpent Eagle** may be seen at the forests of Teluk Dalam*

Agus Nurza Zulkarnain

***Scarlet Minivets** are among the most conspicuous forest birds on Simeulue, often seen in small flocks even in village gardens.*

west of Sinabang between the Km 30 and Km 35 road markers are excellent for birdwatching, especially in the Kuala Baru and Luan Balu areas. The forests here are best visited in the morning and nearly all forest species have been recorded here, including the **Silvery Pigeon** and **Blue-rumped Parrot** (ssp. *abbotti*) (often seen noisily in flight above the forest canopy). A productive way to explore this area is to concentrate on the forests along the roadside and by the edge of rivers. With luck, a good variety of forest species and pigeons may be seen, including both imperial pigeons and the **Barusan** and **Little Cuckoo-Dove**. With luck, one may also see **Common Hill Myna** (ssp. *miotera*), a species in rapid decline across the island due to the illegal bird trade. Although typically found on small forested islands, the spectacular **Nicobar Pigeon** has been sporadically observed in the forests in Luan Balu.

Access & Accommodation

There are regular flights from Medan to Simeulue's Lasikin Airport by Wings (daily flights on larger planes) and Susi Air. Alternatively, Simeulue can be accessed by ferries from Labuhan Haji (Aceh Selatan), Aceh Singkil and Meulaboh on the Sumatran mainland. Meulaboh, Labuhan Haji and Singkil in turn can be reached by bus from Banda Aceh or Medan. Ferries to Simeulue run twice a week and arrives in Sinabang, on the east coast of Simeulue. The key birdwatching sites on the island are best reached by rented car or motorcycles. Accommodation in the form of basic hotels can be found in Sinabang. There are also a number of surf lodges along the beaches close to the airport on the west coast of the island, some of which also offer birdwatching itineraries.

Conservation

Forest clearance for firewood, cultivation and oil-palm plantations pose a major threat Simeulue's forest ecosystem. Large areas of old growth forest have been lost over the years and what is left is increasingly fragmented. Gravel mining also poses a formidable threat to the forests of Simeulue, as does poaching of birds, especially Common Hill Mynas and White-rumped Shamas to supply the pet trade, resulting in depressed densities of both species across much of western Indonesia. There is an urgent need to develop protected areas on Simeulue to secure key sites for the island's forest birds, and tackle the problem of illegal poaching and forest clearance.

Agus Nurza Zulkarnain

*The Simeulue subspecies of the **White-bellied Woodpecker** is substantially smaller than other mainland races.*

Gunung Leuser National Park

Agus Nurza Zulkarnain

Agus Nurza Zulkarnain

Gunung Leuser National Park protects some of the largest and most pristine tracts of rainforest left in the Greater Sundas.

KEY FACTS

Nearest Major Towns
Kutacane, Blangkejeren, Takengon, Meulaboh, Tapak Tuan (Aceh)

Habitats
Lowland and hill rainforests, submontane and montane rainforests, peat-swamp forest, freshwater wetlands

Key Species
White-winged Duck, Roll's* and Red-billed* Partridges, Hoogerwerf's Pheasant*, Bronze-tailed Peacock-Pheasant*, Storm's Stork, Sumatran Green Pigeon*, Rajah Scops Owl, Short-tailed Frogmouth*, Salvadori's Nightjar*, Sumatran Trogon*, Grey-headed Woodpecker, Schneider's* and Graceful* Pittas, Sumatran Drongo*, Sumatran Treepie*, Cream-striped*, Spot-necked* and Orange-spotted* Bulbuls, Sumatran Babbler*, Rusty-breasted* and Sumatran* Wren-Babblers, Chestnut-capped, Black, Sumatran* and Sunda Laughingthrushes, Shiny* and Brown-winged* Whistling Thrushes
Winter Masked Finfoot

Other Specialities
Sumatran Orangutan, Thomas's Langur, Siamang, Lar Gibbon, Binturong, Sun Bear, Sumatran Tiger, Sunda Clouded Leopard, Asian Golden Cat, Asian Elephant, Sumatran Rhinoceros

Best Time to Visit
March–September; rainy season October–January

Gunung Leuser National Park (GLNP) and the wider Leuser Ecosystem is one of the largest and most spectacular areas of rainforest left in Southeast Asia. The Leuser Ecosystem covers over 26,000km^2 and straddles two of Sumatra's provinces, Aceh (22,555km^2) and North Sumatra (3,843km^2). A major part of the Leuser Ecosystem lies within the sprawling Gunung Leuser National Park (10,946km^2). The park is topographically varied, from the peat swamps of Kluet at sea level, rising to its highest point on the summit of the Gunung Leuser massif (3,466m asl), the second highest point in Sumatra. More than 400 bird species have been recorded in the park, including highly sought-after endemics like **Roll's Partridge**, **Hoogerwerf's Pheasant**, **Sumatran Green Pigeon**, **Short-tailed Frogmouth** and the elusive **Sumatran Laughingthrush**. Other highlights include the highly distinctive Sumatran forms of the **Orange-spotted Bulbul** (ssp. *snouckaerti*) and **Silver-eared Mesia** (ssp. *rookmakeri*). In spite of the park's long bird list, much of the Leuser area has been little explored by ornithologists. In fact, **Schneider's Pitta** was only recorded here for the first time in 2013, while the swamps in Kluet have been found to support the **White-winged Duck**, **Storm's Stork** and **Masked Finfoot**. It is possible to see more than 120 species over a visit of 4–5

Agus Nurza Zulkarnain

*The **Silver-eared Mesia** is an uncommon bird in the highland forests of Sumatra due to trapping.*

days, including a number that are more easily seen here than anywhere else in Sumatra.

Birdwatching Sites

Ketambe

The town of Ketambe lies in the heart of the Alas Valley amid a magnificent backdrop of forested peaks. There are some excellent stretches of submontane and hill dipterocarp forests at 500m asl alongside the road that snakes through the valley, particularly near the Orangutan research station at Ketambe, where the road starts to climb. Some of the best forests can only be accessed on foot by taking the many side trails to the east and west of this road. The **Bronze-tailed Peacock-Pheasant** is regular here and best seen in the early morning. Other sought-after species that have been recorded include **Roll's Partridge**, the **Rusty-breasted Wren-Babbler** and the impressive **Helmeted Hornbill**.

Kedah

The settlement of Kedah is situated near a creek on the fringes of virgin rainforest at about 1,400m asl, roughly 15km west of Blangkejeren in the district of Gayo Lues. Starting from Kedah it is possible to make guided day hikes through dense montane forests to more than 1,800m asl. Birdwatching opportunities in this area are excellent, and foraging flocks contain endemics like the **Sumatran Drongo**, **Blue-masked Leafbird** and various laughingthrushes. The highly sought-after **Bronze-tailed Peacock-Pheasant**, **Roll's** and **Red-billed Partridges** are more easily heard than seen here. The rare **Grey-headed Woodpecker**, here of the distinctive Sumatran subspecies (ssp. *dedemi*), has also been regularly recorded in recent years.

Ise-Ise Road

This road connects the towns of Blangkejeren and Takengon and cuts through extensive areas of pristine submontane and montane forests, reaching elevations of nearly 1,900m asl. Birdwatching along the road can be very productive, and noisy flocks of laughingthrushes have been regularly seen here, besides foraging flocks containing **Sumatran Trogons**, **Sumatran Drongos** and **Sumatran Treepies**. Fruiting trees often attract pigeons and barbets, as well as **Cream-striped** and **Spot-necked Bulbuls**. **Rusty-breasted Wren-Babblers**, and **Shiny** and **Brown-winged Whistling Thrushes** all skulk in the dense roadside foliage. At night, the **Short-tailed Frogmouth** is regularly encountered along the road.

Agus Nurza Zulkarnain

*A **Sumatran Babbler** skulks in the dense undergrowth of the forests around Kluet swamp.*

Takengon, Gayo Highlands

The forests in the Gayo Highlands around Takengon and Laut Tawar Lake are excellent for birdwatching and should be thoroughly explored. The Bur Gayo area south of the lake is where **Schneider's Pitta** was recorded for the first time in Leuser, and this may prove to be a reliable site to see this species outside Kerinci in future. In addition, foraging flocks are very active here towards the mid-morning. The forests in the Bur Lintang area further south of Bur Gayo are also excellent for birdwatching and most of the key species, like **Roll's Partridge**, **Hoogerwerf's Pheasant**, **Sumatran Green Pigeon**, **Rajah Scops Owl**, **Short-tailed Frogmouth** and **Salvadori's Nightjar**, can be encountered here.

Agus Nurza Zulkarnain

*The **Sumatran Trogon** is a regular participant of mixed foraging flocks at submontane and montane elevations in Leuser's rainforests.*

Kluet Swamp

The Blok Kluet area is notable for its varied lowland forest communities. Kluet contains extensive areas of some of the last remaining swamp forests on the west coast of Sumatra and the triangular-shaped Danau Laut Bangko. Laut Bangko is where the **Masked Finfoot** has been seen recently, and a population of **White-winged Ducks** is present. Wild Sumatran Orangutans may be encountered as they move along the rivers here, and **Storm's Stork** has also been regularly recorded. The scarce **Sumatran Babbler** can be encountered along the forest edge by the lake.

Nagan Raya-Meulaboh

Although the forest starts from the lowlands along the road out of Meulaboh towards Beutong, it is the extensive submontane and montane forests at Beutong that are of greatest interest to the birdwatcher. The **Sumatran Drongo** and **Sumatran Treepie** are regularly seen in the large mixed flocks that pass through the roadside forest alongside various species of laughingthrushes. Besides most of the Sumatran endemics, other highlights here are the distinctive north Sumatran subspecies of the **Orange-spotted Bulbul** and the **Silver-eared Mesia**. The **Short-tailed Frogmouth** has been seen in the forests during nocturnal rambles.

Access & Accommodation

Ketambe is best reached via Medan, from where you can get to Kutacane by bus or taxi, a journey of 10 hours. Accommodation is available at basic hotels in Kutacane and the many small guesthouses near Ketambe, the most popular being the Wisma Cinta Alam. Takengon in the Gayo Highlands can be accessed from Medan (12 hours) or Banda Aceh by bus (eight hours). Flights are also available from Medan to Rembele Airport in Bener Meriah District in Aceh. The Nagan Raya–Meulaboh area is best reached from Banda Aceh by bus (five hours). From Meulaboh, it is a two-hour drive to the forests of the Leuser Ecosystem by car. The Kluet area can be accessed from the town of Tapak Tuan, which can be reached on a three-hour bus ride from Meulaboh. It is another hour from Tapak Tuan to Kluet, from where boats can be arranged to access and explore Laut Bangko. Entry permits are required for access to all sites in Leuser and can be arranged at the park offices in Kutacane, Ketambe and Tapak Tuan. Entry permits in Leuser Ecosystem areas outside the park can be arranged at Natural Resources Conservation Agency (BKSDA) in Banda Aceh.

Agus Nurza Zulkarnain

*The globally threatened **Sumatran Laughingthrush** is in decline throughout its range due to the pet-bird trade.*

Agus Nurza Zulkarnain

*The endemic subspecies of the distinctive **Grey-headed Woodpecker** is recognised as a full species by some authors.*

Conservation

The Leuser Ecosystem is a high-value biodiversity site in Sumatra and has been recognized as part of the 'Tropical Rainforest Heritage of Sumatra' World Heritage Site by UNESCO. As well as providing representative habitats of Sumatra's biodiversity, the region supports more than four million people who live around it. The broader Leuser Ecosystem is considered to be one of many National Strategic Areas by the Indonesian Ministry of Forestry. Illegal logging and poaching, however, remain a serious threat throughout the area, especially for birds sought after for the songbird trade (like laughingthrushes) and the Helmeted Hornbill, which is nearly extirpated from the area.

Deli-Serdang Coast

Chairunas Adha Putra

Chairunas Adha Putra

The extensive mudflats along the Deli-Serdang coast are key wintering sites for tens of thousands of migratory shorebirds.

The extensive but declining wetlands that fringe Sumatra's long eastern coastline constitute some of the most important wintering habitats for migratory shorebirds in Southeast Asia. The Deli-Serdang Coast is located 30km east of Medan in North Sumatra province. This extensive area of coastal wetlands, comprising large areas of mangroves and intertidal mudflats, has been designated as an Important Bird and Biodiversity Area (IBA) in recognition of its globally important populations of wintering migratory shorebirds, including the endangered **Nordmann's Greenshank**. The wetlands here also hold breeding colonies of waterbirds, including various herons and egrets. In total, at least 124 bird species have been recorded in the area, including 33 migratory shorebird species. Unlike its mountains, Sumatra's coastal wetlands are seldom visited by birdwatchers and there is potential for interesting discoveries.

KEY FACTS

Nearest Major Town
Medan (North Sumatra)

Habitats
Mangroves, coastal mudflats, cultivation and fishponds

Key Species
Milky Stork, Lesser Adjutant, Black-winged Stilt
Winter Black Baza, Pied Harrier, Asian Dowitcher, Black-tailed and Bar-tailed Godwits, Eurasian and Far Eastern Curlews, Nordmann's Greenshank, Great and Red Knots

Other Specialities
Silvered Leaf Monkey, Oriental Small-clawed Otter, Mangrove Pit Viper, Mangrove Snake

Best Time to Visit
October–March; rainy season September–December

Birdwatching Sites

Bagan Percut

Bagan Percut's mudflats can be reached via a one-hour drive from Medan followed by a 15-minute boat ride. Both **Milky Stork** and **Lesser Adjutant** can be seen foraging on the mudflats daily, while during the northern winter tens of thousands of shorebirds may also be seen, including familiar species such as the **Lesser Sand Plover**, **Bar-tailed Godwit**, **Eurasian**

Chairunas Adha Putra

*The globally endangered **Milky Stork** is regularly seen foraging on the mudflats at Percut – and is more reliably seen here than anywhere else in its range.*

Chairunas Adha Putra

*The Endangered **Great Knot** winters in small numbers on the mudflats of Percut.*

Chairunas Adha Putra

*The mudflats along Deli-Serdang support globally important populations of many migratory shorebirds, including the **Eurasian Curlew**.*

Curlew, **Ruddy Turnstone**, **Great Knot** and **Curlew Sandpiper**. Diligent scanning of these flocks may yield rarities such as the **Chinese Egret**, **Asian Dowitcher** and **Nordmann's Greenshank**.

Cemara Asri Residental Area

This site, approximately 20 minutes by car from Medan, contains a waterbird breeding colony within an area of fishponds. During the breeding season between February–May, thousands of **Black-crowned Night Herons**, **Eastern Cattle Egrets**, **Purple Herons** and **Little Egrets** may be seen. The birds are habituated to humans and can be very approachable.

Marshes around Desa Sei Tuan

This site is a one-and-a-half hour drive from Medan, and features coastal marshes surrounded by oil-palm plantations. A different suite of shorebirds can be seen here, including species associated with freshwater habitats like the **Common Snipe**, **Wood Sandpiper** and **Ruff**. There is also a resident population of **Black-winged Stilts**. Besides shorebirds, the area is good for seeing migratory raptors such as the **Black Baza** and **Pied Harrier**.

Access & Accommodation

Bustling Medan, Sumatra's largest city, is only a two-hour flight from Jakarta. There are also many international flights that connect Medan to regional cities, including Singapore and Kuala Lumpur (Malaysia). It then takes a further two hours to reach the key birdwatching sites from Medan. Accommodation and transport by rental cars can be readily arranged from Medan.

Conservation

The Deli-Serdang coast is currently being proposed for designation as an East Asian-Australasian Flyway Network Site. However, many of the wetland areas receive no formal protection and the habitats, in particular mangroves, are being cleared rapidly.

Kerinci-Seblat National Park

Low Bing Wen

Yong Ding Li

Gunung Kerinci (background) is an active volcano and the tallest peak in Sumatra. The montane forest that cloak its slopes is home to the majority of Sumatra's endemic birds.

KEY FACTS

Nearest Major Towns
Padang, Kersik Tua (West Sumatra), Sungaipenuh (Jambi)

Habitats
Lowland and hill rainforests, submontane and montane rainforests, alpine scrub (summit of Gunung Kerinci), freshwater wetlands

Key Species
Montane Red-billed Partridge*, Salvadori's Pheasant*, Bronze-tailed Peacock-Pheasant*, Pink-headed Fruit Dove*, Sumatran Green Pigeon*, Sumatran Ground Cuckoo*, Rajah Scops Owl, Short-tailed Frogmouth*, Salvadori's Nightjar*, Giant Swiftlet, Sumatran Trogon*, Schneider's Pitta*, Sunda*, Cream-striped* and Spot-necked* Bulbuls, Rusty-breasted* and Sumatran* Wren-Babblers, Sumatran Cochoa*, Rufous-vented Niltava, Shiny* and Brown-winged* Whistling Thrushes, ***Submontane*** Graceful Pitta*, Sumatran Drongo*, Sumatran Treepie*, Marbled Wren-Babbler, Sumatran* and Blue-masked* Leafbirds

Other Specialities
Siamang, Mitred Leaf Monkey*, Sumatran Striped Rabbit*, Sun Bear, Binturong, Asian Golden Cat, Sumatran Tiger, Sunda Clouded Leopard

Best Time to Visit
June–September; rainy season October–January

Kerinci-Seblat is the largest national park in Sumatra, spanning four provinces and occupying an area of 13,791km². Gunung Kerinci, Sumatra's highest peak at 3,805m asl, is an active volcano that lies within the boundaries of the park. From an ornithological perspective, the park shot to fame in 1988 after the enigmatic **Schneider's Pitta** was rediscovered on its slopes after a 70-year hiatus. Due to Gunung Kerinci's popularity with local trekkers, it is nowadays the most accessible site in Sumatra for observing the majority of Sumatra's endemic birds, which include the likes of the **Sumatran Cochoa**, **Red-billed Partridge** and **Short-tailed Frogmouth**, to name a few. Additionally, mountain roads that skirt the park's boundaries offer access to pristine submontane forests that feature a different suite of birds, including the **Sumatran Ground Cuckoo**, **Graceful Pitta**, **Blue-masked Leafbird** and **Marbled Wren-Babbler**. No birdwatching trip to Sumatra is really complete without spending some time at Kerinci-Seblat National Park.

Birdwatching Sites

Gunung Kerinci Summit Trail

The focal point of any birdwatching to the park, this trail covers an elevation of 1,800–3,805m asl at the summit of Gunung Kerinci. For birdwatchers, all the key species can be observed below 2,500m, and the trail can be divided into three sections at different elevations, each with its own set of specialities.

Wang Bin

*With a bit of luck the elusive **Rajah Scops Owl** (ssp. solokensis) can be seen on the slopes of Kerinci on a spotlighting trip.*

Base Camp Clearing

The gentle 1km stretch beginning from the forest edge to the base camp clearing is the best area in which to see **Schneider's Pitta**, the most sought-after of the area's avian endemics. The pitta is best seen in the early mornings and at dusk, when it forages unobtrusively on the trail and can be easily flushed by the unwary birdwatcher. A stealthy approach can also yield **Salvadori's Pheasant**, a party of which regularly forages around the clearing. Other diurnal species seen along this stretch include both **Rusty-breasted** and **Sumatran Wren-Babblers**. At dusk, both **Salvadori's Nightjar** and the **Barred Eagle-Owl** may be encountered at the forest edge at the start of the trail, and the **Short-tailed Frogmouth** is frequently seen along this sector.

'Air Minum' Clearing

From the base-camp clearing the trail climbs steeply towards another forest clearing known as 'Air Minum', situated next to a seasonal river. The area around Air Minum is more open than most of the summit trail, offering the opportunity to see arboreal species, notably the Sumatran subspecies (ssp. *sylvaticum*) of the **Collared Owlet**, a potential future split, as well as the **Pink-headed Fruit Dove**. The **Red-billed Partridge** is regularly heard around the clearing, and may reward patient observers with views as it forages near the river. The riverbed is a good place to see both **Shiny** and **Brown-winged Whistling Thrushes**. After dark, the **Rajah Scops Owls** can also be encountered, generally beginning from this sector of the summit trail.

Sam Thuong

*The summit trail of Gunung Kerinci is the most accessible site for the endemic **Schneider's Pitta**.*

Wang Bin

*The stunning **Graceful Pitta** is one of the most colourful of Sumatra's avian endemics and can be regularly encountered along Tapan Road.*

'Camp Cochoa' & Burnt Tree

From 'Air Minum' the trail climbs even more steeply to a collapsed shelter known as Camp Cochoa, and about 250m beyond that to the burned stump of a large tree on the left side of the trail. The area between Camp Cochoa and the burned tree is the best for seeing the **Sumatran Cochoa**, although depending on the presence of fruiting trees the species can be seen at lower sectors of the trail. The trek to this area, while arduous, can be rewarding, as species such as the **Sumatran Wren-Babbler**, **Red-billed Partridge** and **Rufous-vented Niltava** may be seen during the search for the cochoa.

Muaro Sako–Tapan Road

Located two and a half hours by car from Kersik Tua, this paved road starts from the town of Sungai Penuh, then weaves its way to Muaro Sako, finally reaching Tapan (110m

***Sumatran Treepies** are regular participants in the mixed flocks along the Muaro Sako-Tapan Road.*

asl), near Sumatra's west coast. The focal point for birdwatchers lies along a downhill stretch between the mountain pass at Bukit Tapan to a large bridge at the 27km route marker. The **Graceful Pitta**, the primary reason for visiting this site, has been found to be locally numerous and can be readily encountered in the numerous roadside gullies located between the 24km and 27km route markers. These gullies also hold the **Rufous-chested Flycatcher** and **Marbled Wren-Babbler**. The birdwatching in this area is excellent and mixed foraging flocks should be scrutinized for endemics, including **Blue-masked** and **Sumatran Leafbirds**, **Sumatran Trogon** and **Sumatran Drongo**, as well as **Cream-striped** and **Spot-necked Bulbuls**. The **Bronze-tailed Peacock-Pheasant** is regularly heard and occasionally seen on the road or under roadside fruiting trees.

Gunung Tujuh

The forested mountains in the Gunung Tujuh area rise to more than 2,700m asl. The montane forests here can be accessed from the trail that starts near the village of Ulu Jernih not far from Pelompek town, and eventually leads to the Gunung Tujuh Lake at 1,950m asl. The birdlife here is very similar to that at the Kerinci summit trail, and the rare **Grey-headed Woodpecker** of the distinctive Sumatran subspecies (ssp. *dedemi*) has been recorded, but very few birdwatchers have ventured here.

Access & Accommodation

The village of Kersik Tua is reached after a 6–7-hour ride via rented vehicle (double that for public transport) from Padang, the provincial capital of West Sumatra, where an international airport is located. Within Kersik Tua, several basic homestays are available, although most birdwatchers opt to stay at the Keluarga Subandi Homestay, or *losmen*, as Pak Subandi and his family are used to catering to the needs of birdwatchers, including arranging entry permits. Several local guides knowledgeable in the area's avifauna can be hired on request, including Pak Subandi and his sons. For those planning to explore the forests around the Tapan-Muaro Sako area, one possibility is to be based in Sungai Penuh, from where rental vehicles can be arranged.

Conservation

The various threats faced by this site and other protected areas in Sumatra have prompted listing of the 'Tropical Rainforest Heritage of Sumatra' as a UNESCO site in danger. The proliferation of tea and coffee plantations in the region continues to eat away at the perimeter of the national park. Illegal logging is rampant in and around the park, and chainsaws can be regularly heard along both the summit trail and Tapan Road. Poaching is also a serious threat, evident from the paucity of songbirds such as laughingthrushes and Silver-eared Mesias, as well as the low numbers of gamebirds such as partridges and pheasants.

Lim Kim Chuah

Shiny Whistling Thrush *is common along the lower sections of Gunung Kerinci's summit trail.*

Agus Nurza Zulkarnain

Fruiting trees in submontane forest are the best places to see the endemic ***Cream-striped Bulbul****.*

Con Foley

The ***Rufous-vented Niltava*** *is endemic to the montane forests of Sumatra and the Malay Peninsula.*

Bukit Barisan Selatan National Park

Low Bing Wen

Yann Muzika

The riverine submontane forest at Way Titias is home to the mythical Sumatran Ground Cuckoo and a host of other sought-after Sumatran endemics.

KEY FACTS

Nearest Major Town
Bandar Lampung (Lampung)

Habitats
Primary lowland, hill and montane forests, coffee plantations, scrub

Key Species
Montane Sumatran* and Red-billed* Partridges, Bronze-tailed Peacock-Pheasant*, Sumatran Trogon*, Schneider's Pitta*, Sumatran Treepie*, Rusty-breasted* and Sumatran* Wren-Babblers, Sumatran Babbler*, Sumatran Cochoa*
Hill-submontane Long-billed and Ferruginous Partridges, Sumatran Green Pigeon*, Sumatran Ground Cuckoo*, Graceful Pitta*, Sumatran Drongo*, Cream-striped Bulbul*, Marbled Wren-Babbler, Blue-masked Leafbird*

Other Specialities
Mitred Leaf Monkey*, Agile Gibbon, Siamang, Sun Bear, Sumatran Tiger

Best Time to Visit
May–October; rainy season November–April

The extensive and largely pristine rainforests in this 3,568km^2 park were largely overlooked by visiting birdwatchers to Sumatra in favour of the well-trodden Gunung Kerinci. However, the discovery of the mythical **Sumatran Ground Cuckoo** at this national park in 2007 changed all that. Even so, comparatively few birdwatchers have explored Bukit Barisan Selatan, with its limited accessibility, difficult trails and multiple river crossings discouraging many from attempting the hike. However, those with a sense of adventure (and time!) find a magical and largely untouched rainforest ecosystem that is home to many of Sumatra's most sought-after avian endemics, some of which, like **Schneider's Pitta**, **Sumatran Cochoa** and **Bronze-tailed Peacock-Pheasant**, are arguably easier to see here than at Gunung Kerinci.

Birdwatching Sites

Way Titias

This site, where the **Sumatran Ground Cuckoo**, or 'toktor', as it is known to local people, was first discovered in 2007, is still the most reliable site in the world to see this enigmatic species. Access is, however, no mean feat, and requires a 6–7 hour trek that involves at least six knee-high river crossings (in the dry season) and numerous steep sections before reaching the campsite next to a small stream at around 1,000m asl. The ridgelines surrounding the Way

Wang Bin

***Blyth's Hawk-Eagle** is the most regularly seen raptor in Sumatra's submontane forests.*

Jon Foley

The ***Fire-tufted Barbet*** *is heavily trapped for the pet-bird trade but remains fairly common in the forests of Bukit Barisan Selatan.*

Lim Kim Chuah

The endemic ***Red-billed Partridge*** *is arguably easier to see at Danau Ranau than at the more popularly visited Gunung Kerinci.*

Titias campsite are home to several pairs of ground cuckoos, although seeing them well can be exceptionally challenging. Due to the closed forest getting views of arboreal birds is difficult, but there is a diverse array of ground dwellers on offer, so keep your eyes firmly on the ground in the hopes of getting glimpses of Long-billed, Sumatran and Ferruginous Partridges, Bronze-tailed Peacock-Pheasant and Graceful Pitta in addition to the ground cuckoo. The scrubby transition between coffee plantation and rainforest proper is also good for the poorly known Sumatran Babbler.

Danau Ranau

Sumatra's second largest lake is fringed by montane rainforest that cloaks the peaks surrounding the lake, encompassing a wide range of heights, from the foothills at approximately 500m, to almost 2,000m asl. Most birdwatchers base themselves on the Puncak (Summit) Trail, which peaks at a high point of just under 1,400m asl. While steep, this trail lacks any river crossings and is comparatively less arduous than the Way Titias trek. The best area for birdwatching appears to be between 1,000m asl and the high point, where a host of sought-after montane endemics can be seen. These include Schneider's Pitta, Sumatran Cochoa and Sumatran Wren-Babbler. There is noticeable overlap between the species here and at Way Titias, with the notable exception of the ground cuckoo, and this site offers a second chance to see species such as the Sumatran Partridge, Bronze-tailed Peacock-Pheasant and Graceful Pitta. Mixed species foraging flocks are more conspicuous here and feature a range of species, including the Fire-tufted Barbet, Long-tailed Broadbill, Black Laughingthrush and Blue Nuthatch, to name a few.

Lim Kim Chuah

The enigmatic ***Sumatran Cochoa*** *is the most difficult of the four cochoa species to observe.*

Access & Accommodation

Both Danau Ranau and Way Titias are accessed via the town of Liwa, a 3–4 hour drive from the regional transport hub of Bandar Lampung. Danau Ranau requires a further one and a half hour's drive from Liwa to the start of the trail, while a short drive to the village of Landos followed by a 45-minute walk brings you to the start of the trail to Way Titias. Both treks require about 6–7 hours at birdwatching pace to reach the respective campsites, the only available accommodation option in this remote region. Permits, local guides and porters can be arranged in Liwa. Gamal and Toni, in particular, are two well-known individuals who can organize logistics and serve as local guides at these two sites.

Conservation

The continuous and illegal encroachment into the park by shifting cultivation and coffee plantations is a serious threat to the reserve. Habitat clearance from these sources also facilitates the entry of poachers into the reserve.

Way Kambas National Park

Low Bing Wen

Yann Muzika

A visit to the freshwater swamps in the interior of the park may reward you with sightings of the globally threatened **White-winged Duck**.

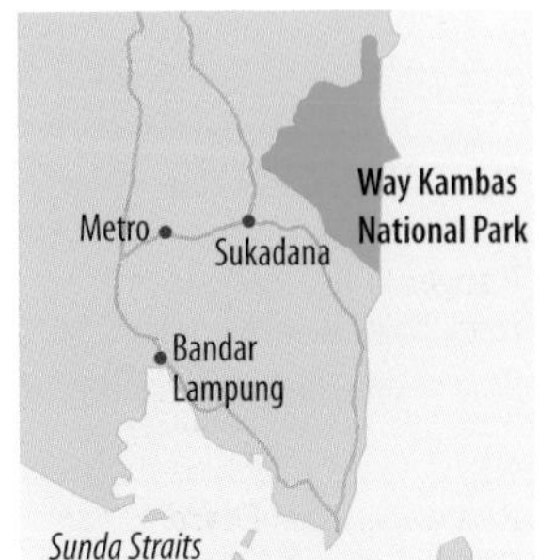

KEY FACTS

Nearest Major Town
Bandar Lampung (Lampung)

Habitats
Lowland rainforest (secondary and primary), freshwater swamp forest and wetlands

Key Species
White-winged Duck, Crested Fireback, Great Argus, Lesser Adjutant, Storm's Stork, Cinnamon-headed Green Pigeon, Oriental Bay Owl, Reddish Scops Owl, Large, Gould's and Sunda Frogmouths, Bonaparte's Nightjar, Cinnamon-rumped and Diard's Trogons, Rufous-collared Kingfisher, Malayan Banded Pitta, Rail-babbler, Black-throated Babbler

Other Specialities
Sunda Slow Loris, Mitred Leaf Monkey*, Siamang, Agile Gibbon, Sun Bear, Marbled Cat, Sunda Clouded Leopard, Sumatran Tiger, Malayan Tapir, Asian Elephant

Best Time to Visit
May–September; rainy season November–February

The brilliant **Malayan Banded Pitta** *is the most regularly encountered pitta in Way Kambas's forests.*

In a region characterized by large-scale deforestation, accessible lowland rainforest sites in southern Sumatra are in very short supply. One such site is Way Kambas National Park in Lampung province, which protects 1,300km² of lowland dipterocarp rainforest. Despite having been logged in the 1970s, the park still supports viable populations of megafauna, including the Malayan Tapir, Sumatran Tiger and even Sumatran Rhinoceros. For birdwatchers, the biggest attraction of the park is the night birds, which occur in relatively high densities – several nights of effort can yield sightings of the majority of the region's night birds, including **Large** and **Sunda Frogmouths** as well as the localized **Bonaparte's Nightjar**. However, the diurnal avifauna is just as diverse, featuring difficult and sought-after species such as the **White-winged Duck**, **Great Argus**, **Cinnamon-headed Green Pigeon** and **Rufous-collared Kingfisher**.

Birdwatching Sites

Road to Way Kanan

The 13km unpaved road to Way Kanan Ranger Station is likely to be the focal point of any birdwatching trip to Way Kambas National Park. The last 2km of the road feature several forest clearings where canopy species, including a variety of bulbuls, sunbirds and pigeons, can be easily seen, while **Bonaparte's Nightjar** can be observed hawking for insects here after dark. Pheasants such as the **Crested Fireback** and **Great Argus** can be seen crossing the road or in the adjacent forest, as can other terrestrial birds such as the **Malayan Banded Pitta** and the highly sought-after **Rail-babbler**. Foraging flocks at the forest edge provide much interest when they occur, and feature a good diversity of babblers, malkohas and trogons. Nocturnal forays along this road can also yield many night birds, including **Oriental Bay Owl** and **Reddish Scops Owl**, **Large**, **Sunda** and **Gould's Frogmouths**, and a range of mammals from flying squirrels to felines such as Leopard Cat and Marbled Cat.

Way Kanan Nature Trail

A short, overgrown loop trail that begins at the Ranger Station and ends a short way down the road, this site offers opportunities for viewing a multitude of understorey and terrestrial species. The trail is best visited in the early morning, when species such as the **Crested Partridge**, **Red-naped** and **Diard's Trogons**, and **Banded** and **Rufous-collared Kingfishers** may be seen along it, alongside a multitude of babblers including the uncommon **Black-throated Babbler**.

Rawa Gajah

A 30-minute boat ride downstream from Way Kanan Ranger Station is needed to reach this area, probably one of the best sites for sightings of **White-winged Ducks** in Indonesia, if not in their entire range. Depending on water levels, accessibility to the site varies, and several clearings may hold the ducks. Most observations involve flight views in the late afternoon, although views of foraging individuals are possible. The other speciality in this area is the **Cinnamon-headed Green Pigeon**, flocks of which can be seen perched on the trees overlooking the river.

Con Foley

*The bizarre **Large Frogmouth** is locally common at Way Kambas, and its haunting calls are frequently heard along the main entry road after dark.*

Gerard Francis

*The scarce **Bonaparte's Nightjar** is best seen along the road to Way Kanan.*

Cheng Heng Yee

***Black-throated Babbler** is one of the most attractive forest babblers that inhabit the forest at Way Kambas.*

Access & Accommodation

Way Kambas is accessed via a 2-hour drive from the town of Bandar Lampung, which itself can be reached by air from Jakarta or regular ferries from Java. Accommodation options at the park are limited, either at the basic Way Kanan Ranger Station within the park (bring your own food and mosquito nets), or at the Satwa Ecolodge just outside the park. A local guide with birdwatching experience named Hari Yono works at the Ecolodge and can be hired if required. As in other national parks in Indonesia, a permit to enter the park is compulsory, as is the requirement to have a park ranger with you at all times within the park due to the possibility of close encounters with megafauna.

Conservation

Poaching is a constant threat to the mammalian megafauna within the park, although incidents appear to have decreased in recent years. As in most lowland rainforests in Indonesia, illegal logging along remote fringes of the park is another major concern. Additionally, nearly all the countryside around Way Kambas is now covered by oil-palm plantations.

Sungai Wain Protected Forest

Low Bing Wen

Low Bing Wen

The low-lying dipterocarp forest at Sungai Wain is home to some of Borneo's most sought-after lowland forest species.

KEY FACTS

Nearest Major Town
Balikpapan (East Kalimantan)

Habitats
Lowland rainforest, freshwater swamp forest

Key Species
Bornean Peacock-Pheasant*, Bornean Ground Cuckoo*, Oriental Bay Owl, Reddish Scops Owl, Large and Sunda Frogmouths, Malaysian Honeyguide, Blue-headed Pitta*, Garnet Pitta, Bornean Bristlehead*, Black Magpie, Grey-breasted Babbler, Rufous-tailed Shama, Grey-chested Jungle Flycatcher, Rufous-chested Flycatcher

Other Specialities
Maroon* and White-fronted* Langurs, Southern Pig-tailed Macaque, Bornean Orangutan, Three-striped Ground Squirrel, Binturong, Sun Bear

Best Time to Visit
March–August; rainy season October–February

Lowland dipterocarp forests are some of the most threatened habitats in Southeast Asia, and Borneo is no exception. The provinces of Indonesian Borneo have lost more than half of their lowland rainforest cover since the 1980s. In East Kalimantan province, the Sungai Wain Protected Forest is one of the last extensive areas of relatively undisturbed lowland forest near the coast (36m asl), and it is protected as a major watershed for Balikpapan city and its oil refineries. Despite its proximity to Balikpapan and relatively small size (100km²), Sungai Wain still supports an impressive diversity of flora and fauna, including Sun Bears and Bornean Orangutans. A combination of local ownership and a passionate local management agency makes Sungai Wain arguably one of Indonesian Borneo's best protected reserves. While the core of the reserve remains inaccessible to tourists, a short visit to the accessible southern sector of the reserve can be very productive, as the site is home to several species that are very difficult to see elsewhere in Borneo, including the **Bornean Peacock-Pheasant**, **Bornean Ground Cuckoo** and **Grey-breasted Babbler**.

Birdwatching Sites

The main birdwatching area in Sungai Wain comprises a single trail that runs from the entrance of the reserve through to Pos 2, a dilapidated outpost in the middle of the forest. There are various side trails, particularly along the first section between the entrance and Pos 1, but most of these trails loop back to the entrance.

Bjorn Olesen

Bornean Ground Cuckoos occur in high densities within Sungai Wain, but seeing them in the dense riverine vegetation can be a challenge.

Low Bing Wen

The drier ridge tops dominated by Licuala palms are the preferred habitat of the Bornean Peacock-Pheasant.

Reservoir & Boardwalk

Access to the best habitat in Sungai Wain first requires you to traverse the edge of a service reservoir, then a stretch of elevated boardwalk through a tract of freshwater swamp forest. These open habitats provide a good introduction to Borneo's lowland forest avifauna, and species likely to be encountered include the Black-and-red Broadbill, White-chested Babbler, Black Hornbill, Hooded Pitta and Bold-striped Tit-Babbler.

End of Boardwalk to Pos 1

At the end of the boardwalk you enter a series of loop trails that meander through excellent lowland riverine forest. The main trail continues uphill to Pos 1, located on a ridge, and features an elevated wooden building with alfresco benches and tables for dining. The riverine forest en route to Pos 1 is home to some of the most sought-after species in Sungai Wain, such as the Bornean Ground Cuckoo, Grey-breasted Babbler, Garnet Pitta and Rufous-tailed Shama. The tall trees around Pos 1 are frequented by the noisy and gregarious Bornean Bristlehead, which is particularly active in the early morning. Spotlighting in this area is also very productive. Large and Sunda Frogmouths are common by call, but seeing them in the dense vegetation can be tough. Various owls, including the Oriental Bay Owl, Reddish Scops Owl and Brown Wood Owl, have also been recorded here.

Con Foley

*Sungai Wain is home to particularly high densities of **Grey-breasted Babbler**, a species that is conspicuous throughout the reserve.*

Pos 1 to Pos 2

The ridgeline upon which Pos 1 is located is home to multiple Bornean Peacock-Pheasants, the males of which are particularly vocal during courtship between March–April. These stunning pheasants prefer the drier forests on the ridge tops and should be looked for along the ridge trail behind Pos 1 and on the descent into the valley between Pos 1 and Pos 2. The ridge trail also holds a known display tree of the scarce Malaysian Honeyguide. Other possible birds along the ridge include the **Crested Partridge**, **Diard's Trogon**, **Banded Kingfisher** and **Great Slaty Woodpecker**. The riverine forest en route to Pos 2 offers a second chance at many of the aforementioned species such as the **Bornean Ground Cuckoo**, as well as the **Blue-headed Pitta** and **Green Broadbill**. The area around Pos 2 is also good for the peacock-pheasants. Mixed foraging flocks in the forests here contain a variety of bulbuls, babblers and flycatchers, including **Puff-backed** and **Grey-cheeked Bulbuls**, **Ferruginous Babbler**, **Chestnut-backed Scimitar Babbler** and **Maroon-breasted Philentoma**.

Access & Accommodation

Sungai Wain is easily accessed via a one-hour drive from the transport hub of Balikpapan, which is connected to Jakarta and Singapore by regular flights. It is no longer possible for tourists to stay within the forest but Kampung Sungai Wain, the main entry point to the reserve, has a basic homestay that is open to foreigners. It is preferable to contact Pak Agusdin, the director of the Sungai Wain protected forest management agency (Badan Pengelola Hutan Lindung Sungai Wain) to facilitate accommodation and permits before arrival. He is also familiar with the birds of the reserve and can act as a guide to visitors upon request.

Mikael Bauer

*The stunning **Bornean Peacock-Pheasant** is likely to be the avian highlight of any trip to Sungai Wain, although getting a good view of it may be difficult, as this photograph shows.*

Conservation

The Sungai Wain Protected Forest is one of the best-managed areas of lowland forest in Indonesia and has its own local management agency. Local ownership, perimeter fencing and its status as a water catchment area has largely spared it from degradation in recent times. Nevertheless, isolated incidents of poaching and illegal encroachment still occur, and parts of the reserve have been damaged by forest fires in recent years.

Con Foley

*The far-carrying calls of the globally threatened **Short-toed Coucal** may be heard at Sungai Wain.*

Tanjung Puting National Park

David Blair

David Blair

Birdwatching is done mainly from a klotok such as this one along the Sekonyer River.

Tanjung Puting National Park in Central Kalimantan is a remarkable park that protects the largest areas of peat-swamp forest left in Southeast Asia. The park is home to Camp Leakey, the research camp of the renowned primatologist Biruté Galdikas, who began her life's work in the park in 1971 studying Bornean Orangutans. Her work continues to this day through Orangutan Foundation International, and is one of the reasons why the forests here are still standing. Tanjung Puting is low lying, with few areas above 50m asl. The birdlife in the vast swamp, heath and lowland forests is spectacular, with more than 230 bird species recorded, including the **Crestless Fireback**, **Storm's Stork**, **Bornean Bristlehead** and enigmatic and highly localized **Black Partridge**. Along with the great diversity of birds, visitors can marvel at a range of other fauna, including seven primate species.

KEY FACTS

Nearest Major Towns
Pangkalan Bun, Kumai (Central Kalimantan)

Habitats
Riverine and peat-swamp forests, lowland rainforest, heath forest (kerangas), mangroves

Key Species
Black Partridge, Crestless Fireback, Storm's Stork, Lesser Adjutant, Great-billed Heron, Oriental Darter, Large Green Pigeon, Barred Eagle-Owl, Rhinoceros, Bushy-crested and Wrinkled Hornbills, Red-crowned Barbet, Great Slaty Woodpecker, Blue-rumped Parrot, Long-tailed Parakeet, Bornean Bristlehead*, Mangrove Whistler, Slender-billed Crow, Puff-backed and Hook-billed Bulbuls, Grey-breasted Babbler, Javan White-eye*, Malaysian Blue Flycatcher, Grey-chested Jungle Flycatcher, Scarlet-breasted Flowerpecker
Austral winter Sacred Kingfisher, Rainbow Bee-eater

Other Specialities
Proboscis Monkey*, Silvery Lutung, Maroon Langur*, Bornean Orangutan*, Bornean Agile Gibbon*, Least and Black-eared Pygmy Squirrels, Bearded Pig*, Sun Bear, Estuarine Crocodile, False Gharial

Best Time to Visit
June–October; wettest months November–December

Birdwatching Sites

Access to the park is primarily by river and there are limited places in which to land, so most of the birdwatching has to be done on motorized house boats, or *klotoks*. When heading ashore, a problem faced by visitors is that access to land-based stations is limited to after 10 a.m. for Tanjung Harapan and Pondok Tanggui, and between midday and 4 p.m. for Camp Leakey. By then, the majority of birds have ceased calling and activity is limited.

Francis Yap

*Parties of noisy **Long-tailed Parakeets** may be seen perched on top of tall trees along the Sekonyer River.*

Con Foley

*The **Great Slaty Woodpecker**, the largest extant woodpecker in the world, can be seen along Tanjung Puting's rivers.*

Untung Sarmawi

*Tanjung Puting is home to a number of bird species characteristic of peat-swamp forests, including the ultra-elusive **Black Partridge**.*

Cheng Heng Yee

***Oriental Darters** are regularly seen along the riverbanks of Tanjung Puting.*

Francis Yap

***Cinnamon-headed Green Pigeons** regularly rest on the bare branches of tall trees along the river.*

Sekonyer River

Birdwatching from a boat is easy and is usually how visiting birdwatchers observe the majority of the area's birds. From Kumai, you soon cross the inlet to the mouth of the Sekonyer River where the **Lesser Adjutant**, herons, shorebirds and terns can be seen. As the boat enters the lower reaches of the Sekonyer River, it passes extensive areas of nipah palms, mangroves and plantations. Further upriver, the landscape becomes dominated by tall riverine swamp forest and it is possible to see Bornean Orangutan nests. This is where to look out for **Storm's Stork**, **Oriental Darter**, and various kingfishers and hornbills. At many points the river is only 10–20m wide, making it easy to spot birds along the forest edge. Tall, exposed trees should be checked for **Large** and **Cinnamon-headed Green Pigeons**, **Long-tailed Parakeets** and even the occasional **Bornean Bristlehead**. It is also possible to see the sought-after **Hook-billed Bulbul** and **Malaysian Blue Flycatcher**. At dusk, both **Buffy Fish Owls** and **Barred Eagle-Owls** may be seen. The boat crew can be asked to stop the boat for better observations of wildlife.

Agata Staniewicz

The peat swamps of Tanjung Puting are also an important habitat for the endangered **False Gharial.**

Michelle & Peter Wong

The threatened **Hook-billed Bulbul** *is a specialist of peat-swamp forests in Borneo and Sumatra.*

Tanjung Harapan, Pondok Tanggui & Camp Leakey

Local guides taking you towards these centres go along the main trail straight to the orangutan feeding stations unless otherwise told. The trails are usually only several hundred metres long, arriving at a feeding platform with nearby seating. One of the highlights here is the highly localized **Crestless Fireback**. Some individuals are used to people and often venture into the orangutan feeding areas. It is possible to request walks along the forest trails, where a variety of lowland birds, including trogons, kingfishers, bulbuls and babblers, can be seen. The swamp forests of Tanjung Puting are home to the elusive **Black Partridge**, one of Southeast Asia's rarest gamebirds, and it should be searched for here.

Access & Accommodation

Tanjung Puting is best accessed from the city of Pangkalan Bun, where you can stay overnight before boarding a *klotok* the following morning from the nearby port town of Kumai. Most visitors spend about three days in the park, investing considerable time at the three major orangutan rehabilitation stations, namely Tanjung Harapan, Pondok Tanggui and Camp Leakey. *Klotoks* tend to vary in quality, but most are fairly comfortable, have shared facilities and carry up to four people. They are anchored to the riverbank during the night. All boats need a guide (for entry into the national park), and it is recommended to have one familiar with the local birdlife. The alternative is to stay at the various lodges in the periphery of the park, including the Rimba Eco Lodge by the Sekonyer.

Conservation

The 4,150km^2 Tanjung Puting National Park was originally set up as a wildlife reserve in 1935. It was subsequently recognized as a UNESCO Biosphere Reserve in 1977, and declared a national park in 1982. As in forests across Borneo, illegal logging, fires and conversion to oil-palm plantations continue to threaten Tanjung Puting, with more than 40% of the park already damaged by these activities.

Carita Nature Park

Khaleb Yordan

Yong Ding Li

The hike from the car park to the waterfall passes through secondary forest and dense groves of bamboo.

KEY FACTS

Nearest Major Towns
Cilegon, Labuhan (Banten)

Habitats
Lowland secondary rainforest, scrub and cultivation

Key Species
Black-naped Fruit Dove, Oriental Bay Owl, Barred Eagle-Owl, Javan Owlet*, Javan Frogmouth*, Banded Kingfisher, Black-banded* and Yellow-eared* Barbets, Rufous Piculet, Freckle-breasted Woodpecker, Yellow-throated Hanging Parrot*, Javan Banded Pitta*, Olive-backed Tailorbird*, White-breasted* and Crescent-chested* Babblers, Grey-cheeked Tit-Babbler*, Fulvous-chested Jungle Flycatcher, Javan Sunbird*, Streaky-breasted Spiderhunter*

Other Specialities
Javan Lutung*, Javan Slow Loris*, Leopard Cat, Sunda Colugo

Best Time to Visit
September–November; rainy season January–March

One of the most accessible lowland forest sites in west Java, Carita Nature Park (Hutan Wisata Carita) in Banten Province is on the itinerary of many visiting birdwatchers. At just 0.95km^2, the park forms a fraction of the larger Carita Forest (Kawasan Hutan Carita), which is fringed by Gunung Aseupan to the east. The vegetation within the park is mostly secondary lowland dipterocarp forest rising to about 50m asl. Despite heavy disturbance and the degraded nature of the forest, Carita is home to a number of lowland Javan endemics that are easily seen here, including the **Black-banded Barbet**, **Grey-cheeked Tit-Babbler** and charismatic **Javan Banded Pitta**. Nocturnal forays here can be very productive, with the **Barred Eagle-Owl** and **Javan Frogmouth** regularly recorded, while the **Oriental Bay Owl** also occurs. As the park is very popular at weekends with visitors to the Curug Gendang Waterfall, the best time to visit is on weekdays. It is possible to see nearly 50 species on a 2–3-day trip.

Birdwatching Sites

Access Road to Car Park

Birdwatching is possible as soon as you leave the coastal road past the entrance arch. On passing the shelter at the T-junction, the left turn leads to the main birdwatching areas. Along this access road, which passes through plantations and secondary forest, both the **Javan Banded Pitta** and the distinctive Javan subspecies (ssp. *capistratum*) of the **Black-capped Babbler** are regularly seen along the forest edge. **Sunda Pygmy** and **Freckle-breasted Woodpeckers** are common in the midstorey, and the **Streaky-breasted Spiderhunter** and **Javan Sunbird** are occasionally seen. The

One of the main reasons for visiting Carita is the stunning **Javan Banded Pitta**, *a species that remains fairly common in the forests here.*

Javan Owlet is regularly heard in the morning, but can be difficult to see. Dense bamboo groves along this road are the best place to see the **Yellow-bellied Warbler** and the pitta. **Sunda Scops Owls** and **Javan Frogmouths** can be encountered during a spotlighting session, and the **Oriental Bay Owl** is also a possiblity.

Forest Around Car Park

The access road eventually leads to a parking area surrounded by secondary forest. The **Black-naped Fruit Dove**, **Olive-backed Tailorbird** and **Grey-cheeked Tit-Babbler** are all easy to find here. A few hundred metres from the car park on the main trail to the waterfall is an open area with dense scrub. From here on, the area is excellent for both **Black-banded** and **Yellow-eared Barbets**, especially where there are fruiting trees. Also keep an eye out for the **Rufous Piculet**, **Crescent-chested Babbler** and **Fulvous-chested Jungle Flycatcher**.

Khaleb Yordan

*The peculiar **Oriental Bay Owl** is one of several species of night bird that may be seen at Carita.*

Ujung Kulon National Park

*The impressive **Great-billed Heron** frequents the remote beaches of the Ujung Kulon Peninsula and its surrounding islands.*

The sprawling Ujung Kulon National Park protects the largest expanse of lowland coastal dipterocarp forests left in western Java, and is best known for supporting the only remaining population of the Critically Endangered Javan Rhinoceros. Few birdwatchers make it this far, but the forest and open habitats here offer some of the best lowland forest birdwatching in west Java. Highlights in the park include the **Green Peafowl**, **Great-billed Heron**, **Nicobar Pigeon**, **Rhinoceros Hornbill** and most of the lowland Javan endemics such as the **Black-banded Barbet**. In addition, the waters around Pulau Peucang and Tanjung Layar support a good diversity of seabirds, with **Wilson's Storm Petrel**, **Streaked Shearwater**, **Brown Booby** and **Aleutian Tern** all recorded recently. The best way to get to the park is from the town of Tamanjaya, which can be reached from Labuhan by public transport. Tours into the park from Pulau Peucang can also be arranged from Anyer, Carita and from travel agents in Jakarta.

Access & Accommodation

While it is possible to get to Carita by public transport via Cilegon or Merak, the best way to reach the birdwatching sites is by car. To get to Carita from Jakarta, take the toll road to Cilegon Barat. From the toll road you can take a shortcut by heading into the Krakatau Steel Industry area. From here follow the signposts towards Carita. After exiting the steel-industry area, the right turn eventually leads to Carita Beach (Pantai Pasir Putih Carita). The nature reserve is on the landward (left) side of the road. There is a good selection of resorts, guesthouses and hotels along the coastal road heading towards Carita Beach and the town of Banjarmasin.

Conservation

Carita has been established as a nature reserve by ministerial decree since 1978. A major conservation problem faced by the park is encroachment at its fringes and, most seriously, the poaching of birds for Indonesia's booming pet-bird trade. Many of the Javan Banded Pittas in the bird shops of surrounding towns have been trapped from the forests in and around Carita.

Gunung Halimun-Salak National Park

Khaleb Yordan & Yong Ding Li

Yong Ding Li

The forest edge by the Nirmala tea plantation is good for seeing raptors and the Brown Prinia.

Although not nearly as famous as the nearby Gunung Gede-Pangrango National Park, Gunung Halimun-Salak National Park is nonetheless one of the most important national parks on Java. Covering nearly 1,000km², it protects some of the most extensive areas of hill and submontane forests left on western Java on the slopes of its two main peaks, Gunung Halimun and Gunung Salak, both rising to nearly 2,000m asl, and a number of lower mountains. Known for harbouring all of Java's endemic primates, including the endangered Javan Silvery Gibbon, Gunung Halimun-Salak is also notable for its birdlife, with over 150 species recorded to date. Several Javan endemics are easier to see here than at Gunung Gede, especially a number of lowland and submontane species. Both the **Javan Trogon** and **Spotted Crocias**, for example, are regularly heard or seen along the main access road from Cikaniki. In addition, the park receives few visitors so the wildlife here is generally more confiding than elsewhere on Java. It is possible to see more than 60 species on a trip of 2–3 days covering the different elevations in the park.

KEY FACTS

Nearest Major Towns
Bogor, Sukabumi (West Java)

Habitats
Submontane and montane rainforests, hill rainforest

Key Species
Chestnut-bellied Partridge*, Javan Hawk-Eagle*, Sumatran Green Pigeon*, Pink-headed Fruit Dove*, Reddish Scops Owl, Javan Frogmouth*, Javan Trogon*, Brown-throated* and Flame-fronted* Barbets, Yellow-throated Hanging Parrot*, Javan Banded Pitta*, Sunda Minivet*, White-bellied Fantail*, Brown Prinia, White-breasted*, White-bibbed* and Crescent-chested* Babblers, Large Wren-Babbler, Spotted Crocias*, Sunda Forktail*, Javan Sunbird*, Streaky-breasted Spiderhunter*
Winter Japanese Paradise Flycatcher, Siberian and Eyebrowed Thrushes, Blue-and-white Flycatcher

Other Specialities
Javan Slow Loris*, Javan Lutung*, Javan Grizzled Langur*, Javan Gibbon*, Red Giant Flying Squirrel, Small-toothed Palm Civet, Leopard, Leopard Cat, Javan Flying Frog*

Best Time to Visit
March–September; rainy season October–January

Birdwatching Sites

Access Road to Cikaniki

The main road that leads to and from the research station at Cikaniki is the best site to see most of Gunung Halimun's specialities. This road passes through a small stretch of good forest immediately before the station, and continues to climb uphill through submontane

Fruiting trees along the access road to Gunung Halimun should be checked for the ***Sumatran Green Pigeon****.*

forest (c. 1,000m asl) for 4km before the vegetation become increasingly open and scrubby near the edge of the park. It is best to explore the road early as it can become busy with traffic later in the day.

The distinctive yellow-billed subspecies (ssp. *flavirostris*) of the **Blue Whistling Thrush** and the **Sunda Forktail** may be seen foraging by the sides of the access road early in the morning, and the song of the **Pale Blue Flycatcher** is often heard, although the singer is only infrequently seen. While walking along the road, you should look out for the mixed foraging flocks that move through the roadside forest. The **Spotted Crocias**, a species hard to see now in Gunung Gede, is a major target, but the flocks should also be checked for the **Sunda Minivet**, **White-bellied Fantail**, **Trilling Shrike-babbler** and **Blue Nuthatch**. Larger birds that regularly join these flocks include the **Red-billed Malkoha** and **Javan Trogon**. Roadside streams, especially those just before the research station, should be checked for the endemic nominate subspecies of the **Blue-banded Kingfisher**, a bird that is now rarely recorded on Java.

Two of Java's endemic barbets, namely **Brown-throated** and **Flame-fronted Barbets**, are easily heard along the road, but seeing them can be tricky. The best way to see the barbets and other fruit-eating birds like pigeons is to locate a fruiting fig tree and wait patiently. With luck, both **Sumatran Green Pigeons** and **Dark-backed Imperial Pigeons** can then be seen.

Con Foley

*The **Javan Trogon** is more easily seen at Gunung Halimun than Gunung Gede.*

Stream Trail

This trail runs for more than 2km along a stream and allows access to excellent submontane forest. Besides the usual mixed flock species, the highlights here are the **Large Wren-Babbler** and **Rufous-chested Flycatcher**, both shy species difficult to see along the roadside. At night the **Javan Frogmouth** has been recorded here, and it may also be possible to see the **Reddish Scops Owl**.

Khaleb Yordan

***Black Eagles** are commonly seen soaring over the forests of Gunung Halimun.*

Khaleb Yordan

The ***Javan Whistling Thrush*** *is uncommon around Cikaniki.*

Nirmala Tea Plantation

This plantation is almost completely surrounded by the forest, so visitors pass the site when entering or leaving Cikaniki. From various vantage points in the plantation overlooking the forest, both the **Black Eagle** and **Javan Hawk-Eagle** are regularly seen soaring overhead. A number of open-country species can also been seen within the plantation, most notably the distinctive Javan subspecies of the **Brown Prinia** (ssp. *polychroa*). Both the prinia and **Striated Grassbird** can be difficult to observe because they tend to skulk in the dense tea bushes.

Access & Accommodation

The best way to access Gunung Halimun-Salak is from Bogor, which is readily accessible from Jakarta via numerous public bus services. From Bogor you can hire vehicles to get to Cikaniki directly. It is possible to stay at the research station in Cikaniki, although arrangements should be made in advance with the park headquarters or via local guides. The alternative is to stay in the village of Citalahab near the park, and make day trips to the key birdwatching areas.

Conservation

Given that much of west Java's lowland and submontane forests have been lost, Gunung Halimun-Salak is undoubtedly one of the most important areas for biodiversity in this part of Java. A narrow corridor of forest links the main block of the park to Gunung Salak, and allows wildlife to disperse between the two major forest blocks. A serious problem faced by the park is encroachment at its fringes and poaching of birds for the pet-bird trade. Some species, including the Javan Green Magpie, are now on the brink of extinction due to the bird trade. Gunung Halimun-Salak used to be a stronghold for this species, but there have been no records here for a number of years.

The ***Javan Slow Loris*** *is regularly seen in the forests of Gunung Halimun.*

The ***White-bibbed Babbler*** *is the most attractive among Java's endemic babblers.*

Gunung Gede-Pangrango National Park

Yong Ding Li & Khaleb Yordan

Yong Ding Li

The forested slopes of Gunung Gede-Pangrango are home to the vast majority of Java's endemic bird species.

KEY FACTS

Nearest Major Towns
Cibodas, Sukabumi (West Java)

Habitats
Hill rainforest, submontane and montane rainforests, alpine meadows, parkland

Key Species
Montane Chestnut-bellied Partridge*, Javan Hawk-Eagle*, Javan Woodcock*, Pink-headed Fruit Dove*, Javan Scops Owl*, Javan Owlet*, Salvadori's Nightjar*, Volcano* and Giant* Swiftlets, Javan Trogon*, Flame-fronted* and Brown-throated* Barbets, Rufous-tailed* and White-bellied* Fantails, Pygmy Bushtit*, Sunda* and Orange-spotted* Bulbuls, Javan Tesia*, Rufous-fronted Laughingthrush*, Spotted Crocias*, White-bibbed* and Crescent-chested* Babblers, Sunda Thrush, Javan Cochoa*, Sunda Robin*, Sunda Forktail*, Javan Whistling Thrush*, White-flanked Sunbird*, Tawny-breasted Parrotfinch
Lowland Javan Frogmouth*, Javan Banded Pitta*, Large Wren-Babbler, White-breasted Babbler*
Winter Siberian and Eyebrowed Thrushes, Blue-and-white and Mugimaki Flycatchers

Other Specialities
Sunda Colugo, Javan Grizzled Langur*, Javan Lutung*, Javan Gibbon*, Leopard, Sunda Stink Badger, Javan Mousedeer*, Javan Flying Frog*, Javan Horned Frog

Best Time to Visit
March–October; rainy season November–April

Rising to over 3,000m asl at the summit of Gunung Pangrango, the twin peaks of Gunung Gede-Pangrango in the namesake national park are one of the most popular sites for birdwatchers visiting Java, a fact helped by its proximity to Jakarta and Bogor. This important park is one of Indonesia's oldest. The park comprises lowland forests on its southern fringes, extensive tracts of montane forest that cover the slopes of the two dormant volcanoes, and alpine meadows at its highest elevations. Adjacent to the park's main entry point is the Cibodas Botanical Gardens, a beautifully landscaped park that merits attention from birdwatchers since many common montane birds can be easily seen here. On a 4–5-day visit covering different heights and habitats, it is possible to see nearly 80 species, including most of Java's endemics. A number of these, particularly the **Chestnut-bellied Partridge**, the fast-declining **Rufous-fronted Laughingthrush** and **Javan Cochoa**, are more easily seen here than anywhere else on Java.

Con Foley

*The **Javan Fulvetta** is common in mixed flocks at the lower elevations of the summit trail.*

Birdwatching Sites

Kebun Raya Cibodas (Cibodas Botanical Gardens)

As many of the montane endemics can be more easily seen here due to its relatively open environment, birdwatchers

Michelle & Peter Wong

*The **Blue-and-white Flycatcher** is one of the wintering passerines that can be seen in the Cibodas Botanic Gardens.*

Khaleb Yordan

*The enigmatic **Javan Cochoa** is regularly seen around the Air Panas area.*

should consider spending at least a full day here before proceeding to the summit trail, where conditions for birdwatching are more difficult. Many mixed flocks pass through the gardens, in particular the patches of montane forest within, and should contain **Pied** and **Trilling Shrike-babblers**, **Rufous-tailed Fantails**, **Pygmy Bushtits** and **Javan Fulvettas**. In recent years the elusive **Salvadori's Nightjar** and **Sunda Thrush** have also been seen in the more wooded areas of the gardens.

Summit Trail to Telaga Biru (Blue Lake)

Climbing steeply into lush montane forest from the entrance arch, this section of the trail is probably the best place to quickly accumulate species on your checklist. Many large mixed flocks pass through the forest throughout the day, with greater species diversity than in the botanical gardens. The lower section of the trail (from the entrance to 500m past the ticket booths) is where the elusive **Sunda Thrush** and **White-bellied Fantail** are most frequently encountered in the early mornings, although seeing them is by no means guaranteed. Notable along this section of the trail are the different ground birds, many of which are unlikely to be seen well without effort and patience. Watch out for the **Eyebrowed Wren-Babbler** and noisy parties of **White-bibbed** and **Crescent-chested Babblers**. With luck, other highlights like the **Chestnut-bellied Partridge**, **Tawny-breasted Parrotfinch** and **Mountain Serin** may also be encountered.

Telaga Biru (Blue Lake) to Air Panas Hot Springs

Besides the usual flocking species, the area around Telaga Biru is especially good for the **Javan Hawk-Eagle**, while spotlighting sessions at night have produced the **Javan Frogmouth** and **Salvadori's Nightjar**. Fruiting trees should be checked for the **Pink-headed Fruit Dove**, as well as for **Flame-fronted** and **Brown-throated Barbets**. Beyond the Blue Lake, the trail enters a boggy section that is raised on a wooden boardwalk. This is the best part of the trail

Khaleb Yordan

*The **Pink-headed Fruit Dove** is not uncommon here, but seeing it in the forest canopy is difficult.*

***Javan Trogons** are often seen quietly sitting on a branch in the forest canopy.*

to encounter **Sunda Forktail** in the morning. Towards mid-morning this section is often frenetic with mixed flock activity, and this is the best time to see the **Checker-throated Woodpecker**, **Sunda Minivet**, **Spotted Crocias** and **Chestnut-backed Scimitar Babbler**, which are regular in the flocks. The trail climbs steeply into dense montane forests after the boardwalk towards Air Panas, where you should listen for the shrill whistling of the enigmatic **Javan Cochoa**. **Javan Trogons** are also regular along this stretch of the trail.

Cibereum Falls

A narrow trail branches off the main summit trail on the right to the Cibereum Falls. The **Javan Tesia** is common in the dense undergrowth along the trail, although the main highlight here is the rare **Giant Swiftlet**, which is most regularly seen in this area of the park. Look out for the **Spotted Kestrel** and **Sunda Forktail**, both regular species here.

Kandang Batu

After the Air Panas hot spring is the Kandang Batu campsite. The **Chestnut-bellied Partridge** regularly visits it at dawn and dusk. The **Javan Scops Owl** is also regularly heard on the trail between here and Kandang Badak, the next campsite.

Kandang Badak – Summit

Few birdwatchers make it this far. Those that do are able to appreciate the transition from upper montane forest into open alpine meadows characterized by the impressive Javan Edelweiss. Besides **Island** and **Scaly Thrushes** (ssp. *horsfieldi*), which are usually not seen below 2,600m asl, the main highlight here is the highly localized **Volcano Swiftlet**, which is confined to the summits of a few of Java's higher volcanic peaks.

Selabintana Gate

Situated on the south side of the park, the forest starts at a lower elevation here than at the main Cibodas Gate. A number of primarily lowland forest species can be expected here, including the **Javan Banded Pitta**, **Large Wren-Babbler** and, with luck, the **White-breasted Babbler**.

Access & Accommodation

Many buses and public vans, the latter known locally as *angkuts*, run between Bogor and Bandung via Cibodas. Once dropped off, you can take the regular van service that shuttles

Khaleb Yordan

*The elusive **Chestnut-bellied Partridge** is often seen foraging for scraps around the campsites along the summit trail.*

James Eaton

*Gunung Gede is the only regular place to encounter the **Rufous-fronted Laughingthrush**.*

between the main road and the Cibodas Botanic Gardens. A number of comfortable homestays and inns occur along this road, the most popular being the 'Freddy Homestay'. Local guides can be hired here. Entry permits are required and can be arranged at the park office in Cibodas.

Conservation

One of Java's best-protected parks, Gunung Gede-Pangrango has also been declared a UNESCO Biosphere reserve in recognition of its high value for biodiversity conservation and cultural significance. The forests on the twin volcanos have been extensively studied since the early 1800s by many leading botanists, like C. G. C. Reinwardt and C. G. G. van Steenis, while the Cibodas Botanical Gardens have a history dating back to the 1830s.

Raden Soerjo Forest Park

Heru Cahyono

Heru Cahyono

Vista of Gunung Welirang as seen from the lower summit trail.

Located in the heart of East Java, the Raden Soerjo Forest Park is a major watershed to many of East Java's cities, and some of its water features, like the Cangar Hot Springs and Watu Ondo Waterfall, are also popular tourist attractions. Encompassing a total area of 279km^2 and rising to a maximum height of 3,339m asl at the summit of Gunung Arjuno, Raden Soerjo provides easy access to remnant montane evergreen forest, home to many of Java's endemic birds. Preliminary surveys have recorded more than 170 bird species, including many that can be hard to see in West Java, such as the **Sunda Thrush**, **Mountain Serin** and subspecies (ssp. *horsfieldi*) of the **Scaly Thrush**, which is endemic to Indonesia. In addition, the regular occurrence of the endangered **Javan Hawk-Eagle** makes Raden Soerjo a difficult place to leave out on a trip to east Java.

Heru Cahyono

*The **Javan Hawk-Eagle**, one of the site's avian attractions, is often seen at Watu Ondo Bridge.*

To Surabaya
Raden Soerjo Forest Park
Gng Arjuna
Gng Arjuno
To Kadangan
Batu
Lawang
Malang

KEY FACTS

Nearest Major Towns
Surabaya, Mojokerto, Malang (East Java)

Habitats
Hill, submontane and montane rainforests

Key Species
Chestnut-bellied Partridge*, Javan Hawk-Eagle*, Pink-headed Fruit Dove*, Orange-breasted Trogon, Black-banded Barbet*, Javan Flameback*, Banded Broadbill, Javan Banded Pitta*, White-bellied Fantail*, Orange-spotted Bulbul*, White-bibbed* and Crescent-chested* Babblers, Sunda Thrush, Rufous-chested Flycatcher, Sunda* and White-crowned Forktails, Javan Whistling Thrush*, Tawny-breasted and Pin-tailed Parrotfinches, Mountain Serin
Passage and winter Eyebrowed Thrush, Grey-streaked, Narcissus, Blue-and-white and Mugimaki Flycatchers, Forest Wagtail

Other Specialities
Sunda Colugo, Javan Lutung*, Javan Ferret-badger*, Red Giant Flying Squirrel, Yellow-throated Marten, a variety of amphibians including Hasselt's Litter* and Javan Horned Frogs

Best Time to Visit
June–December; wettest months November–March

Birdwatching Sites

Cangar Hot Springs

Despite being heavily commercialized and popular with tourists, this area is surprisingly good for birdwatching, with more than 110 species recorded. In the forests around the hot springs, a variety of birds, including the **Pink-headed Fruit Dove**, **Dark-backed Imperial Pigeon**, **Freckle-breasted Woodpecker**, **Sunda Minivet**, **Blue Nuthatch** and **White-flanked Sunbird** can be seen. During the northern winter, a number of wintering flycatchers, including **Yellow-rumped**, **Mugimaki** and **Narcissus Flycatchers**, supplement the resident avifauna. In the early morning and late afternoon, it is possible to see the elusive **Sunda Thrush** on the path to the hall or around the baths at the hot-springs area.

An alternative birdwatching site in the area is along the 'Jogging Track', or around the Japan Cave (Goa Jepang). Here a different suite of birds is present, including the **Orange-breasted Trogon**, **Checker-throated Woodpecker**, **Javan Flameback**, **Rusty-breasted Whistler**, **Crescent-chested** and **White-bibbed Babblers**, **Tawny-breasted Parrotfinch** and **Mountain Serin**. Spotlighting in the area may yield the **Oriental Bay Owl** and **Brown Wood Owl**, here of the distinctive subspecies *bartelsi*.

About 500m from Cangar in the direction of Watu Ondo is a bridge that offers a good vantage point of the surrounding forested landscape and is one of the best places in the area to look for the **Javan Hawk-Eagle**.

Dubi Shapiro

*The **Mountain Serin** can be erratic and difficult to see anywhere in its distribution.*

Watu Ondo Waterfall

Another popular attraction for local tourists, the waterfall is where the **Sunda Forktail** can be regularly seen. The surrounding forest should be checked for species such as the **Sunda Cuckoo**, **Flame-fronted Barbet**, **Sunda** and **Orange-spotted Bulbuls**, and **Javan Whistling Thrush**. Mixed flocks passing this area should be scrutinized for the **Sunda Cuckooshrike**, **Pied** and **Trilling Shrike-babblers** and **Mees's White-eye**.

Heru Cahyono

Subspecies horsfieldi *of the **Scaly Thrush**, which is endemic to Indonesia's mountains, is regularly seen here.*

Heru Cahyono

*The **Sunda Thrush**, like all Zoothera thrushes, is generally shy and retiring, although pairs are occasionally encountered around Cangar Hot Springs between October–December.*

Lemahbang & Watu Lumpang Waterfall

Located about 1km from Watu Ondo, this highland area is relatively flat, making it a great place to look for ground-dwelling forest birds such as the **Chestnut-bellied Partridge** and **Javan Banded Pitta**. At the waterfall itself the **White-crowned Forktail** is regularly encountered, and the difficult **Pin-tailed Parrotfinch** and endemic **White-bellied Fantail** have been seen in the area.

Gajah Mungkur & Sendi

This hilly location provides good viewing opportunities for raptors such as the **Spotted Kestrel** and **Crested Honey Buzzard**, as well as other birds like the **Wreathed Hornbill**. The dense understorey is home to skulkers like the **Brown Prinia** and endemic **Javan Bush Warbler**. Sendi, located at 1,067m asl, features numerous local cafes. A trail heading west that passes through remnant woodland supports the **Violet Cuckoo**, **Banded Kingfisher**, **Rufous-chested Flycatcher**, **Thick-billed** and **Crimson-breasted Flowerpeckers**, the endemic **Javan Sunbird**, and various migratory passerines like the **Narcissus Flycatcher** during the northern winter.

Wonosalam

This region of Gunung Anjasmoro is one of the few known sites for the rarely seen Javan subspecies (ssp. *euryzona*) of the **Blue-banded Kingfisher**. For the more intrepid birdwatcher, reaching this site requires a drive of three and a half hours from Cangar, and it is recommended that Mr Parimun from the nearby Jarak Kebon village be contacted to make arrangements beforehand.

Access & Accommodation

Access to Raden Soerjo is possible via Surabaya in the north or Malang in the south, and a rented/private vehicle is recommended. From Surabaya, access is via Mojokerto and Pacet and takes approximately three hours. From Malang, it takes about two hours via the settlements of Batu and Bumi Aji. There is a variety of accommodation options in both Batu and Pacet. A permit is required for visitors from abroad and can be applied for at the park office (Tahura Raden Soeryo Jl Raden, Simpang Suroso Kav. 144 – Arjosari Malang). Avoid the park on weekends and public holidays due to its popularity with tourists. Local guides are unavailable, although there are birdwatchers based in Surabaya and Malang who can assist prospective visitors.

Conservation

The site was designated a forest park in 1992 and is regularly patrolled by rangers, particularly in the region between Cangar to Sendi. Nevertheless, illegal trapping of birds for the pet-bird trade remains a problem, although such operations are comparatively small in scale.

Heru Cahyono

*The vulnerable **Javan Flameback** is a specialty of east Java.*

Baluran National Park

Heru Cahyono

Baluran is one of the global strongholds of the stunning **Green Peafowl**.

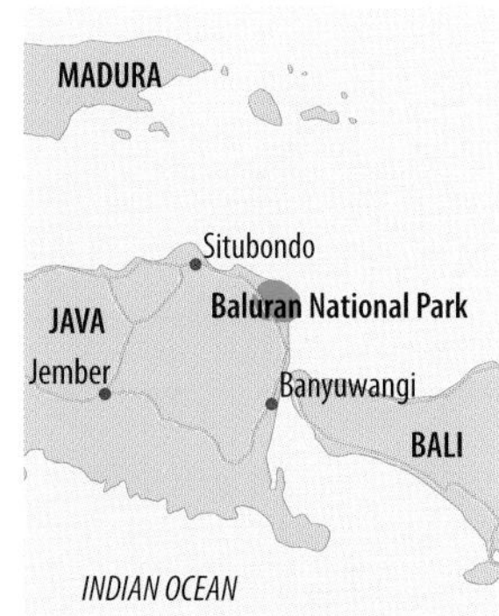

Situated on the northern coast of East Java province, this national park features mixed deciduous (monsoon) forests and extensive savannah grasslands more reminiscent of East Africa than Southeast Asia. Encompassing a total area of 250km² and rising to a maximum elevation of 1,240m asl at Gunung Baluran, the park is home to a diverse array of forest and open-country species. To date more than 230 bird species have been recorded in the park, including three hornbills, 10 woodpeckers and 25 raptors, as well as the endemic and Critically Endangered **Black-winged Starling** (ssp. *tricolor*). In addition, pheasants such as the endemic **Green Junglefowl** and globally threatened **Green Peafowl** (ssp. *muticus*) are readily encountered throughout the park.

Birdwatching Sites

Batangan

The headquarters of the park is located here. Birdwatchers should spend some time around the observation points, where **Streaky-breasted** and **Long-billed Spiderhunters** can be seen. From here, they can opt to visit Bekol (12km away), the focal point of all activities within the park. The route between Batangan and Bekol is good for a variety of lowland species, including the **Woolly-necked Stork**, various pigeons, **Red-billed Malkoha**, **Oriental Dwarf Kingfisher**, the distinctive Javan subspecies (ssp. *silvestris*) of the **Rhinoceros Hornbill**, **Javan Banded Pitta** and **Java Sparrow**.

Bekol

The main tourist hub within the park, Bekol comes complete with waterholes, an observation tower and visitor accommodation. The site is surrounded by savannah and both the **Green Junglefowl** and **Green Peafowl** are easily seen here, particularly in the early morning and late afternoon. The observation tower is a good place for raptor watching, especially during the

KEY FACTS

Nearest Major Towns
Situbondo, Banyuwangi (East Java)

Habitats
Mixed deciduous and evergreen forests, mangrove forest, savannah and scrub

Key Species
Green Junglefowl*, Green Peafowl, Javan Hawk-Eagle*, Lesser Adjutant, Beach Stone-curlew, Cerulean* and Javan* Kingfishers, Rhinoceros, Oriental Pied and Wreathed Hornbills, Javan Flameback*, Black-thighed Falconet, Javan Banded Pitta*, Grey-cheeked Tit-Babbler*, Black-winged Starling*, Java Sparrow*
Winter Short-toed Snake Eagle, Pacific Golden Plover, Wood Sandpiper, Yellow-rumped Flycatcher

Other Specialities
Javan Lutung*, Asiatic Wild Dog, Leopard, Javan Warty Pig*, Javan Rusa, Banteng

Best Time to Visit
May–October

Con Foley

*The **Lesser Adjutant** is often seen soaring over Baluran's plains in the heat of the day.*

Mark J. Villa

A vista of Gunung Baluran as seen from the area around Bekol.

northern winter, when migratory species like the **Short-toed Snake Eagle** can be seen. In the late afternoon small groups of the endangered **Black-winged Starling** may be seen foraging in the savannah or perched on top of Javan Rusa (a deer species). After dark, the **Spotted Wood Owl** has been seen here.

Bama

Located about 3km north of Bekol, Bama offers a change of scenery, featuring dense evergreen coastal forest. Birdwatching along the path towards Manting water source can be particularly productive. Small flocks of **Grey-cheeked Tit-Babblers** are common here, although the east Javan subspecies (ssp. *banyumas*) of the **Hill Blue Flycatcher** may provide more of a challenge to locate. Other species possible here include two of Southeast Asia's largest woodpeckers –**White-bellied** and **Great Slaty Woodpeckers** – as well as mangrove specialists such as the **Mangrove Whistler** and **Mangrove Blue Flycatcher**. At low tide the **Lesser Adjutant** is often found foraging along the beaches alongside the **Cerulean Kingfisher**.

Kacip

Located on the slopes of the dormant Gunung Baluran, the tallest landform within the park, Kacip requires a lengthy 6–7km trek from Bekol to reach, and a stint of overnight camping is required upon arrival. However, the forests at Kacip are the main area in Baluran to see the endangered **Javan Hawk-Eagle**. A good diversity of other forest birds can be seen here, including the **Orange-breasted Trogon**, **Banded Kingfisher**, **Hair-crested Drongo** and **Asian Fairy-bluebird**. In general, the Kacip area is under-surveyed, and more thorough exploration will undoubtedly yield more species.

Francis Yap

*Coastal creeks in Baluran should be checked for the attractive **Cerulean Kingfisher**.*

Bajulmati River

This location is not far from Batangan, and can be reached by following the main path 2km to the south. This sector of riverine evergreen forest is home to a range of lowland forest species, including the endemic **Yellow-throated Hanging Parrot** and **Black-banded Barbet**, as well as other species such as the **Blue-eared Kingfisher**, **Banded Broadbill** and **White-crowned Forktail**.

Access & Accommodation

The park is easily accessed from both Java and Bali. It is a 5–6 hour drive from Denpasar (Bali), or two hours from West Bali inclusive of a one-hour ferry crossing from West Bali to East Java. Alternatively, Baluran can be reached via an eight-hour drive on the northern coastal highway (Jalur Pantai Utara) from Surabaya (Java). Homestay accommodation is available at both Bekol and Bama, and a range of hotels can be found near Batangan. Local guides are available.

Conservation

Despite its protected status as a national park, due to easy accessibility and porous borders poaching is a major threat to the park's wildlife. A range of wildlife is harvested, including the eggs of the globally threatened Green Peafowl, as well as a number of songbirds such as the Javan Banded Pitta and Black-winged Starling for the cage-bird trade. Of particular concern to conservationists has been the rapid decline in the park's Banteng population as a result of poaching and, ironically, predation by the large local population of Asiatic Wild Dogs, another endangered species.

Bali Barat National Park

Low Bing Wen

Low Bing Wen

Bali Barat's picturesque landscape provides the perfect backdrop for observing the area's birdlife.

Located on the north-western tip of Bali, this national park protects some of the last major areas of low-lying wilderness on this otherwise heavily developed island. A variety of habitats can be found within the park boundaries, ranging from open palm savannah to evergreen forest, mangroves and sandy beaches. For many birdwatchers the primary avian attraction here is the Critically Endangered **Bali Myna**, which is endemic to the island. Unfortunately, truly wild populations have been extirpated and virtually all the birds seen within the park nowadays originate from captive breeding programmes. However, a close encounter with this gorgeous starling against the backdrop of its preferred palm savannah habitat will make any birdwatcher's day, with further exploration likely to yield a fine supporting cast of Indonesian endemics that can be difficult to see elsewhere, including the **Black-winged Starling** and **Javan Owlet**, as well as the scarce **Beach Stone-curlew**, which can be regularly seen on Bali Barat's beaches.

KEY FACTS

Nearest Major Town
Gilimanuk (Bali)

Habitats
Open savannah, mixed deciduous and evergreen forests, mangroves

Key Species
Green Junglefowl*, Beach Stone-curlew, Javan Plover*, Javan Owlet*, Javan Banded Pitta*, Crescent-chested Babbler*, Black-winged Starling*, Bali Myna*, Fulvous-chested Jungle Flycatcher, Scarlet-headed Flowerpecker*, Java Sparrow*
Winter Oriental Plover

Other Specialities
Javan Lutung*, Sunda Porcupine, Black Giant Squirrel, Red Muntjac, Javan Rusa

Best Time to Visit
April–October; rainy season November–March

Francis Yap

*The striking **Javan Banded Pitta** is locally common, but can be frustratingly difficult to see in the dense coastal rainforest.*

Birdwatching Sites

Brumbun Ranger Station

Located on the eastern shoreline of the Prapat Agung Peninsula, this is one of the few accessible areas in the

Francis Yap

*The unmistakable **Bali Myna** is Bali's only avian endemic and one of the most beautiful starlings in the world.*

*The charismatic **Beach Stone-curlew** haunts the remote reefs and beaches of Bali Barat.*

park where the **Bali Myna** can be readily seen. Captive-bred individuals are acclimatized here before their release into the wild, and their noisy vocalizations frequently attract free-flying birds to the area. The rolling hills, dotted with scattered palms and thorn scrub surrounding the centre, support small numbers of **Black-winged Starlings** (ssp. *tertius*), which sometimes perch on top of the numerous Javan Rusa, while the **Java Sparrow** is regularly seen in the area especially when the grass is seeding. Brumbun is accessed via a 30-minute boat ride from the jetty at Labuan Lalang, and scanning the sandy beaches from the boat may be rewarded with sightings of the **Beach Stone-curlew**. The nearby Menjangan Island, a famous diving site, supports a population of **Lemon-bellied White-eyes**.

Labuan Lalang Jetty

The monsoon forest located just across the road from the Labuan Lalang Jetty provides a change of scenery and introduces a new suite of birds from those observed on the palm savannah. A network of trails meanders through the area, where the understorey conceals such gems as the **Javan Banded Pitta**, **Oriental Dwarf Kingfisher** and **Fulvous-chested Jungle Flycatcher**. The nearby salt pans support a breeding population of **Javan Plovers**.

Menjangan Resort Road Network

The road network leading to the Menjangan Resort, just down the road from Labuan Lalang jetty, is the best place in the area for getting close views of the endemic **Green Junglefowl**, as well as the Javan Rusa and Red Muntjac. Small numbers of **Bali Myna** have been released here and birdwatchers on a tight schedule might opt to try for them here instead of taking the longer boat trip to Brumbun. The more open forest here allows better viewing opportunities of canopy birds such as the **Yellow-throated Hanging Parrot**, **Grey-cheeked Green Pigeon** and **Black-winged Starling**.

Access & Accommodation

Bali Barat can be reached via a day trip from Bali's capital of Denpasar, although a very early start is required to account for the four-hour drive each way. Alternatively, the nearby town of Gilimanuk offers plenty of accommodation options for those who wish to spend more time in the area. A permit and guide are compulsory, and arrangements can be at the park headquarters in Gilimanuk, or the office at Labuan Lalang. There are also several park rangers (especially Hery Kusumanegara), who are used to dealing with visiting birdwatchers and can be hired as local guides if required.

Conservation

Bali Mynas command a high price in the captive-bird trade and poaching of released birds is a problem the park faces on a regular basis. Additionally, habitat loss through encroachment and illegal woodcutting is a major threat.

*The **Green Junglefowl** adds colour to any morning's ramble around Bali Barat's savannah.*

Komodo National Park

Low Bing Wen

Low Bing Wen

The island of Komodo has a seasonally dry climate that supports large tracts of open palm savannah.

KEY FACTS

Nearest Major Town
Labuan Bajo (East Nusa Tenggara)

Habitats
Mixed deciduous forest, mangroves, open grassy savannah

Key Species
Orange-footed Scrubfowl, Green Junglefowl*, Great-billed Heron, Beach Stone-curlew, Barred Dove*, Flores Hawk-Eagle*, Yellow-crested Cockatoo*, Wallacean Cuckooshrike*, Lemon-bellied White-eye*, Flame-breasted Sunbird*
Austral winter Rainbow Bee-eater

Other Specialities
Javan Rusa, Komodo Dragon*,

Best Time to Visit
May–October; rainy season November–March

Located between Sumbawa and Flores in Indonesia's Nusa Tenggara or south-eastern islands, the Komodo National Park that spans Padar, Rinca and Komodo Islands is famous for its reptilian namesake, the Komodo Dragon, a giant monitor lizard that can grow to more than 3m in length. Geographically, Komodo and the other islands of the Lesser Sundas sit in a region in otherwise tropical Indonesia that experiences a seasonally dry climate with distinct wet and dry seasons. Consequently, the vegetation is characterized by open-palm savannah, thorn scrub and mixed deciduous forest. The island has the added attraction for birdwatchers of being one of the few remaining strongholds of the **Yellow-crested Cockatoo**, a Critically Endangered species that has been extirpated from most of its range and, worryingly, is also showing signs of decline here. Birds are noticeably tame on the island and other species readily observed include the **Orange-footed Scrubfowl**, **Green Junglefowl**, **Barred Dove** and **Lemon-bellied White-eye**, while scanning the seas during boat trips may be rewarded with seabirds such as **Bulwer's Petrel** and the **Red-necked Phalarope**.

Birdwatching Sites

Boat Journey to Pulau Komodo

The return journey to Komodo Island offers the chance for you to see several species of seabird, including **Black-naped Terns** and **Lesser Frigatebirds**. During the winter months the possibilities expand to include species such as **Bulwer's Petrel**, **Red-necked Phalarope** and **Aleutian Tern**. At low tide the **Great-billed Heron** and **Beach Stone-curlew** may be seen on exposed reefs and beaches around the island.

James Eaton

Komodo Island is one of the last strongholds of the Critically Endangered ***Yellow-crested*** ***Cockatoo****.*

Walking Trails

From the jetty at Loh Liang, a series of treks of varying durations lead northwards in a loop around the area. These are identified as 'Short', 'Medium' and 'Long' treks respectively, and it is compulsory to hire a local park guide to accompany visitors on these treks. **Orange-footed Scrubfowl** and several examples of their nesting mounds can be seen en route, and the scattered watering holes are good places to see **Green Junglefowl** and Komodo Dragons during the dry season. Sulphurea Hill, as its name suggests, is a good place to scan for the **Yellow-crested Cockatoo**, which is usually seen in noisy parties foraging in the treetops. While trekking through the scenic palm savannah, keep an eye out for foraging flocks, which can contain the **Wallacean Cuckooshrike**, **Lemon-bellied White-eye** and the striking **Flame-breasted Sunbird**.

Abdelhamid Bizid

*The **Orange-footed Scrubfowl** is relatively easy to see in Komodo's open habitats.*

Restaurant/Souvenir Shop Complex

Ironically, the restaurant complex, a short walk from the park entrance, is probably the best place on the island for close encounters with large Komodo Dragons, which rest during the heat of the day under or around the buildings here. Javan Rusa and **Barred Doves** forage unperturbed among the resting dragons, while the **Beach Stone-curlew** may be seen on the nearby beach.

Access & Accommodation

Komodo National Park is readily accessible from Labuan Bajo, a bustling coastal tourist hub on the nearby island of Flores. Labuan Bajo can be reached via multiple daily flights from Bali, and offers a wide range of accommodation to suit travellers on different budgets. From Labuan Bajo, a boat needs to be chartered to reach Komodo, with more expensive but faster speedboats covering the journey in about an hour each way, while conventional chartered boats take 4–5 hours each way. This can be organized with many tour companies to include organized trips to the park. Once on the island, you need to pay an entrance fee and for a compulsory guide; prices are clearly displayed at the entrance gate. Those opting to stay overnight can do so in basic chalets run by the national park, or on the boats they chartered to reach the island.

Low Bing Wen

*The **Komodo Dragon** is the island's main attraction and can be easily seen around the island's restaurant.*

Conservation

Komodo National Park was established in 1980 and subsequently declared a World Heritage Site by UNESCO in 1991. Illegal fishing and poaching are two of the park's major threats, while its uniquely dry climate makes its terrestrial habitats especially vulnerable to wildfires.

*The **Barred Dove** is endemic to the Lesser Sundas and south-east Maluku.*

Ruteng and Mbeliling Forests

Colin Trainor

The rugged Ruteng Nature Reserve encompasses some of the largest tracts of montane evergreen forest in the Lesser Sundas.

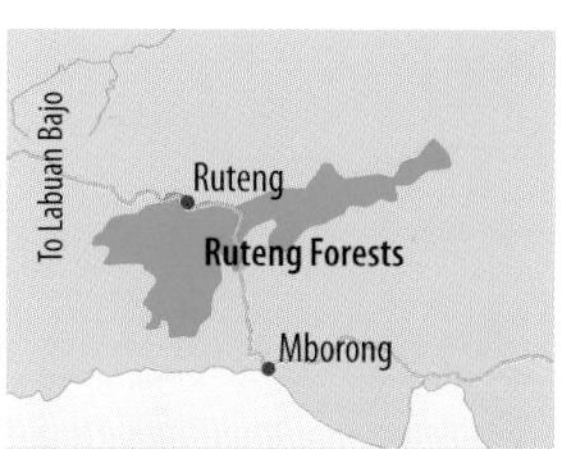

KEY FACTS

Nearest Major Towns
Ruteng, Labuan Bajo (East Nusa Tenggara)

Habitats
Submontane and montane evergreen forests, paddy fields, scrub

Key Species
Montane Flores Hawk-Eagle*, Dark-backed Imperial Pigeon*, Flores* and Wallace's* Scops Owls, Glittering Kingfisher*, Wallace's Hanging Parrot*, Leaf Lorikeet*, Little Minivet*, Bare-throated Whistler*, Russet-capped Tesia*, Timor (Flores) Leaf Warbler, Thick-billed Heleia*, Cream-browed* and Crested* White-eyes, Chestnut-backed Thrush*, Russet-backed Jungle Flycatcher*, White-browed (Flores) Shortwing, Tawny-breasted Parrotfinch
Submontane (Puar Lolo) Flores Hawk-Eagle*, Wallace's Scops Owl*, Glittering Kingfisher*, Wallace's Hanging Parrot*, Leaf Lorikeet*, Little Minivet*, Flores Monarch*, Thick-billed Heleia*, Crested White-eye*, Chestnut-capped Thrush
Passage and winter Crested Honey Buzzard, Chinese and Japanese Sparrowhawks, Swinhoe's and Pin-tailed Snipes, Kamchatka Leaf Warbler

Other Specialities
Long-tailed Macaque (introduced?)

Best Time to Visit
June–September

The Manggarai region of west Flores encompasses the largest mountainous area on the island and contains some of the most extensive tracts of montane forest in the Lesser Sundas. Much of the terrain is rugged, with steep slopes and 'old volcanic' geology. Elevations range from 900m to a high of 2,376m asl. All of Flores's hill and montane bird species occur here, and increasingly some typically lowland species as well, which are colonizing degraded forest edge habitat. However, the endemic **Flores Monarch** and **Flores Crow** are strangely absent. Ruteng's forests occupy an area of about 322km^2 (with a 50km^2 buffer), and include the Danau Rana Mese Tourist Park (5km^2). Access to the forest and forest edge here is easy, providing good opportunities for birdwatching. More than 30 Lesser Sundas endemics are present, including important populations of the **Flores Scops Owl** and **Bare-throated Whistler**. In addition, a distinctive subspecies of **Besra** was described in 1973 based on a specimen from Ruteng, and has been observed recently.

Birdwatching Sites

Danau Rana Mese

The montane evergreen forest fringing this 0.11km^2 crater lake (at 1,220m asl) is home to a variety of birds, while a few waterbirds like the **Pacific Black Duck** are present at the lake. Since the rediscovery of the **Flores Scops Owl** in the 1990s, and more recent documentation of its calls, this has become the prime site to see this little-known endemic. It has been recorded between the park headquarters and the lake, south of the highway and along the Trans-Flores highway for about 2km

Dubi Shapiro

*The **Russet-backed Jungle Flycatcher** is common in the forests around Rana Mese.*

*The **Flores Scops Owl** went missing for nearly 100 years before being rediscovered in the forests here.*

from the headquarters towards Ruteng, but is presumably more widespread. **Wallace's Scops Owl** is another night bird to search for. The **Dark-backed Imperial Pigeon**, **Pale-shouldered Cicadabird**, **Bare-throated Whistler**, **Cream-browed White-eye** and **Russet-backed Jungle Flycatcher** (ssp. *oscillans*) are some of the regional endemics that also occur here. A trail around the lake provides access to forest, as does the main Trans-Flores highway. Swampy habitat on the northern shore could be suitable for **Lewin's Rail**.

Poco Ranaka

The forests on the slopes of the 2,266m asl peak of Poco Ranaka are easily accessible via a poorly maintained but sealed road that almost reaches the summit. A good range of montane forest birds can be seen by walking along the road, including the **Bare-throated Whistler**, **Cream-browed White-eye**, **Russet-backed Jungle Flycatcher** and **Tawny-breasted Parrotfinch**, while **Green Junglefowl** is common by call. The Flores subspecies (ssp. *floris*) of the **White-browed Shortwing** is easy to find here, and may represent a distinct species. The **Flores Scops Owl** may also be present, but efforts to locate it have been limited. There is good potential for seeing migratory raptors.

Golo Lusang

Degraded roadside forests just 8km south of Ruteng allow relatively easy access to a wide range of forest and montane birds, including the **Flores Hawk-Eagle**, **Leaf Lorikeet**, raucous **Bare-throated Whistler** and **Russet-backed Jungle Flycatcher**. The first records of **Wallace's Hanging Parrot** in the Ruteng area were made here as recently as 2014. Recent nocturnal visits have successfully recorded the **Flores Scops Owl** (it was rediscovered on Poco Madasawa a few kilometres west of Golo Lusang) and **Wallace's Scops Owl**.

Pagal

Just 15km north of Ruteng, the roadside secondary forest at Pagal has been the most reliable site to see **Wallace's Hanging Parrot** since the late 1990s. The **Flores Green Pigeon**, **Leaf Lorikeet** and **Rufous-chested Flycatcher** may also be seen here. Pagal lies outside the boundaries of the Ruteng Nature Recreation Reserve.

Puar Lolo

Puar Lolo is a highly accessible tract of secondary hill forest at 900–950 m asl surrounding a Telkom tower along the Trans-Flores Highway about 70 km west of Ruteng. The forest is part of the extensive Mbeiling (or Tanjung Kerita Mese) Forest Reserve, with mountains rising above 1,200 m asl cloaked in primary forest just 6 km to the west. Phillippe Verbelen and Frank Lambert were the first to confirm the presence of the **Flores Monarch** around Puar Lolo in the early 1990s, and it has since become the main site to see this bird.

Flores Crow and **Wallace's Hanging Parrot** have also occasionally been recorded here. The site possibly hosts **Flores Scops Owl**, although few nocturnal surveys have been done here. The key to finding the monarch is knowledge of its vocalizations, although a sighting is not guaranteed. The fig trees along the highway and in the forest attract pigeons, **Wallace's Hanging Parrot, Thick-billed Heleia** and **Crested White-eye** when in fruit.

Access & Accommodation

Ruteng is accessible by daily flights from Kupang (West Timor), although most visitors travel along the Trans-Flores highway by bus or hired car from Labuan Bajo. There are many bungalow and basic hotel options in Ruteng town, and each of the main birdwatching locations is readily accessible by sealed road, although the Poco Ranaka road is in poor condition and some sections may not be driveable. Local motorbike or car hire with a driver can be easily arranged in town.

Dubi Shapiro

One of the dominant members of the dawn chorus at this site is the endemic ***Bare-throated Whistler****, sometimes known as the 'Flores Nightingale' due to its impressive vocalizations.*

The globally endangered ***Flores Monarch*** *was rediscovered at this site in the 1990s, and today it is the most accessible site to see this sought-after species.*

Dubi Shapiro

The handsome ***Chestnut-capped Thrush*** *is perhaps most easily seen on Flores, and is regularly observed in the forest around Puar Lolo.*

Conservation

The Ruteng forests were proposed as a Nature Reserve in 1982; they were identified as a 'terrestrial conservation area of top priority' in the 1993 Indonesian Biodiversity Action Plan, and designated as a Nature Recreation Reserve the same year. This included a ban on timber harvesting in the reserve, but it has not been successful. In the late 1990s a $US25 million Asian Development Bank and Government of Indonesia Integrated Conservation and Development project tackled management of the reserve with some biodiversity surveys and interesting economic and livelihood studies.

The main threat to the Ruteng forests is the ongoing and unregulated collection of timber.

However, the forests at Puar Lolo around the Telkom tower appear well protected, although plots of shifting cultivation are occasionally observed. Mbeliling Forest Reserve has been highlighted as a high conservation priority for more than 30 years. Burung Indonesia (BirdLife International's partner in Indonesia) has worked with communities on sustainable forest management, bird surveys and ecotourism projects since 2002.

Poco Ndeki (Kisol)

Colin Trainor

Low Bing Wen

The remnant lowland forest at Kisol supports the majority of the lowland endemics found on Flores.

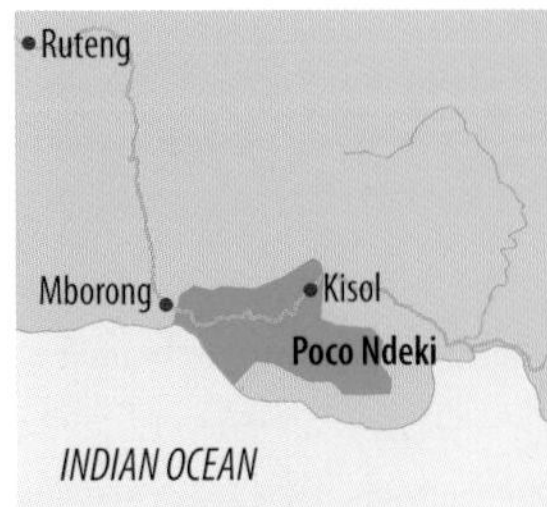

KEY FACTS

Nearest Major Towns
Ruteng, Mborong (East Nusa Tenggara)

Habitats
Coastal and hill mixed evergreen forests, cultivation, sandy beaches

Key Species
Flores Hawk-Eagle*, Flores Green Pigeon*, Wallace's Scops Owl*, Mees's Nightjar*, Glittering Kingfisher*, Leaf Lorikeet*, Great-billed Parrot, Elegant Pitta*, Flores Crow*, Russet-capped Tesia*, Thick-billed Heleia*, Yellow-ringed* and Crested* White-eyes, Chestnut-capped Thrush
Passage and winter Crested Honey Buzzard, Chinese and Japanese Sparrowhawks, Kamchatka Leaf Warbler

Best Time to Visit
June–October; rainy season December–March

Poco Ndeki is a small (*c.* 10km^2) volcanic peak rising to about 950m asl on the south-central coast of Flores. Although the lowlands around it have been cleared for agriculture, especially cocoa, its foothills and slopes remains largely covered in mixed evergreen forests. Located just 3–4km south of the village of Kisol and Mborong on the Trans-Flores highway, the accessibility of this site draws many birdwatchers keen to see some of Flores' rarer endemics, particularly the **Flores Hawk-Eagle** and **Flores Green Pigeon**. Although limited in extent, the forests on Poco Ndeki provide an excellent introduction to Flores' lowland forest bird fauna, and they are usually part of birdwatching itineraries after a visit to the highlands of Ruteng. Many birdwatchers visiting Poco Ndeki also seek out the elusive **Chestnut-capped Thrush** and **Elegant Pitta**, both of which are increasingly hard to see elsewhere on Flores.

Birdwatching Sites

Poco Ndeki

This is the key site for seeing most of Flores' lowland bird species, especially the **Flores Hawk-Eagle**, **Wallace's Scops Owl**, **Mees's Nightjar**, **Glittering Kingfisher**, **Elegant Pitta**, **Flores Crow** and **Chestnut-capped Thrush**. A few **Yellow-crested Cockatoos** were present here during surveys in the 1990s, but these have not been seen for many years and may be extinct.

After a couple of kilometres of cultivation and villages, the rough road heading south from Kisol eventually enters good-quality forest. The shy **Flores Crow** is locally common and easily recognized by its call, but seeing it well can be a tall order. A narrower

*The **Elegant Pitta** (ssp. concinna) occurs in the forests in Kisol and can be particularly vocal between April–July.*

Noisy parties of ***Yellow-ringed White-eyes*** *regularly participate in the mixed flocks at Kisol.*

trail eventually branches off from the road and creeps up along the forested slopes of Poco Ndeki. This is where you should scan the trail sides and forest understorey for the shy **Elegant Pitta** and **Chestnut-capped Thrush**; the latter is becoming increasingly rare across Southeast Asia due to rampant poaching. Mixed species flocks should be checked for **Yellow-ringed** and **Crested White-eyes**, although the **Thick-billed Heleia**, whilst rarer, may also be encountered. A nocturnal ramble should turn up both **Wallace's Scops Owl** and **Mee's Nightjar**, in addition to the more common **Moluccan Scops Owl**.

Nangarawa

Some visitors have explored the coastal forest and beach at Nangarawa, about 12km to the south-east of Kisol. The **Malaysian Plover**, **Red-necked Phalarope**, **Beach Stone-curlew**, **Flores Green Pigeon** and **Flores Crow** have all been recorded here. Another way to see waterbirds and shorebirds is via an excursion to the estuary at Mborong about 4km to the south-west of Kisol.

Access & Accommodation

From Kisol the forest starts about 3–4km to the south and can be reached after a one-hour hike. For those planning to explore the coastal areas, Nangarawa is about 10km south-east of Kisol. There are few accommodation options, but many visitors stay in the Catholic seminary in Kisol, called Wisma Anjas.

Conservation

The status of the forest on Poco Ndeki includes 'Protection forest' and 'Production forest', but there is limited management of either forest type for conservation. Some bird trapping for the pet-bird trade is known to occur here.

One of the main targets for birdwatchers here is the charismatic ***Flores Crow****, a species usually first detected by its baby-like cries.*

Dubi Shapiro

Glittering Kingfishers *call from the canopy and can be easily overlooked.*

Alor & Pantar Islands

Philippe Verbelen & Colin R. Trainor

Philippe Verbelen

Moist forest on Alor, seen here from Kungwera.

KEY FACTS

Nearest Major Towns
Kalabahi (Alor), Baranusa (Pantar)

Habitats
Moist evergreen forest, Eucalyptus woodland, mangroves

Key Species
Flores Hawk-Eagle*, Timor* and Little ('Eucalypt') Cuckoo-Doves, Flores Green Pigeon*, Southern (Alor) Boobook, Mees's Nightjar*, Yellow-crested Cockatoo*, Crimson-hooded (Alor) Myzomela*, Wallacean (Alor) Cuckooshrike*, Javan Bush Warbler*, Common Hill Myna, Chestnut-backed Thrush*
Passage and winter Crested Honey Buzzard, Australian Pratincole, Pallid Cuckoo, Kamchatka Leaf Warbler, Grey Wagtail

Other Specialities
Sperm and Blue Whales (seasonal)

Best Time to Visit
July–October (dry season)

Alor and Pantar lie at the eastern end of the Flores-chain of islands. Both are volcanic islands dominated by rugged terrain covered in *Eucalyptus* woodland on slopes and ridges, while patches of moist evergreen forest occur in the gullies and on mountaintops. Historically, neither island has been on the radar of birdwatchers. This all changed with the discovery of the undescribed **'Alor' Myzomela**, a new population of **Javan Bush Warbler** on Alor in 2009, and the potential recognition of the local subspecies of the **Southern Boobook** (ssp. *plesseni*) as a distinct species endemic to Alor and Pantar. In addition, the local subspecies of the **Wallacean (Alor) Cuckooshrike** may in fact be an Alor-Pantar endemic at the species level, while recent vocalization analysis of the **Little (Eucalypt) Cuckoo-Dove** suggests that it is actually an as yet undescribed pigeon endemic to Alor, Pantar, Sumba and Timor. Among these exciting discoveries, both islands are proving to be interesting biogeographically as the avifauna shows a stronger affinity with that of the Timor region, having many shared species such as the aforementioned bush warbler and **Timor Stubtail**.

The Alor subspecies of the ***Southern Boobook*** *is common and widespread across both islands.*

Birdwatching Sites

Apui-Subo-Manmas

Most, if not all the specialities on Alor island can be seen during a 3–4 day visit focusing on the forests above the village of Apui. From Apui, a road climbs from an elevation of about 800m asl towards a telecom tower at around 1,150m asl on a ridge surrounded by eucalypt woodland. The undescribed **'Alor' Myzomela** was discovered along the track within 500m on either side of the Telekom tower. The myzomela appears closely related to the Crimson-hooded Myzomela, which is otherwise endemic to Wetar. It differs, however, in its ecological preferences, a different song and a red hood that is much less extensive than on the Crimson-hooded Myzomela. The whole area should also be checked for the threatened Flores Hawk-Eagle that can be frequently seen soaring over the forests, but note that the Short-toed Snake Eagle and Bonelli's Eagle also occur here. Areas with high grass and dense bushes below the telecom tower are the habitat for an endemic subspecies of Sunda Bush Warbler (ssp. *kolichisi*) and the Javan Bush Warbler (ssp. *timorensis*). The latter's presence its best detected by its rasping *churr-eep churr-eep song*. Other key birds such as the Little (Eucalypt) Cuckoo-Dove, Timor Stubtail, Lesser Shortwing, Thick-billed Flowerpecker (ssp. *obsoletum*) and Blood-breasted Flowerpecker also occur. Night birding in the woodland on the valley around Apui is good for the Southern (Alor) Boobook, Eastern Barn Owl and Mees's Nightjar. The Yellow-crested Cockatoo is difficult to see on Alor, but there are recent records of birds seen from the main road overlooking deep valleys in the Apui area.

The Critically Endangered Flores Hawk-Eagle *was only recently discovered to persist on both islands.*

Mainang Area

This is a highly accessible area of secondary moist evergreen forest and eucalypt woodland only about 45 minutes' drive from Kalabahi. The Mainang area is a good site for the Flores Hawk-Eagle, Bonelli's Eagle, Javan Bush Warbler and most of the other key species. The 'Alor' Myzomela is present on higher ridges above 1,000m asl, but may be harder to find here than above Apui. The Little (Eucalypt) Cuckoo-Dove, Flores Green Pigeon and Wallacean (Alor) Cuckooshrike (ssp. *alfrediana*) have been recorded but are uncommon. No nocturnal surveys have been done at this site, but both the Southern (Alor) Boobook and Mees's Nightjar can be expected to occur here.

Kungwera, Tanglapui Timur

Birdwatchers with more time and a sense for adventure and exploration are encouraged to visit the forested mountains at the eastern end of the island. The traditional village of Kungwera at Tanglapui Timur is a good base from which to explore the surrounding forests and Mount Koya Koya area, although basic knowledge of Bahasa Indonesia will be necessary. The Southern (Alor) Boobook can be easily located in the forests around Kungwera and Tanglapui Timor; the restricted range Mees's Nightjar occurs in the more open woodland with a lot of grass and stony outcrops. The forests around Kungwera hold the Flores Hawk-Eagle, while the Flores Green Pigeon is fairly common here. Other interesting birds in the area are the Elegant Pitta, Chestnut-backed Thrush, Thick-billed Flowerpecker and scarce Common Hill Myna (ssp. *venerata*).

Mount Wasbila – Baranusa Mangrove (Pantar Island)

The woodland surrounding the bay at Baranusa is a good area to look out for the Eastern Barn Owl, Southern (Alor) Boobook and Mees's Nightjar. There are also fairly extensive mangrove forests in this area with the Oriental Dwarf Kingfisher and Black Bittern present; the bitterns can be heard calling around dusk and throughout the night – this is a species that is little known in the Lesser Sundas. The tropical forest and eucalypt woodland on Mount Wasbila is another good site for the Southern (Alor) Boobook and holds other interesting species such as the Little (Eucalypt) Cuckoo-Dove, Flores Green Pigeon, Bonelli's Eagle and Common Hill Myna. The woodland around the peaks of Wasbila and Sirung should be checked for the Flores Hawk-Eagle and perhaps 'Alor' Myzomela.

Access & Accommodation

Alor is easily accessed by daily flights from Kupang to Mali airport (check for the **Australian Pratincole** and other migrants on the short grass at the airfield), 14km west of Kalabahi, the main town on Alor Island. It is easy to grab a taxi or ojek (motocycle) driver to reach Kalabahi, where there are reasonable hotel options such as Pulo Alor and more basic hotels such as Hotel Pelangi Indah and Hotel Melati. Pantar Island can be easily reached by public ferries that leave almost daily from Kalabahi. These can take 4–6 hours and also provide the opportunity to see seabirds. Accommodation options are limited in Baranusa, a major port town on Pantar, but there is at least one guesthouse near the harbour.

Conservation

Slash-and-burn clearance of forests and scrub across the island remains a perennial threat to remaining areas of moist evergreen forests. Recurrent fires destroy much habitat, leaving poor-quality eucalypt scrubland with a dense grassy understorey favoured by few species. There is very little attempt to manage clearance of vegetation, and the local forestry department is under-resourced. Illegal trapping for the pet trade still occurs, and threatens both the cockatoo and hill myna.

*The unobtrusive **Timor Stubtail** skulks in the dense forest understorey.*

James Eaton

*Flocks of noisy **Olive-headed Lorikeets** may be seen feeding on flowering eucalypts.*

Lewa Forests

Low Bing Wen

Yong Ding Li

The lush monsoon evergreen forests of Sumba are home to important populations of Yellow-crested Cockatoo.

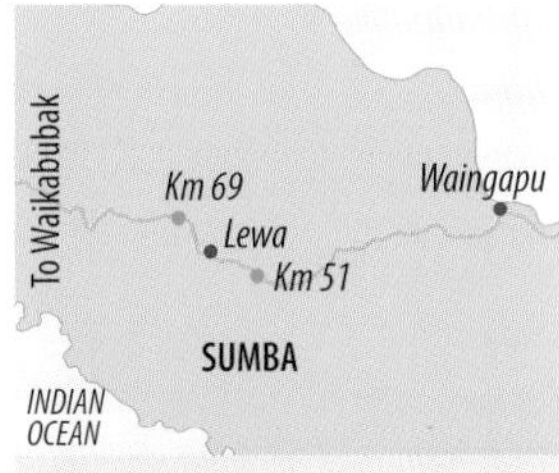

KEY FACTS

Nearest Major Town
Waingapu (East Nusa Tenggara)

Habitats
Hill mixed evergreen forest, scrub

Key Species
Sumba Green Pigeon*, Red-naped Fruit Dove*, Mees's Nightjar*, Sumba Boobook*, Little Sumba Hawk-Owl*, Cinnamon-banded Kingfisher*, Sumba Hornbill*, Yellow-crested Cockatoo*, Marigold Lorikeet*, Elegant Pitta*, Sumba Myzomela*, Chestnut-backed Thrush*, Sumba Brown Flycatcher*, Russet-backed Jungle Flycatcher*, Sumba Flycatcher*, Apricot-breasted Sunbird*

Other Specialities
Black Flying Fox

Best Time to Visit
May–November

Best known for its megalithic traditions and the spear-throwing game of *pasola*, the fascinating island of Sumba features savannah and monsoon forests packed with endemic birds. The town of Lewa, located along the Trans-Sumba Highway, is a small settlement in the island's hilly centre that offers easy access to some of the largest remaining fragments of mixed evergreen forest left in Sumba's interior. All but one of the island's avian endemics can be seen around Lewa, and birdwatching along the relatively quiet highway that winds its way through several tracts of forest can be rewarded with some of the island's most sought-after endemics, including the **Red-naped Fruit Dove**, **Little Sumba Hawk-Owl** and **Sumba Hornbill**. For a change of scenery, there are several side trails east of Lewa that lead to other forest fragments where conditions may be more conducive for viewing secretive understorey and ground-dwelling species such as the **Chestnut-backed Thrush**, **Sumba Flycatcher** and **Elegant Pitta**.

Birdwatching Sites

Km 51 Trail

This trail is located 10km west of Lewa, close to a highway marker bearing the same figure. The trail passes through a section of farmland before entering remnant forest after a 10–15-minute walk. Spending some time along the forest edge, particularly at dawn and dusk, can provide good views of a range of parrots and pigeons, including the **Red-naped Fruit Dove**, **Sumba Green Pigeon** and **Marigold Lorikeet**. The forest also supports specialities such as the **Sumba Brown Flycatcher**, **Chestnut-backed Thrush** and **Cinnamon-banded Kingfisher**. Spotlighting can be productive, with both **Mees's Nightjar** and **Little Sumba Hawk-Owl** possible. The site lies outside the boundaries of any national park, so entry formalities are unnecessary.

James Eaton

*With two endemic owls, including the **Sumba Boobook**, night birdwatching is a necessary component of any birdwatching trip to the island.*

Km 69 Roadside

The nearest stretch of good roadside forest is west of Lewa at Km 68–72, with Km 68 being about 20-minute drive from Lewa. The area lies in the Manupeu-Tanah Daru National Park, the largest of Sumba's national parks, and is best visited in the early morning when traffic is minimal and the forest resonates with the dawn chorus of **Elegant Pittas**. The **Sumba Hornbill**, **Marigold Lorikeet**, **Eclectus Parrot**, **Red-cheeked Parrot**, **Sumba Myzomela**, **Sumba Brown Flycatcher**, **Russet-backed Jungle Flycatcher** and **Apricot-breasted Sunbird** may be seen here, and short forays into the forest via side trails can yield the **Elegant Pitta**, **Chestnut-backed Thrush** and **Sumba Flycatcher**. After dark the **Sumba Boobook** is regularly seen here.

Manurara

Located on the western boundary of the Manupeu-Tanah Daru National Park near the village of Waikabubak, Manurara is reached after a two-hour drive from Lewa. The site is one of the best places on the island to see the distinct Sumba subspecies (ssp. *citrinocristata*) of the Critically Endangered **Yellow-crested Cockatoo**. A short trek up one of the hills in the area offers panoramic views of the forested valley below. Parrots are particularly conspicuous, and an early morning vigil should provide views of multiple species, including **Eclectus**, **Great-billed** and **Red-cheeked Parrots**, as well as **Marigold Lorikeets**, in addition to the cockatoo. This site also offers a good back-up for the **Sumba Hornbill** and **Sumba Green Pigeon**. The **Broad-billed Flycatcher** may be spotted on the scrubby hillsides.

Access & Accommodation

The island of Sumba is served by regular flights from Bali. These arrive in Waingapu, the island's capital city, from where it is a drive of an hour and a half along paved roads to Lewa town in the island's interior. A good variety of hotels can be found in Waingapu. There are a couple of basic homestays in Lewa (like Mama Riwu Homestay) where you can spend the night. A permit and local ranger are required to birdwatch within the boundaries of the national park.

Conservation

Forest clearance for cultivation and grazing, and bush fires are a major threat to the long-term existence of the forests across Sumba. Much of the original forest has been lost over the years, and what is left is now fragmented.

James Eaton

*The globally threatened **Sumba Hornbill** is one of the most sought-after and elusive endemics on the island.*

James Eaton

*Given its unobtrusive habits, the endemic **Red-naped Fruit Dove** is easy to miss.*

James Eaton

*The hill evergreen forest around Lewa is home to several regional endemics, such as this **Pale-shouldered Cicadabird**.*

Rote Island

Philippe Verbelen & Colin Trainor

Yann Muzika

The dry forests of Rote extend to the island's coastline.

KEY FACTS

Nearest Major Towns
Ba'a, Kupang (Timor)

Habitats
Dry forest, freshwater and saline wetlands, mangroves, savannah woodland

Key Species
Black Cuckoo-Dove*, Timor Green Pigeon*, Southern (Rote) Boobook, Jonquil Parrot*, Rote Myzomela*, Olive-brown Oriole*, Northern (Rote) Fantail, 'Rote' Leaf Warbler (undescribed)*, Orange-sided Thrush*, Timor Blue Flycatcher*, White-bellied Bush Chat*
Wet season Elegant Pitta*, Island Monarch

Best Time to Visit
June–October (dry deason); the relatively short wet season occurs in November–May.

Yann Muzika

Formerly lumped with Sumba Myzomela, the ***Rote Myzomela*** *was finally recognized as a distinct species in 2017.*

Lying about 12km south of Timor, Rote is the southernmost island in Asia and is only 145km north of Ashmore Reef at the edge of the Australian continental shelf. The Rote landscape is flat and dominated by coralline limestone and karst. The substrate, low rainfall and mostly sparse savannah vegetation gives the impression of a harsh landscape, particularly during the long dry season. Rote's avifauna is essentially a subset of nearby Timor, but some Timor endemics such as the **Timor Green Pigeon** and **Jonquil Parrot** are more readily observed on Rote than in West Timor. Presently, only one endemic is recognized for Rote, the recently described **Rote Myzomela**. However, recent field observations indicate that there are probably at least three additional endemic species, including the undescribed **'Rote' Leaf Warbler**, as well as the distinctive and endemic subspecies of the **Southern (Rote) Boobook** and **Northern (Rote) Fantail**. For visiting birdwatchers, the most interesting areas are the small patches of dry tropical forest, particularly on the northern Pakuafu peninsula, and the relatively intact patches of evergreen forest on Mount Musaklain, near Seda, the highest point on the island. Other interesting habitats include mangroves, saline and freshwater wetlands, particularly on the northern peninsula and the satellite islands off south Rote.

Birdwatching Sites

Sotimori-Daurendale-Dead Sea

The endemic taxa, including the **Rote Myzomela**, the **Timor (Rote) Leaf Warbler** and **Northern (Rote) Fantail**, are all common in woodland and tropical dry forest throughout the Tapuafu peninsula in northern Rote. The **Black Cuckoo-Dove** and **Jonquil Parrot** are more of a challenge to locate, but are likely to be seen

Philippe Verbelen

The Rote subspecies of the ***Northern Fantail*** *differs substantially from the other subspecies in the Moluccas.*

while quietly walking through the evergreen forest or woodland. There are very few Rote records of the **Timor Green Pigeon** to date, but it is certainly more likely to be seen here than in Timor. Knowledge of their calls, and spending time in areas at numerous fruiting fig trees would certainly help in locating these rare pigeons. Other Timor specialities such as the **Olive-brown Oriole**, **Timor Stubtail**, **Timor Blue Flycatcher** and **White-bellied Bush Chat** are likely to be encountered quite regularly. Lake Oendui can be an easy spot to see a variety of Australian waterbirds, such as the **Green Pygmy Goose**, **Australasian Grebe**, **Royal Spoonbill**, **White-faced Heron**, **Australian Pelican**, **Australasian Darter**, and possibly even **Masked Lapwing** and **Black-fronted Dotterel**. Saline lakes on Rote, including the 'Dead Sea', can host large numbers of resident and migratory shorebirds, including the **Red-capped Plover** and **Red-necked Stint**.

Bolatena

This area of tropical dry forest is probably the most convenient place to look for the **Southern (Rote) Boobook**. The local people are now very used to visiting birdwatchers wanting to see the owl and thus are keeping track of its whereabouts. It is only a one hour (5km) walk between Bolatena and Sotimori villages, and there is contiguous woodland between the two areas. The local teacher (Pak Rens Maku) usually accommodates visiting birders at his house and guides them to the boobook

Philippe Verbelen

The Rote subspecies of the ***Southern Boobook*** *has distinctive vocalizations*

Currently lumped with the Timor Leaf Warbler, the ***'Rote' Leaf Warbler*** *has a distinctively longer bill, among other traits.*

territories found around the village. Like other Ninox owls, the boobooks are most vocal around dusk and just before dawn, and they are very responsive to play-back (this should, however, be used sparsely considering the potential rarity of the species).

Satellite Islands around Rote

These islands host unusual bird assemblages, including 'small-island' and 'supertramp' species such as the **Beach Stone-curlew** (Ndana, also Nuse Island), **Rose-crowned Fruit Dove** and **Island Monarch** (Ndana Island). There is a good chance to see seabirds while travelling to and from the islands, including migratory **Red-necked Phalaropes**.

Access & Accomodation

A visit to Rote can easily be combined with a birdwatching trip to Timor. From Kupang, there are now almost daily boats to Rote. Pitoby Tour and Travel in Kupang are a good source of information on flights and ferry services. The most convenient option is a fast (about two hour) catamaran that runs from Kupang's Tenau Harbour to Ba'a, Rote's main town. Alternatively, there is a slow ferry. Ferry crossings between Rote and Timor can be productive for a variety of seabirds, including frigatebirds, boobies, Bulwer's Petrel and the Sooty Tern. Ferries can be erratic due to heavy seas in the wet season. If they are cancelled, Wings Air has daily flights from Kupang to Ba'a's airport. Minibuses, bemos or ojek can be arranged from the port to Sotimori for the 1.5–2 hours' drive to the main birding areas. There is little tourist infrastructure but visitors can stay at family houses in villages like Sotimori and Bolatena. It is customary to first introduce yourself to the village chief and/or local authorities to explain the purpose of your visit. Basic food supplies are available locally but it is best to buy supplies before heading to the villages. Rote is nowadays on the international tourism map mainly because of surfing opportunities in the south of the island, with beaches in the Nemberala area being popular with surfers.

Conservation

The Rote government authorities and Rotenese people have been mostly unaware that their island harbours unique wildlife species and consequently conservation efforts have been limited. New information on the taxonomic status of Rote's birds highlights the global significance of the island. Tropical forest loss or conversion and tree cutting may be a significant threat to some of the island endemics, particularly the loss of mature, hollow-bearing trees used by owls and parrots. Hunting is clearly the major threat to populations of forest pigeons, Yellow-crested Cockatoo and other parrots. Visiting birdwatchers could help raise awareness about the importance to protect wildlife on Rote by telling their hosts, guides and any government officials they meet that the main reason they are visiting Rote is to see wild birds found only on Rote or the region.

Bipolo Forest

Colin Trainor

Colin R. Trainor

A small patch of evergreen and semi-evergreen tropical forest remains at Bipolo.

The largest island in the Lesser Sundas, Timor is also the driest and its remaining natural vegetation is dominated by palm savannah and mixed evegreen forest. Much of West Timor's original forests have been cleared, leaving small, scattered patches. One of these, Bipolo, is a small patch (1.29km² in 2014) of lowland alluvial semi-evergreen forest near the mouth of Kupang Bay. Although the forest is small and increasingly degraded, the majority of Timor's lowland forest birds can still be seen here. A key feature of Bipolo is the many large fruiting fig trees, which attract a lot of pigeons. The selective loss of large forest trees has probably increased the ease of viewing birds at this site. It is possible to see as many as 35 species in one morning's walk in and around Bipolo forest, including a fine selection of endemics such as the **Timor Stubtail** and **White-bellied Bush Chat**.

Birdwatching Sites

Bipolo Forest

A walk along the main road or numerous side tracks that wind through the forest can yield most of the lowland frugivores and nectarivores on Timor, as well as the attractive **Orange-sided Thrush**. The **Timor Green Pigeon** is possible but there have been very few records since 1999, and only 1–2 individual **Yellow-crested Cockatoos** have been present since around 2000. The **Great-billed Parrot** was recorded in the early 1990s but has also become irregular. Pigeons such as **Rose-crowned** and **Banded Fruit Doves** and **Pink-headed Imperial Pigeon** should be reasonably easy to see, but the **Black Cuckoo-Dove** and **Timor Cuckoo-Dove** are less common. Understorey birds like the skulking **Timor Stubtail**, **Sunda Bush Warbler** and **Buff-banded Thicketbird** are present but difficult to see. The **White-bellied Bush Chat** and **Black-banded Flycatcher** are present but locally uncommon.

KEY FACTS

Nearest Major Town
Kupang (West Timor)

Habitats
Lowland evergreen forest, ricefields, fishponds

Key Species
Timor Green Pigeon*, Black Cuckoo-Dove*, Timor Cuckoo-Dove*, Cinnamon-banded Kingfisher*, Yellow-crested Cockatoo* (rare), Jonquil Parrot*, Timor Friarbird*, Streak-breasted* and Flame-eared* Honeyeaters, Plain Gerygone*, Green Figbird*, Timor Stubtail*, Buff-banded Thicketbird*, Timor Leaf Warbler*, Spot-breasted Heleia*, Orange-sided Thrush*, Timor Blue Flycatcher*, Timor Sparrow*, Tricolored Parrotfinch*
Passage Crested Honey Buzzard
Regional migrant Elegant Pitta* (present in wet season)

Other Specialities
Northern Common Cuscus

Best Time to Visit
June–October; rainy season November–May

*The **Black-banded Flycatcher** is restricted to clumps of bamboo in the forest understorey.*

*The skulking **Buff-banded Thicketbird** is one of the harder Timor endemics to get good views of.*

*The increasingly rare **Jonquil Parrot** is threatened by the pet-bird trade but can still be seen at Bipolo.*

Ricefields & Fishponds

These are located 1–2 km south of the forest. An unusual mix of Asian and Australian waterbirds frequents the fishponds and intertidal mudflats, including the Royal Spoonbill, White-faced Heron, Australian Pelican, resident Red-capped Plover, Black-tailed Godwit, Little Curlew, Australian Pratincole and Gull-billed Tern. This area is also a good spot for Bonelli's Eagle and other raptors, as well as seed eaters like the Timor Sparrow and Five-colored Munia.

Pariti-Kuka

Located 5km south-west of the forest, this is a key site for shorebirds including the Beach Stone-curlew, Broad-billed Sandpiper and Far Eastern Curlew.

Access & Accommodation

From Kupang it is a 50-km (50-minute) drive to the Bipolo forest, a further five-minute drive or 30-minute walk to the ricefields and fishponds, and a 20–30-minute drive to Pariti-Kuka. There are numerous accommodation options in Kupang, and if you bring provisions it is easy to arrange accommodation in villages near the forest (check with the *kepala desa*, or village chief).

Conservation

The Bipolo area is classified as a Recreation Reserve (Taman Wisata Alam) but received little on-the-ground management. Following the 'Bipolo war' in 1912, the forest was divided between two local groups – the Rotinese and Timorese. The forest had been considered sacred and was managed by local communities until 1960. Since its management by government, many areas have been legally logged (when classed as production forest) and converted to rice fields. Consequently, the area of forest has declined rapidly from 716ha in 1972 to 176ha in 2006 and 129ha in 2014. It was designated as protection forest in 1999, which has slowed logging and forest loss.

*The attractive **Orange-sided Thrush** is a highlight of any birdwatching trip to Bipolo.*

James Eaton

*The striking **White-bellied Bush Chat** is endemic to Timor and nearby Rote.*

Gunung Mutis & Gunung Timau

Colin Trainor

Yong Ding Li

Uniform stems of Eucalyptus forest at the summit of Gunung Mutis.

KEY FACTS

Nearest Major Towns
Soe, Lelogama (East Nusa Tenggara)

Habitats
Montane evergreen forest, eucalyptus (*Eucalyptus urophylla*) forest, open grasslands

Key Species
Metallic Pigeon, Timor Green Pigeon*, Timor Imperial Pigeon*, Black Cuckoo- Dove*, Southern (Timor) Boobook, Cinnamon-banded Kingfisher*, Olive-headed* & Iris* Lorikeets, Flame-eared Honeyeater*, Pygmy Wren-babbler (ssp. *timorensis*), Timor Leaf Warbler*, Javan Bush Warbler*, Buff-banded Thicketbird*, Spot-breasted Heleia*, Chestnut-backed*, Sunda* and Island (ssp. *schlegelii*) Thrushes, Timor Blue Flycatcher*, Blood-breasted Flowerpecker*, Tricolored Parrotfinch* and undescribed 'Timor Parrotfinch' *Erythrura* sp.*
Passage Crested Honey Buzzard

Other Specialities
Timor Forest Rat

Best Time to Visit
August–October; rainy season November–June

Gunung Mutis (2,400m asl) is the highest mountain in West Timor, and together with Gunung Timau (1,774m asl), hosts the most extensive areas of mixed and evergreen montane forests left in West Timor. Gunung Mutis contains more than 12km² of mixed montane forest, tall *Eucalyptus urophylla* forest, and heavily grazed grasslands. Much of the forest lies at *c.* 1,700–2,400m asl. Gunung Timau is rarely visited by birdwatchers, but supports both montane evergreen forest above *c.* 1,200m asl and extensive semi-evergreen forest down to 500m asl. The **Javan** (formerly Timor) **Bush Warbler** was first collected on Gunung Mutis at 1,900m asl by Georg Stein in 1931, but has not been seen since. The rediscovery of this bird on Mutis would make an interesting focus for the intrepid birdwatcher.

The montane birds on Gunung Mutis and Timau are the main focus for visiting birdwatchers, but a good variety of Timor endemics is also present, including birds more typical of lowlands. All of Timor's montane species are present except the **Pygmy Flycatcher** (only in Timor-Leste). Sightings of the undescribed **'Timor Parrotfinch'** on Gunung Mutis should also be a major target – it may also occur on Timau, but its presence there has yet to be confirmed. The very rare **Wetar Ground Dove** and **Timor Green Pigeon** have been reported from lowland sections of Gunung Timau, but only the most intrepid visitor is likely to see these extreme Timor rarities.

*The **Timor Imperial Pigeon** is regularly seen in the forests of Gunung Mutis.*

Birdwatching Sites

Fatumnasi

The extensive *Eucalyptus* (or *ampupu*, as it is known locally) forest at the edge of the village along the road to Nenas gives access to birds such as the **Timor Imperial Pigeon**, **Banded Fruit Dove**, and **Chestnut-backed** and **Island Thrushes**. To see skulking understorey birds such as the **Pygmy Wren-babbler**, **Sunda Thrush** and **Lesser Shortwing**, you need to find montane evergreen forest, which often occurs on or around large limestone outcrops. Other specialities like the **Timor Leaf Warbler** (ssp. *presbytes*) and **Timor Blue Flycatcher** can occur in evergreen forest as well as *Eucalyptus* forest, where they may be easier to observe.

Nenas to Gunung Mutis Summit

Nenas village allows easy to access to the *Podocarpus*-dominated montane evergreen forests on the upper slopes of Gunung Mutis, as well as tracts of montane woodland and grassland. Agricultural fields near the village can be good for munias. All montane species can be seen in forest above c. 1,700m asl, including the undescribed **'Timor Parrotfinch'**.

Gunung Timau

A vehicle track gives easy access to evergreen forest and *Eucalyptus* forest to within 1.2km of the summit. Most of Timor's endemic and montane birds are readily seen in the tropical forest. Be on the lookout for easy views of specialities in *Eucalyptus* forest, including the **Timor Imperial Pigeon**, **Black Cuckoo-Dove**, **Island Thrush** and **Snowy-browed Flycatcher**.

Access & Accommodation

Gunung Mutis is a 4–5 hour drive east of Kupang via the town of Soe. Access to the most extensive tropical forest on the Mutis massif is easiest and closest from Nenas, but also possible from Fatumnasi. A vehicle track between Fatumnasi and Nenas provides easy access to *Eucalyptus* forest, although it can be very rough and requires a four-wheel drive vehicle. There are no clearly marked tracks so it is advisable to walk with a local guide. The terrain is not overly steep except on the upper slopes of the Mutis massif. A walk to the Gunung Mutis summit from Nenas is c. 5km and may take around three hours. Access to Gunung Timau is more difficult: from Kupang turn north at Takari after c. 90 minutes, then travel to Lelogama, then a further c. 18km on tracks to the base of Gunung Timau. There is no formal accommodation at Gunung Timau, but it may be possible to camp or stay with local villagers (or government officials in Lelogama). This is most easily arranged by local driver-guides. Cattle tracks allow easy access to forest and the rocky summit.

Conservation

There were plans in the early 2000s for a national park covering both Gunung Mutis and Timau, but they were rejected by local communities. The main part of Gunung Mutis is a Strict Nature Reserve of 120km^2 with an additional 200km^2 managed as Protection Forest and Limited Production Forest. Both Mutis and Timau were identified as Important Bird Areas. Montane evergreen forest and *Eucalyptus* forest boundaries appear steady, but annual fires and intensive livestock grazing limit the distribution of tropical forest largely to the rockiest locations. Fuel-wood collection is also an issue. These land-management conflicts are likely to be a greater issue in the future as the local population increases and land use intensifies. Hunting of birds and trapping for trade are ongoing problems, particularly for large pigeons, doves and parrots.

James Eaton

*The **Timor Blue Flycatcher** is easily seen in the forests above the village of Fatumnasi.*

James Eaton

*The **Banded Fruit Dove** is regularly encountered in the montane forests on Gunung Mutis.*

Sangihe Island

Yann Muzika

Yann Muzika

The Critically Endangered **Cerulean Paradise Flycatcher** *is now confined to a handful of forested gullies on the island.*

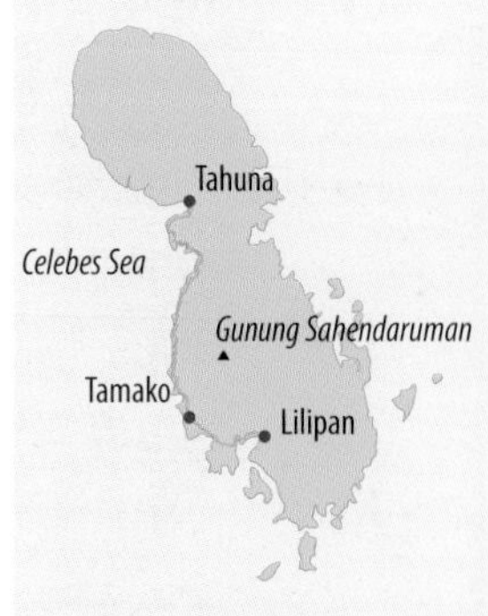

KEY FACTS

Nearest Major Town
Tahuna (North Sulawesi)

Key Species
Elegant Imperial Pigeon*, Sangihe Scops Owl*, Lilac Kingfisher*, Red-and-blue Lory* (possibly extinct), Sangihe Hanging Parrot*, Sangihe Pitta*, Sangihe Shrikethrush*, Hair-crested Drongo (undescribed endemic taxon), Cerulean Paradise Flycatcher*, Northern Golden Bulbul*, Sangihe White-eye* (extinct?), Yellow-sided* and Grey-sided Flowerpeckers*, Elegant Sunbird*
Passage Chinese and Japanese Sparrowhawks, Grey-faced Buzzard

Other Specialities
Sangihe Tarsier*

Best Time to Visit
Dry season May–September; wet season November–March

The Sangihe (or Sangir) Islands are a group of oceanic islands located in the Celebes Sea midway between Sulawesi and Mindanao (the Philippines). Sangihe became known to the ornithological world in 1873 when a bluish bird was collected by a local hunter on behalf of the noted German naturalist A. B. Meyer on Sangihe Besar, who then described it for science. This bird was first thought to be related to the paradise flycatchers, and was given its own genus *Eutrichomyias*. Despite subsequent collection expeditions, no further specimens were obtained, nor were there any reliable records of the **Cerulean Paradise Flycatcher** until 1998, when surveys led by Jon Riley finally obtained definitive sightings of the species. Recent genetic studies have revealed that it is closely related to the silktails of Fiji. Other recent studies have clarified the taxonomic positions of many of Sangihe's birds. For instance, the **Sangihe Shrikethrush**, once thought to be a shrikethrush with a disjunct distribution, has been revealed to be closely related to Sulawesi's Maroon-backed Whistler. With six endemic species that can be seen over a 2–3 day visit, Sangihe is becoming an increasingly attractive destination for serious birdwatchers.

Birdwatching Sites

Gunung Sahendaruman

Located in the southern half of Sangihe, Gunung Sahendaruman is the highest point of a volcanic caldera's ridge on which the last remnants of Sangihe's original forests can be found. Sitting at the bottom of the ridge inside the caldera is the village of Lilipan, which is the starting point of most Sangihe birdwatching adventures. There is still remnant degraded forest around the village, and this is where some of the target birds can be found. The **Lilac Kingfisher** (ssp. *sanghirensis*) is usually seen sitting motionless in the middle storey, while the **Sangihe Hanging Parrot**, **Elegant Sunbird** and both flowerpeckers are best located

The Sangihe subspecies of the **Lilac Kingfisher** *is more richly coloured than its congeners on Sulawesi.*

Yann Muzika

*The **Elegant Sunbird** is the most common of the Sangihe endemics.*

by their calls. At night, the **Sangihe Scops Owl** may be seen. As you climb the valley out of Lilipan the main road becomes increasingly narrow, and after the last houses, it shrinks into a trail that follows a river. This river has to be crossed at several points on the trail to access better quality forest. Although the vegetation here is dense, it is still essentially kebun, and offers little habitat for most forest birds except the adaptable **Elegant Sunbird**. The trail becomes steeper as it approaches a ridge, and it is in patches of degraded forest here where you should start looking for the **Sangihe Pitta**. An undescribed form of the **Hair-crested Drongo** can also be found here, and **Elegant Imperial Pigeons** are often seen in flight overhead. From the top of the first ridge a secondary trail leads into a steep narrow valley, where the **Cerulean Paradise Flycatcher** can still be found. As you follow the main ridge up on a very steep and narrow trail, you eventually enter the remnants of Sangihe's primary forest, and this is where the **Northern Golden Bulbul** (ssp. *platenae*) and Critically Endangered **Sangihe Shrikethrush** can be found. The whole hike takes the better part of the day, and the conditions can be rather slippery and treacherous on a rainy day.

Access & Accommodation

Sangihe is easily accessed by a daily flight from Manado in North Sulawesi, with Naha Airport located in its main town of Tahuna. Alternatively, a daily ferry leaves the port of Manado at 6 p.m. and arrives at Tahuna Harbour the next morning at 6 a.m. To access Lilipan from Tahuna, you can either take a scheduled public bus (the journey takes roughly two hours), or hire a car easily found around the airport or in town. At Lilipan, the Rainbow Losmen run by Frets Pangimangen is the natural choice for accommodation. Fret's son, Wesley, can be hired as a birdwatching guide.

Francis Yap

*Many tens of thousands of **Chinese Sparrowhawks** migrate through Sangihe annually.*

Alternatively, Pak Niu, who can be found in his house higher up the road near the start of the trail, can be your guide, as he owns the land where the remnants of forest are found. Pak Niu can also arrange for you to stay in his house, which is beautifully located at the top of the first ridge, but then you will have to arrange logistics at the village before setting out.

Conservation

Sangihe has lost nearly all but a tiny portion of its original forest cover, and the Sangihe White-eye, a species recently separately from the Black-crowned White-eye, has not been seen for more than 20 years and is probably extinct. A similar fate may have befallen the nominate subspecies of the Red-and-blue Lory. Clearance of forest for agriculture and firewood continues to threaten the island's remaining forests. Efforts are ongoing by various stakeholders to engage the local community to conserve Sangihe's forest, with the option of establishing protection forests or nature reserves.

Tangkoko Nature Reserve & Minahassa Highlands

Ch'ien C. Lee

The coastal forests of Tangkoko provide the perfect introduction to Sulawesi's endemic birdlife.

With some of the most famous coral reefs (like Bunaken) in Indonesia and an impressive landscape dominated by towering volcanoes, the northern tip of the Minahassa Peninsula is undoubtedly Sulawesi's most visited region among ecotourists. Although scuba diving opportunities attract most foreign tourists to this area, the proximity of both the Minahassa Highlands and Tangkoko Nature Reserve, with its abundant and easily seen wildlife, also serves as an important attraction for naturalists in general, particularly birdwatchers.

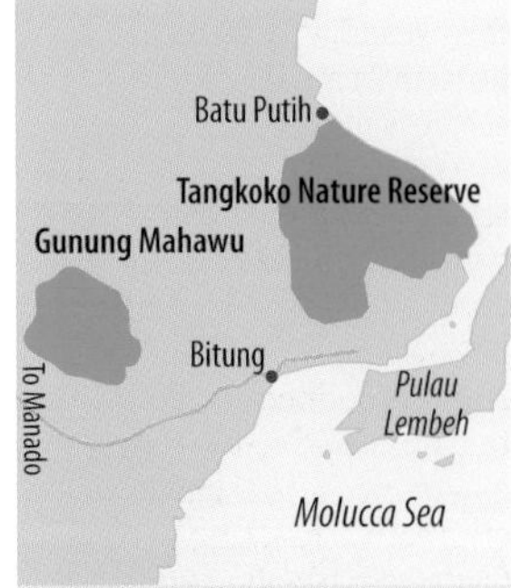

KEY FACTS

Nearest Major Town
Manado (North Sulawesi)

Habitats
Lowland rainforest, submontane and montane rainforests

Key Species
Lowland Maleo*, Philippine Megapode, Blue-faced Rail*, Minahassa Masked* and Sulawesi* Masked Owls, Sulawesi Scops Owl*, Ochre-bellied* and Speckled* Boobooks, Sulawesi Nightjar*, Purple-winged Roller*, Green-backed*, Lilac* and Great-billed* Kingfishers, Sulawesi Dwarf Kingfisher*, Knobbed* and Sulawesi* Hornbills, Ashy Woodpecker*, Yellow-breasted Racket-tail*, Great* and Pygmy* Hanging Parrots, Sulawesi Pitta*, Red-backed Thrush*, Sulawesi* and White-necked* Mynas
Montane Metallic Pigeon, Superb Fruit Dove, Scaly-breasted Kingfisher*, Sulawesi Pygmy Woodpecker*, Sulawesi Myzomela*, Sulphur-vented Whistler*, Chestnut-backed Bush Warbler*, Red-backed Thrush*, Sulawesi Blue Flycatcher*, Rufous-throated Flycatcher*, Crimson-crowned Flowerpecker*

Other Specialities
Sulawesi Rousette, Sulawesi Bear Cuscus*, Spectral Tarsier*, Sulawesi Black-crested Macaque*

Best Time to Visit
All year round

Birdwatching Sites

Tangkoko Nature Reserve

This is the best place to get acquainted with Sulawesi's endemic birds. It is not only home to an impressive range of species, including numerous endemics, but many of the mammals and birds here have become so habituated to the presence of people that they often allow for remarkably close viewing. The park entrance is located near the village of Batu Putih, where both accommodation and guides can be found. From here the main trail (a dirt road) runs south-east, parallel to the beach. The lowland forest of Tangkoko is seasonally dry,

A highlight of any visit to Tangkoko, the ***Spectral Tarsier*** *here is now recognized as a distinct species.*

with occasional large trees (like figs) and a fairly open understorey, making for easy walking and wildlife spotting. Species such as the **Philippine Megapode**, **Yellow-billed Malkoha**, **Green-backed Kingfisher**, **Lilac Kingfisher** (ssp. *cyanotis*) and **White-necked Myna** are regularly found here, and scrutiny of the forest floor may reveal the **Sulawesi Pitta** and **Red-backed Thrush**. Roving bands of Critically Endangered Sulawesi Black-crested Macaques are frequently encountered, and their playful antics and lack of shyness make them a joy to watch.

Trails heading uphill to the south lead into somewhat denser forest, where visitors often gather at dusk to watch the emergence of Spectral Tarsiers from their sleeping spots in hollow fig trees. In this area, birds such as the **Ashy Woodpecker**, **Yellow-breasted Racket-tail** and **Knobbed Hornbill** can be seen. The experienced local guides are also able to point out regular roosting sites for the **Sulawesi Scops Owl**, **Ochre-bellied Boobook** and **Great Eared Nightjar**. Occasional sightings of the **Maleo** and **Blue-faced Rail** have been made in this area.

An active **Maleo** nesting ground can be found further east along the coast – it is accessible by boat. However, the birds at this site are exceedingly shy and infrequently

Chi'en C. Lee

*A beautifully camouflaged **Great Eared Nightjar** roosts during the day on a low branch near the forest floor.*

Chi'en C. Lee

*A **Sulawesi Pitta** shows off its vivid plumage on the shady forest floor of Tangkoko.*

*Although they are generally silent and unobtrusive, **Lilac Kingfishers** can be very approachable once located.*

Wang Bin

*The **Green-backed Kingfisher**, a large and robust forest-dwelling species, is frequently encountered in Tangkoko's understorey.*

seen. Boats can also be hired to visit the mangrove swamp to the north of Batu Putih, where the **Great-billed Heron**, **Great-billed Kingfisher** and **White-rumped Cuckooshrike** are regularly seen.

Spotlighting sessions may yield **Sulawesi Nightjars** along the open, grassy areas of the main trail. **Sulawesi Scops Owls** are vocal and conspicuous, while forays into the forest may be rewarded with **Ochre-bellied** and **Speckled Boobooks**, and with luck the rare **Minahassa Masked Owl**.

Gunung Mahawu

Located just south of Manado in the Minahassa Highlands, Gunung Mahawu is a forested volcano with a summit at 1,324m asl. Less geologically active than neighbouring volcanoes, Mahawu is frequented by visitors due to its easy access and the breathtaking views of its impressive crater. A paved road winds up the south-eastern slope of the mountain, ending at a small parking lot where hikers proceed up a stairway to the viewpoint. Birdwatchers visit the mountain primarily because it offers one of the most accessible habitats to look for the **Scaly-breasted Kingfisher** (ssp. *princeps*), Sulawesi's most difficult endemic kingfisher.

Most birdwatching is done along the forested sections of the road at 1,150–1,250m asl, and a walk here in the morning often yields many of the most common montane endemics, including the **Sulawesi Pygmy Woodpecker**, **Sulawesi Myzomela**, **Sulphur-vented Whistler**, **Streak-headed White-eye** and occasional **Metallic Pigeon**. To observe the kingfisher it is usually necessary to enter the forest via any number of trails, and forays may potentially

Chi'en C. Lee

*Tangkoko is one of the last remaining wild localities for the **Sulawesi Black-crested Macaque**, now listed as Critically Endangered.*

Chi'en C. Lee

*The main avian draw of Gunung Mahawu's montane forests is the shy **Scaly-breasted Kingfisher**.*

yield the **Red-backed Thrush**, **Sulawesi Blue Flycatcher** and rare **Rufous-throated Flycatcher**. In addition, the thickets and stunted montane evergreen forest on the path to the summit are home to the **Chestnut-backed Bush Warbler** (ssp. *castanea*) and scarce **Crimson-crowned Flowerpecker**.

Access & Accommodation

Most visitors hire vehicles in Manado as public transportation is not a reliable means of accessing these sites. Tangkoko can be reached in less than two hours, and a host of small lodges can be found just outside the entrance of the reserve. Both lodge proprietors and park rangers are helpful in arranging local guides, which are required for entering the reserve. Gunung Mahawu is just over an hour's drive south of Manado, but it can also be reached via a 3–4-hour drive from Tangkoko. There are several lodges and hotels in the nearby town of Tomohon, but visitors can also opt to do day trips from Manado.

Conservation

Although both Tangkoko and Mahawu are protected forests, the Minahassa Peninsula is such a densely populated region that threats such as illegal logging and hunting still remain. During drought seasons there is always a considerable risk of forest fires, particularly around the borders of Tangkoko, and much of the secondary forest in this area has been lost to burning in recent years. Donations are collected at the entrance to the parking lot at Gunung Mahawu for the purpose of planting trees on the deforested slopes of the mountain.

Bogani Nani Wartabone National Park

Abdelhamid Bizid & Yong Ding Li

*The **Sulawesi Hawk-Eagle** may be encountered within Bogani Nani Wartabone National Park, here photographed with a captured cuscus.*

KEY FACTS

Nearest Major Town
Kotamobagu (North Sulawesi)

Habitats
Lowland and hill rainforests, freshwater wetlands, cultivation

Key Species
Maleo*, Vinous-breasted Sparrowhawk*, Spot-tailed Sparrowhawk*, Spotted Harrier, Isabelline Bush-hen*, Blue-faced Rail*, Metallic Pigeon, Oberholser's Fruit Dove*, Stephan's Emerald Dove, Black-billed Koel*, Bay Coucal*, Yellow-billed Malkoha*, Speckled Boobook*, Purple-winged Roller*, Green-backed Kingfisher*, Knobbed* and Sulawesi* Hornbills, Ashy Woodpecker*, Yellow-breasted* and Golden-mantled* Racket-tails, Great* and Pygmy* Hanging Parrots, Sulawesi Pitta*, Pale-blue Monarch*, White-necked Myna*, Red-backed Thrush*, Sulawesi Streaked Flycatcher*
Winter Grey-streaked Flycatcher

Other Specialities
Bear Cuscus*, Spectral Tarsier*, Gorontalo Macaque*, Sulawesi Palm Civet*, Lowland Anoa*, Sulawesi Babirusa*, Sulawesi Warty Pig*

Best Time to Visit
June–October

The vast Bogani Nani Wartabone (formerly Dumoga Bone) National Park (2,871km^2) straddles extensive areas of montane and lowland rainforests between North Sulawesi and Gorontalo provinces on the Minahassa Peninsula, and forms one of the largest protected areas on Sulawesi. Its diversity of habitats ranges from lowland riverine rainforest to montane forest that cloaks the slopes of its many mountains, which rise to 1,950m asl at Gunung Bulawa. Despite the large area of pristine habitats, access to most areas of the park is very poor, and most birdwatchers visit only one of the famous **Maleo** breeding grounds (such as Tambun) and a handful of other spots to see lowland specialities such as the **Blue-faced Rail**, **Oberholser's Fruit Dove** and **Pied Cuckooshrike** in the disturbed lowland forests on the eastern border of the park, near the villages of Pinonobatuan, Toraut, Dumoga Ketjil and Doloduo. Other areas of the park, like the well-forested Hungoyono camp in the west of the park in Bone Bolango (Gorontalo province) are more difficult to access, but can also be rewarding for the intrepid birdwatcher.

Birdwatching Sites

Tambun

The popular site of Tambun lies near Pinonobatuan village (East Dumoga district) on the eastern boundary of the park, and is one of the best and most convenient places to see the charismatic **Maleo** at its nesting grounds. Maleos should be relatively easy to see here at dawn, when adults descend from the hills to lay their eggs in soil

Abdelhamid Bizid

*The **Red-backed Thrush** may be observed in the lowland forests around Toraut.*

heated by volcanic hot springs. Besides seeing the adults, it is also possible to view their eggs, which are relocated by the wardens from the nesting grounds to safe areas far from poachers and egg collectors, as well as newly hatched chicks, which are released within a day or two after hatching. The small trails around the sandy areas can yield a good number of widespread forest species such as the **Yellow-billed Malkoha**, **Bay Coucal**, **Sulawesi Pitta** and even the **Red-backed Thrush**. Very occasionally, one may be treated to the rare sighting of a **Blue-faced Rail** crossing a forest trail. Tambun is also a very good area to see **Oberholser's Fruit Dove**, while the more open areas at the forest edge may hold the **Black-billed Koel**, **Purple-winged Roller**, **Sulawesi Cicadabird**, **White-rumped Triller**, **White-necked Myna** and distinctive white-eyed subspecies (ssp. *leucops*) of the **Hair-crested Drongo**.

James Eaton

Sulawesi has more endemic Accipiter *hawks than anywhere else in Asia, including this* ***Sulawesi Goshawk****.*

Toraut & Doloduo

The Toraut River near the villages of Toraut and Doloduo roughly follows the eastern boundary of the park. While degraded by human activity, the lowland forest across the river can still yield interesting species such as the **Yellow-breasted Racket-tail** and **Ornate Lorikeet**, both endemic hornbills and hanging parrots, and the **Pied Cuckooshrike**. Large numbers of pigeons may be seen when some of the trees are in fruit, including various fruit doves, imperial pigeons and green pigeons. The cultivated areas and secondary growth support different species like the **Spotted Harrier**, **Barred Rail** and **White-faced Cuckoo-Dove**, while at dusk the **Isabelline Bush-hen** and **Sulawesi Nightjar** may been encountered.

The small lakes and wetlands along the access road to Toraut may yield, among common wetland species, the **Dusky Moorhen** and various rails, as well as the distinctive subspecies (ssp. *hispidoides*) of the **Common**

Wang Bin

The ***Ochre-bellied Boobook*** *can be seen at dusk in the forests at Tapakulintang.*

Knobbed Hornbills *are widespread across Sulawesi and is common in the park's forest.*

*The **Maleo** is the highlight of any trip to Sulawesi's lowland forests.*

Kingfisher. At night more common night birds like the **Sulawesi Scops Owl**, **Sulawesi Masked Owl**, and **Speckled** and **Ochre-bellied Boobooks** have been recorded in the forest.

Tapakulintang

This area of hilly rainforest by the south-east edge of the park can be accessed from the Trans-Sulawesi Highway, south of Doloduo and in the direction of Molibagu. **Oberholser's Fruit Dove**, **Purple-winged Roller**, **Sulawesi Dwarf Kingfisher**, **Sulawesi Hornbill** and various owls have been seen here. The area holds a well-known day roost of the **Speckled Boobook**.

Access & Accommodation

The best way to access the park is from Kotamobagu city. This is where permits can be arranged at the national park office (Jalan AKD Mongkonai Barat) to enter the park. Hotels and lodges in Kotamobagu cater to varying budgets. Most of the key sites can be accessed by car along the Trans-Sulawesi Highway leading out of Kotamobagu and in the direction of Molibagu. For easy access to the park, many birdwatchers base themselves in the 'Tante Mien Ujang' homestay (Losmen) in Doloduo.

Conservation

The Dumoga valley just outside the park was the site of a major transmigration scheme to resettle rice farmers. The park's fringes are subject to deforestation, cultivation and illegal gold-mining activities, which have resulted in contamination of some rivers. The Maleo breeding sites are vulnerable to egg poaching, and only a handful of sites are now under the protection of local conservation agencies (BKSDA, PHPA) in collaboration with the Wildlife Conservation Society. Mammals, especially fruit bats, babirusa and wild pigs, are heavily hunted by locals.

*Sulawesi's **Purple-winged Roller** is probably most closely related to the Indian Roller of mainland Asia.*

Nantu Wildlife Reserve

Ch'ien C. Lee

Chi'en C. Lee

The forests of Nantu represent some of the last remaining pristine lowland forest in northern Sulawesi.

KEY FACTS

Nearest Major Town
Gorontalo (Gorontalo)

Habitats
Lowland rainforest

Key Species
Snoring* and Blue-faced* Rails, White-faced Cuckoo-Dove*, Stephan's Emerald Dove, Sulawesi Ground Dove*, Oberholser's Fruit Dove*, Ochre-bellied Boobook*, Purple-winged Roller*, Lilac Kingfisher*, Knobbed* and Sulawesi* Hornbills, Ashy Woodpecker*, Golden-mantled Racket-tail*, Sulawesi Dwarf Kingfisher*, Pied Cuckooshrike*

Other Specialities
Heck's Macaque*, Sulawesi Babirusa*, Sulawesi Warty Pig*, Lowland Anoa*

Best Time to Visit
July–October; rainy season January–March

Chi'en C. Lee

So small that it can sometimes be mistaken for an insect in flight, a ***Sulawesi Dwarf Kingfisher*** *perches quietly on the branch of a small tree near the entrance to Nantu.*

Situated approximately midway along Sulawesi's arching Minahassa Peninsula, Nantu comprises a 310km^2 expanse of forest along the Paguyaman River and the slopes of Gunung Boliohuto. The reserve is best known for its population of Sulawesi Babirusa (the best known of the three species of babirusa), the enigmatic 'pig-deer' whose bizarre appearance piqued the interest of 19th-century naturalist Alfred Russel Wallace and which is extremely difficult to see elsewhere on the island. At Nantu large numbers of babirusa and other wildlife, including a number of endemic pigeons, congregate at a natural salt lick where they can be observed from a concealed hide. In addition to its mammals, the forest reserve also offers excellent birdwatching opportunities and is home to a large number of Sulawesi's endemic lowland species, including the seldom-seen **Snoring** and **Blue-faced Rails**. Nantu is, however, seldom visited compared with other Sulawesi reserves, primarily due to its remote location.

Birdwatching Sites

Saritani & Panguyaman River

Nantu is approached by road or river from the east, with the last settlement before the reserve being the village of Saritani. Nearly all the forest on the relatively flat southern side of the river has been cleared for agriculture and timber. Nevertheless, birdwatching here can be productive for species that benefit from the open habitats, such as the **Spotted Harrier**, **Buff-banded Rail**, **White-rumped Triller** and occasional **Purple-winged Roller**.

*Seeing the bizarre **Sulawesi Babirusa** is worth a visit to Nantu in itself. Here, a mature male sports his unique upturned tusks.*

Nantu Forest

Intact lowland rainforest occurs only on the northern side of the river, within the borders of the reserve. Most visitors base their hikes from the ranger station, which is accessible by fording the river near Saritani village. Although there are few trails, the forest has an open understorey and is relatively easy to walk in. Nevertheless, caution must be exercised to avoid getting lost. The forest around the station is good for birds, with the noisy calls of **Bay Coucals** and **Hair-crested Drongos** a constant feature. Shyer species such as **Lilac** and **Sulawesi Dwarf Kingfishers** and the occasional **Sulawesi Pitta** can also be spotted here with some patient searching of the forest floor. Both **Knobbed** and **Sulawesi Hornbills** are frequently seen in the forest, the former being particularly abundant and noisy with its loud wingbeats.

Salt Lick

The highlight for most visitors, the salt lick consists of a wide, muddy clearing that is seeping with mineral-laden water, reached after a 30-minute walk into the forest from the ranger station. Here, large herds of Sulawesi Babirusa, perhaps mainland Sulawesi's most

*Although less conspicuous than the larger and more common Knobbed Hornbills, **Sulawesi Hornbills** are not difficult to find in the forests around Nantu.*

Wang Bin

*The bubbly chatter of the **Ashy Woodpecker** often betrays its presence.*

Chi'en C. Lee

White-faced Cuckoo-Doves can often be seen congregating at the muddy salt spring.

charismatic large mammal, gather throughout the day to ingest the spring water and bathe in the mud. Other mammals, including Heck's Macaque and the rare Lowland Anoa (a species of dwarf wild buffalo), also occasionally visit the clearing, as do a number of pigeons such as the **White-faced Cuckoo-Dove**, **Stephan's Emerald Dove** and **Sulawesi Ground Dove**. A small hide, albeit currently in a state of disrepair, has been constructed at the edge of the clearing to offer views of visiting wildlife. The babirusa are exceptionally wary of people, and care must be taken to approach the hide silently with the animals upwind.

Access & Accommodation

All visitors to Nantu are required to first obtain permits from the Kantor Seksi Bilaya Dua Gorontalo (in Gorontalo) before their trip. The permit process can usually be completed in half a day (weekdays only) although delays are not uncommon, and it is advisable to inform the office in advance of any visit. Trips can be arranged by contacting info@nantuforest.org.

Before the completion of the road to Saritani village, all visits to Nantu required an approach by boat up the Paguyaman River, which was costly and time-consuming. The road from Gorontalo to Saritani now takes under four hours, and private cars for this trip can be hired in town.

Overnight stays within Nantu Wildlife Reserve are currently prohibited (except for researchers), so visitors have the option of lodging in Saritani village, where basic homestay-style accommodation is available upon request. Alternatively, it is possible to camp in tents or hammocks in the secondary forest directly across the river from the ranger station. Hiking into Nantu each day from Saritani takes about 30 minutes and involves crossing the river, which is usually shallow, although filled with slippery stones. Occasional heavy storms, which can occur at any time of the year, may make the river temporarily impassable.

Conservation

Nantu is designated a wildlife reserve and regulations are strictly enforced. For example, rules require that no vegetation is cut even to make trails, and visitors are not allowed to bring *parangs* (machetes) into the forest. Despite this numerous threats persist, including illegal logging, rattan collecting, hunting, gold mining along the Paguyaman River and the continuing encroachment of agricultural plots.

Lore Lindu National Park

Yong Ding Li

Lore Lindu National Park protects a large expanse of montane rainforest.

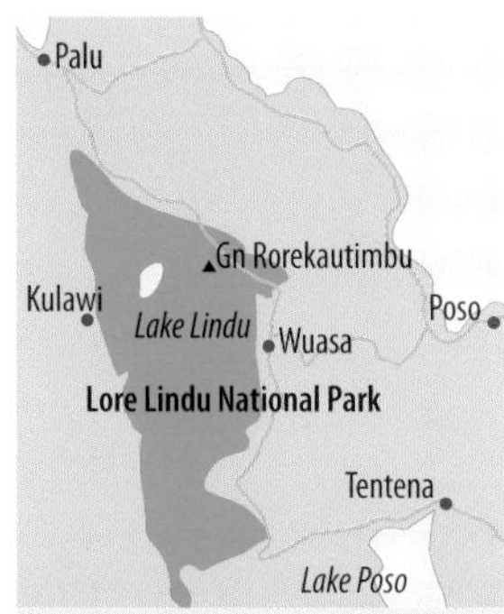

KEY FACTS

Nearest Major Towns
Palu, Wuasa (Central Sulawesi)

Habitats
Hill rainforest, submontane and montane rainforests, cultivation

Key Species
Lowland Sulawesi Goshawk*, Oberholser's Fruit Dove*, Green-backed Kingfisher*, Knobbed* and Sulawesi* Hornbills, Red-backed Thrush*, Sulawesi Streaked* and Rufous-throated* Flycatchers ***Montane*** Dwarf Sparrowhawk*, Sulawesi Woodcock*, Grey-headed* and White-bellied* Imperial Pigeons, Sombre Pigeon*, Red-eared Fruit Dove*, Sulawesi Ground Dove*, Minahassa Masked Owl*, Cinnabar Boobook*, Satanic Nightjar*, Scaly-breasted Kingfisher*, Purple-bearded Bee-eater*, *all* endemic woodpeckers, Golden-mantled Racket-tail*, Ivory-backed Woodswallow*, Piping Crow*, Hylocitrea*, Malia*, *all* endemic starlings, Geomalia*, Sulawesi Thrush*, Great Shortwing*, Crimson-crowned Flowerpecker*, Mountain Serin ***Winter*** Grey-streaked Flycatcher, Pechora Pipit

Other Specialities
Sulawesi Bear* and Dwarf* Cuscuses, Pygmy* and Dian's* Tarsiers, Tonkean Macaque, Sulawesi Ground Squirrel*, Sulawesi Palm Civet*, Sulawesi Babirusa* (rare), Mountain Anoa*

Best Time to Visit
June–October; wettest months January–May

Covering a vast 2,180km², the mountainous Lore Lindu National Park reaches a maximum elevation of 2,500m asl at Gunung Nokilalaki and protects some of the largest expanses of montane rainforests left on Sulawesi. Of equal significance are the lowland forests that remain on the hill slopes of the park's northern and eastern frontiers, and along the Napu Valley. Besides its importance to biodiversity, Lore Lindu also features significant archaeological sites and a number of geologically interesting tectonic lakes like Danau Lindu. For the birdwatcher, the main draw here are Sulawesi's montane endemics, most of which occur in this park. These include the recently described **Sulawesi Streaked Flycatcher** and many of Sulawesi's most elusive species, like the mysterious **Sulawesi Woodcock**. Most visitors concentrate on the forests along the Anaso Track and around Lake Tambing, which is home to a variety of species, from the **Great Shortwing** to the highly sought-after **Geomalia**. In the lower elevations around the Napu Valley, it is also possible to observe a number of lowland species.

Birdwatching Sites

Anaso Track (Gunung Rorekautimbu)

Probably the single most important birdwatching site in Lore Lindu, most birdwatchers spend at least three days here as most of the key species can be seen along the track, which climbs the slope of Rorekautimbu at the eastern fringe of the

*The nominate subspecies of the **Great Shortwing** is regularly heard, but seeing it is another matter.*

Yong Ding Li

*One of the most sought-after species from Lore Lindu is the **Satanic Nightjar**, which is known to occur at only a handful of sites on Sulawesi.*

Abdelhamid Bizid

*The local form of the **Cinnabar Boobook** is heavily spotted on its underparts, and may represent an undescribed subspecies.*

Con Foley

*The **Purple-bearded Bee-eater** often nests in holes dug into eroded banks along the Anaso Track.*

park. The increasingly eroded Anaso Track, particularly along the lower 2km, rises to nearly 2,000m asl at the turn-off to Puncak Dingin, an open lookout point from which pristine montane forest can be accessed via a narrow mountain trail. Otherwise, the forest along the track ranges from scrubby, fern-dominated clearings to good-quality montane forest. The lowest kilometre of the track is probably the best place to see the sought-after **Sulawesi Thrush**, which tends to follow large mixed flocks led by the noisy **Malia** and joined by many whistlers, drongos and cuckooshrikes. The **Great Shortwing** skulks in dense gullies and may take a bit of effort to see. The **Geomalia**, an aberrant ground thrush, is best seen in the forests past the large clearing otherwise known as the 'Helipad'. Also keep an eye out for the **Hylocitrea** (formerly Olive-flanked Whistler), an unusual relative of the waxwings formerly classified as a whistler. The **Satanic Nightjar** can often be located roosting in dense fern clearings. At higher elevations an undescribed subspecies of the **Mountain Serin** can be seen. It is also possible to see the central Sulawesi form of the **Cinnabar Boobook** and **Minahassa Masked Owl** on a night ramble, although the latter is infrequently reported.

Danau Tambing

Situated at an elevation of 1,600m asl, the forests around the lake and along the road in the direction of Sedoa or Palu are excellent for birdwatching. In the early morning the **Sombre Pigeon** and **Scaly-breasted Kingfisher** (ssp. *erythrorhamphus*) may be seen from the roadside. Besides that, large mixed flocks containing **Yellow-billed Malkohas**, **Cerulean** and **Pygmy Cuckooshrikes**, **Sulawesi Drongos** and **Piping Crows** often pass through the forests, as well as smaller flocks containing warblers, white-eyes and whistlers. The tall trees around the lake are often favoured by the **Ivory-backed Woodswallow** and imperial pigeons. It is possible to see the secretive Scaly-breasted Kingfisher and **Sulawesi Ground Dove** on the trail leading to the lake.

Kulawi

Located on the western boundary of the park, it is possible to access forests at lower elevations compared to the Napu Valley to the east. Few birdwatchers make it to this part of the park, but recent bird surveys show that many lowland species occur here, including **Oberholser's Fruit Dove**, **Knobbed** and **Sulawesi Hornbills**, and the **Sulawesi Myna**.

*The **Red-eared Fruit Dove** (*ssp. centralis*) is locally common in the montane forests of Lore Lindu, but is best seen in fruiting trees at the forest edge.*

Benjamin Schweinhart

*The unobtrusive **Sombre Pigeon** often perches in the dense canopy of the rainforest.*

Gunung Nokilalaki

The highest mountain entirely within the park, Nokilalaki is seldom visited by birdwatchers. The main point of access for hikers and climbers at Palolo can be reached via a 45-minute drive from Palu, from which a narrow forest trail leads to the summit. The area is densely forested with few clearings, and the birdlife here is very similar to Anaso's, although birdwatching conditions may prove more challenging.

Access & Accommodation

Palu is the main city closest to the park, from where entry permits can be arranged at the Lore Lindu Park headquarters. From Palu most birdwatchers charter vehicles to reach Wuasa, and are based there for most of their trip. For the more adventurous there are public buses running between Palu's main bus station (Petobo terminal) to Sedoa and Wuasa. Basic homestays (like Sendy's Homestay) are available in the town of Wuasa and Kulawi if you plan to enter the park from the west. There are local bird guides based in Palu and Wuasa, such as Idris Tinulele. It is also possible to camp at clearings along Anaso Track.

Conservation

Given its biodiversity and archaeological significance, Lore Lindu National Park has received international recognition as a UNESCO Biosphere Reserve. However, encroachment and illegal hunting remain a problem on its fringes. Small- to medium-scale cultivation, especially cocoa plantations, continue to eat into the park's northern fringes, and conflicts have occurred in the past due to illegal settlers. Poaching of babirusa, anoa and fruit bats for the illicit bush-meat trade has decimated these animals throughout the more accessible areas in the park.

Abdelhamid Bizid

*The **Sulawesi Thrush** (*ssp. abditiva*) is best seen around Lake Tambing and the lower parts of Anaso Track.*

Bantimurung-Burusaraung National Park

Abdelhamid Bizid

Mikael Bauer

The restricted-range **Lompobattang Flycatcher** *is only found in the montane forests of south-west Sulawesi.*

KEY FACTS

Nearest Major Town
Makassar (South Sulawesi)

Habitats
Hill and submontane rainforests, karst forest, paddy fields

Key Species
Woolly-necked Stork, Sulawesi Hawk-Eagle*, Sulawesi Goshawk*, Green-backed Kingfisher*, Knobbed* and Sulawesi* Hornbills, Pygmy Hanging Parrot*, Blue-backed Parrot, Golden-mantled Racket-tail*, Pale-blue Monarch*, Piping Crow*, Black-ringed White-eye*, White-necked Myna*, Pale-bellied Myna*, Sulawesi Streaked Flycatcher*

Other Specialities
Sulawesi Dwarf Cuscus*, Spectral Tarsier*, Moor Macaque*

Best Time to Visit
June–October; wettest months March–July

The Bantimurung-Bulusaraung (BaBul) National Park unites a number of protected areas in South Sulawesi Province (such as Bantimurung, Karaenta and Bulusaraung), and is home to some of the most significant areas of limestone landscapes in Indonesia. Much of the park is hilly, and there are large sections of impressive karst hills and a number of large cave chambers, all of which are major visitor attractions. A number of rivers, such as the Walanae, flow from the park's mountains, and one drains into the Danau Tempe lake system further north. For the birdwatcher BaBul is popular not only because of its accessibility from Makassar, but also due to a number of localized endemics that are best seen here. Key birdwatching targets include the **Black-ringed White-eye**, a south Sulawesi speciality, and the recently described **Sulawesi Streaked Flycatcher**. In addition to charismatic species such as **Knobbed** and **Sulawesi Hornbills**, **Golden-mantled Racket-tail**, **Blue-backed Parrot**, **Piping Crow** and many lowland forest specialities, the distinctive subspecies of the **Green-backed Kingfisher** (ssp. *capucinus*), restricted to south Sulawesi, is also of interest. Birdwatchers usually spend at least a full morning at this site before heading north to central Sulawesi.

Birdwatching Sites

Karaenta Area (Poros Maros-Soppeng Road)

Referred to in many birdwatching reports simply as 'Karaenta forest', the hill forest along this winding road is probably the most popular birdwatching area in the park, and it is usually not long before one

Abdelhamid Bizid

The **Sulawesi Hornbill** *is regularly seen along the roadside forests.*

The restricted-range ***Black-ringed White-eye*** *is easily seen in the forest of BaBul National Park.*

Con Foley

The high-pitched shrieks of the bizarre ***Piping Crow*** *are an integral part of the park's soundscape.*

obtain views of the restricted-range **Black-ringed White-eye**. Most birdwatchers spend the early morning exploring the hill forest beside the winding road as they slowly walk downhill from a high point near the rest stop. The **Yellow-billed Malkoha**, **Green-backed Kingfisher**, **Sulawesi Hornbill**, **Piping Crow** and a good selection of parrots and pigeons can be seen here. Unfortunately, due to traffic disturbance bird activity can drop rapidly from mid-morning onwards.

Bantimurung Area

The extensive karst hills here are some of the largest in Southeast Asia. As the Bantimurung area is easily reached from Makassar Airport and contains a number of ecotourism attractions, like waterfalls, caves and butterfly farms, it is very popular with domestic tourists. Many birdwatchers visit the site to see the recently described **Sulawesi Streaked Flycatchers** in the forest remnants near the main entrance to the park. Other birds of interest include the **Woolly-necked Stork**, which is increasingly difficult to see in Southeast Asia, and the scarce **Pale-bellied Myna**. The storks usually forage in the fields near the park entrance and may be seen while passing by. Species that have been recorded in the forest include the **Sulawesi Hawk-Eagle**, **Ruddy Kingfisher** and **Sulawesi Babbler**.

Malino

Another site not far from Makassar, this is the only place to see the endangered **Lompobattang Flycatcher**, which occurs in the montane forests on the ridges of Mount Lompobattang. Birdwatchers usually base themselves at the hill town of Malino and make morning excursions to the forests on the lower slopes of the mountain at 1,200m asl, south of the town. From the roadside and the many forest trails, one can also observe the **Black-ringed White-eye**, **Red-eared Fruit Dove** (ssp. *meridionalis*), **Hylocitrea** (ssp. *bonthaina*) and **Sulawesi Thrush** (ssp. *turdoides*).

Access & Accommodation

Makassar, the capital of South Sulawesi province, is the city closest to the key birdwatching areas. Accommodation in the city comprises hotels of all standards. There is no need to arrange accommodation outside the city as the sites are readily accessible from it. As most of the main birdwatching areas are on the fringes of the park on roads and at major tourist spots, a permit is not necessary for access.

Conservation

Like all lowland forests in Sulawesi, BaBul's forests are increasingly threatened by human activities, particularly illegal logging and poaching, despite the protection afforded to them.

Peleng Island

Lim Kim Chuah & Alpian Maleso

Philippe Verbelen

Peleng's rugged and well-forested interior supports many of the Sula archipelago's endemic birds.

KEY FACTS

Nearest Major Town
Luwuk (Central Sulawesi)

Habitats
Coastal forest (secondary), hill and submontane rainforest

Key Species
Sula Megapode*, Banggai Fruit Dove*, Sulawesi Scops Owl*, Great-billed Kingfisher*, Ruddy Kingfisher, Sula Hanging Parrot*, Sula Pitta, Slaty Cuckooshrike, Drab Whistler*, 'Peleng Fantail'*, Banggai Crow*, Northern Golden Bulbul*, 'Peleng Leaf Warbler'*, Helmeted Myna*, Red-and-black Thrush*, Henna-tailed Jungle-Flycatcher*
Winter Gray's Grasshopper Warbler

Other Specialities
Peleng Tarsier*, Peleng Cuscus*

Best Time to Visit
August–December

*The **Sula Pitta** is one of the most beautiful of the Sula archipelago's endemic birds.*

Lying off the eastern peninsula of Sulawesi are the thinly populated Banggai Islands. The largest among them, Peleng, has the shape of a distended 'M', and covers about 2,406km^2. Peleng has a rugged interior, with mountains rising to more than 1,000m asl to the west (such as Gunung Tombia). Peleng caught the attention of the international ornithological community when the supposedly extinct **Banggai Crow** was rediscovered on the island by Indonesian conservationist Mochamad Indrawan in 2007, after an absence of confirmed sightings since the first specimens were collected more than 100 years earlier. The Critically Endangered crow is now confined to the submontane forest of Peleng. Besides the crow, Peleng is home to a fantastic supporting cast of endemics, including the **Sula Megapode** and the beautiful **Red-and-black Thrush**, as well as many taxa represented by distinctive subspecies (and thus avenues for future splits) in the archipelago.

Birdwatching Sites

Around Salakan

There are extensive fragments of degraded coastal and lowland forest, broken by farmland (mostly 'kebun') and plantations, around the main town of Salakan. About 10km north of Salakan near the village of Kawalu is a small patch of coastal secondary forest frequented by birdwatchers. This is a good site to look for the Vulnerable **Sula Megapode**, which is also known to nest here. Other interesting species include widespread Sulawesi endemics such as the **Ornate Lorikeet**, **Ivory-backed Woodswallow** and **Grosbeak Starling**, as well as endemics restricted to the Banggai and Sula islands, like the **Banggai Fruit Dove**, **Sula Hanging Parrot**, **Red-and-black Thrush**, **Sula Pitta**, **Henna-tailed Jungle-Flycatcher** and **Helmeted Myna**. In addition, birdwatchers should also look

Philippe Verbelen

*Perhaps one of the most narrowly distributed corvids in the world, the **Banggai Crow** has drawn birders from far and wide to the island of Peleng.*

out for several taxa represented by distinctive subspecies here, notably the **Ruddy Kingfisher** (ssp. *pelingensis*) and **Sulawesi Scops Owl** (ssp. *mendeni*), the latter split by some authorities as 'Banggai Scops Owl'.

Kokolomboi

The compact village of Kokolomboi can be reached from Tataba by an uphill walk (about 6km) through agricultural land and patches of degraded forest. The relatively undisturbed submontane forest around Kokolomboi is well worth the hike and is probably the best site in which to see the Critically Endangered **Banggai Crow**. Other birds of interest in the montane forests here include the **Banggai Fruit Dove**, **Sula Pitta**, **Northern Golden Bulbul** and two undescribed taxa, namely the '**Peleng Fantail**' and '**Peleng Leaf Warbler**'. Both taxa are now the focus of phylogenetic studies. In recent years, the **Sula Pitta** and other sought-after birds like the **Red-and-black Thrush** and **Yellow-throated (Banggai) Whistler** have been seen in a patch of privately own cinnamon/cocoa plantation not far from Kokolomboi. Birdwatchers interested in visiting should contact local guides, who can make arrangements to visit this.

*The striking **Red-and-black Thrush** may be seen in the forests around Kokolomboi.*

Philippe Verbelen

*The Banggai subspecies of the **Sulawesi Scops Owl** differs markedly in plumage and vocalizations.*

Access & Accomodation

Peleng can be reached from the port of Luwuk in eastern Sulawesi. Luwuk is in turn served by regular flights from Ujung Pandang. From Luwuk, there are regular air-conditioned ferries leaving for Salakan, a three-hour journey. Another ferry departs for Tataba three times a week and takes about three hours.
A road follows the coast on Peleng, and it is possible to drive along and explore the remnant forests between Salakan and Tataba. Hotels catering for different budgets are available at Salakan. However, at Kokolomoboi, homestays are the only option. Local bird guides Pak Maleso and his son Alpian manage a basic homestay near Kokolomboi that is frequented by many visitors.

Conservation

Much of the lowland forests of Peleng have been cleared for farming, especially cocoa plantations, leaving mostly patches of remnant forests. In recent years, the remaining submontane forests around Kokolomboi have been increasingly threatened by slash-and-burn agriculture. There is an urgent need to gazette protected areas in order to conserve the island's remaining forests.

Weda Bay

Ch'ien C. Lee

Chi'en C. Lee

Thick and impenetrable Pandanus *swamps are the domain of one of Halmahera's most elusive birds, the Invisible Rail.*

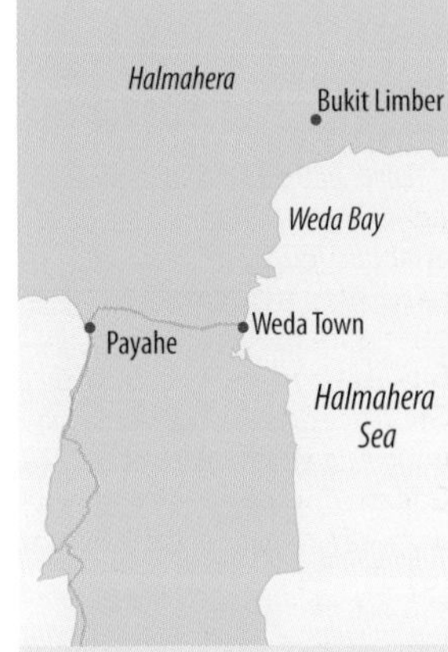

KEY FACTS

Nearest Major Towns
Weda, Sofifi (North Maluku)

Habitats
Lowland rainforest (primary and logged), submontane forest (Bukit Limber), mangroves, cultivation

Key Species
Lowland Gurney's Eagle, Invisible Rail*, Barking Owl, Halmahera Boobook*, Moluccan Scops Owl*, Moluccan Owlet-nightjar*, Azure Dollarbird*, Blue-and-white* and Sombre* Kingfishers, Blyth's Hornbill, White Cockatoo*, Great-billed Parrot*, Violet-necked Lory*, Moluccan King Parrot*, Ivory-breasted Pitta*, Rufous-bellied Triller*, Moluccan and Halmehera* Cuckooshrikes, Standardwing*, Paradise-crow*
Submontane Moluccan Cuckoo*, Great Cuckoo-Dove, Island Leaf Warbler, Rufous Fantail

Other Specialities
Ornate Cuscus*, Masked Flying Fox*, Sugar Glider, Tricolored Monitor*, Weber's Sailfin Lizard, Moluccan Scrub Python

Best Time to Visit
March–November; wettest months May–August

Weda Bay is located on the eastern shore of central Halmahera and is fast becoming one of the island's most popular birdwatching destinations, being much more accessible than some of the alternative sites in the north. Nearly all the island's specialities (aside from the Moluccan Megapode), and perhaps the most accessible lekking site for the **Standardwing**, can be found in the rainforests around Weda. Birdwatchers visiting this area usually stay at the Weda Resort (see below), which offers excellent accommodation and easy access to most of the important birdwatching sites.

Birdwatching Sites

Weda Bay Standardwing Lek

A 20-minute drive from the resort, this area comprises undulating lowland rainforest on limestone rocks. The main attraction here is the lek site of the iconic **Standardwing**, which is a further 15-minute walk from the main road, making it one of the most accessible sites for this bird on the island. On most mornings at least 3–5 male birds are in attendance at the lek, their raucous calls being audible from a great distance. As the birds display at dawn, it is best to reach the site in the dark so a headlamp is recommended for the walk.

Most visitors head to the Standardwing lek site first; after the display has stopped (usually at around 7–8 a.m.), they continue through the trails in the surrounding

Chi'en C. Lee

The iconic ***Standardwing*** *bird-of-paradise is the highlight of any birdwatching trip to Halmahera.*

forest to search for other species. The loud, taunting calls of **Ivory-breasted Pittas** are often heard, but the birds are quite shy and often considerable effort must be spent to see them. Also present in the vicinity are the **Dusky Megapode**, **Nicobar Pigeon**, **Common Paradise Kingfisher**, **Sombre Kingfisher**, **Dusky-brown Oriole** and **Paradise-crow**. Before dawn, **Moluccan Scops Owls** and **Moluccan Owlet-nightjars** are often heard calling here.

Birdwatching along the forested road near the entrance to the Standardwing trail enables you to see a variety of other local birds, particularly frugivores. Large fruiting fig trees often yield several pigeon species, including **Blue-capped**, **Superb**, **Scarlet-breasted** and **Grey-headed Fruit Doves**, along with the imposing **Blyth's Hornbill**. The aptly named **Goliath Coucals** are frequently sighted resting on the fronds of coconut palms that line the road.

Blue-and-white Kingfishers are regularly seen along the roads in the Weda area.

Bioyu

This swampy area located very close to Kobe Village (about a 20-minute drive from the resort) offers a chance to see one of Halmahera's most elusive birds, the aptly named **Invisible Rail**. Although sightings are not guaranteed, the bird has been spotted with fair frequency in the mangroves and *Pandanus* swamps not far from the roadside.

Bukit Limber

Comprising mineral-rich ultramafic soils, this mountain (c. 1,200m asl) is viewable from Weda across the bay to the north. The entire formation has been slated for nickel mining, but as of the time of writing, all operations have been indefinitely suspended and the forests remains largely intact. An excellent gravel track leads to the summit of

Chi'en C. Lee

*The submontane forests of Bukit Limber offer opportunities to see a different suite of birds, including the **Island Leaf Warbler*** (ssp. henrietta).

Limber and is easily navigable by a four-wheel drive vehicle, making this the only accessible highland area within the region. The forests here comprise largely trees and other vegetation adapted for the unusual soils, resulting in a habitat very different from the typical lowland Halmahera rainforest, and a chance to find species (including a number of submontane birds) that are scarce or absent elsewhere around Weda.

From Weda Resort, this track can be reached via vehicle in about an hour and a half by following the coastal road, then turning sharply left uphill. Birdwatching is good around the foot of the mountain, yielding various lowland species, including the **Moluccan King Parrot**, which is sighted with some frequency here. From here the road gradually climbs up the mountain slope, dipping several times before petering out at around 900–1,000m asl. Here the forest is composed of smaller trees growing on orange ultramafic soils, and the large strangling figs so predominant in the lowlands of Halmahera are absent. Walking along the road offers a good opportunity to meet both lowland and submontane species, including the **Moluccan Cuckoo** and the distinctive subspecies of the **Rufous Fantail** (ssp. *torrida*) and **Island Leaf Warbler** (ssp. *henrietta*).

Payahe-Weda Road

As the main highway connecting the east and west coasts of the island on the southern peninsula, this road is well paved and crosses the hilly interior to an elevation of about 400m asl. Although much of the original forest on either side of the road has been cleared, undisturbed habitats still remain at some distance, and there are some spectacular views over lowland rainforest on the higher reaches of the road. Aside from the occasional danger of speeding vehicles, walking along some sections of the road provides opportunities to see a large variety of birds, including the **Azure Dollarbird**, which is occasionally spotted perched on the tops of dead tree snags. **Blyth's Hornbills** are also particularly numerous here, often congregating in large numbers at dusk to roost in the crowns of large trees.

Chi'en C. Lee

*A young **Gurney's Eagle** soars over the forests of Weda.*

Wang Bin

*The **Ivory-breasted Pitta** is on the must-see list of many birdwatchers visiting Weda.*

Access & Accommodation

Weda Resort is a quiet and comfortable complex of bungalows located directly on the coast. Originally catering primarily to divers, in recent years the resort has become increasingly popular with birdwatchers, who comprise the bulk of its visitors in May–September during the diving off-season. The resort provides transportation to/from Sofifi or Tobelo, as well as guides and vehicles to visit all the nearby birdwatching sites.

Conservation

At the time of writing Halmahera's forests are disappearing rapidly, due primarily to clearance for agriculture and timber. Most roadside forest has been felled, making it necessary for visitors to venture through cultivation on foot to reach undisturbed habitats. Fortunately, roads have yet to penetrate the rugged interior of this part of the island, and a considerable amount of forest remains intact in these inaccessible regions. In the Weda area the Sawai Eco Tourism Foundation, established in 2012 by the owners of Weda Resort, has been responsible for the purchase of 750ha of lowland forest, including the Standardwing lek site, all of which is designated as a permanent forest reserve.

Chi'en C. Lee

*The **Sultan's Cuckoo-Dove** is common in the forests at Weda.*

*The **Moluccan Owlet-nightjar** is the only representative of the family in Wallacea.*

Aketajawe-Lolobata National Park

Yann Muzika & Hanom Bashari

*The hulking **Goliath Coucal** is often seen perching motionless in the forest canopy.*

KEY FACTS

Nearest Major Town
Sofifi (North Maluku)

Habitats
Lowland and lower montane rainforests (including forest on limestone)

Key Species
Dusky Megapode, Moluccan Goshawk*, Rufous-necked Sparrowhawk*, Invisible Rail*, Nicobar Pigeon, all endemic fruit doves, Spectacled and Cinnamon-bellied Imperial Pigeons, Goliath Coucal*, Moluccan Cuckoo*, Halmahera Boobook*, Moluccan Scops Owl*, Moluccan Owlet-nightjar*, Azure Dollarbird*, Common Paradise Kingfisher, Blue-and-white* and Sombre* Kingfishers, Moluccan Dwarf Kingfisher*, Moluccan King Parrot*, Chattering* and Violet-necked* Lories, Moluccan Hanging Parrot*, Ivory-breasted Pitta*, Moluccan* and Halmahera* Cuckooshrikes, Rufous-bellied Triller*, Black-chinned* and Drab* Whistlers, Moluccan Monarch*, Long-billed Crow*, Paradise-crow*, Standardwing*

Other Specialities
Ornate Cuscus*, Masked Flying Fox*

Best Time to Visit
May–October; rainy season December–March

Covering more than 1,600km^2 in two widely separated blocks of forest, this is one of the newest national parks in Indonesia. It protects a vast area of lowland and lower montane rainforests in north-eastern and central Halmahera. Nearly all the 30 or so species endemic to Halmahera and nearby islands can be found here. The park has gained ornithological attention in recent years as a reliable site for at least hearing the mythical **Invisible Rail**, one of Halmahera's most sought-after endemics. Like most other biodiversity conservation areas on Halmahera and the Moluccas (Weda Bay being an exception), its facilities and infrastructure for visitors remain underdeveloped. However, for birdwatchers with time and a sense of adventure, the little-visited forests of this beautiful park offer a viable alternative for seeing the majority of Halmahera's avian specialities.

Birdwatching Sites

Binagara

This is presently the only site that allows birdwatchers to easily access the Aketajawe-Lolobata National Park. The house of Pak Roji (a park ranger with good knowledge of the birds) is the last one on the road and is located right at the edge of the forest; from his front door visitors can enjoy views of flocks of **Blyth's**

Wang Bin

*The **Common Paradise Kingfisher** is the westernmost representative of the principally Papuan genus* Tanysiptera.

Hornbills and various parrots flying to and from their roost sites at dawn and dusk. A small river demarcates the boundary between cultivation and the forest, and here some lucky birdwatchers have been able to glimpse the extremely elusive **Invisible Rail**. The forest edge is also the best place to find fruiting trees, and once located patient observation will yield various pigeons, fruit doves and parrots, as well as cuckooshrikes, **Dusky-brown Orioles** and **White-streaked Friarbirds**. Look for movement among the vines to find foraging **Paradise-crows**, as well as the impressive **Goliath Coucal**. The **Rufous-bellied Triller** and **Blue-and-white Kingfisher** are regularly seen in the cultivated areas, especially during the rainy season.

Crossing the river brings you into the rainforest proper, and a few trails well known to Pak Roji allow sightings of the most elusive species. The splendid **Ivory-breasted Pitta** is quite common and surprisingly showy for its genus, while the **North Moluccan Pitta** is noticeably more retiring. **Black-chinned Whistlers**, **Moluccan Monarchs** and **Moluccan Flycatchers** forage in the middle storey and are best located by their songs. One sought-after bird is the beautiful **Common Paradise Kingfisher** from New Guinea, which reaches the western limit of its distribution in Halmahera; it is extremely hard to spot as it tends to sit unobtrusively in the tangles.

Wang Bin

*The **Sombre Kingfisher** is most vocal in the early hours of the morning.*

Abdelhamid Bizid

*Threatened by poaching, **Moluccan King Parrots** are increasingly uncommon.*

Abdelhamid Bizid

*The **Scarlet-breasted Fruit Dove** is the least regularly seen of the three endemic fruit doves.*

One morning should also be dedicated to a two-hour trek into the forest, leading to the **Standardwing** lek. Such an excursion involves a pre-dawn start as the birds only display for a short period at dawn. On the trail back to Pak Roji's house, look out for the **Chattering Lory**, **Moluccan Hanging Parrot** and **Dusky Megapode**. With a bit of luck the charismatic **Moluccan King Parrot**, **Azure Dollarbird** and **Moluccan Goshawk** may also be seen.

Spotlighting sessions are also recommended as they provide an opportunity to see the bizarre **Moluccan Owlet-nightjar**, the only representative in Asia of an otherwise Australasian family. The **Halmahera Boobook** and **Moluccan Scops Owl** are other possibilities on these sessions.

Access & Accommodation

Visiting the park first involves getting a permit from the national park office in Sofifi, the administrative capital of North Maluku Province, something that can be easily done since the friendly staff are supportive of tourists and birdwatchers. Access to the park is usually from the village of Binagara, located an hour's drive from Sofifi (a car with driver can be arranged easily from Sofifi). There, Pak Roji can offer a very basic homestay with his family, and guiding services provided he has no other commitments.

Conservation

The park is subject to the same conservation issues as many of the other large parks in Indonesia, particularly illegal logging and clearance of land for cultivation. In addition, bird trapping targeting parrots for the pet trade remains relatively rampant in Halmahera. The Chattering Lory and White Cockatoo are of special concern, as they are especially popular in the international pet trade.

Obi Island

Yann Muzika

Yann Muzika

The enigmatic ***Moluccan Woodcock*** *was thought to be extinct until its rediscovery in 2010.*

KEY FACTS

Nearest Major Town
Jikotamo (North Maluku)

Habitats
Lowland, hill and montane rainforest

Key Species
Pygmy Eagle, Bare-eyed and Invisible Rails*, Moluccan Woodcock*, Barking Owl, Scarlet-breasted* and Carunculated Fruit Doves*, Spectacled* and Cinnamon-bellied Imperial Pigeons*, Moluccan Cuckoo*, Common Paradise Kingfisher, Blue-and-white Kingfisher*, Moluccan Dwarf Kingfisher*, Chattering Lory* and Violet-necked Lory*, North Moluccan Pitta*, Dusky* and Sulawesi Myzomelas*, Pale Cicadabird*, Black-chinned* and Cinnamon-breasted Whistlers*, Moluccan Monarch*, Paradise-crow*, Northern Golden Bulbul*, Island Leaf Warbler, Cream-throated White-eye*, Halmahera Flowerpecker*
Winter Matsudaira's Storm Petrel, Gray's and Sakhalin Grasshopper Warblers

Other Specialities
Obi Cuscus

Best Time to Visit
July–September (active displays by Moluccan Woodcock)

Oval shaped and roughly 85km long, Obi is a mountainous island in the northern Moluccas that rises to an elevation of 1,500m asl. Human settlements on Obi are concentrated along the coasts, leaving the interior devoid of roads. Until recently, Obi was difficult to access and had been sparingly explored by ornithologists. Because of this, the island's avifauna was known only from a handful of collecting expeditions in the late 19th and early 20th centuries. In 2010, a survey led by Marc Thibault rediscovered the **Moluccan Woodcock**, a species previously known only from old museum specimens, and the **Carunculated Fruit Dove**, an Obi endemic that was last seen in the 20th century. Obi has now become a fixture on the birdwatching circuit because it supports three endemic species and several endemic subspecies likely to be elevated to full species (such as the **Sulawesi Myzomela**, **Paradise-crow** and **Cream-throated White-eye**). However, access to Obi for the independent birdwatcher remains particularly challenging for now due to its remoteness, and limited accommodation and infrastructure.

Birdwatching Sites

There are two lowland forests sites located fairly close to the main village and harbour of Jikotamo, and both are known to hold the main target species. Accessing forests at higher elevations can only be done after obtaining a permit to enter one of the logging concessions, and it is also necessary to arrange a vehicle to drive a

Yann Muzika

The ***Obi Cuscus*** *is common in the forests of Obi and surrounding islands.*

logging road up to 350–400m asl at least. A minimum of five days (to and from Ternate) are needed to cover the main sites on the island; more time is needed to explore higher elevation forests.

Jikotamo–Sembiki Road

This is an area of degraded lowland forests and plantations south-east of Jikotamo. About 3km from the village, the road passes through a swampy area where observations of **Moluccan Woodcock** in display flight (at dawn or dusk) are regular outside the dry season. Continuing along the road uphill leads you to a lightly wooded area broken with plantations and orchards (kebun) where many of the target species can be seen, including the **Carunculated Fruit Dove** (look for fruiting trees), **Moluccan Cuckoo**, **Common Paradise Kingfisher**, **Violet-necked Lory**, **North Moluccan Pitta**, **Pale Cicadabird** (ssp. *hoogerwerfi*), **Paradise-crow** (ssp. *obiensis*) and **Northern Golden Bulbul** (ssp. *lucasi*). From the swampy area, a dirt road leads into a logging concession and enables access to the higher elevations. An access permit is required in advance to enter these forests.

Yann Muzika

*The **Moluccan Cuckoo** is frequently heard at the forest edge.*

Cabang River Area

Directly south of Jikotamo, a small road meanders through agricultural land and after 2.5km reaches the Cabang River. After crossing the river, a trail runs through rice paddies and remnant forest with swampy areas along the river. The hill slopes on either side of the valley still hold good forest. Here, there is a slim chance of seeing two elusive rails – the **Invisible** and **Bare-eyed Rails**. However, the area is good for a a selection of forest species including the **Moluccan Dwarf Kingfisher**, **Red-cheeked Parrot**, **Red-flanked Lorikeet**, **North Moluccan Pitta**, **Dusky Myzomela**, **Cinnamon-breasted Whistler** and **Moluccan Monarch**, while displaying **Moluccan Woodcocks** have also been recorded. With the help of local guides, more forested areas can be accessed and it is possible to access areas above 400m elevation, where distinctive forms of the **Sulawesi Myzomela** (subspecies unclear), **Island Leaf Warbler** (ssp. *waterstradti*) and **Cream-throated White-eye** (ssp. *atriceps*) can all be seen.

Access & Accommodation

Obi is only accessible by a daily ferry that leaves from Ternate (which is well connected with several major Indonesian cities) in the evening and reaches Jikotamo harbour in north Obi the next afternoon after an 18-hour crossing. At Jikotamo and in nearby villages of Buton and Laiwui, basic supplies can be obtained. At Laiwui, the Celebes Inn is possibly the best accommodation on the island. A local guide named Pak Sabar has been catering to the needs of visiting birdwatchers, and can provide a spot in his back garden to pitch a tent and serve meals. He can also guide you to the birdwatching locations, or arrange logistics for a camp if needed.

Conservation

Although the interior of Obi Island remains mostly forested and sparsely inhabited, a major part this forested landscape falls under one of several selective logging licences operated by three logging companies. In addition, small scale mining has also damaged old-growth forest in various parts of the island. There is a need to designated new protected areas on Obi to protect its resident birdlife.

Yann Muzika

*The beautiful **North Moluccan Pitta** is a highlight of any birding trip to Obi.*

Manusela National Park

Mikael Bauer & Hanom Bashari

Hanom Bashari

Manusela National Park is the largest national park in the Moluccas and is home to all of Seram's endemic birds.

KEY FACTS

Nearest Major Towns
Wahai, Sawai, Masohi (South Maluku)

Habitats
Lowland and montane rainforests, freshwater swamp forest, coastal beach forest

Key Species
Dusky Megapode*, Seram Mountain Pigeon*, Moluccan Masked Owl*, Hantu Boobook*, Lazuli Kingfisher*, Salmon-crested Cockatoo*, Purple-naped Lory*, Red Lory*, Blue-eared Lory*, Red-breasted Pygmy Parrot, Moluccan King Parrot*, Seram Friarbird*, Drab* and Wakolo* Myzomelas, Seram Honeyeater*, Moluccan Cuckooshrike*, Pale Cicadabird*, Drab Whistler*, Grey-collared Oriole*, Streak-breasted Fantail*, Black-chinned Monarch*, Seram Golden Bulbul*, Rufescent Darkeye*, Grey-hooded White-eye*, Long-crested Myna*, Seram Thrush*, Cinnamon-chested Flycatcher*
Winter Chinese Crested Tern (rare)

Other Specialities
Common Spotted Cuscus, Seram Flying Fox*

Best Time to Visit
September–January; rainy season January–April

Manusela is the largest national park in the Moluccas. Its 1,890km² encompass lowland rainforest, swamps and montane forest up to the barren peak of Gunung Binaiya (3,027m asl), the highest peak in the Moluccas. The park covers roughly 10% of the island of Seram and is home to all the endemic birds of the island. As in other parts of eastern Indonesia, parrots are well represented. The charismatic **Salmon-crested Cockatoo** is, despite heavy trapping pressure, not too difficult to connect with, while the **Purple-naped Lory** has suffered badly and is now very difficult to see. Most of the endemic birds of Seram can be seen on a 3–4-day visit while birdwatching along the Masohi-Sawai stretch of the Trans Seram road that traverses the island. However, to have a fair chance at the **Blue-eared Lory** and the near-mythical **Seram Thrush**, a strenuous week-long trek up the slopes of Gunung Binaiya is needed. The recently discovered endemic subspecies (ssp. *almae*) of the **Moluccan Masked Owl**, a likely future split, is the ultimate prize for those keen on night birds. Visiting birdwatchers will be amazed at the breathtaking scenery and avian treasures that Seram has to offer.

Birdwatching Sites

Trans-Seram Road

The northern stretch of the highway between Masohi and Sawai passes through Manusela National Park and offers an easily accessible site for seeing most of the endemic birds of Seram. Traveling from Masohi, the first stretch passes through small patches of forest and agricultural areas. As you go further north, it might be worthwhile stopping in the disturbed forests to look for the **Long-crested Myna**, which is more easily seen in degraded areas outside the national park. About 90km north of Masohi the road starts to climb up the hills of the park, and the next 15km

James Eaton

*The endemic **Long-crested Myna**, with its funky hairdo, is regularly seen in degraded forest at the fringes of the park.*

*The **Lazuli Kingfisher** is best seen in the fast-disappearing lowland forests of Seram, and neighbouring Ambon.*

*The eye-catching **Salmon-crested Cockatoo** is a major target for any birdwatcher visiting Seram.*

*Mixed flocks should be checked for the **Streak-breasted Fantail**.*

*The **Seram Mountain Pigeon** is one of the many species on the island with Papuan affinities.*

or so offer access to superb forest. The road largely undulates at 800–1,200m asl, peaking at a mountain pass at 1,250m asl. After the pass it descends into a spectacularly steep passage down to Sawai. Most of the high-altitude endemics, like the **Streak-breasted Fantail**, **Seram Honeyeater**, **Rufescent Darkeye** and **Grey-hooded White-eye**, are fairly easy to see in this area. This is also a great place to find the **Salmon-crested Cockatoo** and **Moluccan King Parrot**. There are claims of small groups of **Blue-eared Lories** being seen at the pass, and note that the **Red Lory** is numerous. The lowland forest around the junction to Sawai is good for the **Lazuli Kingfisher** and **Hantu Boobook**. With luck you might also find the **Salmon-crested Cockatoo** and **Long-crested Myna** in this area. Check the langsat (*Lansium domesticum*) plantations for the diminutive **Red-breasted Pygmy Parrot**.

Gunung Binaiya Trek

To access proper montane forest a hike up the slopes of Gunung Binaiya is needed. Starting in Wahai, this is a strenuous, week-long, circular trek. The disturbed lowland forests around Wahai are good for the **Lazuli Kingfisher**, while the likelihood of seeing a **Purple-naped Lory** increases as you move further away from civilization. Continuing the ascent up Gunung Binaiya or Gunung Kobipoto you enter the habitat of the **Blue-eared Lory**, which rarely ventures below 1,350m asl. This is also the area to look for the rarely seen **Seram Thrush**, while the type specimen of the **Moluccan Masked Owl** (ssp. *almae*) was collected at 1,350m asl in the park.

Offshore Islands

Just off the north coast at Sawai is Pulau Lusaolate, a tiny island that hosts a good population of the **Olive Honeyeater**, a true small-island specialist. Nearby, the island of Pulau Sailoemania has hosted wintering **Chinese Crested Terns** in recent winters. Boats to reach these two areas can be hired from Sawai.

Off the north-western end of Seram is the small island of Boano, which is home to the Critically Endangered **Black-chinned Monarch**. This species is still quite common in the remaining forest patches at the northern end of the island. In order to reach the site, plan a two-day excursion and try to charter a boat from Masika on Seram to the small village of Huhua on the northern coast of Boano to save the hike across the island. There is also a good chance to see the endemic subspecies of the **Dusky Megapode** (ssp. *forsteni*) in the coastal forest on Boano.

Access & Accommodation

Masohi is a good point of entry into Seram, with several daily boat connections with Ambon. As the regional capital, there is good infrastructure and numerous accommodation options, and for the visiting birdwatcher this is a good place to stock up before venturing to the more remote northern areas. Manusela National Park's office, where permits to enter the park can be arranged, is also located here. Sawai is a small village situated in a stunning cove on the northern coast about 3–4 hours from Masohi. This is the ideal base for birdwatching along the Masohi-Sawai road as there is an excellent ecotourism hotel in the village. Further along the Trans Seram road is the town of Wahai, the starting point for treks up the slopes of Gunung Binaiya.

Conservation

The avifauna of Seram is under heavy pressure. About half of the remaining forest on Seram has been divided into forestry concessions, and trapping of parrots for the pet trade is rampant. Manusela National Park is the last stronghold for most of Seram's montane species, and keeping levels of trapping inside the park low will be necessary to ensure the long-term survival of species such as the Salmon-crested Cockatoo.

Mikael Bauer

Spotlighting sessions at the park are likely to yield the endemic ***Hantu Boobook.***

Wang Bin

Blyth's Hornbills *are not uncommon in the forests of Seram, and their whooshing wingbeats are regular sounds in the rainforests on the island.*

Buru Forests

Robert Hutchinson

Robert O. Hutchinson

Forested ravines along the Wamlana logging road are the abode of the Buru Thrush.

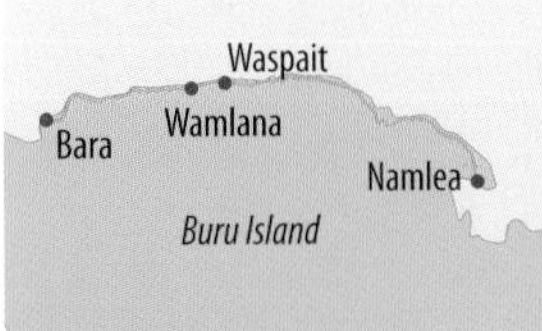

KEY FACTS

Nearest Major Towns
Waspait, Namlea (South Maluku)

Habitats
Lowland rainforest and mixed evergreen forest, montane forest, scrub

Key Species
Lowland, hills Dusky Megapode*, Buru Green Pigeon*, White-bibbed & Claret-breasted Fruit Dove, Spectacled Imperial Pigeon*, Moluccan Masked Owl*, Hantu Boobook*, Buru Dwarf Kingfisher*, Moluccan King Parrot, Buru Racket-tail*, Black-lored Parrot*, Blue-fronted Lorikeet*, Red Lory*, South Moluccan Pitta, Wakolo Myzomela*, Black-faced Friarbird*, Buru Cuckooshrike*, Pale Cicadabird*, Black-eared Oriole*, Spangled Drongo, Tawny-backed Fantail*, Buru Golden Bulbul*, Buru White-eye*, Streak-breasted Jungle Flycatcher*, Flame-breasted Flowerpecker*
Montane Buru Mountain Pigeon*, Buru Honeyeater*, Island Leaf Warbler, Chestnut-backed Bush Warbler*, Buru Thrush*, Madanga*
Winter Grey-streaked Flycatcher

Other Specialities
Moluccan and Seram Flying Fox, Buru Babirusa (rare)

Best Time to Visit
August–November

As the third largest island in Maluku at more than 9,600km^2, well-forested Buru is home to at least 20 endemic bird species including the enigmatic and nocturnal **Black-lored Parrot**. Despite this, Buru is one of the least-visited of all of Indonesia's islands by virtue of the once difficult access and tough trekking needed to reach the key birdwatching sites in its rugged interior. All this has fortunately changed in recent years with daily ferries providing access from the major transport hub of Ambon and regular flights scheduled to start soon. On top of this, the highland birdwatching sites in Buru's interior are now easily accessible via an old logging road from Wamlana on the north coast where one can stay in a remarkably comfortable beach resort. Given these positive developments, a trip to Buru will now become increasingly irresistible for any serious birdwatcher with an interest in Wallacea's avian endemics.

Birdwatching Sites

Wamlana Logging Road

This logging road rises from sea level near the settlement of Waspait and gradually climbs up the mountainous slopes of the Kapalat Mada massif in western Buru. This road provides the visiting birdwatcher an opportunity to see the vast majority of Buru's endemics with the **Buru Cuckooshrike**, **Black-eared Oriole**, which closely mimics the **Black-faced Friarbird**, **Tawny-backed Fantail**, **Buru Golden Bulbul**, **Buru White-eye**, **Streak-breasted Jungle Flycatcher** and **Flame-breasted Flowerpecker** possible at most elevations, while the **Buru Racket-tail** remains pleasantly common here. In addition, there are many sought-after species shared only with adjacent

*The enigmatic **Black-lored Parrot** has the distinction of being tropical Asia's only nocturnal parrot.*

islands of Seram or Halmahera regularly found here, notably the **White-bibbed Fruit Dove**, **Spectacled Imperial Pigeon, Moluccan King Parrot**, **Red Lory**, **Pale Cicadabird, Black tipped** and **White-naped Monarch** as well as the **Cinnamon-chested Flycatcher**. A number of endemic subspecies of wide-ranging Wallacean species are highly likely to receive taxonomic attention in the future and should be looked for, including **Wakolo Myzomela** (ssp. *wakoloensis*), **Drab Whistler** (ssp. *examinata*), **Northern Fantail** (ssp. *bouruensis*), **Spangled Drongo** (ssp. *buruensis*), and for pitta fans the Buru subspecies (ssp. *rubrinucha*) of the **South Moluccan Pitta**. The evergreen forests at the mid-elevation sections of the logging road have also produced recent sightings of two of Buru's most enigmatic species, namely **Black-lored Parrot** and **Blue-fronted Lorikeet**. However, both appear to be nomadic and unpredictable with the nocturnal Black-lored Parrot perhaps following its favoured fruiting trees, and the lorikeet being attracted to particular flowering trees.

At the higher elevations of the Wamlana logging road above 900 m asl, a different suite of endemics are present, with the uncommon **Buru Honeyeater** attracted to flowering trees while both the **Chestnut-backed Bush Warbler** (ssp. *disturbans*) and **Buru Thrush** are common inhabitants of roadside gullies. The **Buru Mountain Pigeon** is a common sight perched in open trees or passing overhead, and **Island Leaf Warbler** (ssp. *everetti*) is ubiquitous. The only difficult species in the montane forests along the logging road is the **Madanga**, a fascinating species recently and sensationally shown by Per Alstrom and his colleagues to be a highly aberrant Old World pipit. Finding it will require a much longer expedition into the high mountains south-west of Danau Rana. At night the distinctive vocalizations of **Hantu Boobook** make this recent split easy to track down.

Bara

The scrub and degraded forest at Bara are easily accessible along the coastal road heading west from Wamlana and are the best place to see the locally common **Black-tipped Monarch** as well as small numbers of **Buru Green Pigeons,** which are apparently restricted to the lowlands and very rare. Finding fruiting trees seems to be the key to locating one of these pigeons and they have even been spotted occasionally in villages on the coastal road when suitable trees are fruiting. Buru's other endemic nightbird, the **Moluccan Masked Owl** (ssp. *cayelii*), is also more reliably searched for in the lowland forests here.

Access & Accommodation

Daily overnight ferries shuttle between Ambon and the city of Namlea on Buru, usually arriving on Buru in the late morning or early afternoon. From Namlea, a 4WD vehicle is needed to access the key birdwatching sites, many of which can only be reached via rugged logging roads. Visiting birdwatchers can now base themselves at the comfortable Buru Island

Robert O. Hutchinson

*Due to its unusual taxonomic status, the **Madanga** is one of the most sought-after of Buru's avian endemics.*

Dubi Shapiro

*The attractive **Cinnamon-chested Flycatcher** skulks in dense roadside thickets.*

Resort at Waspait, from where it is easy to access the Wamlana logging road during the dry season. Given the difficulty in organizing logistics for Buru, most birdwatchers visit the island on organized birdwatching tours.

Conservation

While Buru's montane forest is largely undisturbed, much of the island's coastal lowland forest has been cleared for timber and agriculture. Additionally, small-scale gold mining operations across the island have increased exponentially in recent years, resulting in adverse environmental impacts such as mercury contamination that have yet to be adequately documented.

Yamdena Forests

Yong Ding Li

Yong Ding Li

The remote Tanimbar Islands are still cloaked in dense forest and contain their own suite of endemics found nowhere else in Indonesia.

KEY FACTS

Nearest Major Town
Saumlaki (South Maluku)

Habitats
Mixed evergreen forests, mangroves, coastal mudflats, cultivation, scrub

Key Species
Tanimbar Megapode*, Wallace's* and Rose-crowned Fruit Doves, Elegant and Pink-headed* Imperial Pigeons, Tanimbar Corella*, Blue-streaked Lory*, Tanimbar Boobook*, Moluccan Masked Owl*, Elegant Pitta*, Long-tailed Fantail*, White-naped* and Black-bibbed* Monarchs, Tanimbar Oriole*, Tanimbar Bush Warbler*, Fawn-breasted* and Slaty-backed* Thrushes, Tanimbar Flycatcher*, Tanimbar Starling*, Mistletoebird, Tricolored Parrotfinch*

Other Specialities
Black-bearded Flying Fox*

Best Time to Visit
June–December

Located at the furthest south-western corner of Wallacea, the Tanimbar Islands are as remote as it gets for any intrepid birdwatcher travelling in Indonesia. Flights to this part of Indonesia are limited, but the reward for coming this far are forests full of endemic birds. While much of the Tanimbar Islands is yet to be gazetted as formal conservation areas, they still retain a large proportion of their forest cover due to the rugged limestone terrain that renders cultivation challenging.

Although administered as part of Maluku province, the Tanimbars are zoogeographically part of the Lesser Sundas, being the easternmost extent of this long chain of islands. Low-lying Yamdena is the largest of the Tanimbar Islands and supports all the key species desired by visiting birdwatchers. Much of Yamdena's interior and its north-eastern coast remain covered in dense monsoon evergreen forests, while the south-eastern coastal belt along the island's main road has been largely cleared for agriculture. This road, however, provides some excellent opportunities to access Yamdena's forests further inland, and at a number of points tracks that branch off the road into the forest are also worth exploring to see shy species. Most of the area's endemics can be seen on a trip lasting 3–4 days.

Birdwatching Sites

Saumlaki – Larat Road

This major road traverses the entire length of Yamdena and provides access to the island's varied habitats. Most birdwatchers concentrate on the stretch of road from the 20–23km mark, and base themselves at a local property known as 'Kebun 45'. The

***Wallace's Fruit Dove** is arguably the region's most attractive-looking pigeon and is easily seen on Yamdena.*

James Eaton

Spotlighting sessions on Yamdena may yield sightings of the mythical ***Moluccan Masked Owl*** (ssp. sororcula), *a species seen by very few birdwatchers until recent years.*

habitat here is mostly open roadside scrub, broken by patches of forests and creeks. Many of the common endemics like the **Tanimbar Corella**, **Rufous-sided Gerygone**, **White-browed Triller**, **Cinnamon-tailed Fantail** and **Golden-bellied Flyrobin** can be easily seen in the forest patches adjacent to the roadside. To see some of the shyer species like **Fawn-breasted** and **Slaty-backed Thrushes**, and the **Tanimbar Flycatcher**, one needs to explore the smaller trails and logging roads that branch off the main road. Heading further north, a side track that leads pass a military camp is where some lucky birdwatchers have seen the rare **Moluccan Masked Owl**, on top of the **Tanimbar Boobook** and many forest species.

Lermatang

A mosaic of cultivation, scrub and forest, Lermatang can be accessed by boat from Saumlaki and is best tackled as a day trip. Exploring the recently constructed roads, **Bonelli's Eagle** (ssp. *renschi*) may be seen, as well as various parrots and pigeons in flight or resting on high, open perches. A network of narrow trails enters the forest here and it is possible to see most of the forest species, including the **Black-bibbed Monarch**, **Tanimbar Bush Warbler**, **Slaty-backed Thrush** and – for the lucky few – the shy **Tanimbar Megapode**.

Saumlaki

Remnant patches of mangroves and open mudflats line some parts of Saumlaki Bay near the city. Waders, terns, cormorants and migrant **White-faced Herons** and **Australian Pelicans** are regularly seen.

Dubi Shapiro

The endemic subspecies (ssp. castus*) of the* ***White-naped Monarch*** *dwells in the evergreen forests of the Tanimbar Islands.*

James Eaton

Unlike the more adaptable Cinnamon-tailed Fantail, the ***Long-tailed Fantail*** *prefers closed canopy forests.*

With two endemic thrushes up for grabs, including this ***Fawn-breasted Thrush****, a trip to the Tanimbar Islands is a must for serious birdwatchers.*

Access & Accommodation

The Tanimbars can be reached on daily flights between Pattimura Airport (Ambon) and Saumlaki. Flights also connect Saumlaki and Langgur in the Kai Islands, and many birdwatchers plan trips to visit both island groups. A variety of accommodation is available in Saumlaki, but most birdwatchers base themselves at the Harapan Indah Hotel, whose owner also owns the 'Kebun 45' property. Cars, pick-ups and motorcycles can be rented in town, while public buses that regularly ply the main Yamdena route can provide a budget option to access some of the forest sites. If attempting to access new areas of forest, seek local advice on whether there are any customary requirements to enter forested areas on the island, as the forest is held in high regard by many traditional local communities.

Conservation

Compared with many islands in the Lesser Sundas, the Tanimbars still retain a high proportion of their forest cover. However, due to the paucity of formally gazetted conservation areas in the archipelago, much of these forests remain at the mercy of loggers. Logging at a small scale, usually by community-based operations for a few species of commercially valuable timber, is gradually degrading Yamdena's forest. Logging also opens up areas of forest that can be accessed by farmers or poachers who trap various parrots, especially the Tanimbar Corella and Blue-streaked Lory.

Dubi Shapiro

The handsome ***Slaty-backed Thrush*** *is fairly common in the forests on the Tanimbar Islands.*

Kai (Kei) Islands

Colin R. Trainor & Yann Muzika

Much of low-lying Kai Kecil has been cleared, leaving many remnant patches of forest and woodland.

KEY FACTS

Nearest Major Town
Tual (South Maluku)

Habitats
Tropical dry forest, secondary forest, mangroves, beach and coastal scrub

Key Species
Pheasant Coucal (ssp. *spilopterus*), Yellow-capped Pygmy Parrot, Papuan Pitta, Tanimbar Friarbird*, Kai Cicadabird*, Island and Grey Whistlers, Australasian Figbird, Wallacean Drongo (ssp. *megalornis*)*, Northern Fantail, White-tailed*, Island and White-naped Monarchs, Island Leaf Warbler, Pearl-bellied* and Golden-bellied White-eyes*, Cinnamon-chested Flycatcher* and Mistletoebird (ssp. *keiense*)
Winter Oriental Plover, Little Curlew

Best Time to Visit
May–October (dry season); rainy season November–April

Dominated by Kai Kecil and Kai Besar and encompassing 50 other smaller islands, the Kai (Kei) Islands lie at the easternmost margin of Wallacea about 560km south-east of Ambon and 200km north-east of the Tanimbars. The islands are dominated by Kai Kecil which is flat and low-lying, and Kai Besar which is very hilly. With the current trend of elevating endemic subspecies of islands and archipelago to species level, the Kai islands are increasingly recognized for their distinctive avifauna with several endemic taxa. The endemic birds of Kai are fairly easy to see, with the endemic subspecies of the **Pheasant Coucal** being found in secondary forest and scrub on Kai Kecil, the **Pearl-bellied White-eye** in most habitats on Kai Besar and the **Golden-bellied White-eye** in many habitats on Kai Kecil. The **White-tailed Monarch** occurs on both islands in woodland and forest, from sea level to the hills. Additionally, small satellite islands such as Tayandu support a number of 'small-island specialists' such as the **Island Monarch**, **Elegant Pitta** and **Scaly-breasted Honeyeater**. For visiting birdwatchers, taking special notes on the endemic subspecies may prove worthwhile as many of these taxa could eventually be raised to species level.

Birdwatching Sites

Kai Kecil

Some of the regularly visited sites by birdwatchers on Kai Kecil include Ohoililir Beach, Taman Anggrek Lake, Danau Ngadi, Samawi and Polisi forest, as well as the remnant forest and scrub in and around Tual Airport. However, secondary forest and woodland occur widely across the island, so

*The distinctive local subspecies of the **Pheasant Coucal** is considered by some authorities as a full species, the Kai Coucal.*

Yann Muzika

seeing the endemics and other forest birds can be fairly straightforward. Freshwater wetlands such as Taman Anggrek support a number of Australasian waterbird species, including the **Green Pygmy Goose**, **Pied Heron** and **Australian Pelican**. The larger areas of forest further away from Tual, such as those in the south, can be expected to support the full complement of forest species, including the **Orange-footed Scrubfowl**, imperial pigeons and fruit doves.

Tual Airport

From September to November, open areas and the scrub around the old Dumatubun Airport can be good spot to search for uncommon Palearctic shorebirds such as the **Oriental Plover** (more than 50 birds have been recorded) and **Little Curlew** (up to 160 birds have been recorded). Since 2014, birdwatchers have concentrated their efforts at the new Karel Sadsuitubun Airport, which is located about 10km to the south. Patches of remnant woodland on the new 'airport road' provide easy viewing opportunities for a good range of woodland and forest birds, including the endemic subspecies of the **Pheasant Coucal**.

Kai Besar

The most accessible site on Kai Besar is the Bukit Indah Mission and its surrounding environs, which contains good-quality evergreen forests. The **Pearl-bellied White-eye** is usually easy to locate. On the other hand, it would require a hike to about 400m asl before the **Island Leaf Warbler** (ssp. *avicola*, or 'Kai Leaf Warbler') would be likely to be found. The **White-tailed Monarch** and the endemic form of the **Northern Fantail** (ssp. *assimilis*, or 'Kai Fantail'), as well as the **Eclectus Parrot** and **Great-billed Parrot** may also be spotted at this site. Kai Besar is relatively little explored and it is likely that any site with reasonable forest cover will harbour the endemics. The main challenge most birdwatchers face is likely to be getting easy access to forest over 400m in elevation.

Yann Muzika

*The **Pearl-bellied White-eye** is the more commonly encountered of the two endemic white-eyes of the Kai Islands.*

Myron Tay

*The **Little Curlew** may be encountered in open grassy areas such as the airport.*

Access & Accomodation

The Kai islands are easily reached from Ambon, with several flights daily. Less reliable and far slower are the PELNI ferries from Ambon. Ferry rides might be particularly enticing for those keen on observing seabirds. A variety of basic accommodation is available in Tual, but a better option on Kai Kecil are the cottages at Ohoililir or Ohoidertawun, which also provide great opportunities for birding. Accessing Kai Besar is more difficult, with boats leaving Tual-Langgur and landing at Elat on Kai Besar. A variety of public boats may head for Elat, depending on demand. If these options do not suit your schedule, then chartering a boat may be necessary, costing in excess of 100 USD. There is a local hotel near the port at Elat, but there are few other accommodation options on the island, apart from camping, or asking permission from the village heads or kepala desa to stay in villages.

Conservation

In 1993, a 142km^2 nature reserve covering Gunung Daab on Kai Besar was gazetted by the Indonesian government. There is, however, little recent information on conservation activities and success on the islands.

Raja Ampat Islands

Shita Prativi and Low Bing Wen

The Raja Ampat Islands are one of the most visually spectacular birdwatching sites in the region.

KEY FACTS

Nearest Major Towns
Sorong, Waisai (West Papua)

Habitats
Lowland and hill rainforests; forested limestone islets

Key Species
Waigeo Brushturkey*(rare), Dusky Megapode*, Red-necked Crake, Western Crowned Pigeon*, Spice Imperial Pigeon*, Marbled and Papuan Frogmouths, Hook-billed, Common Paradise, Beach and Yellow-billed Kingfishers, Papuan Pitta, Spotted Honeyeater, Raja Ampat Pitohui*, Brown-headed Crow, Wilson's Bird-of-paradise*, Red Bird-of-paradise, Black-sided Robin
Winter Gray's Grasshopper Warbler

Other Specialities
Waigeo Cuscus, Northern Common Cuscus

Best Time to Visit
October–April is the drier time of year; but most groups visit from July onwards

The Raja Ampat Islands are an island archipelago comprising four main islands (Misool, Salawati, Batanta and Waigeo) and more than 1,500 smaller islands and islets scattered across the Bird's Head Seascape, an area widely regarded to have the richest marine biodiversity on Earth. These islands are also cloaked in dense forest that is home to a diverse array of species found nowhere else on Earth, including the exquisite **Wilson's Bird-of-paradise**, widely regarded as the best bird in the world. As is the case with many sites in Indonesian New Guinea, infrastructure improvements have made these islands much more accessible and comfortable for the visiting birdwatcher in recent years. Most birdwatchers now visit the island of Waigeo, the largest of the four main islands, and its tourist hub of Waisai. From here, it is easy to arrange visits to the surrounding forests to see the birds-of-paradise from specially constructed hides, along with a fine supporting cast of regional endemics including the **Western Crowned Pigeon**, **Spice Imperial Pigeon** and **Raja Ampat Pitohui**.

Birdwatching Sites

Bird-of-paradise Hides

A network of hides has been built by locals overlooking active leks for the island's two endemic birds-of-paradise – **Wilson's** and **Red Birds-of-paradise**. These hides can be reached after a drive of 30 minutes to an hour from Waisai, followed by a short walk. For the best results, these hides should be visited in the early morning. Most visiting birdwatchers allocate three mornings to visit the various hides for the best experience. While waiting for the birds-of-paradise to make an appearance, the **Dusky Megapode**, **Red-necked Crake** and **Black-sided Robin** may be seen

Chi'en C. Lee

*The striking **Wompoo Fruit Dove** may be seen in the rainforests of Waigeo.*

Arman A

The stunning ***Wilson's Bird-of-paradise*** *is widely regarded as one of the world's best birds.*

An endemic bird-of-paradise on these islands is the spectacular ***Red Bird-of-paradise****.*

from or around the hide. The other big prize in this area, the **Western Crowned Pigeon**, may be observed along the road or trails leading to the various hide. The scarce **Brown-headed Crow** may be seen flying over the road, while fruiting trees attract a variety of fruit doves, including **Wompoo**, **Beautiful** and **Claret-breasted Fruit Doves**. Mixed flocks in these forest comprise a variety of insectivores, including the **Yellow-breasted Boatbill**, **Raja Ampat Pitohui** and **Spot-winged Monarch**. The hides should be visited in the company of local guides from the Sapokren village, who are the owners of the land and have been actively involved in conserving the area's birdlife.

Kabui Bay

A popular tourist attraction featuring dozens of forested limestone islets located between Gam Island and south-western Waigeo, a boat ride around this area provides a nice change of pace from rainforest birding and a good way to spend an afternoon. The main draw here for birdwatchers is the localized **Spice Imperial Pigeon**, which returns to these islets to roost in the evening. The islets are also the roosting site for large numbers of **Great-billed Parrots**, which can also be seen together with the pigeons. During the boat trip, there is a chance to see coastal birds such as the **Great-billed Heron** and **Beach Kingfisher**.

Bjorn Olesen

The pretty ***Frilled Monarch*** *is a regular participant of mixed flocks here.*

*The **Spotted Honeyeater** is one of many species of honeyeater found on New Guinea.*

Mount Danai

One of the tallest peaks on the island at 950m asl, Mount Danai is located in remote eastern Waigeo and is seldom visited by either locals or tourists. However, the mountain supports most of the global population of the Endangered **Waigeo Brushturkey**. Birdwatchers with a lot of time and a sense of adventure can contemplate organizing a multi-day expedition into the highlands of this region in search of this mythical creature.

Roadside Birding near Sorong

Birdwatchers with time to spare between their forays to Waigeo can consider exploring the lowland forest near Sorong. A 1–2 hour drive east of the city provides access to lowland forest where species like the **Moluccan King Parrot**, **Black Lory**, **Large Fig Parrot** and **Lowland Peltops** are present. For more intrepid birdwatchers, spending a few nights in the Malagufuk village located in the Klasow Valley can be very rewarding for species such as the **Red-billed Brushturkey** and **Western Crowned Pigeon**, and the exquisite and rarely seen **Red-breasted Paradise Kingfisher**, as well as an active **Lesser Bird-of-paradise** lek, to name a few.

Access & Accommodation

Most birdwatchers access Waigeo via a 2–3 hour speedboat or ferry ride from the coastal city of Sorong, which is connected by air to other major Indonesian cities. However, it is also possible to take a direct flight from the city of Manado on Sulawesi to the town of Waisai on Waigeo. Upon arrival in Waisai, foreign visitors have to pay 1,000,000 IDR for an entry permit that is valid for one year from the date of purchase. Waisai is becoming an increasingly popular destination for divers and a range of accommodation options is now available. Transport arrangements to visit the various bird-of-paradise hides or offshore islands can be made with the accommodation provider. There are also accommodation options around Sapokren village.

Conservation

The entire island of Waigeo is divided into two nature reserves – West Waigeo Nature Reserve and East Waigeo Nature Reserve. The key birdwatching sites are in the West Waigeo Nature Reserve, and the Forestry Department is working towards making the area easier to visit for tourists. Although the island remains densely forested, logging concessions have been granted for the island's forests and hunting is a local threat for specific birds such as the brushturkey, crowned pigeon and cuscus.

Chi'en C. Lee

*The bizarre **Papuan Frogmouth** is widely distributed across New Guinea and its satellite islands.*

Arfak Mountains

Shita Prativi & Marc Thibault

Chi'en C. Lee

The rugged mountains of the Vogelkop Peninsula remain densely forested and support many regional endemics found nowhere else.

Perhaps the most recognizable geographical feature of New Guinea, the Bird's Head or Doberai Peninsula is low lying and swampy on its southern coast, yet extremely rugged and mountainous in the north, being capped by the formidable Arfak and Tamrau ranges. Just south-west of Manokwari, the imposing Arfak Mountains (Pengunungan Arfak) rise steeply out of the narrow coastal plain and reach nearly 3,000m asl at their highest ridges. Although development is beginning to encroach into the foothills, much of the Arfaks remains densely forested. Due to this, combined with its relative isolation from the larger 'torso' of New Guinea, the Arfaks support a large and distinctive bird fauna, including a number of localized endemics such as the **Vogelkop Owlet-nightjar**, **Vogelkop Whistler** and **Western Parotia**, along with sought-after montane endemics such as the **Feline Owlet-nightjar**, **Mottled Berryhunter** and **Black-billed Sicklebill**. The mid-montane forests here are very species rich, and it is possible to see a fantastic representation of pigeons, honeyeaters and parrots, as well as uniquely Papuan families such as the distinctive berrypeckers and painted berrypeckers. The jewel in the crown of the birding experience here is undoubtedly the opportunity to see the extravagant displays of numerous birds-of-paradise, ranging from the toy-like trot of the Western Parotia, to the more conventional dance of the **Magnificent Bird-of-paradise**, or simply to admire the superlative-defying **Black Sicklebill**. This experience is complemented by examining the impressive bowers built by the modest-looking **Vogelkop Bowerbird** and the fiery **Masked Bowerbird** at lower elevations. It is nearly impossible to omit the Arfaks in any field trip to Indonesian New Guinea.

Tamrau Mountains
PACIFIC OCEAN
Manokwari
Syoubri
Mt Mebo
Mupi Gunung
Anggi Lakes
Doberai (Bird's Head) Peninsula
Cenderawasih Bay

KEY FACTS

Nearest Major Town
Manokwari (West Papua)

Habitats
Hill, submontane and montane rainforests, scrubland, grassland

Key Species
Dwarf Cassowary, Wattled Brushturkey, Papuan Eagle, Black-mantled Goshawk, White-striped Forest Rail*, Cinnamon and Bronze Ground Doves, Ornate Fruit Dove, Papuan Mountain Pigeon, Rufous-throated and White-eared Bronze Cuckoos, Papuan Boobook, Feline, Mountain and Vogelkop Owlet-nightjars*, Pesquet's Parrot, Yellow-capped and Red-breasted Pygmy Parrots, Brehm's and Modest Tiger Parrots, Papuan Lorikeet (ssp. *papou*), Arfak Catbird*, Vogelkop* and Masked Bowerbirds, Papuan Treecreeper, Orange-crowned Fairywren, Rufous-sided Honeyeater*, Arfak Honeyeater*, Cinnamon-browed, Vogelkop* and Ornate Melidectes, Goldenface, Mountain Mouse-warbler, Vogelkop Scrubwren*, Papuan Logrunner, Mid-mountain and Fan-tailed Berrypeckers, Tit Berrypecker, Spotted and Chestnut-backed Jewel-babblers, Black-breasted Boatbill, Mountain Peltops, Mottled Berryhunter, Rufous-naped Bellbird, Black Pitohui, Vogelkop* and Regent Whistlers, Crinkle-collared Manucode, Long-tailed Paradigalla*, Arfak Astrapia*, Western Parotia*, Superb Bird-of-paradise (ssp. *superba*) Black and Black-billed Sicklebills, Magnificent Bird-of-paradise, Ashy, Smoky, Slaty and Green-backed Robins, Garnet and Lesser Ground Robins, Grey-banded Mannikin*

Other Specialities
Black Tree-kangaroo*, Reclusive Ringtail Possum*, Striped Possum

Best Time to Visit
May–October (best time for displaying birds-of-paradise); December–April are the wettest months

Arman A

Seeing the captivating display of the ***Western Parotia*** *is arguably the biggest attraction for visiting birdwatchers.*

Birdwatching Sites

Sioubri Area

Most birdwatching trips to the Arfak Mountains start from the settlement of Syoubri, which sits at around 1,500m asl, after climbing the undulating foothills south of Manokwari. Made convenient because of a number of accommodation possibilities, the key sites here are the various established blinds set up to observe birds-of-paradise and other species. The access road to Syoubri itself is also excellent for seeing a good selection of species, including pigeons, lorikeets (like the **Fairy Lorikeet**) and, with luck, **Pesquet's Parrot**. At Syoubri, most visitors can expect to spend some time at the blinds scattered in the forest around the village. The aptly named 'Magnificent Bird-of-paradise hide', accessed from the Syoubri road, is a must visit, and most trips here result in observing the displays of the **Magnificent Bird-of-paradise**. Other species that may be seen in and around the hide include the **White-striped Forest Rail**, **Cinnamon Ground Dove** and **Green-backed Robin**.

Yann Muzika

The skulking ***White-striped Forest Rail*** *is one of the region's many avian endemics.*

A steep hike into the montane forest above Syoubri leads on to the most frequented of the Parotia hides, another focal point of most visits. Not only are visitors able to see the bizarre and somewhat mechanical display of the **Western Parotia** with regularity; the **Wattled Brushturkey** and **Greater Sooty Owl** are also possible, along with more common species such as the **Vogelkop Whistler**.

Minggrei & Maibri Area

Minggrei (1,500m asl) is located to the right of the road just before the turn-off to Mokwam and Syoubri. Maibri sits slightly lower and before Minggrei on the road at an elevation of 1,100m. The mid-montane forests at Maibri contain stakeouts for the **Western Parotia**, **Black-billed Sicklebill** and **Magnificent Bird-of-paradise**. The **Black-eared Catbird**, **Masked Bowerbird**, **Papuan Scrub Robin** and very rare **Vogelkop Owlet-nightjar** are all possible here. In the forests below Maibri, the **Papuan Boobook**, **Papuan Frogmouth**, **Feline Owlet-nightjar** and **Magnificent Riflebird** have all been seen. Minggrei village is surrounded by excellent forest and many of the key species occur here, including the endemic **White-striped Forest Rail**, **Pesquet's Parrot**, **Papuan Logrunner** (ssp. *novaeguineae*), **Spotted Jewel-babbler**, **Lesser Melampitta** and **Black-billed Sicklebill**, and the endemic subspecies of the **Superb Bird-of-paradise** (ssp. *superba*), which can be seen from a hide.

Trails above Syoubri to Camp 'German', Camp 'Japan'

To see the remaining birds-of-paradise and other specialities, a strenuous hike into the high-elevation forest is necessary. Most birdwatchers base themselves at the 'German' and 'Japan' Camp area, both extensive set-ups prepared by local people to enable access to the cloud forest, where most eventually connect with displaying **Black Sicklebills**, **Arfak Astrapias** and – with a bit of luck – the elusive **Long-tailed Paradigalla**. About 20 minutes' hike from 'German' Camp past a couple of the impressive bowers of the rather drab-looking **Vogelkop Bowerbird** is the frequented display post of the Black Sicklebill, while seeing Arfak Astrapia requires a hike to the forests above

Chi'en C. Lee

There are several observation hides in the area from which to see displaying ***Magnificent Bird-of-paradise****.*

'Japan' Camp. The hike to 'German' Camp from Syoubri passes through excellent forest, and birdwatchers should look out for specialities such as the unusual **Mottled Berryhunter**, **Rufous-naped Bellbird**, **Tit Berrypecker**, **Garnet Robin**, **Canary Flyrobin**, both tiger parrots in flight and – with help from the local guides – both the **Feline** and **Mountain Owlet-nightjar** sitting at their daytime roosts. The viewpoint at 'German' Camp should be checked for soaring birds-of-prey such as the **Black-mantled Goshawk**.

Minyambou Area

Covered with cloud forests at 2,350m, this is a newly opened site for birdwatchers and is where **Vogelkop Bowerbirds** and their bowers, among other high-elevation species, can be seen. The bowers can be accessed from a short but steep hike from Minyambou.

Anggi Lakes Area

This area is known for its beautiful view of two high-elevation lakes, Anggi Gigi and Anggi Gida. At an elevation of about 2,500m asl, the grassy areas and reedbeds surrounding the lakes are where the endemic **Grey-banded Mannikin** can be found. The Anggi lakes can be reached after a two-hour drive on rough roads (best on 4WD) from the Syoubri area.

Mupi Gunung Area

Located about 30km south of Manokwari, this little settlement is now proving to be a viable, if not better alternative to the traditionally visited sites around Syoubri. Access involves a hike starting from the lowlands near the coast to well over 1,100m asl, allowing access to lower-elevation forest. Besides a number of species shared with Syoubri, including the birds-of-paradise, visitors here have reported lowland species such as the **Yellow-billed Kingfisher** and **Papuan Pitta**. Forests at mid-

Chi'en C. Lee

The male ***Vogelkop Bowerbird*** *constructs a highly elaborate bower in a bid to attract prospective females.*

Chien C. Lee

The local villagers have an uncanny ability to locate roosting owlet-nightjars, such as this ***Feline Owlet-nightjar****.*

montane elevations (1,100–1,300m asl) support the **Pheasant Pigeon**, **Wallace's Fairywren**, **Chestnut-backed Jewel-babbler**, **Frilled Monarch**, **Papuan Scrub Robin**, **Crinkle-collared** and **Trumpet Manucodes**, and **Magnificent Riflebird**, among others. The Black Tree-kangaroo is regularly encountered around the settlement.

Access & Accommodation

Most trips to the Arfaks start from Manokwari. Given the rough quality of the road, 4WD vehicles are needed for the journey from Manokwari to Syoubri, where most birdwatchers spend a number of days. Visitors mostly stay at the Syoubri guesthouse, but accommodation is also available at the Papua Lorikeet Guesthouse further down the road towards Manokwari. Most trips in the Syoubri area are arranged through Zeth Wonggor, who runs the main guesthouse and sorts out vehicle arrangements, porters and local guides. The trek to the remote Mupi Gunung area should be organized with the help of Hans Mandacan. At Minggrei, villagers have converted their houses to homestays to cater to tourists. Currently six homestays are available, each with two double bedrooms. Visitors to Minggrei are guided by the village head and his sons, who all know the area very well. Homestays are not available at Maibri and visitors may have to stay with villagers here.

Conservation

A significant portion of the Arfaks Mountains is protected within the Arfak Range (Pengunungan Arfak) Nature Reserve, which covers about 683km^2 of cloud forest above 2,000m elevation, but few, if any, of the key birdwatching sites fall within the reserve. However, since much of the forests are inaccessible, there is little encroachment at the moment. Bird and butterfly poaching is a localized problem along the fringes of the reserve. Local people across various villages are contributing to bird conservation by protecting key areas of forests themselves and encouraging local ecotourism. The Papua Bird Club works closely with local communities to support livelihoods and promote sustainable ecotourism.

Arman A

The beautiful ***Spotted Jewel-babbler*** *is usually glimpsed as it moves along the forest floor.*

Biak & Numfor Islands

Carlos Bocos & Marc Thibault

Daniel López Velasco

The Geelvink Islands are densely forested and support many endemic species.

KEY FACTS

Nearest Major Town
Kota Biak (Numfor Regency)

Habitats
Lowland rainforest (primary and secondary), mangroves, scrub

Key Species
Biak Scrubfowl*, Yellow-bibbed (Geelvink) Fruit Dove*, Spice (Geelvink) Imperial Pigeon*, Biak Coucal*, Biak Scops Owl*, Biak Paradise Kingfisher*, Numfor Paradise Kingfisher*, Geelvink Pygmy Parrot*, Black-winged Lory*, Biak Lorikeet*, Hooded Pitta, Dusky (Red-brown) Myzomela*, Biak Gerygone*, Common (Geelvink) Cicadabird*, Black-browed (Biak) Triller*, Biak Monarch*, Biak Black Flycatcher*, Island Leaf Warbler, Biak White-eye*, Long-tailed Starling*

Other Specialities
Biak Spotted Cuscus, Biak Glider

Best Time to Visit
All year round

Daniel López Velasco

*The striking **Biak Monarch** is one of the more difficult endemics to observe.*

Located off northwestern New Guinea in the sprawling Cendrawasih Bay, the relatively low-lying islands of Biak and Numfor support an impressive array of endemic species and are the main birdwatching hotspots in the Geelvink archipelago. Being the more densely populated of the two islands, extensive clearance for agriculture means that little primary rainforest is left on Biak. In general, most of Biak's specialities are relatively widespread and easy to find, but there are a few species that are scarce and are localized to remnant forest patches in the north of the island. Unlike Biak, Numfor is sparsely populated and the heart of the island still contains some excellent areas of undisturbed rainforest. A number of the Geelvink endemics are more common here than on Biak, especially the **Biak Scrubfowl** and **Geelvink Pygmy Parrot**. Seeing a good selection of the Geelvink endemics would thus necessitate a trip to both islands.

Birdwatching Sites

Remnant Forests on Biak

Most of the remnant forest patches to the east of Kota Biak host the main target species and can be reached within 1–2 hours from town. Areas with tall fruiting or flowering trees can be expected to have good numbers of **Black-winged Lory**, **Biak Lorikeet** and the relatively common **Yellow-bibbed (Geelvink) Fruit Dove**. While the **Spice (Geelvink) Imperial Pigeon** is reasonably common, it tends to be very shy due to hunting pressure. The **Biak Coucal** is frequently heard but this shy skulker is usually not easy to see. The sought-after **Biak Paradise Kingfisher** is common in most of the forest on the island. The endemic subspecies of the **Hooded Pitta** (ssp. *rosenbergii*) is also common, and frequently heard at dawn and dusk. Feeding flocks on Biak usually contain the **Black-browed (Biak) Triller**,

Daniel López Velasco

The elegant ***Biak Paradise Kingfisher*** *is pleasingly common in Biak's remaining forest.*

Bjorn Olesen

The highly distinctive ***Raja Shelduck*** *is present in the coastal wetlands of Biak.*

Biak Gerygone, **Biak Black Flycatcher** and sometimes the very scarce **Biak Monarch** and **Island Leaf Warbler** (ssp. *misoriensis*). One of the rarest endemics is the **Geelvink Pygmy Parrot**, which nests in old termite mounds on trees and is more common on Numfor. At night, with some luck the uncommon **Biak Scops Owl**, a species possibly more common than previously thought, may also be seen.

Scrub

The **Biak White-eye** is the main target in this habitat. Small parties of this big *Zosterops* can be located easily by their contact calls. The rare **Dusky (Red-brown) Myzomela** also frequents this habitat and the forest edge. Sometimes, the **Biak Scrubfowl** forages in scrub in the late afternoon.

Coasts & Mangrove Forest

Along the coasts, the mangrove forests are home to the **Raja Shelduck**, **Claret-breasted Fruit Dove**, **Torresian Imperial Pigeon**, **Beach Kingfisher** and good numbers of waders and terns, especially during the northern winter.

Forests on Numfor

On Numfor, both the **Numfor Paradise Kingfisher** and **Island Leaf Warbler** (ssp. *maforensis*) are readily seen as soon you enter the forest. Other interesting targets there are the local forms of the **Hooded Pitta** (ssp. *mefoorana*) and the **Barred Cuckooshrike** (ssp. *maforensis*). Most specialities can be seen in forest patches that are readily accessible from the main road circumnavigating the island.

Daniel López Velasco

The distinctive subspecies of the ***Island Leaf Warbler*** *on Numfor is a potential future split.*

although the rare Cuckooshrike is best searched for from a track leading to the island's interior that can be accessed just south of Namber village, on the south-west of the island.

Access & Accommodation

Biak has daily flights from several airports in Indonesia. Many hotels and hostels are to be found in the main town, Kota Biak, although they are sometimes full and are best booked in advance. It is fairly easy to rent a car with driver but no one speaks English. There is also the option of public transport, although it does not cover all the birding areas. Numfor is much more difficult to access. Flights are irregular and sometimes unreliable. Another option to visit Numfor is by sea. There is a PELNI ferry service from Manokwari, but most visitors opt for the rather expensive option of chartering a speedboat from Biak. The sea is often rough and travellers have to assess that the conditions are safe enough to cross. Accommodation on Numfor is limited to a few homestays and guesthouses that are to be found in the main town, Yenburwo (Jenboeroeo). There are not many vehicles and renting a car in advance is highly unlikely, but it might be arranged on the spot with the help of one of the guesthouses' owners, in order to access the good-quality forest to the south of the island. Alternatively, exploring the small patches of forest accessible by foot from Yenburwo might yield a good selection of Numfor's specialities.

Conservation

There is no effective protection at any site on both islands. Logging and farming are widespread activities, while hunting threatens medium-sized birds like scrubfowls, pigeons and fruit doves. There is also incipient trapping for the cage-bird trade, mainly associated with the Javanese community. This is already affecting species like the globally threatened Biak Scops Owl, Hooded Butcherbird and Biak White-eye.

Daniel López Velasco

*The stunning **Numfor Paradise Kingfisher** is easily seen in suitable habitat on Numfor.*

*The globally threatened **Biak Scops Owl** inhabits remnant forest fragments on Biak.*

Maoke (Snow) Mountains

Marc Thibault & Shita Prativi

The rainforest and grassland around Lake Habema provide a stunning backdrop for observing the spectacular birds of the region.

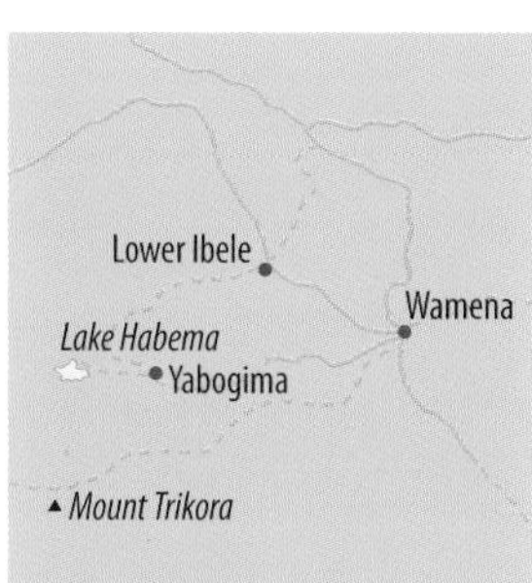

KEY FACTS

Nearest Major Town
Wamena (Papua)

Habitats
Montane rainforest and scrubland, alpine grassland and shrubland, alpine scree

Key Species
Salvadori's Teal, Snow Mountain Quail*, Chestnut Forest Rail, New Guinea Woodcock, Brehm's and Painted Tiger Parrots (ssp. *lorentzi*), Papuan Lorikeet (ssp. *goliathina*), Archbold's Nightjar, Mountain Kingfisher, Archbold's and MacGregor's Bowerbirds, MacGregor's Honeyeater, Sooty and Belford's Melidectes, Orange-cheeked Honeyeater*, Mountain Mouse-warbler, Papuan Logrunner, Loria's and Crested Satinbirds, Crested Berrypecker (ssp. *olivacea*), Great Woodswallow, Hooded Cuckooshrike, Black Sittella, Wattled Ploughbill, Baliem* and Lorentz's Whistlers, Lesser Melampitta, Blue-capped Ifrit, Short-tailed Paradigalla, Splendid Astrapia, King of Saxony and Superb Birds-of-paradise, Brown Sicklebill, Snow Mountain Robin*, Mountain Robin, Greater Ground Robin, Mountain Firetail, Black-breasted Mannikin*, Western Alpine Mannikin, Alpine Pipit

Other Specialities
Speckled Dasyure, Silky Cuscus, New Guinea Quoll

Best Time to Visit
June–November are slightly cooler with lower rainfall

The Maoke (Snow) Mountains are part of the New Guinea's Central Cordilleras and extend across roughly 700km of Indonesian New Guinea, with several peaks rising to well over 4,000m asl. Puncak Jaya, Indonesia's tallest mountain at 4,884m asl, lies within the Maoke Mountains. In birdwatcher lingo, the Snow Mountains collectively refers to several birdwatching sites around the Baliem Valley. First discovered by Richard Archbold in 1938, the densely populated (by Papuan standards) valley and the associated tourist hub of Wamena offer easy access to a range of elevations and habitats in the surrounding mountains. With a variety of regional endemics and sought-after species such as the **Snow Mountain Quail**, **MacGregor's Honeyeater**, **Wattled Ploughbill** and the mythical **Snow Mountain Robin**, set against a backdrop of breathtaking scenery, the Baliem Valley has become a cornerstone of birdwatching trips to Indonesian New Guinea.

Birdwatching Sites

Pondok Tiga (Camp 3) Area & Lake Habema

The Pondok Tiga campsite (now no longer in use) is located close to the start of Ibele Trail (more below) and has hosted birdwatchers for decades now. It offers a convenient base to explore both the upper sections of the Ibele Trail and Lake

Daniel López Velasco

One of the more underrated birds-of-paradise found on New Guinea is the ***King of Saxony Bird-of-paradise****.*

Habema. The grassland and shrubland around the lake are home to the **Papuan Harrier**, **Snow Mountain Quail**, uncommon **Sooty Melidectes**, **Orange-cheeked Honeyeater**, **Mountain Firetail**, **Western Alpine Mannikin** and **Alpine Pipit**. The lake itself should be checked for **Salvadori's Teal**, while the shy **Spotless Crake** may be seen in the tall marsh vegetation surrounding the eastern side of the lake. Patient scanning at wooded clearings around the lake may be rewarded with good, albeit usually distant, views of the bizarre **MacGregor's Honeyeater**. Another noteworthy species that should be looked for in this area is the endemic subspecies of the **Painted Tiger Parrot** (ssp. *lorentzi*), a potential future split.

Ibele Trail to Yabogima Camp

The focal point of any birdwatching trip to the Snow Mountains, early groups of birdwatchers used to do a multi-day trek downhill from the top of the Ibele Trail near Lake Habema all the way back to Wamena. Due to improved accessibility, it is more productive to now concentrate efforts within a 5km stretch that runs from the top of the trail (3,300m asl) to the area around Yabogima Camp (2,810m asl), which passes through excellent montane rainforest. The open area at the start of the trail is the place to look for the roding **New Guinea Woodcock** and **Archbold's Nightjar** at dusk. Just further down the trail, the forest is notable for many sought-after skulkers, including the **Chestnut Forest Rail**, **Papuan Logrunner**, **Lesser Melampitta** and **Greater Ground Robin**. Other species to look out for here include the **Crested Satinbird**, monotypic **Wattled Ploughbill**, **Lorentz's Whistler** and **Splendid Astrapia**. The forest around Yabogima Camp, located at a lower elevation, has its own suite of birds. These include **Archbold's Bowerbird**, **Loria's Satinbird**, **Black Sittella**, **Torrent-lark**, the monotypic **Blue-capped Ifrit** and the exquisite **King of Saxony Bird-of-paradise**, to name a few.

Puncak Trikora

One of the tallest peaks on the Maoke Mountains at 4,750m asl, this strenuous multi-day hike is not for the faint of heart. The singular reward is seeing the mythical **Snow Mountain Robin**, a regional endemic confined to the alpine scree zone above 4,000m asl. On average, a 2–3 day return expedition is required from Pondok Tiga, depending on fitness levels. The trek passes through large tracts of alpine grassland and shrubland where local specialities such as the **Snow Mountain Quail**, **Mountain Robin** and **Western Alpine Mannikin** may be seen.

The rainforest around Lake Habema is home to the unusual ***MacGregor's Honeyeater****, once thought to be a bird-of-paradise.*

*The local subspecies of the **Painted Tiger Parrot** is recognized as a full species by some authorities.*

Roadside Birding along Wamena & Lake Habema

Birding along the road between Wamena and Lake Habema can be very productive and a nice change of pace, although logging in the lower elevation areas is a problem. Not far from the start of the Ibele trail, the top of the road is good for the **Mountain Robin** and **Mountain Firetail**. The scrubby areas and remnant forest along the highway close to Wamena are good places to search for some of the regional endemics such as the **Baliem Whistler** and **Black-breasted Mannikin**, as well as the **Superb Bird-of-paradise**. Further along the highway there are some trails that enter degraded forest by the road. Exploration of these trails may be rewarded with species not found at higher elevations, such as the **Mountain Kingfisher**, **MacGregor's Bowerbird**, **Short-tailed Paradigalla** and **Brown Sicklebill**.

Access & Accommodation

Access to Wamena, the area's main town, has improved significantly in recent years – it is now a tourist hub with amenities such as a new airport terminal serviced by a variety of airlines, a range of accommodation options and vastly improved roads. As such, it is now possible for birdwatchers to base themselves within the township and make daily forays to many of the birdwatching sites (approximately an hour's drive on paved roads). However, for the serious birdwatcher seeking the area's montane endemics, camping remains the preferred option. There is a range of local companies based in Wamena that can facilitate this, and arrangements can be made either in advance or on arrival. Arrangements should include a permit to visit the area around Lake Habema and parts of the access road, which are within the boundaries of the massive Lorentz National Park, Indonesia's largest protected area.

Conservation

The Snow Mountains and its associated ecosystems have remained relatively untouched due to the traditional lifestyle and low density of the area's human population. In more recent times, infrastructure projects like the construction of the Trans-Irian Highway have cleared large areas of habitat, and greater accessibility to the areas's habitats brings with it other threats such as illegal logging and hunting.

Daniel López Velasco

*The aptly named **Splendid Astrapia** is restricted to the montane rainforest along New Guinea's rugged Central Cordillera.*

Daniel López Velasco

*The **Western Alpine Mannikin** is regularly seen in alpine grassland around Lake Habema.*

Nimbokrang Forests

Marc Thibault & Shita Prativi

Daniel Lopez Velasco

*The **Victoria Crowned Pigeon** is one of the most difficult of Nimbokrang's specialities to observe.*

Swarms of Malaria-carrying mosquitos, extreme heat and humidity, birds that are impossible to see – these were some of the phrases birdwatchers have used to describe Nimbokrang. And yet, with large tracts of lowland rainforest close to the bustling and well-connected city of Jayapura, Nimbokrang is also arguably one of the more accessible sites in Indonesian New Guinea to come to grips with the island's spectacular lowland rainforest bird fauna. The birding here can be very difficult but features a long list of possible species, including **King** and **Twelve-wired Birds-of-paradise** displaying at leks, a supporting cast featuring the likes of the **Papuan Eagle**, **Victoria Crowned Pigeon**, **Shovel-billed Kookaburra** and **Blue Jewel-babbler**, and the chance to glimpse the mythical **Northern Cassowary**. Consequently, it is not hard to see why Nimbokrang has become a must-visit destination for many birdwatchers exploring Indonesian New Guinea.

Birdwatching Sites

Secondary Forest around Nimbokrang Town

There are still significant areas of accessible secondary forest around the town of Nimbokrang. Although degraded, many of the area's specialities can still be encountered in these forests. Part of the forest around the southwestern entrance to the town is the property of Alex Waisimon and contains active display leks of **King**, **Twelve-wired** and **Lesser Birds-of-paradise**. These are best visited in the early morning when male birds are likely to be seen at close range displaying from observation towers. While exploring the area, birdwatchers may also encounter species such as **Yellow-billed** and **Papuan Dwarf Kingfishers**, **Buff-faced**

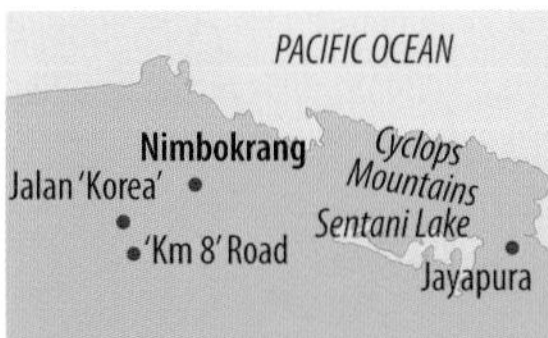

KEY FACTS

Nearest Major Town
Jayapura (Papua)

Habitats
Lowland rainforest and swamp forest

Key Species
Northern Cassowary, Collared Brushturkey, New Guinea Scrubfowl, Papuan Eagle, Victoria Crowned Pigeon, Papuan Hawk-Owl, Shovel-billed Kookaburra, Blue-black and Hook-billed Kingfishers, Palm Cockatoo, Brown and Black-capped Lories, Salvadori's Fig Parrot*, Papuan King-Parrot, Buff-faced Pygmy Parrot, White-crowned Cuckoo, Tan-capped Catbird, Blue Jewel-babbler, Lowland Peltops, Golden Cuckooshrike, Papuan Babbler, Ochre-collared Monarch, Sooty and White-bellied Thicket Fantails, Jobi Manucode, Pale-billed Sicklebill, Magnificent Riflebird, Pale-billed Sicklebill, King Bird-of-paradise, Twelve-wired Bird-of-paradise, Lesser Bird-of-paradise

Best Time to Visit
June–October; but rain is possible at any time

*The tiny **Buff-faced Pygmy Parrot** is the world's smallest species of parrot.*

Daniel López Velasco

*Nimbokrang is one of the few places in New Guinea where the enigmatic **Papuan Hawk-Owl** is seen with some regularity.*

Pygmy Parrot, **Blue Jewel-babbler**, **Ochre-collared Monarch** and **Pale-billed Sicklebill**. A late-afternoon foray might even yield the shy **Collared Brushturkey**. This area also features several low hills where birdwatchers can spend a relaxing afternoon scanning the surrounding areas for various pigeons and parrots, including the **Brown Lory** and **Salvadori's Fig Parrot**. To the town's east lie wetter forest where the enigmatic **Shovel-billed Kookaburra** is known to occur. This species is best detected when it calls just before dawn. Other specialities in this area include the **White-crowned Cuckoo**, **Hook-billed** and **Blue-black Kingfishers**, and **Tan-capped Catbird**. Night birding in the area may yield both the rarely seen **Papuan Hawk-Owl** and the more common **Marbled Frogmouth**.

Jalan 'Korea'

So named for the Korean loggers who used this logging road in the past, much of the forest here has been selectively logged, with local logging

*The attractive **Yellow-billed Kingfisher** is one of several kingfisher species that inhabit Nimbokrang's forests.*

Chi'en C. Lee

*Birdwatchers can enjoy the spectacular display of the **King Bird-of-paradise** from well-positioned observation towers.*

operations still being carried out. Nevertheless, it is still a good area for birdwatching for the time being. There are several active mounds of the elusive **Collared Brushturkey** in this area but seeing one requires a great deal of luck. This site is also a good alternative for **Pale-billed Sicklebill**. Mixed foraging flocks in the area can contain the **Black Berrypecker**, **Grey Whistler** and a variety of monarchs, including the **Spot-winged Monarch**, and the exquisite **Golden** and **Ochre-collared Monarchs**.

Km 8

The low hills in this area are cloaked in dense, largely primary rainforest, which might be the best area to look for the **Papuan King Parrot**. The **Magnificent Riflebird** and **Magnificent Bird-of-paradise** can be seen displaying from hides conveniently set up near their leks. Mixed species flocks should be searched for a variety of arboreal species, including **Hooded** and **Ochre-collared Monarchs**. Ground-dwellers frequenting this area include the **Cinnamon Ground Dove**, **Papuan Pitta** (ssp. *habenichti*) and **Blue Jewel-babbler**; active mounds of the **Collared Brushturkey** are also found here. Further along the road, a viewpoint at Km 9 overlooking large tracts of lowland forest is best visited in the early morning or late afternoon for pigeons, mynas and parrots including **Salvadori's Fig Parrot**.

'Northern' Swamp Forest

This area is where many birdwatchers try their luck for two of the most difficult and sought-after birds in Nimbokrang – the **Victoria Crowned Pigeon** and **Northern Cassowary**. The usual strategy for finding both these species is dedicating at least one full day to search the area's forest in the company of local guides. The pigeons are terrestrial but have a tendency to flush and perch in nearby trees when disturbed. In the case of the cassowary, fresh footprints and droppings are usually the closest most people come to seeing one. However, camping in the forest overnight, especially close to an active fruiting tree, will dramatically increase your chances of encountering this mythical bird. While searching for these ground-dwellers, birdwatchers are also liable to be distracted by a suite of lowland species such as the **New Guinea Scrubfowl**, **Papuan Eagle**, **Blue-black Kingfisher**, **Palm Cockatoo**, **Black-capped Lory**, **Golden Cuckooshrike** and **Magnificent Riflebird**.

Access & Accommodation

Nimbokrang is usually accessed via a three-hour drive from Jayapura, the provincial capital of Papua province, which in turn is easily accessed from many Indonesian cities by air. Accommodation options are limited and most birdwatchers opt to stay either at the homestay of Pak Jamil, who also acts as a local guide for the area, or at another homestay recently set up by Alex Waisimon. The latter's homestay is increasingly popular and provides direct access to his property. An alternative would be to camp in the forest at one of the abandoned logging camps. Transport to other birding sites from Nimbokrang are either by motorbikes or vehicles that can be arranged through the homestays.

Conservation

Although industrial-scale logging of Nimbokrang's rainforest ceased some time ago, local logging operations are ongoing and continue to chip away at the area's forests. Birdwatchers are likely to utilize the same trails made of planks that local loggers use to transport timber out of the forest. Hunting for large birds and mammals for the bush-meat trade is widespread in the area and probably has an effect on large-bodied, ground-dwelling birds, in particular the Northern Cassowary, Collared Brushturkey and Victoria Crowned Pigeon.

*The elegant **Pink-spotted Fruit Dove** is best observed by staking out fruiting trees in the forest.*

*A key avian attraction around Nimbokrang is observing displaying **Twelve-wired Bird-of-paradise** on top of a bare stump in the early morning.*

LAOS

Lao People's Democratic Republic (PDR) is the only landlocked country in Southeast Asia. Much of Laos is very rugged, with its eastern frontier with Vietnam defined by the long Annamite (Truong Son) Mountains, which rise to more than 2,800m asl at Phou Bia, and on its north-west border with Thailand by the Luang Prabang Range. In the far south near the border with Cambodia is the Bolaven Plateau, an upland region that rises to more than 1,100m asl. The middle and lower courses of the Mekong River wind through much of the country, flowing through rapids and limestone country, and form much of the international border with Thailand.

For many years, Laos has been largely outside the radar of global birdwatchers due to a general lack of information on its native biodiversity, and the fact that many species are more easily seen in other Southeast Asian countries. Biodiversity surveys carried out in remote parts of the country, like the pristine forested mountains of Nam Theun in the Annamites, have revealed species like the Crested Argus, White-winged Duck and Rufous-necked Hornbill, but these remain difficult to access due to the poor infrastructure. However, the recent discovery of the Bare-faced Bulbul, a country endemic and localized resident of wooded limestone outcrops, brought Laos into the ornithological limelight and has attracted an increasing number of birdwatchers in recent years. Nevertheless, the country is still under-visited, with opportunities for future discoveries. Although birdwatching at many sites can be slow going due to low bird density, the birdwatcher who perseveres is rewarded with a surprising number of species, including a suite of limestone-forest specialities that are difficult to see elsewhere in the region.

Climate

Laos experiences a tropical monsoon climate with distinct wet and dry seasons. The driest time of the yeaar occurs between November–April, during which temperatures gradually rise from around 20° C in December to more than 40° C. During the rainy season between May–October, the Mekong River may flood, inundating settlements along its banks.

Access, Transportation & Logistics

There are two main international gateways into Laos – the sleepy capital of Vientiane and the UNESCO-listed old city of Luang Prabang, both located in the country's north. A tourist visa is required for most foreign visitors, and is available on arrival at the airport in either city. On the ground, roads linking major cities are generally paved but poorly maintained, while those leading to remote villages are unpaved and potentially inaccessible during the rainy season. Some English is spoken around urban centres.

Health & Safety

Vaccinations for Hepatitis A and B and typhoid are recommended for visiting birdwatchers. Malaria is prevalent in remote areas and prophylaxis should be taken, although this is less of a problem in the dry season when most birdwatchers visit. Unexploded ordnance, a dangerous legacy of the Laotian Civil War and Vietnam War, litters some parts of the Laotian countryside and local advice should be sought when exploring new areas in the country.

KEY FACTS

No. of Endemics
1 (Bare-faced Bulbul)

Country List
c. 700 species

Top 5 Birds
1. Bare-faced Bulbul
2. Sooty Babbler
3. Red-collared Woodpecker
4. Rufous-necked Hornbill
5. Blyth's Kingfisher

Francis Yap

*Important populations of the brilliant **Rufous-necked Hornbill** remain in protected areas like the Nakai-Nam Theun National Biodiversity Conservation Area.*

Birdwatching Highlights

Visiting birdwatchers spending 2–3 nights at Ban Na Hin can expect to record upwards of 50 species during their stay. This figure, although comparatively low, is packed with quality, and features the endemic Bare-faced Bulbul as well as both the Sooty Babbler and Limestone Leaf Warbler, a trio of limestone forest specialities that are often encountered here. There are also tracts of excellent rainforest in the area that, despite being subject to intense hunting pressure, still hold sought-after species including the Silver Pheasant, Spot-bellied Eagle-Owl, Red-collared and Pale-headed Woodpeckers, and occasional Green Cochoa.

Le Manh Hung

*The montane forests of the Annamites are a stronghold for the **Short-tailed Scimitar Babbler**.*

Besides Ban Na Hin, there remain large areas of pristine evergreen and mixed evergreen deciduous forests, especially on the long and remote border with Vietnam on the Annamites (Sai Phou Louang in Laos). The protected area network spans 20 sites and covers 14 per cent of the country's area. The extensive Nakai-Nam Theun National Protected Area (NPA) covers a diversity of habitats from riverine forest to *Fokienia*-dominated cloud forests and supports important populations of the Crested Argus, Rufous-necked Hornbill and other Indochinese specialties. East of Luang Prabang, the Nam Et-Phou Louey NPA overlaps with large areas of evergreen and mixed forests, which support a range of species such as Blyth's Kingfisher and the localized Rufous-vented Laughingthrush.

Nam Et-Phou Louey National Protected Area

Janina Bikova

The higher elevations of Nam Et-Phou Louey are cloaked in beautiful mossy forest.

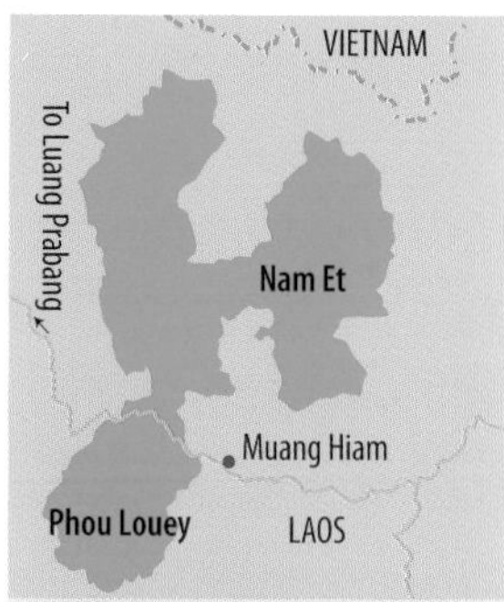

KEY FACTS

Nearest Major Towns
Muang Hiam (HQ), Sam Neua (Houaphan Province), Phonsavan (Xienkhouang Province), Luang Prabang (Luang Prabang province)

Habitats
Broadleaved evergreen forest (from lowland to montane), mixed deciduous forest, bamboo thickets, grassland

Key Species
Rufous-throated and Bar-backed Partridges, Silver Pheasant, Grey Peacock-Pheasant, Oriental Bay Owl, Blyth's and Crested Kingfishers, Great and Rufous-necked Hornbills, Pale-headed and Crimson-breasted Woodpeckers, Pied Falconet, Blue-naped and Rusty-naped Pittas, Rufous-throated Fulvetta, Spot-throated Babbler, Grey, Rufous-vented and Red-tailed Laughingthrushes, Rufous-headed Parrotbill, Beautiful Nuthatch, Jerdon's Bush Chat, Brown Dipper
Winter Yellow-vented Warbler

Other Specialities
Indochinese Tiger, Leopard, Clouded Leopard, Marbled Cat, Asian Golden Cat, Asiatic Black Bear, Sun Bear, Asiatic Wild Dog, Northern White-cheeked Gibbon, Bengal Slow Loris, Sambar, Chinese Serow, Gaur

Best Time to Visit
October–May; rainy season in June–September

Located in north-east Laos, the Nam Et–Phou Louey National Protected Area (NPA) currently covers 4,107km² and is effectively the largest protected area in the country. This ecologically diverse NPA spans nine districts across three provinces (Houaphan, Luang Prabang, Xieng Khouang), and is marked by a rugged, mountainous topography, with elevations ranging from about 336 to 2,267m asl. A number of major rivers, such as the Nam Nern, Nam Khan and Nam Seng, have their sources in this park. Nam Et–Phou Louey is well surveyed for its mammals, and is known to support important populations of Northern White-cheeked Gibbons, Asiatic Wild Dogs, a rich suite of other carnivores, and large ungulates. Less well known, however, is the park's rich bird diversity, and surveys have revealed nearly 300 bird species, including sought-after Southeast Asian specialities such as the **Grey Peacock-Pheasant**, **Rufous-necked Hornbill** and **Blyth's Kingfisher**, and localized populations of the **Beautiful Nuthatch** and the poorly known **Rufous-vented Laughingthrush**. Despite this, few birdwatchers have visited Nam Et–Phou Louey in recent years, and there is much potential for interesting discoveries.

Birdwatching Sites

To access the Totally Protected Zone of the Nam Et-Phou Louey NPA, it is mandatory for visitors to be accompanied by Protected Areas guides (a local English-speaking guide and local village guides who are former

Cheng Heng Yee

*The **Oriental Bay Owl** can be seen on the river cruises offered at Nam Et-Phou Louey.*

hunters). Visitors can select between the different tours organized by the protected area's ecotourism unit.

The Night Safari

This component of the NPA's standard ecotours involve a 24-hour boat-based trip into the core of the protected area with a combination of birdwatching and night-time wildlife spotlighting. A good variety of mammals, such as Sambar deer, Muntjac, Binturong, Bengal Slow Loris and various civets, as well as the **Oriental Bay Owl**, may be spotlighted at night while travelling along the river. During the day, both **Crested** and **Blyth's Kingfishers** are possible, as is a good variety of widespread Southeast Asian species.

Jason Wong

Riverine forests in Nam Et-Phou Louey are a stronghold of the ***Blyth's Kingfisher****.*

Wildlife Treks

The 2–5-day trekking tours allow access to excellent broadleaved and mixed evergreen forests from the hills to montane elevations. The longest five-day trek goes up to the summit of Phou Louey (2,257m asl) – the third highest peak in Laos. The **Silver Pheasant** and **Grey Peacock-Pheasant** are regular along the trail in the lower hills, while the **Pale-headed Woodpecker** is regular in tall stands of bamboo and secondary growth at around 600m asl. Dense forest undergrowth above 800m is where the attractive and localized **Rufous-throated Fulvetta** occurs. On these tours, visitors are offered an opportunity to participate in the NPA's conservation work by maintaining the wildlife camera traps set up along the trails. At higher elevations, it is possible to see the **Rufous-necked Hornbill**, best detected by its heavy wingbeats as it flies over the forested valleys. The **Rufous-throated Partridge** is common from above 1,600m; at this elevation, large, mixed flocks become regular, and this is where they should be scrutinized for the striking **Beautiful Nuthatch**.

James Eaton

The ***Crimson-breasted Woodpecker*** *occurs in the montane evergreen forests of Phou Louey.*

Access & Accommodation

Maung Hiam (also called Viengthong) is where the National Protected Area's headquarters and visitor centre are located. Daily buses link Muang Hiam from Sam Neua (four hours), Nong Khiaw (five hours) and Luang Prabang (eight hours). If you travel from Phonsavan (four hours), it will be necessary to change buses at the Koa Hing junction. Several guesthouses are located in Muang Hiam and there is also the option of village homestays at the NPA's ecotourism villages.

Conservation

The Nam Et–Phou Louey National Protected Area (NEPL NPA) was established in 1993. The International Union for the Conservation of Nature (IUCN) provided technical and financial support from 2000 to 2002. The Wildlife Conservation Society (WCS) has been supporting the NEPL Management Unit since 2003, and has assisted with the development of ecotourism products commencing in 2009–2010. Living inside or immediately adjacent to the NPA are 30,000 villagers from 98 communities, many in some of the poorest districts of the country. There is a long history of human settlement in and around NEPL, with local people relying heavily on natural resources for their subsistence.

The ***Grey Peacock-Pheasant*** *is commonly heard in the broadleaved forests of Nam Et-Phou Louey.*

Nakai-Nam Theun National Protected Area

Souvanhpheng Phommansane & Yong Ding Li

Yong Ding Li

Submontane evergreen forest still persists on the rugged slopes of the Annamites at the Laotian-Vietnamese border.

The vast Nakai-Nam Theun National Protected Area (3,710km^2) and its extensions constitute one of the largest protected areas in Laos, and cover a diverse complement of ecosystems, from dry deciduous forests to montane evergreen forests. Much of Nakai-Nam Theun overlaps with the Annamites (Sai Phou Louang) mountains. Many of the forested mountains and ridges soar to well over 1,000m asl, and reach a maximum elevation of 2,200m asl at Phou Laoko. A number of rivers, including the meandering Nam On and Nam Xot, flow west from the Annamites into the Nakai Plateau. Nakai-Nam Theun is one of the most important forested landscapes for biodiversity in Southeast Asia, and surveys have revealed more than 400 species of bird, including many Indochinese specialities such as the **Crested Argus** (ssp. *ocellata*), **Coral-billed Ground Cuckoo**, **Red-collared Woodpecker**, **White-winged Magpie** and **Short-tailed Scimitar Babbler**. Riverine forests on the lower and middle elevations of the park support the endangered **White-winged Duck**, and formerly the **Green Peafowl**. Evergreen forests at higher elevations are important to a number of species better associated with the Eastern Himalayas, such as the **Rufous-necked Hornbill**, **Red-tailed Laughingthrush** and **Beautiful Nuthatch**. Despite this enormous

Abdelhamid Bizid

The raucous screeches of the ***White-winged Magpie*** *can be regularly heard in riverine forest at the lower elevations of Nakai-Nam Theun.*

KEY FACTS

Nearest Major Towns
Nakai District Town (Khammouane), Lak Xao (Bolikhamsai)

Habitats
Lowland, hill and montane evergreen forests, pine forest, grassland, riverine scrub, cultivation

Key Species
White-winged Duck, Siamese Fireback, Crested Argus, Coral-billed Ground Cuckoo, Tawny Fish Owl, Blyth's and Crested Kingfishers, Great, Austen's Brown and Rufous-necked Hornbills, Red-collared and Pale-headed Woodpeckers, Pied Falconet, Blue-rumped Pitta, White-winged Magpie, Spotted Elachura, Short-tailed Scimitar Babbler, Red-tailed Laughingthrush, Beautiful Nuthatch, Green Cochoa, Jerdon's Bush Chat

Other Specialities
Red-shanked Douc, Pygmy Slow Loris, Annamite Striped Rabbit, Asian Elephant, Gaur, Saola, Large-antlered and Truongson Muntjacs, Indochinese Tiger, Asiatic Wild Dog, Mainland Clouded Leopard, Asian Golden and Marbled Cats, Owston's Civet

Best Time to Visit
November–April; the south-west monsoon starts in May

Yong Ding Li

*The **Red-tailed Laughingthrush** is arguably the most attractive of the laughingthrushes in the montane forests of the Annamites.*

diversity, as a result of its remoteness and limited infrastructure, Nakai-Nam Theun is seldom visited by birdwatchers.

Birdwatching Sites

Nam Xot River

Upstream of the enclave of Ban Namxot, the slow-flowing Nam Xot River meanders through some excellent areas of dry evergreen forests as it snakes through the western part of the Nakai Plateau before eventually joining the Nam Theun. The riverine forest here still holds small numbers of **White-winged Duck** and **Lesser Fish Eagle**, and is a known stronghold of the **River Lapwing**, which may be seen in forested riverbanks and sandbars. The dry evergreen forest along the lower to middle Nam Xot supports high bird diversity, and past surveys have reported sought-after species such as the **Coral-billed Ground Cuckoo**, **Tawny Fish Owl**, **Blyth's Frogmouth**, **Pied Falconet**, **Yellow-vented Warbler** and other forest birds.

Nam Pheo

The middle courses of the Nam Pheo snakes through dry evergreen forest at an elevation of more than 550m. The forested banks of the river should be checked for both **Blyth's** and **Crested Kingfishers**.

Montane Forest above Ban Navang

If time permits, you should attempt to access the montane forests, including the Fokienia-dominated forests above the enclave of Ban Navang. These forests are extremely bird rich and support an essentially eastern Himalayan avifauna, including significant populations of the threatened **Rufous-necked Hornbill**. At above 1,300m, mixed flocks should be checked for the **Beautiful Nuthatch**, various minlas and fulvettas.

Nam Kwai

Located north-west of the Nakai-Nam Theun, this area falls under a proposed extension to the conservation area. There remain excellent stands of wet evergreen forests, some of which can be accessed from tracks and old logging roads to well over 1,000m asl. Past surveys have revealed these forests to be important to the **Crested Argus**, a spectacular pheasant restricted to submontane evergreen forests in the Annamites, and with a disjunct population in Peninsular Malaysia.

Access & Accommodation

The enclave of Nakai can be accessed from Thakek on Route 12, followed by Route 8B. The road becomes very rough in the wet season due to erosion and rutting. A road (in poor condition) enters the conservation area from Route 8B (around Ban Kengdaeng, Khamkeut District), allowing limited access to Ban Thamuang and Ban Navang. The western side of the conservation area skirts Routes 8A, 8B and 12. Given the limited infrastructure and accommodation options, a visit to Nakai-Nam Theun could be expected to involve a lot of camping.

Conservation

The completion of the Nam Theun II dam resulted in the inundation of more than 400km of low-lying, valley-bottom forest on the fringes of the conservation area. Forest clearance by shifting cultivators continues unabated throughout the NBCA, especially along rivers and streams. Poaching, both by local hunters and those from Vietnam, poses a severe threat to the wildlife in Nakai-Nam Theun, and ranger patrols have uncovered widespread evidence of poaching, including thousands of snares and hunting camps.

Ah Kei

*The **Pied Falconet** is regularly encountered in the dry forests of the Nakai Plateau.*

Ban Na Hin Forests

Low Bing Wen

Low Bing Wen

The road leading to Ban Na Hin offers birdwatchers easy access to a large expanse of limestone karsts and their associated birdlife.

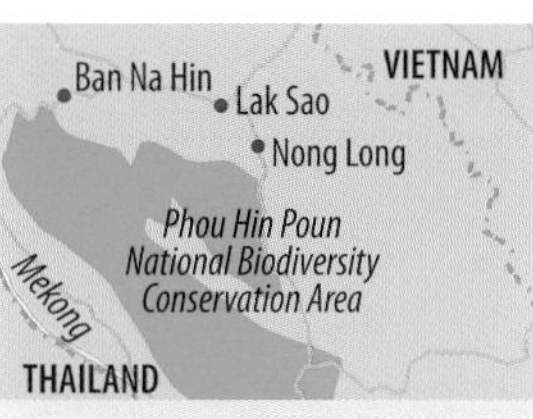

KEY FACTS

Nearest Major Town
Ban Na Hin

Habitats
Mixed deciduous and hill evergreen forests, limestone karsts, scrub

Key Species
Silver Pheasant, Spot-bellied Eagle-Owl, Red-vented Barbet, Red-collared and Pale-headed Woodpeckers, Eared Pitta, Bare-faced Bulbul*, Limestone Leaf Warbler, Sooty Babbler, Streaked Wren-Babbler, Black-browed Fulvetta, Indochinese Yuhina, Green Cochoa, White-tailed Flycatcher, Fork-tailed Sunbird
Winter White-throated Rock Thrush, Siberian Blue Robin, Hainan Blue Flycatcher

Other Specialities
Assam Macaque, Lao Langur*, Black Giant Squirrel, Kha-nyou*

Best Time to Visit
November–March; rainy season May–October

Benjamin Schweinhart

*The **Spot-bellied Eagle-Owl** has been reported in the forests of Ban Na Hin in recent years.*

For birdwatchers, Laos remained largely under the radar due to a general lack of information and widespread reports of large-scale hunting of biodiversity. All this, however, changed with the discovery of the **Bare-faced Bulbul**, a country endemic and localized resident of wooded limestone outcrops, now placed in its very own genus *Nok*. Situated within the extensive Phou Hin Poun National Biodiversity Conservation Area (1,801km^2), Ban Na Hin is readily accessible from the capital Vientiane, and now features regularly on many birdwatching trips to Indochina. Although illegal hunting and logging are rife, other restricted-range avian specialities associated with limestone karst ecosystems, such as the **Limestone Leaf Warbler** and **Sooty Babbler**, are frequently encountered, while exploration of the surrounding forests can yield sought-after species such as the **Spot-bellied Eagle-Owl** and **Green Cochoa**. The **Red-collared Woodpecker**, perhaps Southeast Asia's most elusive woodpecker, has also been recorded here.

Birdwatching Sites

Highway 8 Route Marker Km 31 to 35

This 4-km stretch of road heading towards Ban Na Hin passes through visually spectacular limestone karsts and is the main site for avian limestone specialities. The viewpoint at Km 31.5 provides a convenient landmark highlighting the start of the best habitat. The recently described **Bare-faced Bulbul** is frequently seen in fruiting trees or perched on top of the limestone karsts, while parties of localized **Sooty Babblers** forage unobtrusively in the understorey. The **Limestone Leaf Warbler** and **Streaked Wren-Babbler** may be seen, and the **Green Cochoa** has been recorded foraging on fruiting trees here on a number

James Eaton

*Small parties of **Sooty Babblers** are regularly seen foraging in low vegetation on the limestone karsts.*

James Eaton

*One of the specialities of the limestone forests here is the **Limestone Leaf Warbler**.*

Low Bing Wen

There are still remnant tracts of excellent hill evergreen forest in the hills surrounding Ban Na Hin.

of occasions. This stretch is also the best area to spot the globally threatened Lao Langur, which regularly sits on top of the limestone karsts in the early morning and late afternoon.

Highway 8 Route Marker Km 48 Forest Trail

Along Highway 8 heading east out of Ban Na Hin towards Lak Sao and the Vietnam border is a trail beginning at the Km 48 route marker, which leads to an apparently abandoned weather station. It is a moderately steep ascent and takes about 30 minutes to complete. Just before the end, the trail turns sharply to the left, with a less conspicuous branch heading off to the right. Taking the right-hand branch leads to excellent hill evergreen forest with a dense understorey of gingers and bamboo. The trail runs level for a further 100m or so before descending sharply into a seasonal riverbed, where the open nature of the understorey makes it easy to get lost – it is recommended that birdwatchers turn back at this point.

This site provides some of the best forest birdwatching in the area with an extensive list of possibilities, including the **Silver Pheasant**, **White-browed Piculet**, **Red-vented Barbet**, **Red-collared Woodpecker**, **Silver-breasted** and

James Eaton

*The recently described **Bare-faced Bulbul** is frequently seen perched on top of the limestone karsts on the way to Ban Na Hin.*

Long-tailed Broadbills, Red-headed and Orange-breasted Trogons, Eared Pitta, Spot-necked Babbler, Black-browed and Rufous-throated Fulvettas, White-tailed Flycatcher, Rufous-tailed Robin and Fork-tailed Sunbird.

Namsanam Waterfall Trail

This trail starts from the only temple in Ban Na Hin, with the entrance to the trail 'demarcated' by a pile of granite boulders and a bare area of soil in front of the temple. The trail crosses a small stream shortly thereafter, before entering a tract of lowland monsoonal evergreen forest that can be explored via a network of side trails. After about 1.5km a riverbank is reached, and the waterfall is a further 2km from this point. Exploratory visits to this area have yielded species such as the Spot-bellied Eagle-Owl, Collared Owlet, Ruddy Kingfisher, Blue-bearded Bee-eater and Grey-backed Shrike.

Access & Accommodation

Ban Na Hin is a pleasant five-hour journey from Vientiane along paved roads heading east along Highway 13 before turning off to Highway 8 at the crossroad town of Vieng Kham. The village itself offers a range of accommodation of varying quality. No entrance fees or permits are needed to explore the area, and motorbikes can be hired from the village if private transportation is not an option.

Conservation

The forests and limestone karsts around Ban Na Hin are generally located within the Phou Hin Poun National Biodiversity Conservation Area, a large protected area that spans more than 1,800km^2. Phou Hin Poun has been the scene of two major zoological discoveries in recent years, namely the Bare-faced Bulbul and the Kha-nyou, a rodent so distinct that it has been classified in its own family, the Diatomyidae.

Widespread hunting and logging throughout the protected area poses a serious threat to its animal life. Another significant threat to the area and many of Laos's most important protected areas is forest clearance for hydroelectric power projects funded by neighbouring countries Vietnam and Thailand.

Abdelhamid Bizid

*The **Hainan Blue Flycatcher** has been recorded wintering in the forests around Ban Na Hin.*

MALAYSIA

1. Taman Negara
2. Tasik Kenyir
3. Langkawi Archipelago
4. Teluk Air Tawar
5. Fraser's Hill
6. Kuala Selangor
7. Panti Forest
8. Kinabalu Park
9. Danum Valley
10. Mantanani Islands
11. Crocker Range
12. Kabili-Sepilok
13. Kinabatangan
14. Tawau Hills
15. Payeh Maga
16. Bakelalan
17. Buntal Bay
18. Borneo Highlands
19. Kubah

One of the world's most biodiverse countries, equatorial Malaysia straddles the southern half of the Thai-Malay Peninsula and northern Borneo. Peninsular Malaysia is backboned by the long Titiwangsa Mountains, which contain some of the largest areas of montane forest in the peninsula, and a number of smaller ranges like the Tahan and Bintang. The Malaysian states of Sabah and Sarawak on Borneo are extremely rugged, and contain the highest peak in the country, Gunung Kinabalu (4,090m asl). While there are few large natural lakes in the country, a number of major river systems drain the interior of Malaysian Borneo and the peninsula, the best known being the Rajang, Kinabatangan and Pahang Rivers.

Malaysia's extensive tropical rainforests offer some of the most exciting birdwatching opportunities in Southeast Asia, with species-filled forests and mountains that feature charismatic Asian bird families such as trogons, pittas, frogmouths and broadbills. Due to the generally high standard of infrastructure and accommodation, Malaysia is one of the most hassle-free countries in the region for the independent birdwatcher to visit. However, because of the dynamics of rainforest birdwatching birds are more often heard than seen, and a significant investment of time is generally needed at each site to come to grips with the most secretive rainforest specialities like pittas and pheasants. To showcase the entirety of Malaysia's rich birdlife, 19 of the best sites have been selected from both East and West Malaysia. Besides the well-trodden sites like Taman Negara and Danum Valley, a number of lesser-known sites like Payeh Maga and Langkawi have been chosen – each has its own specialities and great potential for interesting discoveries.

KEY FACTS

No. of Endemics

4 (West Malaysia including 3 near-endemics), 6 (Malaysian Borneo)

Country List

805 species

Top 5 Birds

1. Bornean Bristlehead (Malaysian Borneo)
2. Great Argus
3. Blue-headed Pitta (Malaysian Borneo)
4. Rail-babbler
5. Storm's Stork

Climate

Malaysia's climate can be described as equatorial, with uniformly high temperatures, high humidity and rainfall throughout the year. Mountain ranges, in particular the Titiwangsa (Main) Range, which runs along the length of much of Peninsular Malaysia, play a significant role in rainfall distribution. Rainfall is most intense in East Malaysia and the north-eastern areas of Peninsular Malaysia during the north-east monsoon between November–February, while the south-west monsoon between April–September brings considerable rainfall to the west coast of Peninsular Malaysia.

Access, Transportation & Logistics

The country's main transport hub is its capital Kuala Lumpur. Many of Peninsular Malaysia's best birdwatching sites can be reached within a few hours' drive from the capital city. Frequent domestic flights connect numerous cities in East Malaysia to Kuala Lumpur, including the state capitals Kota Kinabalu (Sabah) and Kuching (Sarawak). Visitors from many countries do not require tourist visas to visit, and can stay in the country for between

14 days to three months, depending on their nationality. On the ground, transport infrastructure is comprehensive and of a good standard. Many birdwatching sites in Peninsular Malaysia and Malaysian Borneo can be easily reached by car, while the most popular destinations like Gunung Kinabalu and Taman Negara are served by taxis and even public transport. By and large, English is widely spoken across Malaysia. However, outside the major cities and tourist areas, learning some basic Malay goes a long way towards a more enjoyable trip.

Mikael Bauer

*The **Large Wren-Babbler** is a speciality of pristine lowland forests in Peninsular Malaysia.*

Health & Safety

There are numerous natural annoyances in rainforests, particularly in the lowlands. In addition to terrestrial leeches, especially the infamous 'tiger leeches' of Borneo's rainforests, ticks, mosquitoes and various species of biting fly occur. The use of long sleeves and insect repellent is recommended for rainforest birdwatching. Vaccinations are also recommended for diseases such as Hepatitis A and B and typhoid. The risk of malaria is comparatively low throughout the country, although visitors can still opt to take prophylaxis.

Birdwatching Highlights

A two-week trip to Peninsular Malaysia can yield around 250 species, including the endemic Malayan Whistling Thrush and near-endemic Malaysian Partridge in the highlands (Fraser's Hill), as well as the two near-endemic peacock-pheasants and Malayan Laughingthrush, the ranges of which extend marginally into the southernmost Thai province of Narathiwat. Other avian highlights include the charismatic Rail-babbler (now a monotypic family), up to six species of trogon, seven species of broadbill and ten species of hornbill.

Similar returns can be expected from a two-week foray into Sabah with regard to number of species, but such a trip also offers the tantalizing prospect of seeing more than 40 of the avian endemics found only on the island of Borneo. These include the bizarre Bornean Bristlehead, skulking Bornean Ground Cuckoo, four species of endemic pitta, Storm's Stork and a host of montane endemics, including the 'Whitehead's trio' (broadbill, trogon and spiderhunter). For a more comprehensive experience, a visit into the mountains of Sarawak offers a very good chance of observing the enigmatic Black Oriole, Dulit Frogmouth and Hose's Broadbill, three Bornean endemics that are unlikely to be seen in Sabah, as well as an excellent selection of montane and lowland species.

*No less than a dozen nightbird species occurs in Malaysia, including the uncommon **Reddish Scops Owl**.*

Choy Wai Mun

*The forests of Peninsular Malaysia supports important populations of the **Chestnut-necklaced Partridge**.*

Taman Negara National Park

Yong Ding Li

Yong Ding Li

The Sungei Tembeling, a tributary of the Sungei Pahang, skirts the eastern boundaries of the park and provides access for visitors heading to the park by boat.

KEY FACTS

Nearest Major Towns
Jerantut (Pahang), Gua Musang (Kelantan)

Habitats
Lowland and hill rainforests, lower and upper montane rainforests, subalpine scrub, caves

Key Species
Lowland Crestless and Crested Firebacks, Great Argus, Crested and Long-billed Partridges, Malayan Peacock-Pheasant*, Storm's Stork, Bat Hawk, Wallace's Hawk-Eagle, Large Green Pigeon, Moustached Hawk-Cuckoo, Reddish Scops Owl, Large, Gould's and Blyth's Frogmouths, *all* Malaysian hornbills (except Plain-pouched), Malaysian Honeyguide, Malayan Banded, Garnet and Giant Pittas, Crested Jay, Black Magpie, Rail-babbler, Finsch's and Straw-headed Bulbuls, Large and Striped Wren-Babblers, Chestnut-capped Thrush
Montane Crested Argus, Mountain Peacock-Pheasant*, Malaysian Partridge*, Fire-tufted Barbet, Grey-headed Woodpecker, Hill Prinia, Malayan and Chestnut-capped Laughingthrushes
Winter Masked Finfoot (rare), Slaty-legged Crake, Eyebrowed and Siberian Thrushes, Brown-chested Jungle Flycatcher, Green-backed Flycatcher, Siberian Blue Robin

Other Specialities
Sunda Colugo, Sunda Pangolin, Sunda Slow Loris, White-handed Gibbon, Siamang, Dusky Langur, White-thighed Surili, Sun Bear, Malayan Tiger, Leopard, Asian Golden Cat, Asiatic Wild Dog, Asian Elephant, Malayan Tapir, Sambar, Gaur

Best Time to Visit
March–August; rainy season November–January

Straddling the Malaysian states of Pahang, Terengganu and Kelantan, the 4,343km² Taman Negara protects some of the largest intact forest landscapes in Sundaic Southeast Asia. This vast national park contains nearly the full complement of vegetation types in the Thai-Malay Peninsula, from lowland riverine forests (80m asl) to subalpine scrub on Peninsular Malaysia's highest peak, Gunung Tahan at 2,183m asl, and boasts a diversity of flora numbering more than 3,000 species. It is thus of no surprise that more than 400 bird species have been documented within Taman Negara, including two that are largely confined in terms of their distribution in the peninsula to the park, the **Crested Argus** and **Hill Prinia**. As most birdwatchers tend to focus on the forests around the Kuala Tahan and Sungei Relau section of the park, much of it remains virtually unexplored. However, even the rainforests around the park headquarters at Kuala Tahan are enough to keep the most dedicated of birdwatchers busy for 3–4 days, as the majority of the lowland forest species can be seen here, including the highly sought-after **Malayan Peacock-Pheasant**, **Malayan Banded Pitta**, **Rail-babbler** and groups of the bold **Crested Fireback**. Night rambles may yield a variety of owls and frogmouths, including the unusual **Large Frogmouth**.

Birdwatching Sites

Kuala Tahan & Tabing Area

A fairly extensive trail system here allows access to pristine lowland forest, the most popular trails being the Swamp Loop, Jenut Muda and Tabing Trails.

Wang Bin

*The **Garnet Pitta** is one of six pitta species that may be seen in the park.*

*Small parties of **Crested Firebacks** can be regularly seen around the Kuala Tahan section of the park.*

The Jenut Muda Trail is very popular, and many encounter their first peacock-pheasant, trogon, pitta, hornbill and malkoha, and the bizarre **Crested Jay**, along this trail. Careful scanning of the riverbank may reveal the **Blue-banded Kingfisher**, **Chestnut-naped Forktail** and, very occasionally, a **Masked Finfoot** during the winter months (November–January). Other sought-after species possible here include the **Moustached Hawk-Cuckoo**, **Reddish Scops Owl**, **Malaysian Honeyguide**, **Great Slaty Woodpecker**, **Maroon-breasted Philentoma** and **Chestnut-capped Thrush**. The canopy walk at Bukit Teresek is recommended for close-up views of the many mixed feeding flocks that move through the area, including flycatcher-shrikes, nuthatches, minivets, woodshrikes, **Bar-bellied** and **Lesser Cuckooshrikes**, the **Green Iora** and numerous woodpecker species.

Kumbang Hide & Trail

This trail and associated hide is best reached from Kuala Trenggan, which can be accessed by a 30-minute ride on a boat or a three-hour hike. As the trail is infrequently used, it can get very muddy and overgrown in some sections. However, those who make it this far can appreciate the diversity of lowland forest birds in the presence of minimal human traffic. Highlights here include all the lowland trogons, the **Malaysian Honeyguide**, **Rail-babbler**, **Chestnut-capped Thrush** and elusive **Crestless Fireback**, which is infrequently seen on the trail. Although **Storm's Stork** has been reported here, it appears to be rare and is probably best seen during a boat trip up the Tembeling River. The Kumbang Hide is an excellent place for watching mammals, and is often visited by the Malayan Tapir and Sambar Deer and, very occasionally, the Gaur and Malayan Tiger after dark.

Cheng Heng Yee

*The **Chestnut-naped Forktail** is frequently encountered along the Sungei Tahan.*

Yong & Belau Hides

Both these hides can be reached via a short 15-minute boat trip downstream on the Tembeling River. The access trails leading to the hides feature excellent riverine forest, and the **Malayan Peacock-Pheasant**, **Garnet Pitta** and **Rail-babbler** are all regularly heard, although seeing them takes some effort. The **Giant Pitta** is very scarce, but recent records are from this area.

Sungei Relau (Merapoh)

Less visited but increasingly popular among birdwatchers, this site can be reached from the western boundaries of the park. The 13km-long access road through lowland forest is excellent for forest birds, including the **Malayan Peacock-Pheasant**, **Large Green Pigeon**, **Jambu Fruit Dove**, **Garnet** and **Malayan Banded Pittas**, and a good variety of hornbills, bulbuls and babblers. Shy **Striped** and **Large Wren-**

Cheng Heng Yee

*The whistles of the **Rufous-collared Kingfisher** are part of the dawn chorus of Malaysia's lowland rainforest.*

Babblers skulk in the dense undergrowth, and may require some patience before showing themselves. Night walks here are good for the full complement of Malaysian night birds, including the **Oriental Bay Owl**, and **Gould's** and **Large Frogmouths**.

Gunung Tahan Hike

This strenuous seven-day hike can now be cut back to 3–4 days if you start from the Sungei Relau entrance. While hardly any birdwatcher has hiked to the summit of Gunung Tahan or any of the adjoining peaks, those that do are able to see montane forest species otherwise seen only in the Titiwangsa (Main) Range, including the rare **Mountain Peacock-Pheasant**, as well as mixed flocks of laughingthrushes, woodpeckers, warblers and babblers. The **Hill Prinia** (ssp. *waterstradti*) occurs in the peninsula only in the alpine scrub at the summit of Gunung Tahan. The ultimate prize, however, is a glimpse of the ultra-localized and highly distinctive subspecies (ssp. *nigrescens*) of the **Crested Argus**, one of the world's largest and most elusive pheasants. The **Crested Argus** is usually heard from across the mountain slopes, but it is possible to see it in the lower montane forests at the elevation of Camp II on the Tahan hike.

Access & Accommodation

In the past most visitors caught a motorized boat from the jetty at Kuala Tembeling for the 3–4-hour upriver trip to Kuala Tahan. While this option still exists, you now have the option of driving straight into Kuala Tahan from Jerantut if coming from the direction of Kuala Lumpur. Alternatively, many birdwatchers now drive to the Sungai Relau entrance from the town of Gua Musang on the western boundary of the park, where there are fewer tourists. Access to the more remote Kuala Koh entrance in the north is possible by road from Gua Musang or Kota Bharu. A variety of accommodation, ranging from basic dormitories to air-conditioned chalets, is available in Kuala Tahan, Sungei Relau and Kuala Koh. Camping facilities are also available.

Conservation

As a national park and one of the leading tourist destinations in Peninsular Malaysia, Taman Negara is relatively well protected. Some illegal poaching of animals occurs on the remote and less well-patrolled boundaries of this vast park. A controversial project to start sturgeon farming has been proposed near the park's boundaries, but has been temporarily halted pending a review.

Cheng Heng Yee

*The **Olive-backed Woodpecker** is the most elusive of Malaysia's lowland woodpeckers.*

*Four trogon species coexist in the lowland dipterocarp forests of Malaysia, including this **Red-naped Trogon**.*

Tasik Kenyir

Anuar McAfee

Anuar McAfee

Hill dipterocarp forests at the edge of Tasik Kenyir shrouded in mist after a rain.

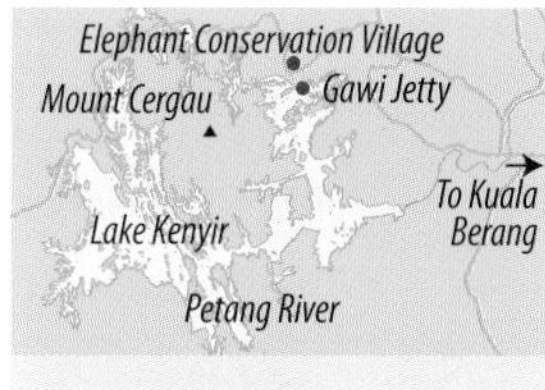

KEY FACTS

Nearest Major Town
Kuala Terengganu (Terengganu)

Habitats
Lowland and hill rainforests, montane forest

Key Species
Bat Hawk, Wallace Hawk-Eagle, Lesser and Grey-headed Fish Eagles, Large Green Pigeon, White-crowned, Rhinoceros, Great, Helmeted, Black, Bushy-crested, Wreathed and Wrinkled Hornbills, Crested Jay, Black Magpie, Straw-headed Bulbul, Puff-throated Babbler
Montane Malaysian Partridge, Crested Argus, Malayan Laughingthrush

Other Specialities
White-handed Gibbon, White-thighed Surili, Asiatic Wild Dog, Sun Bear, Malayan Tiger, Clouded Leopard, Marbled Cat, Asian Golden Cat, Asian Elephant, Malayan Tapir, Sumatran Serow, Gaur

Best Time to Visit
March–October; rainy season November–February

Con Foley

The metallic calls of the ***Black Magpie*** *are one of the more familiar sounds of the forests around Kenyir.*

Tasik Kenyir is a massive, man-made lake formed in the late 1980s after the completion of a hydroelectric dam, the Sultan Mahmud Power Station. The lake covers an area of 368km^2 and is surrounded by extensive lowland and hill dipterocarp forests. While parts of the forest reserves adjoining the lake have been selectively logged over the years, the forest landscape surrounding the lake remain largely intact, supporting some 296 species of bird as well as a healthy assemblage of large mammals, including the Malayan Tiger, Asian Elephant and Malayan Tapir. The forests adjacent to the south side of the lake form the Terengganu portion of the Taman Negara, which is accessible only by boat. Tasik Kenyir is a wonderful place to observe many of the lowland forest birds in Malaysia, and is exceptionally good for seeing a good diversity of raptors, including the uncommon **Lesser Fish Eagle**. Tasik Kenyir is also one of the best places to see hornbills in the country, with nine species possible, including the scarce **Helmeted** and **Wrinkled Hornbills**. In recent years, the **Crested Argus** has been seen in the submontane forests in the mountains south of the lake, and this may prove to be a good site to observe this elusive species in future, alongside other submontane specialities.

Birdwatching Sites

Kelah Sanctuary

The Petang River, which drains into Tasik Kenyir, can be accessed on a half-hour boat trip from Gawi Jetty. After passing a wildlife department checkpoint, the boat will take you further upriver right to excellent lowland forest at the fringes of Taman Negara. Here, a canopy walkway allows access to the forest canopy. A good variety of woodpeckers, malkohas, barbets, bulbuls and trogons can be seen here. The walkway runs for several hundred metres, then crosses the Petang River on a large suspension bridge connected to a watchtower. The best areas for birding are along

*The **Rhinoceros Hornbill** is perhaps the most iconic of the Sundaic hornbills.*

*The handsome **Red-bearded Bee-eater** is regularly encountered along the forest edge.*

the canopy walkway, suspension bridge and watchtower. Access to the canopy walk is only allowed up to 12 noon.

Sungai Buweh Recreational Road

This area can be reached from Gawi Jetty following Route T185 towards Kelantan, followed by a left turn into the road opposite the 'Elephant Conservation Village'. This road runs for several kilometres along the edge of Tasik Kenyir at the base of Lawit Hill. Most birdwatchers drive along the road and stop every now and then to access the roadside forest on foot. At the more open areas, various raptors can be seen, including both **Blyth's** and **Wallace's Hawk-Eagles**, and the **Black-thighed Falconet**. Along sections of the road in closed forest, where you can see bulbuls, barbets, babblers, cuckoos, **Large Green Pigeons** and the retiring **Rufous-collared Kingfisher** in the early morning. Primates regularly seen here include the White-handed Gibbon and Dusky Langur.

Road T185

This is the only major road connecting Terengganu with Kelantan, and it skirts the northern side of Tasik Kenyir through extensive forest that is part of the Kenyir Wildlife Corridor. It is fringed by logged forest for much of its length, and many raptors and hornbills can be seen along it, including **Rhinoceros**, **Great** and **Helmeted Hornbills**. Old logging trails run off both sides of T185 and are good areas to explore on foot to see many lowland species.

Access & Accommodation

Kuala Terengganu, the capital of Terengganu State, can be reached by air from Kuala Lumpur (one hour) or by road (five hours). Gawi Jetty, the main entry point for Tasik Kenyir, is 60km from Kuala Terengganu. A rented car provides the easiest way to reach Tasik Kenyir, and is most convenient for exploring the various roads. Travel agents in Kuala Terengganu or at Tasik Kenyir can assist with boats and entry permits for those wanting to visit the National Park or Kelah Sanctuary. Petang Island Resort is the only hotel operating at the moment, located 30 minutes by boat from Gawi Jetty. Hotels are under construction at Gawi Jetty and should be open in the near future. Many groups visiting the area also rent 'houseboats' to explore the lake.

Conservation

The forested landscape of Kenyir is particularly difficult to patrol given its many access points, and poaching remains a serious problem. Wildlife surveys are conducted here regularly by local NGOs (like Rimba, www.rimbaresearch.org), universities and the wildlife department, Perhilitan. While forests on the southern fringe of the lake are contiguous with Taman Negara (also an Important Bird Area), much of the remaining forests are unprotected. Recognizing the significance of the wider Kenyir landscape for biodiversity, including many large mammals, lobbying of the Terengganu state government is ongoing to gazette sections of the forests around Tasik Kenyir into a state park.

*The **Lesser Fish Eagle** forages mainly along rivers in well-forested areas.*

Langkawi Archipelago

Sofian Zack

The forested slopes of Gunung Raya, Langkawi's tallest landform, support the majority of the island's avian specialities.

KEY FACTS

Nearest Major Towns
Kuah, Padang Matsirat, Pantai Cenang (Kedah)

Habitats
Lowland and submontane rainforests (including forest on limestone), mangroves, cultivation

Key Species
Mountain Hawk-Eagle, Large Green Pigeon, Orange-breasted Trogon, Brown-winged and Ruddy Kingfishers, Chestnut-headed Bee-eater, Great and Wreathed Hornbills, Great Slaty Woodpecker, Blue-winged and Mangrove Pittas, Black-hooded Oriole
Winter Jerdon's Baza, Oriental Scops Owl, Asian House Martin, Yellow-browed and Pale-legged Leaf Warblers, Brown-streaked Flycatcher

Other Specialities
Sunda Colugo, Sunda Slow Loris, Dusky Langur, Black Giant Squirrel, Hairy-nosed Otter, Lesser Mousedeer

Best Time to Visit
November–May

The Langkawi Archipelago consists of a group of 104 islands about 30km off the state of Perlis on the Malaysian mainland. The largest, Langkawi Island, is dominated on its north-east by the rugged Machincang Hills, which geologically date back to the Cambrian Period, while large parts of its coast are fringed by either steep cliffs or mangroves. However, the most prominent feature of Langkawi Island is the forested peak of Gunung Raya, which rises to nearly 900m asl and is the highest point in the archipelago. Recognized for its complex and geologically ancient rock formations, Langkawi has been afforded Geopark status by UNESCO. Langkawi is rich in birds and nearly 330 species have been recorded there, including a number with highly restricted distributions in Peninsular Malaysia (**Mountain Hawk-Eagle**, **Brown-winged Kingfisher** and **Black-hooded Oriole**). In addition, Gunung Raya is noted for being one of the easiest places to see large numbers of hornbills, in particular the **Great Hornbill**, in Peninsular Malaysia. It is possible to see nearly 100 species on a trip lasting 3–4 days.

Birdwatching Sites

Gunung Machinchang Forest Reserve

Some of the most pristine coastal lowland rainforests can be found at this site located on the remote north-eastern corner of Langkawi. Access to the forest is possible along the road (Jalan Datai and Jalan Teluk Datai) that leads to the major resorts at Datai Bay. In addition, there are small forest trails that snake into the forest from the Datai resort. Besides various bulbuls and flycatchers, the **Banded Kingfisher** and **Blue-winged Pitta** can be seen here. Large fruiting trees should be checked for green pigeons, especially the localized **Large Green Pigeon**.

Con Foley

*The striking **Banded Kingfisher** is more often heard than seen at Machinchang's forests.*

Sofian Zack

The patchily distributed ***Brown-winged Kingfisher*** *is readily observed in the extensive mangroves along Langkawi's coastline.*

Francis Yap

The whistles of the ***Mangrove Whistler*** *are a distinctive sound of Southeast Asian mangroves.*

Gunung Raya Forest Reserve

The 881m-tall Gunung Raya is the most popular birdwatching spot on Langkawi. The lowland and hill forests on the mountain's slopes can be accessed from a road that snakes its way for 25km from the mountain's foot to its peak. Most of Langkawi's forest birds can be seen in the forest along the road, including the Large Green Pigeon, Orange-breasted Trogon, Banded Kingfisher, Great Slaty Woodpecker, Blue-winged Pitta, White-bellied Erpornis, and Black-crested and Ochraceous Bulbuls. Migratory raptors include Jerdon's Baza and the Grey-faced Buzzard.

One of Gunung Raya's highlights is the Mountain Hawk-Eagle, a species found only on Langkawi within Malaysia and often seen soaring over the forested slopes. Another is the hornbills. While the Oriental Pied Hornbill is generally common throughout Langkawi, Great and Wreathed Hornbills are most easily seen on the upper slopes in the evenings after 5 p.m.

Cenang Area

This site is dominated by paddyfields with isolated patches of scrub and secondary forest. A good diversity of open-country species can be expected here. Flooded areas are good for whistling ducks and Little Grebes, while denser patches should be checked for Greater Painted-snipes and rails like the Slaty-breasted Rail and White-browed Crake.

Kisap & Kilim Area

Some of the most accessible and pristine mangroves on Langkawi can be found here. The best way to explore the mangroves is to charter a boat and explore the mangroves along the Kilim River for the various mangrove specialities. For many visiting birdwatchers the restricted-range Brown-winged Kingfisher is high on the agenda (although it is reasonably easy to see), and numerous other mangrove specialists can be found here, including the Greater Flameback, Mangrove Pitta, Mangrove Whistler and Mangrove Blue Flycatcher. Also look out for the Black-hooded Oriole, another bird confined only to Langkawi in Malaysia. The Oriental Scops Owl and Black-capped Kingfisher may be seen in winter.

Pulau Tuba & Pulau Dayang Bunting

These two islands are a little further off the beaten track, and can be reached from Kuah via a 10-minute ferry journey. Both islands remain densely forested, and it is best to explore their coastal forests by bike. The Greater Flameback, Mangrove and Blue-winged Pittas, Mangrove Whistler and Black-hooded Oriole can all be found here.

Access & Accommodation

Langkawi can be accessed by regular ferries that set off from the jetty at Kuala Perlis (one hour each way). The island is also served by regular domestic flights from Penang and Kuala Lumpur. When on the island it is best to be based in the area around Pantai Cenang, given the good variety of budget guesthouses and motels as well as an abundance of eateries. Luxury resorts are found around Burau Bay and Datai, the latter being particularly good for birdwatching. As public transport is limited, taxis and rented cars or scooters are best for exploring the sites on the island. Alternatively, you can drive onto Langkawi using the RORO ferry service at Kuala Perlis.

Conservation

Most of the low-lying areas on Langkawi Island have been cleared for agriculture. However, the forests on the rugged hills and tallest peaks have been gazetted as forest reserves, while the entire Langkawi archipelago has been afforded 'Geopark' status in recognition of its outstanding geological formations.

Teluk Air Tawar-Kuala Muda

Choy Wai Mun

Large flocks of sandpipers and plovers gather to feed on the mudflats along the Teluk Air Tawar coast during the northern winters.

KEY FACTS

Nearest Major Towns
Georgetown, Butterworth (Penang)

Habitats
Intertidal mudflats, mangroves and coastal scrub, orchards, ricefields

Key Species
Lesser Adjutant, Little Cormorant, Grey Heron, Pacific Reef Heron, White-bellied Sea Eagle, Little Bronze Cuckoo, Streak-eared Bulbul
Winter Indian and Javan Pond Herons, Chinese Egret, Far Eastern Curlew, Black-tailed Godwit, Great and Red Knots, Broad-billed and Spoon-billed Sandpipers, Little Stint, Asian Dowitcher, Nordmann's Greenshank, Brown-headed and Black-headed Gulls, Gull-billed and Lesser Crested Terns, Black-capped Kingfisher

Other Specialities
Malayan Water Monitor, Dog-faced Water Snake, Mangrove Pit Viper

Best Time to Visit
September–April

Located along the northern fringe of mainland Penang (Seberang Perai) south of the estuary of the Muda River, the intertidal mudflats and mangrove forests of the Teluk Air Tawar-Kuala Muda coastline cover more than 7,000ha and have been recognized as an Important Bird and Biodiversity Area (IBA) since 2007. From August onwards, thousands of migratory shorebirds arrive to stage on the extensive mudflats, with many continuing further south. More than 200 species have been recorded at the site so far, including the globally threatened **Chinese Egret**, **Far Eastern Curlew**, **Great Knot** and **Nordmann's Greenshank**. The mangrove belt fringing the mudflats support a number of resident species, including the **Lesser Adjutant** and increasing numbers of the **Little Cormorant**, a species formerly rare in Peninsular Malaysia. On a visit timed with the right tidal conditions, you can expect to see thousands of shorebirds of multiple species on the mudflats, and a good selection of residents.

Birdwatching Sites

Teluk Air Tawar-Kuala Muda Mudflats

The extensive mudflats along the coastline from Teluk Air Tawar north to Kuala Muda bordering Kedah state remain one of the most important sites for migratory shorebirds in Peninsular Malaysia, as well as large numbers of resident waterbirds. However, there are only a few sites here where the coastline is accessible due to the fringing mangrove forest belt. To access more areas along the coast, one viable option is to hire a small boat. Regularly seen species such as the **Lesser Sand Plover**, **Eurasian Curlew**, **Red-necked Stint** and **Common Redshank** congregate in huge numbers on the mudflats. Rarities such as the **Javan Pond Heron**, **Chinese Egret**, **Far Eastern Curlew**, **Spoon-billed Sandpiper**, **Little Stint**, **Asian Dowitcher**, **Nordmann's Greenshank** and **Black-headed Gull** have all been found here during the winter months. The enormous **Lesser Adjutant** and the striking **Pacific Reef Heron** are among the resident species to look out for.

Choy Wai Mun

*The captivating **Black-and-red Broadbill** is best detected by its cackling calls.*

Choy Wai Mun.

*The **Slender-billed Gull** typically winters on the coasts of the Middle East and India, but has showed up along the Penang coast*

Mangroves & Scrub

The fringing mangroves and adjoining coastal scrub on the coastline between Teluk Air Tawar to Kuala Muda support a fairly rich birdlife, including a number of species that do not occur further south in Peninsular Malaysia. These include Streak-eared and Red-whiskered Bulbuls, with the latter in decline across the region due to trapping pressure. Recently, small numbers of the formerly rare Little Cormorants have been observed nesting in the coastal vegetation. Migratory species such as the stunning Black-capped Kingfisher and Blue-tailed Bee-eater occur in the mangroves and scrub during the winter months.

Air Hitam Dalam Educational Forest

Recognized as one of the best birdwatching sites in Penang, this relatively compact site consists of remnant freshwater swamp forest along the Perai River, patches of scrub and adjoining areas of paddy fields. Located near Sungai Dua town on mainland Penang, the site sports a list of over 140 bird species, including the distinctive Asian Openbill, which has recently expanded its range into Peninsular Malaysia from populations further north in Thailand. Other species of interest to birdwatchers include the Green-billed Malkoha, Spotted Wood Owl, Ruddy Kingfisher, Streak-breasted Woodpecker (a northern specialty in Peninsular Malaysia), Black-and-red Broadbill, Blue-winged Pitta and Mangrove Blue Flycatcher. At dusk, the impressive Great Eared Nightjar, a localized specialty, is regularly encountered. From November onwards, a number of migratory cuckoos occur here, including Hodgson's Hawk-Cuckoo, as well as various flycatchers and warblers.

Access & Accommodation

The coastal mudflats and mangroves of Teluk

Choy Wai Mun.

*At certain times of the year, the resident **Ruddy Kingfisher** can be quite conspicuous.*

Air Tawar can be easily reached on a 45-minutes drive by car from Georgetown or Penang International Airport. An alternative is to stay in Butterworth. Car hire can be easily arranged in town or at the airport. It is also possible to take public buses to Teluk Air Tawar or Kuala Muda from Butterworth town. As Penang is a major tourist destination, there is a good variety of accommodation ranging from budget homestays to luxury hotels to choose from.

Conservation

Although listed as an Important Bird Area (IBA), the coastal wetlands of Teluk Air Tawar – Kuala Muda currently receive no formal protection and remain vulnerable to pollution, encroachment and future development of the coastline. The site also has notable economic value, and provides important ecosystem services, especially in supporting fisheries, coastal protection and ecotourism. The Malaysian Nature Society, Birdlife International and Cemex are working closely to advocate for the long-term conservation of this important wetland through engaging various stakeholders, including the state government.

*Confined mostly to coastal mangroves, the **Mangrove Blue Flycatcher** is regularly encountered in Air Hitam Dalam.*

Fraser's Hill

Cheong Weng Chun & Yong Ding Li

Yong Ding Li

Fraser's Hill has been largely spared from development and still provides easy access to excellent montane rainforest and associated birdlife.

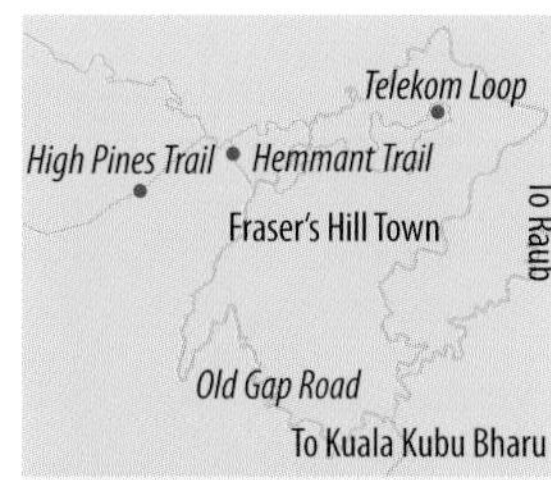

KEY FACTS

Nearest Major Towns
Kuala Kubu Bharu, Bentong (Selangor), Raub (Pahang), Kuala Lumpur

Habitats
Hill, submontane and montane rainforests

Key Species
Long-billed, Ferruginous and Malaysian Partridge*, Mountain Peacock-Pheasant*, Blyth's Hawk-Eagle, Black Eagle, Yellow-vented and Wedge-tailed Green Pigeons, Mountain Scops Owl, Brown Wood Owl, Orange-breasted and Red-headed Trogons, Rhinoceros and Great Hornbills, Fire-tufted, Red-throated and Yellow-crowned Barbets, Bamboo and Bay Woodpeckers, Speckled Piculet, Long-tailed and Silver-breasted Broadbills, Rusty-naped Pitta, Rail-babbler, Scaly-breasted Bulbul, Marbled and Streaked Wren-Babblers, Black and Malayan* Laughingthrushes, Himalayan Cutia, Rufous-browed Flycatcher, Malayan Whistling Thrush*
Winter Mugimaki Flycatcher, Eyebrowed and Siberian Thrushes

Other Specialities
Siamang, Southern Pig-tailed Macaque, White-thighed Surili, Sunda Slow Loris, Yellow-throated Marten, Red Giant Flying Squirrel

Best Time to Visit
April–July; rainy season November–January

Fraser's Hill is one of the most famous birdwatching sites in Peninsular Malaysia. Located along the densely forested Titiwangsa Range north of Kuala Lumpur and being largely spared from the wave of development affecting other hill stations in the peninsula, Fraser's Hill offers access to excellent submontane and lower montane forest across a range of elevation from as low as 500m to 1,310m asl at its highest point. With a comprehensive network of paved roads and hiking trails, visiting birdwatchers can opt to drive around in search of foraging flocks or hit the trails for the more elusive species. The cooler temperatures also permit activity throughout the day and can be a refreshing change from the humidity of the lowlands. All of West Malaysia's montane endemics have been recorded here, along with a host of difficult Sunda specialties including **Ferruginous Partridge**, **Red-throated Barbet**, **Rail-babbler**, **Marbled Wren-Babbler** and **Scaly-breasted Bulbul**. It is possible to see more than 120 species on a trip of 3-4 days.

Birdwatching Sites

Telekom Loop

Formally known as Jalan Girdle, this 10km loop road is best traversed by vehicle. Mixed foraging flocks are a regular sight in the forests along the road, and feature a range of species, notably the **White-throated Fantail**, **Blyth's Shrike-babbler**, **Mountain Fulvetta**, **Silver-eared Mesia** and **Long-tailed Sibia**. Careful scanning of these flocks may reward you with more

Con Foley

*Parties of **Long-tailed Broadbills** are regular in the forests at Fraser's Hill.*

Con Foley

*The near-endemic **Mountain Peacock-Pheasant** has been recorded in recent years at the High Pines Trail.*

uncommon flock participants such as the **Long-tailed Broadbill** and **Himalayan Cutia**. The gullies by the road are home to the shy **Rusty-naped Pitta** (ssp. *deborah*), although seeing it requires a great deal of luck.

Hemmant Trail

Arguably the best-marked and least demanding hiking trail on Fraser's Hill, the Hemmant Trail provides an excellent introduction to birdwatching in a closed-canopy setting. In addition to the foraging flocks that can be seen anywhere along the trail, the forest understorey is home to other species such as the **Rufous-browed Flycatcher** and **White-tailed Robin**. The roadside forest between this trail and Bishop's Trail is particularly good for the **Malaysian Partridge**, which is often seen either crossing the road or foraging on the slope adjacent to Silverpark Resort.

High Pines Trail

Located close to the summit of Fraser's Hill at 1,310m asl, this is the longest hiking trail at the site, requiring a 6–7-hour journey to the end at Pine Tree Hill and back (c. 5.5km). Birdwatchers do not need to walk quite so far. The first few hundred metres of the trail should be checked at dawn and dusk for a foraging **Mountain Peacock-Pheasant**, while the **Bay Woodpecker** and **Himalayan Cutia** have been recorded in

Con Foley

*The **Malayan Laughingthrush** is one of three possible laughingthrushes that can be seen.*

Sreedharan Gopalsamy

*TThe near-endemic **Malaysian Partridge** is best seen at Fraser's Hill.*

Abdelhamid Bizid

*The elusive **Marbled Wren-Babbler** is a key target for many birdwatchers visiting Fraser's Hill.*

Con Foley

*The **Common Green Magpie** is a favourite among birdwatchers.*

foraging flocks along the trail. The **Long-billed Partridge** has been regularly heard about 2km from the start of the trail, but seeing it is an entirely different matter.

The Gap Area & Old Gap Road

Situated at the border between the states of Selangor and Pahang, the Gap and the 'Old Road' is the only entry point into Fraser's Hill (the 'New Road' is for exit only). As it is located at a lower elevation (average of 700–800m asl) than Fraser's Hill, a different set of birds inhabits the area, including a variety of lowland and submontane Sundaic specialities. The site is best explored in the early morning, with roadside birdwatching potentially yielding such specialities as the **Yellow-vented Green Pigeon**, **Bamboo Woodpecker**, **Scaly-breasted Bulbul** and **Black Laughingthrush**. The numerous densely vegetated gullies should be checked for the shy **Marbled Wren-Babbler**, while both **Ferruginous** and **Long-billed Partridges**, as well as the **Rail-babbler**, have been recorded along the 'Petak Fenologi' forest trail at the 2km marker. The upper 1km of the trail, especially the area immediately below the gate, is where the **Malayan Whistling Thrush** has been seen in recent years, usually only around daybreak.

Access & Accommodation

Fraser's Hill is 100km from Kuala Lumpur and can be easily accessed after a two-hour drive, the last 60km of which comprises an ascent up a winding road (Jalan Kuala Kubu Bharu-Teranum) to the hill station proper at 1,165m asl. For those not driving, the option is to take the *Komuter* train to Kuala Kubu Bahru and to charter a taxi up to Fraser's Hill. There is a variety of accommodation options around the town centre, the most popular being the Puncak Inn. Although a permit and guide are not required to explore the area, several hiking trails are overgrown and/or lengthy, and it is advisable to inform the local police station or hotel staff of your hiking plans if a long stint in the forest is needed.

Conservation

The Titiwangsa Range, of which Fraser's Hill is a part, encompasses the largest block of Peninsular Malaysia's remaining montane forest cover, but much of it does not receive formal protection. Parts of the Fraser's Hill area are protected as a wildlife sanctuary. Around Fraser's Hill, landslides, aggravated by development in the area as well as hunting, are local threats. The hill station has also experienced noticeable warming in recent years, with an increase in average temperature by 2–3° C in the last two decades. This may have resulted in the disappearance of several species of upper montane forest bird, such as the Brown Bullfinch.

Con Foley

*Mixed flocks are regular in the forests of Fraser's Hill, and are regularly attended by the **Black-and-crimson Oriole**.*

Kuala Selangor Nature Park

Cheong Weng Chun

Cheong Weng Chun

Kuala Selangor National Park is a readily accessible site from Kuala Lumpur that offers visitors a chance to see various mangrove specialities.

KEY FACTS

Nearest Major Towns
Kuala Selangor, Kuala Lumpur

Habitats
Mangrove forest, secondary forest and scrub, brackish lakes, coastal mudflats

Key Species
Milky Stork (rare), Lesser Adjutant, Chestnut-bellied Malkoha, Barred Eagle-Owl, Buffy Fish Owl, Stork-billed Kingfisher, Greater Flameback, Mangrove Pitta, Mangrove Whistler, Cinereous Tit, Mangrove Blue Flycatcher, Ashy Tailorbird, Copper-throated Sunbird
Winter Black Baza, Chestnut-winged Cuckoo, Ruddy and Black-capped Kingfishers, Blue-winged Pitta, Tiger Shrike, Yellow-rumped Flycatcher, Forest Wagtail

Other Specialities
Silvered Leaf Monkey, Long-tailed Macaque, Smooth-coated Otter

Best Time to Visit
September–May

Situated about 70km north-west of Kuala Lumpur, Kuala Selangor Nature Park is a 2.97km² nature park located within the north-central Selangor coast Important Bird and Biodiversity Area. It was established in 1987, and since then more than 168 bird species have been recorded in and around it. Kuala Selangor comprises three main habitats, namely coastal secondary forest mainly consisting of mature strangling fig trees with climbers and ferns, a 0.1km² open brackish water lake system and a remnant patch of mangrove forest. There are several trails within the secondary forest and a boardwalk through the mangroves. For the visiting birdwatcher, Kuala Selangor's main attraction is the convenience of observing the majority of Peninsular Malaysia's mangrove specialists and a good selection of widespread waterbirds. The park supports several globally threatened waterbirds, notably both the **Milky Stork** and **Lesser Adjutant**, as well as several mangrove specialities with a patchy global distribution, including the **Mangrove Pitta**, **Mangrove Whistler** and **Mangrove Blue Flycatcher**. At low tide various shorebirds occur in the coastal mudflats during the northern winter, including godwits and **Asian Dowitchers**.

Birdwatching Sites

Secondary Forest

The secondary forest around the car park contain a number of mature fig trees that attract various fruit-eating birds. In winter, migratory species such as the **Blue-winged Pitta**, **Ashy**

Con Foley

*The striking **Greater Flameback** is generally found around mangroves and coastal forests in Peninsular Malaysia.*

Minivet and **Yellow-rumped Flycatcher** have all been seen here. It is worth scanning the trees and stumps in the area, as well as the trail just past the visitor centre, for roosting **Barred Eagle-Owls**.

Coastal Bund & Brackish Water Lake

This habitat starts after crossing the bridge from the secondary forest. A nearby watch tower offers excellent viewing of waterbirds such as **Purple** and **Grey Herons**, which often perch on the tops of the trees along the bund. Other species recorded in the area include the **Little Bronze Cuckoo**, **Lineated Barbet** and **Greater Flameback**. The **Crested Serpent Eagle** and **Brahminy Kite** can often been seen soaring above the lake.

Mangroves

The boardwalk that loops around the mangrove forest fringing the coast provides excellent opportunities to see a number of mangrove specialities, especially the **Mangrove Pitta** (rare), **Mangrove Whistler**, **Cinereous Tit** (ssp. *ambiguus*) and **Mangrove Blue Flycatcher**. During the northern winter, the **Chestnut-winged Cuckoo** and **Ruddy Kingfisher** are possible here.

Access & Accommodation

The park is readily accessed from Kuala Lumpur via the Kuala Lumpur–Kuala Selangor Highway, and can be reached after an hour and a half's drive. Chalet accommodation in the park is currently unavailable, but there is a good range of hotels available in Kuala Selangor town.

Conservation

The park is managed by the Malaysian Nature Society. Water pollution from surrounding aquaculture farms, as well as their uncontrolled expansion along the coastline, is a constant threat to the integrity of the park's mangrove forests and inter-tidal habitats.

*The unobtrusive **Chestnut-bellied Malkoha** can be seen in the coastal forest at Kuala Selangor.*

Sreedharan Gopalsamy

*With luck, **Buffy Fish Owls** may be seen in the secondary forest within the park.*

Panti Forest Reserve

Yong Ding Li & Con Foley

Yong Ding Li

Gunung Panti as seen from the foothills at the Bunker Track.

KEY FACTS

Nearest Major Town
Kota Tinggi (Johor)

Habitats
Lowland and hill rainforests, freshwater swamp forest

Key Species
Crested Partridge, Great Argus, Short-toed Coucal, Blyth's and Gould's Frogmouths, Cinnamon-rumped, Red-naped and Diard's Trogons, Banded, Rufous-collared and Blue-banded Kingfishers, Rhinoceros, Black and Wrinkled Hornbills, Banded, Black-and-yellow, Dusky and Green Broadbills, Giant and Malayan Banded Pittas, Rail-babbler, Grey-breasted, White-necked and Black-throated Babblers, Rufous-tailed Shama, Brown-backed Flowerpecker
Winter Green-backed, Mugimaki and Yellow-rumped Flycatchers, Eyebrowed Thrush, Brown-chested Jungle Flycatcher

Other Specialities
Sunda Slow Loris, Southern Pig-tailed Macaque, Dusky and Banded Langurs, White-handed Gibbon, Leopard, Banded Linsang, Flat-headed Cat, Asian Elephant, Malayan Tapir, Sambar, Bearded Pig

Best Time to Visit
March–September

The relatively low-lying state of Johor in the southern quarter of Peninsular Malaysia was once covered with vast expanses of rainforest and freshwater swamps. However, as Malaysia's economic development took off, the species-rich rainforests of these lowlands were soon replaced by endless stretches of oil-palm and rubber plantations. Situated along the upper reaches of the Johor River is Panti Forest Reserve, one of the larger rainforest patches (*c.* 100km^2) remaining that is close to the state capital of Johor Bahru. The defining feature of this reserve is its series of tall hills, notably the ridges of Gunung Panti and Sumalayang. Although its foothills consist mainly of a mosaic of largely secondary and logged forests, it continues to draw birdwatchers from afar, many of whom come in search of the remarkable **Rail-babbler**, perhaps best seen here than anywhere else in its range. The reserve sports a list of nearly 300 species, and with a bit of effort it is possible to see more than 120 in a 3–4-day trip, including a large set of Sundaic lowland forest specialities like trogons, babblers and flowerpeckers.

Birdwatching Sites

Bunker Track

An old logging road stretching nearly 8km along the foothills of Gunung Panti, Bunker Track has remained the heart of the birdwatching action in the reserve for more than a decade. The track allows for easy roadside birdwatching and a frequent strategy among visitors is to drive slowly, stopping whenever bird activity is detected. A good variety of woodpeckers, malkohas, bulbuls, babblers, sunbirds and trogons can be seen this way, as well as mixed

*The shy **Cinnamon-rumped Trogon** may be seen when birdwatching in Panti.*

Con Foley

Panti Forest Reserve is one of the easiest place in the world to see the monotypic and highly sought-after ***Rail-babbler****.*

Con Foley

Blyth's Frogmouth *is common by call at Panti.*

Con Foley

The ***White-necked Babbler*** *is one of the rarest of the lowland babblers in Malaysia, but has been encountered in Panti in recent years.*

flocks of minivets, ioras and flowerpeckers. Hornbills appear to be increasingly rare, with **Black** and **Rhinoceros Hornbills** being the species most often seen. Flowering trees along the track usually attract many bulbuls and leafbirds, including the scarce **Finsch's** and **Black-and-white Bulbuls**. There are numerous forest trails that branch off the main track, and these are where you should keep an eye out for shyer species like the **Crested Partridge**, **Rufous-collared Kingfisher**, **Malayan Banded Pitta**, **Rail-babbler**, **Rufous-tailed Shama** and, very occasionally, the **Giant Pitta**.

About 5km from the start of the track is a large clearing where a stream crosses the track. Frequented by both birdwatchers and campers, it is often rewarding to check the stretch of forest along the track on both sides of the stream for sunbirds, leafbirds and flowerpeckers coming low to feed in *Melastoma* and *Clidemia* bushes. Both the rare **White-necked Babbler** and **Olive-backed Woodpecker** have been encountered here, besides the usual supporting cast of bulbuls and babblers. Just before the forest fizzles out near the 8km mark, the few standing fig trees attract barbets, hornbills, bulbuls and, in recent years, the rare **Brown-backed Flowerpecker**.

Panti Recreational Forest

Although less frequently visited than the above, the recreation forest provides easy access to some swamp forest and has good access to facilities for camping, including public toilets and parking grounds. A narrow trail penetrates the forest for a couple of hundred metres before fading into the swamp. Besides the usual set of lowland specialities, the **Rail-babbler**, **Crested Jay** and rare **Cinnamon-rumped Trogon** have all been encountered here.

Gunung Arong Forest

About 80km or roughly one hour's drive from Bunker Track is the Gunung Arong Forest, a relatively undisturbed patch of lowland dipterocarp and swamp forest north of Mersing. As in the case of the Bunker Track, you can drive along Jalan Tanjung Resang and stop to look at fruiting or flowering trees. The main draw for birdwatchers here is the stunning **Scarlet-breasted Flowerpecker**, a scarce speciality of lowland and nutrient-poor forests in the region. Other species of note include the **Spotted Fantail**, **Brown Fulvetta** and **Striped Wren-Babbler**.

Access & Accommodation

Due to the proximity of Kota Tinggi town, you can be based at the range of hotel accommodation in town and make daily trips to the birdwatching spots in the morning. The hotels can arrange taxis. Many birdwatchers also make day trips while being based in Singapore or Johor Bahru. Visitors planning to spend some time in Gunung Arong Forest may want to consider staying overnight in the range of hotels located at the nearby town of Mersing instead. Currently, a permit must be obtained from the forestry department in advance to access the forest at Bunker Track.

The ***Malayan Banded Pitta*** *is the most regularly seen member of the Pittidae at Panti Forest Reserve.*

Conservation

Panti Forest Reserve consists of both production forests and protected areas. Much of the lowlands around the foothills of Gunung Panti were logged, and are likely to be relogged in the future. Logged-over parts of the reserve are now planted with oil palm. Sand quarrying is still being carried out in the reserve, with trucks carrying sand passing the Bunker Track regularly. Several sections along the Bunker Track were gazetted as the Panti Bird Sanctuary a few years ago, and some attempt has been made to provide basic visitor facilities along the track. Poaching of animals for the bush-meat trade, as evidenced by snare traps, remains a problem in parts of the reserve.

Con Foley

Various Sundaic lowland rainforest birds can be easily encountered, including ***Chestnut-rumped Babbler*** *(right) and the charming* ***Banded Broadbill*** *(left).*

Kinabalu Park

Yong Ding Li

The forested slopes of Gunung Kinabalu, one of Southeast Asia's tallest mountains, are home to the vast majority of Borneo's endemic montane birds.

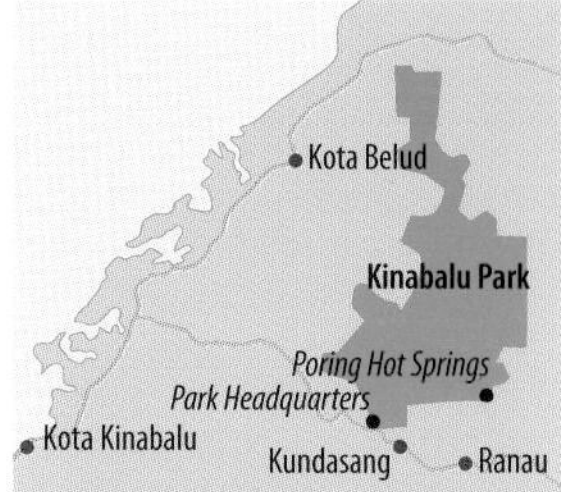

KEY FACTS

Nearest Major Towns
Kota Kinabalu, Kundasang

Habitats
Hill dipterocarp forests, lower and upper montane evergreen forests, alpine scrub

Key Species
Lowland Rufous-collared and Blue-banded Kingfishers, Bornean Barbet*, Hose's Broadbill*, Blue-banded* and Bornean Banded* Pittas, White-necked Babbler, Chestnut-capped Thrush, Chestnut-naped Forktail, Bornean Blue* and White-tailed Flycatchers
Montane Crimson-headed* and Red-breasted* Partridges, Mountain Serpent Eagle*, Mountain and Rajah* (rare) Scops Owls, Collared Owlet, Bornean Swiftlet*, Giant Swiftlet, Whitehead's Trogon*, Golden-naped* and Mountain* Barbets, Whitehead's* and Hose's* Broadbills, Bornean Green Magpie*, Bornean Stubtail*, Friendly Bush Warbler*, Mountain Wren-Babbler*, Bare-headed*, Sunda and Chestnut-hooded* Laughingthrushes, Mountain Blackeye*, Everett's Thrush*, Fruithunter*, Eyebrowed Jungle Flycatcher*, Whitehead's Spiderhunter*
Winter Blue-and-white, Mugimaki and Narcissus Flycatcherers, Pechora Pipit

Other Specialities
Mountain Treeshrew, Kinabalu, Jentink's, Bornean Mountain Ground and Whitehead's Pygmy Squirrels

Best Time to Visit
March–September; rainy season November–January

Rising to more than 4,090m asl, Gunung Kinabalu is perhaps Malaysia's best-known geological feature, visited annually by thousands of climbers, hikers and birdwatchers. Many biologists also recognize its importance to biodiversity, and describe it as the single most significant biogeographic feature on Borneo. Technically, Kinabalu is not a single mountain, but a raised granite massif on which sits a number of peaks, near the northern end of the Crocker Range. Given its broad altitudinal range, Gunung Kinabalu supports nearly the full complement of vegetation zones on Borneo, from lowland rainforests around its northern fringes to the high-elevation alpine scrub (above 3,000m asl) that occurs on few other mountains in the Greater Sundas. The 750km² park has a bird list in excess of 300 species, although many are lowland birds that occur at the fringes of the park seldom visited by tourists. From a birdwatching perspective, a visit to Kinabalu Park is a must on any itinerary to Sabah. The park's extensive trail network provides easy access to a wide range of habitats, and many of Borneo's montane endemic birds are easier to see here than anywhere else on the island.

Abdelhamid Bizid

The plumage of the hulking ***Whitehead's Broadbill*** *offers superb camouflage among the dense foliage of the montane rainforest.*

Birdwatching Sites

Power Station Road

Besides the buses that ply between the summit trail and the park headquarters, there are usually few vehicles on this road. Walking along the Power Station Road in the early morning provides an excellent introduction to many of the common forest species in the park. On top of the regularly seen mixed flocks of warblers, babblers, flowerpeckers and white-eyes, also look out for fruiting or flowering trees, which often attract **Golden-naped Barbets**, **Fruithunters**, cuckoo-doves, and with luck the elusive **Whitehead's Spiderhunter**. Many birds, like the **Sunda Cuckooshrike**, **Bornean Green Magpie** and **Chestnut-hooded Laughingthrush**, are often drawn to the large insects attracted to the lamp posts in the mornings. **Bornean Whistling Thrushes** are also best seen along the roadside in the early mornings, while the **Mountain Serpent Eagle** can sometimes be seen soaring from vantage points along the road.

Silau-Silau Trail

Most tall mountains in Southeast Asia are often inaccessible or have very limited visitor infrastructure, but this is not the case for Gunung Kinabalu. The park headquarters is not only easily accessed via a major road, but also contains a well-managed network of forest trails that provide unparalleled access to excellent forests for birdwatching. The most popular trail among birdwatchers is the Silau-Silau Trail, which runs parallel to the Power Station Road; its southern end links to the Liwagu river trail. The trail is best walked in the mornings, when mixed flocks of warblers, fantails, white-eyes and the exquisite **Temminck's Sunbird** can often be encountered. **Bornean Stubtails** are easy to hear along the trail, while **Bornean Forktails** haunt the stream that flows alongside the trail. In the evenings, small flocks of **Temminck's** and **Grey-throated Babblers** can be seen bathing in the stream. With a bit of effort both **Whitehead's Trogon** and **Broadbill** can be encountered on the trail.

*The diminutive **Bornean Stubtail** utters such a high-pitched call that it is inaudible to many visiting birdwatchers.*

Bukit Ular Trail

This zigzagging, densely forested trail starts near the upper quarter of the Power Station Road, and ends behind the power station near the start of the summit trail at Timpohon Gate. Both **Red-breasted** and **Crimson-headed Partridges** are regularly heard and seen on the

James Eaton

*The fiery **Whitehead's Trogon** glows in the dim understorey of Mount Kinabalu's montane forests.*

*The **Bornean Treepie** is common around the park headquarters.*

trail. The main draw is **Everett's Thrush**, which is occasionally seen here just after first light.

Liwagu River Trail

Stretching nearly 6km from its upper end near Timpohon Gate to where it meets the Silau-Silau Trail, this trail can be physically challenging as it is very steep along some sections, and landslips have occurred in a number of areas. Besides the usual species, the Collared Owlet, Whitehead's Trogon, Bornean Green Magpie, Fruithunter and Tawny-breasted Parrotfinch are present here. Also pay attention to the large mixed flocks that forage along the trail as these are often joined by the scarce **Bare-headed Laughingthrush**.

*The **Black-sided Flowerpecker** is frequently observed around fruiting trees.*

*The vegetation around Laban Rata is a good starting point to search for the **Friendly Bush Warbler**.*

*One of the rewards for exploring the forests at Poring is a chance to see the **Bornean Banded Pitta**.*

Summit Trail

To see the higher elevation species, it is necessary to hike up this trail. However, doing so first requires registration at the park headquarters. **Mountain Blackeyes** are common along the trail and should be readily encountered, but seeing the **Friendly Bush Warbler** requires a hike to at least the Layang-Layang Hut (2,700m asl). This trail can also be accessed from the Mesilau Nature Resort, which starts at a higher elevation (2,000m asl) than at Timpohon.

Poring Hot Springs

The main attraction for birdwatchers at Poring is the canopy walk and the 5km-long Langanan Waterfall Trail. Poring is nearer to the town of Ranau and getting there from the park headquarters involves at least a 45-minute drive. The trail rises gently, but eventually becomes more rugged a kilometre from the start. It is possible to see a good selection of lowland and hill forest species here, including the shy **Rufous-collared Kingfisher**, **Blue-banded Pitta** and **Chestnut-capped Thrush**. To see **Hose's Broadbill**, you need to first find a fruiting tree to increase the chances of an encounter with this stunning endemic.

Access & Accommodation

Kinabalu Park headquarters is readily accessed via the main road east from Kota Kinabalu, with the drive taking about two hours. Public buses to towns further east, like Ranau, Lahad Datu and Sandakan, ply this road and you can ask to be dropped off here on the eastbound buses. A wide range of accommodation is available within the park (Sutera Sanctuary Lodges) and on the main road outside the park headquarters. Chalets and dormitory-type accommodation are also available at Poring.

Conservation

Due to the significance of Gunung Kinabalu as a geological and biological feature, on top of its tourism value, the park has received strong protection from the Sabah state government. The park was also designated as Malaysia's first UNESCO World Heritage Site in 2000.

Danum Valley Conservation Area

Yong Ding Li

Low Bing Wen

The pristine dipterocarp forests of Danum Valley offers some of the best forest birdwatching in Southeast Asia.

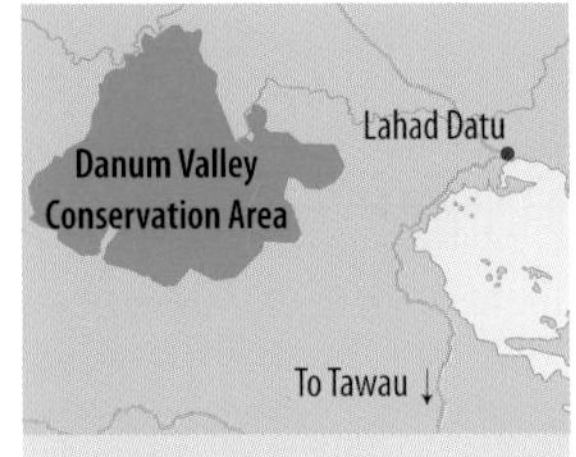

KEY FACTS

Nearest Major Town
Lahad Datu

Habitats
Lowland and hill rainforests; lower montane rainforests (Gunung Danum), caves

Key Species
Crested Fireback, Bulwer's Pheasant* (rare), Great Argus, Crested Partridge, Chestnut-necklaced Partridge, Great-billed Heron, Bat Hawk, Wallace's Hawk-Eagle, Large Green Pigeon, Bornean Ground Cuckoo*, Moustached Hawk-Cuckoo, Large and Gould's Frogmouths, Bonaparte's Nightjar (rare), Helmeted Hornbill, Bornean Banded*, Blue-headed*, Black-crowned*, Blue-banded* and Giant Pittas, Bornean Bristlehead*, Crested Jay, Black Magpie, Finsch's and Straw-headed Bulbuls, Black-throated*, Bornean* and Striped Wren-Babblers, Pygmy White-eye*, Chestnut-capped Thrush, Sunda, Malaysian and Bornean Blue* Flycatchers
Winter Von Schrenck's Bittern, Blue-winged and Fairy Pittas, Blue-and-white and Narcissus Flycatchers

Other Specialities
Sunda Colugo, Sunda Pangolin, Horsfield's Tarsier, Bornean Slow Loris*, Bornean Gibbon*, Maroon* and Hose's* Langurs, Bornean Orangutan*, Sun Bear, Sunda Clouded Leopard, Asian Golden, Marbled and Flat-headed Cats, Asian Elephant, Sambar, Banteng

Best Time to Visit
March–September; rainy season November–January

Stretching from Sabah's rugged interior to the coast, the vast Yayasan Sabah Forest Concession contains some of the most pristine lowland forest left in Borneo, including the fabled Danum Valley Conservation Area and its associated Field Centre (DVFC). Comprising about 438km^2 of mostly hill and lowland dipterocarp forests, the reserve protects parts of the headwaters of the Segama River, which rises to a maximum elevation of 1,090m asl at Gunung Danum. Gazetted since 1995, the conservation area is surrounded by a larger mosaic of logged and secondary forest, and smaller patches of primary forest at its core. Given its relatively undisturbed condition and wide elevation range, this area is exceptional for Borneo's biodiversity, and is known to support nearly 300 bird species as well as a healthy community of forest mammals, including important populations of Asian Elephants and Bornean Orangutans. Besides the charismatic **Bornean Bristlehead**, Danum's forests support notably high densities of birds, and many shy ground birds like the **Great Argus**, **Crested Fireback** (ssp. *nobilis*) and various pittas are probably more easily seen here than anywhere else in Borneo.

Birdwatching Sites

Danum Valley Grid Trails

Accessed by a suspension bridge across the Segama River, this network of trails is connected to the main artery, the East (Clive Marsh Trail), which continues onto the hilly Rhino Ridge Trail. Although

Bjorn Olesen

The skulking ***Black-throated Wren-Babbler*** *often betrays its presence with its melodious and far-carrying song.*

*The **Chestnut-necklaced Partridge** (ssp. graydoni) is the most common of the three species of partridge found around Danum Valley.*

*The Critically Endangered **Helmeted Hornbill** is still locally common in the forest at Danum Valley.*

viewing conditions may be challenging, birdwatchers should consider spending a day or two on this trail to get acquainted with many of Borneo's lowland species, as well as a number of scarce ground birds like the **Giant Pitta** and **Bornean Wren-Babbler**. The section along the riverbank is comparatively more open, and here is where the **Chestnut-necklaced Partridge** (ssp. *graydoni*), various pittas and **White-crowned Shama** can be seen. In the past, **Bulwer's Pheasant** has been encountered on the Rhino Ridge Trail, but trekkers should note that this trail is not well marked.

Danum Valley Waterfall Trail

This trail starts from behind the field centre and gradually increases in altitude, allowing access to excellent hill forest. While the **Bornean Ground Cuckoo** has been sporadically encountered here, other sought-after birds can also be found, including various trogons, the **Rufous-collared Kingfisher**, **Rufous-tailed Shama**, and flycatchers and pittas. The grating mew of the **Bornean Bristlehead** is regularly heard, and birdwatchers should scrutinize mixed flocks of malkohas and woodpeckers, which the bristleheads regularly join.

Access Road to Danum Valley Field Centre

Much of the forest along the access road has been logged in the past and some sections are relatively open, facilitating the scanning of tree crowns for forest birds. Besides the **Crested Fireback**, various bulbuls, babblers, malkohas and woodpeckers can be easily seen by slowly walking along the access road in the direction of the Borneo Rainforest Lodge, and a small viewing tower about 1.5km from the field centre provides great vistas of the surrounding forest. **Bat Hawks** are regularly seen here. Birdwatchers should also look out for fruiting fig trees, which often draw many pigeons and hornbills, including the desirable **Large Green Pigeon**.

Borneo Rainforest Lodge (BRL) Trails

The trails around the lodge support a variety of sought-after birds within easy reach of the accommodation. Even the trees within and around the lodge attract birds, including leafbirds, woodpeckers, flycatchers and the occasional fireback, while the river behind is where the **Great-billed Heron** is sometimes seen. One of the best trails for birdwatchers is the Sapa Babandil Trail, which hugs the Segama River's western edge, and can be good for the **Chestnut-necklaced Partridge**, **Blue-headed Pitta**, **Crested Jay**, trogons, hornbills, broadbills, blue flycatchers and babblers. The trail then crosses the BRL access road and continues to the Hornbill Trail, which is probably the best place to see the **Great Argus**. **Bornean** and **Black-throated Wren-Babblers** have also been recorded on the trail. The Coffin Cliff Trail, named for archaeological relics found there in the past, is where the scarce **Blue-banded Pitta** has been seen regularly.

Borneo Rainforest Lodge Canopy Walk

Accessed from the Sapa Babandil Trail, the canopy walkway not only provides excellent views of the surrounding virgin forests, but allows birdwatchers to observe the many mixed flocks that move through the area, comprising minivets, ioras, woodshrikes, nuthatches and woodpeckers. Broadbills and hornbills, including the magnificent **Helmeted Hornbill**, are regularly seen here. The **Bornean Bristlehead** is also often sighted from the walkway.

Borneo Rainforest Lodge Access Road

Besides the usual forest birds, **Bornean Banded Pittas** and **Black-throated Wren-Babblers** have been regularly reported from the forest along the access road. The road is also excellent for spotlighting sessions at night, with the sought-after **Large Frogmouth** and **Brown Wood Owl** possible.

Access & Accommodation

Transport to access Danum Valley is only available from Lahad Datu, but prior arrangements must be made. At Danum Valley a range of dormitory rooms and chalets is available at the field centre. The only other option is to stay at the Borneo Rainforest Lodge, which is situated 30km away. All accommodation and access permits to Danum Valley Field Centre and the Borneo Rainforest Lodge must be pre-arranged. More details are available at www.borneonaturetours.com.

*The **Black-crowned Pitta** is common in Danum's forests.*

Conservation

Although part of a large logging concession, the Danum Valley Conservation Area is currently secure from exploitation, given the special status allocated to it. Due to the large buffer area of logged and secondary forest there is fairly little human disturbance. Poaching has been reported in the periphery of the area, and the small population of Sumatran Rhinoceros that used to occur has since been extirpated.

Cheng Heng Yee

*The elusive **Giant Pitta** is one of Danum's star birds.*

*The mournful whistles of the **Rufous-tailed Shama** are regularly heard on the Waterfall Trail.*

Mantanani Islands

Yong Ding Li

Christmas Frigatebirds are best seen when they return to roost at dusk.

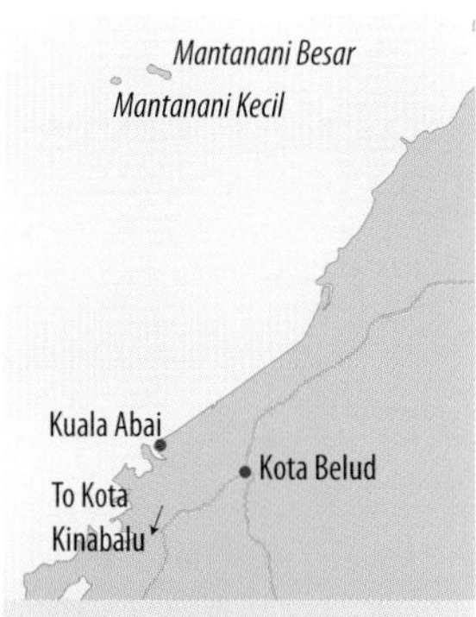

KEY FACTS

Nearest Major Towns
Kuala Abai, Kota Belud

Habitats
Coastal forest, *Casuarina* groves, sandy beaches

Key Species
Philippine Megapode, Pacific Reef Heron, Great, Lesser and Christmas Frigatebirds, Grey and Pied Imperial Pigeons, Metallic Pigeon, Nicobar Pigeon, Mantanani Scops Owl, Blue-naped Parrot
Passage Streaked Shearwater, Japanese Sparrowhawk, Northern Boobook, White-throated Needletail, Chestnut-cheeked Starling, Blue-and-white, Grey-streaked and Dark-sided Flycatchers, Pechora Pipit

Other Specialities
Island Flying Fox, Dugong

Best Time to Visit
April–November

This group of forested islands 50km off the coast of Sabah may not at first seem significant from a birdwatching perspective, but it gave its name to the **Mantanani Scops Owl**, a species confined to this and a few islands in the Philippines. Comprising three small rocky islands, Mantanani Besar is the largest and most interesting for visiting birdwatchers as it supports a few different habitats, home to a number of species hard to see elsewhere, notably the threatened **Grey Imperial Pigeon**. However, the main highlight is the large groups of frigatebirds of three species, including the Critically Endangered **Christmas Frigatebird**, which roosts nearby and covers the skies of Mantanani every evening.

Birdwatching Sites

Mantanani Besar

The north-western end of the island is covered with coastal forests, and fringed by coconut groves and plantation gardens. These remnant forests can be accessed via a steep trail that starts near the site of the old Mantanani Dive Resort. With luck, the **Mantanani Scops Owl** (ssp. *mantananensis*) can be seen here during the day, although it is known to occur throughout the island and is regularly heard after dark. You may also be able to find the **Metallic Pigeon** and threatened and localized **Grey Imperial Pigeon** by checking the flocks of pigeons present, although the **Pied Imperial Pigeon** is usually the dominant species. Early mornings are the best time to see the secretive **Philippine Megapode**, of which a few pairs still occur.

From late afternoon onwards large groups of frigatebirds starts arriving to roost on Lugisan, a small forested islet off the northern coast of Mantanani Besar. You should be able to see these frigatebirds from anywhere on the island, although the best vantage point is probably on the beaches leading to the old resort. More than 5,000 frigatebirds have been counted, the majority

Frigatebirds circling over Pulau Lungisan in the evening.

Yann Muzika

Uthai Treesucon

*With luck, the **Nicobar Pigeon** can be found in the coastal forests of Mantanani Kecil.*

comprising **Lesser Frigatebirds** and smaller numbers being **Christmas Frigatebirds**.

Due to the position of the islands relative to the island of Borneo many migratory land birds move through the islands during autumn passage. Recent visitors have included the **Northern Boobook**, migrating **Japanese Sparrowhawks**, and a number of passerines such as various flycatchers, the **Pechora Pipit**, and regional rarities like the **Willow Warbler** and **Black-headed Bunting**. Birdwatchers to the islands in October–November should concentrate on the scrub, woodland and plantations to find these temporary visitors. Also look out for seabirds during the sea crossing between Kuala Abai and the mainland, as the **Streaked Shearwater**, various skuas and the **Red-necked Phalarope** have been recorded.

Mantanani Kechil

This small forested island is about 10 minutes by boat from the dive resort on Mantanani Besar. The **Nicobar Pigeon** has been recorded here recently and may occur anywhere on the island. Other key species include the **Philippine Megapode** and various arboreal pigeons.

Access & Accommodation

Currently, the only places to stay are the Rasa Sayang Mantanani Resort and the Mari Mari Backpackers Lodge, a comfortable resort catering mostly for visiting divers. A ferry service shuttles between the jetty to the islands, but is mainly arranged for visiting divers. Visitors should contact the lodge for more details.

Conservation

The islands are protected as part of the Mantanani Islands Wildlife Sanctuary. Poaching of marine animals may still occur sporadically, and some of the reefs around the island were damaged in the past by dynamite fishing.

*The **Mantanani Scops Owl** is a common species in the island's forests.*

Crocker Range National Park

Wong Tsu Shi

Low Bing Wen

A paved road through the Crocker Range National Park offers easy access to mid-montane birds that are very difficult to see at Kinabalu Park.

KEY FACTS

Nearest Major Towns
Tambunan, Penampang, Keningau

Habitats
Hill rainforest, submontane and montane rainforests

Key Species
Mountain Serpent Eagle*, Blyth's Hawk-Eagle, Besra, Sunda Cuckoo, Dark and Moustached Hawk-Cuckoos (rare), Collared Owlet, Bornean Swiftlet*, Bornean*, Mountain* and Golden-naped* Barbets, Whitehead's* and Long-tailed Broadbills, Bornean Whistler*, Bornean Banded Pitta*, Flavescent and Bornean* Bulbuls, Bare-headed*, Chestnut-hooded* and Sunda Laughingthrushes, Pygmy White-eye*, Mountain Blackeye*, Everett's* and Orange-headed Thrushes, Fruithunter, Hill Blue Flycatcher, Bornean Forktail*, Bornean Leafbird*, Whitehead's Spiderhunter*
Winter Blue-and-white, Mugimaki and Narcissus Flycatchers

Other Specialities
Rafflesia keithii (seasonal)

Best Time to Visit
March–July; rainy season November–February

The Crocker Range National Park is the largest terrestrial national park in Sabah, covering an area of 1,399km^2 and spanning an altitudinal range of 200–2,050m asl, reaching its highest point at Gunung Alab. The park encompasses eight districts and is an important water catchment for 12 major rivers that supply water to the west coast of Sabah. The park has become a permanent feature of birdwatching tours to Sabah in recent years, as a major road cutting through the park offers easy access to lower montane forest at around 1,000m asl, and an associated suite of birds that may be very difficult to find at Gunung Kinabalu. A day trip from the city of Kota Kinabalu can yield sightings of a range of Bornean endemics, including the diminutive **Pygmy White-eye**, **Fruithunter** and elusive **Whitehead's Spiderhunter**.

Birdwatching Sites

Crocker Range Park Headquarters

Located above Keningau town at an elevation of around 950m asl, this site offers easy access to good-quality submontane forest and associated birdlife. While inside the park headquarters follow the cemented footpath to the Insectarium, located next to the Crocker Nature Centre. Birdwatching around the Insectarium can yield endemics such as the **Bornean Banded Pitta** and **Pygmy White-eye**. Alternatively, a walk on the 2km Crocker Nature Trail offers

Wong Tsu Shi

*The elusive **Whitehead's Spiderhunter** is more regularly observed around the Crocker Range National Park than at Kinabalu.*

additional opportunities to connect with the above species and a range of lowland and montane ones. The distinctive subspecies (ssp. *montanus*) of the **Hill Blue Flycatcher** may be seen here as well.

Rafflesia Information Centre & Gunung Alab

The Rafflesia Forest Reserve is a 3.6km^2 reserve that was established to protect the spectacular *Rafflesia* plants, which are locally common on the hill slopes within the reserve. The Rafflesia Information Centre is the main gateway to the reserve, and is located about 15km from Tambunan on the Penampang-Tambunan Road. While there are hiking trails leading from the information centre down the hill, most birdwatching is done along the roadside around the Information Centre, at an elevation of around 1,200m asl. The road does become busier as the day wears on, and birdwatchers should always be wary of passing traffic while walking along the roadside. Birdwatching here can be very productive, especially if there are fruiting trees close to the road. **Bornean Bulbuls** and **Bornean Leafbirds** are both fairly conspicuous; **Whitehead's Spiderhunters** often perch on the tops of dead snags, and the endemic **Mountain** and **Bornean Barbets** are common by call and regularly seen. The **Fruithunter** is sometimes seen in the trees around the Information Centre. As the day warms up, the **Mountain Serpent Eagle** may be seen soaring over the Information Centre.

Further up the road, the summit of Gunung Alab is accessible via a narrow paved road from the highest point of the Tambunan-Penampang Road. As the highest point of the park it features some species more characteristic of Mount Kinabalu, including the **Bornean Whistler**, **Bornean Treepie** and **Mountain Blackeye**. The dilapidated Gunung Alab Resort is also worth a look as a colony of **Bornean Swiftlets** nests there.

Chi'en C. Lee

*The attractive **Bornean Bulbul** is easily seen in roadside forest along the Crocker Range National Park.*

Abdelhamid Bizid

*The difficult **Mountain Serpent Eagle** can sometimes be seen soaring over the Rafflesia Information Centre.*

Mahua Waterfall

Located 26km away from Tambunan along the Ranau-Tambunan Road, the waterfall is located at around 1,000m asl and offers an opportunity to observe submontane species. The car park at the waterfall has an unobstructed view of the sky and surrounding canopy, making it a good place to watch for raptors such as **Blyth's Hawk-Eagle**. The trail down to the waterfall is good for species associated with slope forest, including the **Orange-breasted Trogon** and **Hill Blue Flycatcher**. The waterfall and its environs are worth checking for the **Bornean Forktail**.

Access & Accommodation

The Crocker Range National Park can be visited as a day trip from Kota Kinabalu. It takes about an hour and a half of driving along the winding Jalan Penampang-Tambunan to reach the Rafflesia Information Centre. Alternatively, for birdwatchers who wish to spend more time here, accommodation is available at both the park headquarters and Mahua Waterfall, although arranging transport to and from the birdwatching sites may be difficult.

Conservation

As the largest terrestrial park in Sabah and due to its rugged terrain, enforcement of its boundaries is a challenge. As in many parts of Borneo, illegal logging and the over-exploitation of non-timber forest organisms such as insects and orchids are arguably the main threats faced by park authorities.

Kabili-Sepilok Forest Reserve

Wong Tsu Shi

Chien C. Lee

Some of Sabah's most accessible virgin lowland rainforests can be seen at Kabili-Sepilok Forest Reserve.

KEY FACTS

Nearest Major Town
Sandakan

Habitats
Lowland rainforest, mangrove forest

Key Species
Bat Hawk, Diard's, Scarlet-rumped and Red-naped Trogons, Banded, Ruddy and Rufous-collared Kingfishers, Rhinoceros, Black, Bushy-crested and Wreathed Hornbills, Crimson-winged, Orange-backed, Maroon and Grey-and-buff Woodpeckers, Black-and-yellow, Black-and-red, Dusky and Banded Broadbills, Hooded, Blue-headed* and Black-crowned* Pittas, Bornean Bristlehead*, Crested Jay, Black Magpie, Grey-cheeked, Black-and-white, Streaked, Yellow-bellied and Spectacled Bulbuls, Yellow-vented, Brown-backed and Yellow-rumped* Flowerpeckers, Van Hasselt's, Copper-throated and Red-throated Sunbirds, Bornean, Thick-billed and Long-billed Spiderhunters

Other Specialities
Southern Pig-tailed Macaque, Maroon Langur*, Bornean Gibbon*, Bornean Orangutan*, Cream-coloured Giant Squirrel, Red Giant Flying Squirrel, Bearded Pig

Best Time to Visit
March–July; rainy season November–February

Named after the two rivers that flow through the reserve into Sandakan Bay, Kabili-Sepilok Forest Reserve protects about 43km^2 of lowland dipterocarp forest and mangroves, two habitats that are fast disappearing in Borneo. While the Orangutan Rehabilitation Centre is a major tourist attraction, the reserve also features an excellent network of forest trails and a canopy bridge, which provide opportunities to see most of Borneo's lowland birds. The monotypic **Bornean Bristlehead** is seen here regularly, and the sight of a flock of these bizarre birds foraging in the canopy will remain in the memories of visiting birdwatchers long after their birdwatching trip.

Birdwatching Sites

Sepilok Orangutan Rehabitilitation Centre

Most visitors come here to view the free-roaming Bornean Orangutans. However, there is also a network of boardwalks around the visitor centre leading to various orangutan feeding platforms that are conducive for birdwatching. Alternatively, birdwatchers can register and hire local guides to hike along the 5km Mangrove Forest Trail, which leads from the Rehabilitation Centre through lowland dipterocarp forests and sandstone ridges into mangrove swamps at Sepilok Bay, where the Sepilok Laut

Con Foley

Whiskered Treeswifts are regularly seen from the canopy bridge.

Reception Centre is situated on the reserve's boundary. Many Sundaic lowland birds can be seen en route, as well as the localized **Copper-throated Sunbird**. However, as the trail network is not as well maintained as around the Rainforest Discovery Centre (see below), this area is less visited by birdwatchers.

Rainforest Discovery Centre (RDC)

There is a well-maintained trail network around the RDC, with each trail being named after a special bird family that has been recorded there. Examples include the Pitta Trail and Kingfisher Trail. Even the three observation towers connected by the canopy bridge are named Bristlehead, Trogon and Hornbill Tower respectively. These trails meander through regenerating forest, and are therefore not only conducive for birdwatching, but also well marked and inter-connected, so the chances of getting lost are minimal and a local guide is not needed in this area. The main target here for most birdwatchers is the **Bornean Bristlehead**, which can sometimes be seen at eye level from the canopy bridge. As its name suggests, the Kingfisher Trail is a good place to look for kingfishers such as **Rufous-collared** and **Banded Kingfishers**, while the Pitta Trail is frequented by the endemic **Black-crowned Pitta**. A host of other birds can be encountered, including various bulbuls, babblers and sunbirds, as well as the **White-crowned Forktail**.

Access & Accommodation

Both the Sepilok Orangutan Rehabilitation Centre and RDC are accessible from Sepilok Road, which is 24km from Sandakan town along Labuk Road. While on Sepilok Road, turn right onto Sepilok Arboretum Road to reach the RDC, while the Orangutan Rehabilitation Centre is at the end of Sepilok Road. Transport can be arranged from Sandakan or through various tour operators. There are a few accommodation options in the vicinity of the reserve.

Con Foley

__Orange-backed Woodpeckers__ are one of 12 species of woodpecker that may be seen here.

Conservation

In addition to protecting natural habitats, the Orangutan Rehabilitation Centre located within the reserve helps rehabilitate orphaned or injured Bornean Orangutans in the hope of releasing them back into the wild in the future. Similarly, the Sun Bear Conservation Centre, opened in 2014 and located in the reserve, was set up to raise public awareness about the bears' plight.

Bjorn Olesen

The highly sought-after __Bornean Bristlehead__ is the star avian attraction at Kabili-Sepilok Forest Reserve.

Lower Kinabatangan & Gomantong Caves

Wong Tsu Shi

Chien C. Lee

The Kinabatangan River at dawn.

KEY FACTS

Nearest Major Towns
Sandakan, Kota Kinabatangan

Habitats
Lowland rainforest, peat-swamp and riverine forests

Key Species
Chestnut-necklaced Partridge, Storm's Stork, Great-billed Heron, Oriental Darter, Grey-headed and Lesser Fish Eagles, Bat Hawk, Wallace's Hawk-Eagle, Jerdon's Baza, Moustached Hawk-Cuckoo, Bornean Ground Cuckoo*, Oriental Bay Owl, Large Frogmouth, Diard's, Scarlet-rumped and Red-naped Trogons, Helmeted, Rhinoceros, Bushy-crested, White-crowned, Wrinkled and Wreathed Hornbills, Red-throated Barbet, White-bellied and Great Slaty Woodpeckers, White-fronted Falconet*, Black-and-red, Dusky and Banded Broadbills, Giant, Hooded, Black-crowned* and Blue-headed* Pittas, Black-throated Babbler

Other Specialities
Bornean Slow Loris*, Horsfield's Tarsier, Silvered Leaf Monkey, Maroon Langur*, Proboscis Monkey*, Bornean Gibbon*, Bornean Orangutan*, Sunda Otter Civet, Flat-headed Cat, Leopard Cat, Asian Elephant, Estuarine Crocodile, Reticulated Python

Best Time to Visit
March–July; rainy season November–February

The Kinabatangan Wildlife Sanctuary, comprising 270km² of lowland riverine forest, is an oasis for wildlife in a sea of oil-palm plantations. It encompasses much of the lower reaches of the Kinabatangan River, which at 560km is Malaysia's second longest river after Sarawak's Rajang. Flora and fauna in this sanctuary are concentrated in relatively narrow tracts of riparian forest that line the river and its tributaries, making it one of the most productive wildlife-watching experiences in the region. From the comfort of a sturdy boat, you can enjoy close encounters with a wide variety of birds, reptiles and mammals. The cream of the crop among the birds here is undoubtedly the **Bornean Ground Cuckoo**, a secretive ground dweller of these riverine forests that is endemic to Borneo. Due to the structure of the riverine forests many birds of the canopy that can be difficult to see at other sites such as Danum Valley are easier to locate here, and include a variety of specialities ranging from regal **Storm's Storks** and majestic **Rhinoceros Hornbills** to tiny **White-fronted Falconets**.

The nearby Gomantong Caves host some of the largest swiftlet colonies in Borneo and comprise several species of this perplexing family, which can otherwise be very difficult to identify in flight. Additionally, the daily exodus of hundreds of

Con Foley

Storm's Stork, a globally threatened species that is difficult to see elsewhere in Southeast Asia, is locally common along the Kinabatangan River.

James Eaton

*An encounter with the secretive **Bornean Ground Cuckoo** is one of the key motivations behind planning a visit to the Kinabatangan River.*

Chi'en C. Lee

*The Gomantong Caves offer visitors an opportunity to see a variety of swiftlets on their nests, such as this **Black-nest Swiftlet**.*

thousands of Wrinkle-lipped Free-tailed Bats at dusk is both a spectacle in itself and a buffet for raptors such as **Bat Hawks**, which perform feats of aerial agility to capture the bats on the wing as they leave the cave.

Birdwatching Sites

Sukau & Kinabatangan Wildlife Sanctuary

Sukau is a small village along the Kinabatangan River where many tour operators maintain lodges and homestays. From here you generally go on up to two diurnal and one nocturnal wildlife-watching cruises based on your accommodation choice. There are also limited opportunities to explore the riverine forest on foot using self-guided nature trails at certain lodges.

Birdwatching from the boat along the main channel of the Kinabatangan River should yield most of the area's specialities, including **Storm's Stork** and the **White-fronted Falconet**, which regularly perches on dead trees, **Grey-headed** and **Lesser Fish Eagles**, and a variety of kingfishers and hornbills. Seeing the **Bornean Ground Cuckoo**, however, requires a concerted effort along the various quiet tributaries such as the Menanggol and Tenagang, where the forest comes right up to the narrow river channel on both sides. Its far-carrying calls often give it away, but seeing one requires a good deal of luck, fieldcraft and potentially jockeying for position on the boat. These channels are also a good place to listen out for other ground dwellers such as the **Chestnut-necklaced Partridge** and mythical **Giant Pitta**. Note that depending on local conditions, it is possible to disembark from the boat and enter the riverine forest in search of calling pittas and pheasants, although the dense understorey makes getting good views challenging. Primates, notably the endangered Proboscis Monkey, are also a permanent feature of the landscape, and herds of Asian Elephants may be encountered.

The action continues after the sun goes down and nocturnal cruises can yield sightings of **Buffy Fish Owls** and various roosting birds, including groups of **Black-and-red Broadbills**. **Oriental Bay Owls** and **Large Frogmouths** are also regularly seen in the riverine forests, and mammals such as the poorly known Flat-headed Cat may be encountered.

Gomantong Caves

This intricate network of caves can be visited as a day trip from lodges around the Kinabatangan Wildlife Sanctuary. From the entrance building there are two options – visiting the Simud Hitam (Black Cave) or Simud Putih (White Cave). For birdwatchers these caves offer the opportunity to add **Edible-nest**, **Black-nest** and **Mossy-nest Swiftlets** to their lists due to the colour of the nests the individual birds perch on.

The forest around the caves provides a good introduction to Southeast Asia's lowland forest birdlife, and possibilities here include the charming **Black-and-yellow Broadbill** and **Black-crowned Pitta**. Timing the trip to coincide with the exodus of bats at dusk yields several raptor species, including **Bat Hawks** and the globally threatened **Wallace's Hawk-Eagle**.

Access & Accommodation

Accommodation is available around the Gomantong Caves and in Sukau Village. It caters to a range of budgets and is often marketed as a complete package comprising transfers to and from Sandakan Airport, accommodation and food, and daily river cruises. The area can also be accessed overland via the Sandakan-Lahad Datu Road, as well as by boat from Sandakan through the mouth of the Kinabatangan River.

James Eaton

*The **Wrinkled Hornbill** is not uncommon in the riverine forests along the Kinabatangan.*

Con Foley

***Oriental Darters** are frequently seen along the Kinabatangan River.*

Conservation

Despite having status as a wildlife sanctuary, the forests in the lower reaches of the Kinabatangan are surrounded by oil-palm plantations that continue to encroach into its boundaries. Illegal logging and poaching are also major threats that continue to persist within the boundaries of the reserve.

Cheng Heng Yee

*In the evenings the spectacular hunting manoeuvres of the **Bat Hawk** may be seen.*

Tawau Hills Park & Maliau Basin

Wong Tsu Shi

Wong Tsu Shi

A 4WD vehicle is needed to access the headquarters of the Maliau Basin Conservation Area.

KEY FACTS

Nearest Major Town
Tawau

Habitats
Lowland, hill and submontane rainforests

Key Species
Chestnut-necklaced and Crested Partridges, Great Argus, Bulwer's Pheasant*, Crested Fireback, Jerdon's Baza, Bonaparte's Nightjar, Dark Hawk-Cuckoo, Diard's and Red-naped Trogons, Blue-banded and Banded Kingfishers, Helmeted, Bushy-crested and White-crowned Hornbills, Maroon Woodpecker, Black-and-yellow, Dusky and Banded Broadbills, Black-crowned*, Blue-headed* and Blue-banded* Pittas, Spotted Fantail, Crested Jay, Black Magpie, Grey-cheeked, Black-and-white, Streaked and Yellow-bellied Bulbuls, Horsfield's and White-necked Babblers, Striped, Eyebrowed and Bornean* Wren-Babblers, Rufous-tailed and White-crowned* Shamas, Bornean*, Sunda and Malaysian Blue Flycatchers, Chestnut-naped Forktail, Bornean* and Long-billed Spiderhunters

Other Specialities
Maroon Langur*, Bornean Gibbon*, Hose's Langur*, Sun Bear, Sunda Clouded Leopard, Bearded Pig, Bornean Yellow, Red Muntjac

Best Time to Visit
March–July; rainy season November–February

Tawau Hills Park is located 24km north-west of the town of Tawau and comprises more than 280km^2 of lowland rainforest largely surrounded by oil-palm plantations. The rugged landscape is dominated by two notable peaks, Gunung Magdalena (1,310m asl) and Gunung Lucia (1,201m asl). The park also has an ancient volcano crater, Bombalai Hill (530m asl), and boasts the second-tallest known tropical forest tree, a 88.1m-tall Yellow Meranti (*Shorea faguetiana*). What the park lacks in size it makes up for in accessibility, with the remnant rainforest supporting several sought-after Bornean endemics, including the stunning **Blue-banded Pitta** and **Bornean Wren-Babbler**. A further 170km north-west of Tawau Hills lies the Maliau Basin Conservation Area, a wild area widely known as 'Sabah's Lost World'. This site comprises more than 580km^2 of pristine lowland and lower montane forests, rising to a height of 1,675m asl. Aside from spectacular natural scenery, the main target for visiting birdwatchers is the equally spectacular **Bulwer's Pheasant**.

Birdwatching Sites

Tawau Hills Park

Birdwatching is generally carried out along the main trail, which eventually leads to the peaks of Gunung Magdalena and Lucia. This trail has two main forks leading to the Tallest Tree Trail after 300m and the Hot Spring Trail after 1.5km.

John Mathai

*Few birdwatchers have laid eyes on the legendary **Bulwer's Pheasant**.*

Main Trail Before Tawau River Crossing & Tallest Tree Trail

The main trail hugs the banks of the Tawau River, and diligent scanning may be rewarded with various kingfishers perched along the banks, including the rare **Blue-banded Kingfisher**. Several species of blue flycatcher, including the endemic **Bornean Blue Flycatcher**, are possible in the riverine forest. Mixed species foraging flocks contain various malkohas, woodpeckers, babblers and occasionally trogons. Careful scrutiny of the forest floor may be rewarded with a **Great Argus** crossing the trail, or a breathtaking **Blue-headed Pitta** glowing in the dim understorey.

The Tallest Tree Trail forks from the main trail after 300m and leads to the aforementioned *Shorea faguetiana*. You can either retrace your steps to exit this trail or follow it to return to the Main Trail at the 1km marker. This trail is a good place to observe various hornbills, including **Rhinoceros** and **Bushy-crested Hornbills**.

Hot Spring Trail

This trail forks from the Main Trail after about 1.5km and leads to a number of sulphur springs after a roughly 1.7km walk. The streams along the trail are home to wary **White-crowned** and **Chestnut-naped Forktails**, and **Black-crowned Pittas** are also possible. **Striped** and **Bornean Wren-Babblers** and various mixed flock species including the **Spotted Fantail** should be looked for here.

Gunung Lucia Trail

This trail starts after the Main Trail crosses the Tawau River after about 2km, and a further 10km hike leads to the summit of Gunung Lucia. A reasonable degree of fitness is required to manage this rarely visited, uphill ascent. However, due to a lack of human disturbance, close encounters with a variety of sought-after slope specialists is possible here. These include **Bulwer's Pheasant** and the **Blue-banded Pitta** at the upper reaches of the trail, and even mammals such as the Sun Bear are possible.

Maliau Basin Conservation Area

Access Road

The 20km unpaved road (four-wheel drive recommended) leading to Agathis Research Station is flanked by pristine forest and provides excellent birdwatching opportunities. Various pheasants, including the **Crested Fireback**, **Crested Partridge** and the mythical **Bulwer's Pheasant** may be seen foraging along the road or flushed at close range. The whole suite of Sundaic rainforest birds inhabits the surrounding forests, and various trogons, hornbills and babblers can be expected. Encounters with Asian (Bornean Pygmy) Elephants are possible on the road.

Trails Around Agathis Research Station

Agathis Research Station, the closest and best equipped of all the stations, serves as the ideal base camp for any exploration of Maliau Basin. A network of trails leading out from this area includes a short nature trail and numerous trails leading to other remote research stations beyond, including Ginseng Research Station (5–6-hour hike), Nepenthes Research Station (1,005m asl) and Lobah Research Station, to name a few. In addition to providing spectacular views of the region's landscape and waterfalls, the treks provide ample opportunity to get acquainted with Borneo's endemic birds, including the bizarre **Bornean Bristlehead** and difficult Sundaic species such as the **Large Green Pigeon**, **Bonaparte's Nightjar** and **Black-and-white Bulbul**. Further exploration will undoubtedly expand the area's bird list significantly. **Bulwer's Pheasant** has been recorded in the vicinity of Agathis Research Station with some regularity, and should be expected anywhere along these treks. Listen for their harsh cackles, a vocal feature unique among its genus.

Wong Tsu Shi

The gorgeous ***Blue-headed Pitta*** *may well be the prettiest member of its family and can be seen at Tawau Hills Park.*

Access & Accommodation

Tawau Hills Park can be accessed via a 30-minute drive from Tawau town on the Tawau-Kalabakan Highway. Taxis can be arranged to transport birdwatchers there as well, but be sure to make return arrangements due to a lack of public transport. Accommodation and food are available at the park headquarters and at a hostel on Gunung Lucia.

In order to visit Maliau Basin, permission must be obtained in advance from Yayasan Sabah (Sabah Foundation), and registration is conducted at the entry gate. A four-wheel drive vehicle is highly recommended to navigate the rugged terrain. Accommodation and meals are provided at the various research stations, but these must be booked in advance.

Conservation

Both Tawau Hills Park and Maliau Basin Conservation Area were gazetted to protect water resources for the various settlements in the area, as many of the region's rivers have their headwaters within these two sites. Local authorities recognize the significance of the area for conservation and are attempting to nominate Maliau Basin as a UNESCO World Heritage Site. Despite the area's remoteness, the over-exploitation of the natural resources at both sites, in particular its unique orchids, pitcher plants, reptiles and amphibians for the exotic plant and wildlife trade, is a major problem.

Bjorn Olesen

The resplendent ***Great Argus*** *is locally common at Maliau Basin and may be seen maintaining a lek along the trails - a large, open scrap on the ground devoid of leaves and twigs.*

Payeh Maga & Meligan Highlands

Ch'ien C. Lee

A disused logging road offers easy access to montane forest on the Payeh Maga Plateau.

KEY FACTS

Nearest Major Town
Lawas

Habitats
Lowland and hill rainforests, montane forests

Key Species
Lowland Blue-banded* and Bornean Banded* Pittas, Rufous-tailed Shama, Bornean Spiderhunter*
Submontane Bornean Frogmouth*, Bornean Barbet*, Whitehead's* and Hose's* Broadbills, Black Oriole*, Rail-babbler, Bornean Bulbul*, Bare-headed Laughingthrush*, Pygmy White-eye*, Bornean Leafbird*
Montane Mountain Serpent Eagle*, Whitehead's Trogon*, Golden-naped* and Mountain* Barbets, Bornean Whistler*, Bornean Green Magpie*, Bornean Treepie*, Flavescent Bulbul, Bornean Stubtail*, Chestnut-hooded Laughingthrush*, Mountain Blackeye*, Black-sided Flowerpecker*, Whitehead's Spiderhunter*

Other Specialities
Kayan Slow Loris*, Bornean Gibbon*, Maroon* and Hose's* Langurs, Whitehead's Pygmy Squirrel*, Leopard Cat, Sunda Clouded Leopard, Lesser Mousedeer

Best Time to Visit
March–November

Located at the remote north-eastern edge of Sarawak, Payeh Maga is a mountainous region that encompasses the forested slopes rising from Sungai Tuyo (300m asl) to the summit of Gunung Matalan (1,851m asl). Much of the area was selectively logged before 2002, but a large portion of pristine forest still remains, especially on the steep upper slopes, and is now protected. The site was thrust into the limelight after a Heart of Borneo biological expedition to the area in 2010 rediscovered the **Black Oriole**, a rare endemic that has never been recorded in Sabah, and has an inexplicably patchy distribution in northern Sarawak. Subsequent surveys have revealed a very high avian diversity with more than 250 species, including more than half of Borneo's endemics.

Birdwatching Sites

Gunung Do'a

Located at the lower elevations of the Payeh Maga Track (*c.* 600m asl), Gunung Do'a consists of a small cluster of houses and an auditorium that are utilized by the villagers of Long Tuyo as an occasional religious retreat. For most of the year these buildings are completely vacant, and can be used (with permission) as a convenient base camp for exploring the nearby habitats.

Chi'en C. Lee

Payeh Maga is presently the only reliable location in the world to see the enigmatic **Black Oriole**.

Wang Bin

A shy resident of steep hill forest, the ***Blue-banded Pitta*** *is more often heard than seen.*

Although much of the forest in this area comprises old secondary growth, some steep slopes such as the one immediately above the buildings retain large mature trees. Avian diversity is very high here, comprising mostly lowland species and some hill/submontane birds such as **Whitehead's Broadbill** (rare), the **Flavescent Bulbul** and the **Pygmy White-eye**. The **Oriental Bay Owl** has occasionally been recorded around the buildings.

Payeh Maga Track

This trail consists of an old disused logging road that begins near Gunung Do'a and winds its way up onto the Payeh Maga Plateau above 1,600m asl. The hike is fairly easy, with a few moderately steep sections. The wide track allows easy viewing into the forest canopy, and it is here that the bulk of Payeh Maga's birds can be seen. Babblers, bulbuls and flycatchers are particularly common in the lower sections, and Bornean endemics such as the **Bornean Barbet** and **Bornean Leafbird** can be found on the ascent. The **Black Oriole** is most reliably encountered at around 900–1,200m asl along the track, a 3–4-hour walk from Gunung Do'a. After dark the sought-after **Bornean Frogmouth** is commonly heard at this elevation, but can be frustratingly difficult to see. Above 1,400m asl, montane birds begin to appear, including the **Mountain Serpent Eagle**, **Golden-naped Barbet** and **Whitehead's Spiderhunter**. It takes about 4–6 hours of hiking from Gunung Do'a to reach the montane sections of the road, so those wishing to spend more time here should plan to camp at the various shelters.

Merarap

Spanning an elevation range of 400–700m asl, the access road to Merarap Hot Spring Lodge is an excellent and convenient location for observing many lowland and hill-forest birds. Although walks can be started directly from the lodge at the bottom of the hill, there are several steep uphill sections and the going is much more leisurely if you are dropped off at the top of the hill and walk slowly down. Hornbills, including **Rhinoceros**, **Bushy-crested**, **Wreathed** and **Black**, are shy but commonly seen in the area. The steep forested slopes near the upper end of the road are good habitats for **Blue-banded** and **Bornean Banded Pittas**, and their calls are frequently heard in the morning.

Access & Accommodation

Visitors arriving from Sarawak can reach Lawas via multiple daily flights from Miri, whereas those arriving from Sabah can make the drive from Kota Kinabalu in about four hours. There is an immigration crossing between Sabah and Sarawak so it is advisable to reach the border during working hours. From Lawas, Payeh Maga is a drive of two and a half hours inland, following a solid dirt road. Public vehicles regularly take villagers along this road, but to avoid delay it is suggested that visitors charter their own four-wheel drive vehicles from Lawas as this will also allow for an easier approach to the Payeh Maga trail head.

The village of Long Tuyo, which is located near the beginning of the Payeh Maga Track, offers simple homestay accommodation. Alternatively, you can stay further up the main road (about 20 minutes by car) at Merarap Hot Spring Lodge, which offers clean, affordable rooms and excellent meals. From either of these locations you can visit Payeh Maga on day hikes, but this limits the amount of time available for birdwatching at the upper elevations due to the length of the trail. Birdwatchers wanting to spend enough time in the montane areas to have good chances of finding most of the endemics need to camp along the Payeh Maga Track at either of two cabins: Black Oriole Camp (c. 800m asl) or Payeh Maga Camp (c. 1,600m asl). Each of the cabins provides a simple shelter, an area to cook and access to water. All sleeping gear and food must be brought in. Guides and porters for the Payeh Maga Trail can be hired in Long Tuyo.

Conservation

In recognition of its rich biodiversity, the Forestry Department of Sarawak has classified Payeh Maga as a High Conservation Value Forest and formally listed the area as a proposed national park. In addition, in 2008 Payeh Maga was included within the Heart of Borneo Initiative as part of a network of reserves that form a contiguous link across the interior of Borneo. Although the future of Payeh Maga's land usage remains uncertain, such recognition may serve to secure the protection of its forests permanently. As in much of rural Sarawak, poaching is very common in the area. It is hoped that increased revenue from future ecotourism activities such as birdwatching will encourage more sustainable practices among the local community.

Chi'en C. Lee

__Mountain Blackeye__ (ssp. fusciceps*) occur in the Payeh Maga Mountains at lower elevations compared with Sabah's mountains.*

Chi'en C. Lee

The distinctive subspecies (ssp. leucops*) of the __Flavescent Bulbul__ may be seen at the higher elevations of the Payeh Maga Track.*

Bakelalan & The Bario Highlands

Ch'ien C. Lee

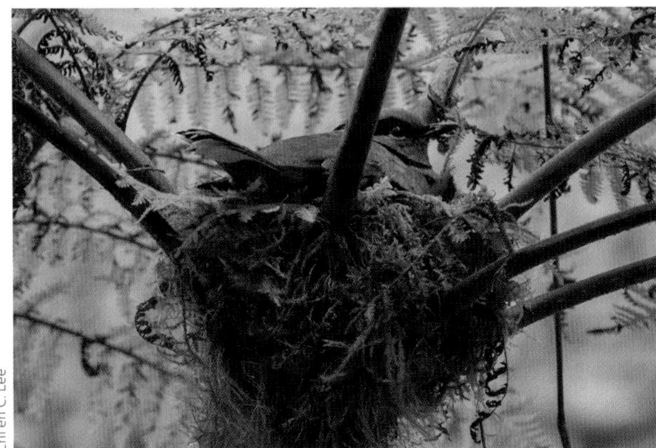

Chi'en C. Lee

*A female **Fruithunter** sits on its camouflaged mossy nest situated in the crown of a tree fern.*

KEY FACTS

Nearest Major Towns
Lawas, Bario

Habitats
Hill submontane and montane rainforests, cultivation

Key Species
Montane Crimson-headed* and Red-breasted* Partridges, Mountain Serpent Eagle*, Dulit* and Bornean* Frogmouths, Golden-naped Barbet*, Long-tailed, Hose's* and Whitehead's* Broadbills, Bornean Stubtail*, Fruithunter*, Whitehead's Spiderhunter*
Submontane Ferruginous Partridge, Bulwer's Pheasant*, Sunda and Reddish Scops Owls, Hose's Broadbill*, Blue-banded* and Bornean Banded* Pittas, Black-thoated Wren-Babbler*

Other Specialities
Bornean Leafbird* Bornean Gibbon*, Maroon* and Hose's Langur*, Cream-colored Giant Squirrel, Hose's Civet*, Bearded Pig, Sambar, Red Muntjac

Best Time to Visit
March–October

Located in the far north-eastern corner of Sarawak, Bakelalan and the surrounding valleys are the homeland of the Lun Bawang people. This site consists of extensive forest-covered mountains separated by lush cultivated valleys. The region has come to the attention of birdwatchers due to recent sightings of the **Dulit Frogmouth**, one of Borneo's most sought-after endemics, and field trips to the area have confirmed an abundance of many other prized species.

Trips to the Bario Highlands are almost invariably centred around the village of Bakelalan, the largest and most accessible settlement. At 1,000m asl, the days here are warm and the nights pleasantly cool. Most birdwatchers base themselves in Bakelalan and take day trips out to several sites in the vicinity, which allows them to cover a variety of elevations and habitats.

The nearby Pulong Tau National Park, one of Sarawak's largest protected areas, covers nearly 700km² of both pristine and partially logged montane forest. Near the northern end of the park, and within a two-day walk from Bakelalan, is the state's highest peak, Gunung Murud (2,423m asl). Intrepid birdwatchers and naturalists who wish to explore this mountain for the opportunity to see upper montane flora and fauna can hire porters and guides in Bakelalan for the trek.

Chi'en C. Lee

The forest at the higher elevations of the Payeh Lam Road is frequently saturated with mists, and the trees are covered with moss and epiphytes.

Birdwatching Sites

Bakelalan Village

The extensive wet rice fields and meadows around Bakelalan offer easy birdwatching and a chance to see species that cannot be found in the forest. Waterbirds including the **Common Moorhen**, **Yellow** and **Cinnamon Bitterns**, and **Greater Painted-snipe** can be found in the fallow wet fields. Mistletoes, which are abundant in the trees at the edges of the fields, attract small birds including **Yellow-vented** and **Plain Flowerpeckers**. The **Malaysian Eared Nightjar** and **Brown Hawk-Owl** can often be heard calling after nightfall at the village edges.

The forested hills around Bakelalan can be reached by a network of trails, some leading to houses or small gardens. **Red-breasted** and

Chi'en C. Lee

*Before 2013 the **Dulit Frogmouth** was known from only a handful of specimens and sightings, mostly in northern Sarawak.*

Ferruginous Partridges can be found here, although both are usually exceedingly shy and more often heard than seen. This is also a good place to look for the vocal **Black-throated Wren-Babbler**.

The main focus here for most birdwatchers is the legendary **Dulit Frogmouth**. To see this bird it is necessary to hire a local guide from Bakelalan (Mr Sang Sigar), who monitors several sites and is updated on where the birds have been most recently spotted, with sites generally being within a short drive from the village. Like other frogmouths, the Dulit Frogmouth is most responsive shortly after nightfall and before dawn.

Payeh Lam Road

Beginning just south of Bakelalan, this newly constructed road is part of a development project to bring easy vehicle access to the town of Bario, situated in the Kelabit Highlands further south. Unfortunately for the developers, the unusual contour of the Indonesian border at this point has required that the road make an impractical detour and excessively steep climb over a 1,700m asl ridge before descending into the valley beyond. This is actually a boon for birdwatchers, since the road bisects a convenient gradient through beautiful lower and upper montane forests, and a morning or afternoon walk here yields a great diversity of species.

Chi'en C. Lee

*The endemic **Golden-naped Barbet** is the only barbet occurring in upper montane forests.*

The lower sections of the road, at 1,100–1,400m asl, can be productive for forest birds such as the **Orange-breasted Trogon**, **Mountain Barbet**, **Long-tailed Broadbill**, some of Borneo's seldom-seen submontane and hill forest endemics such as **Hose's Broadbill**, and both **Bornean Banded** and **Blue-banded Pittas**. Further up, where the road passes over the ridge, the forest becomes stunted and mossy, and these birds are replaced by their upper montane equivalents, such as **Whitehead's Trogon**, **Golden-naped Barbet** and **Whitehead's Broadbill**. Other montane endemics can be found with some luck, including **Whitehead's Spiderhunter**, **Bornean Stubtail** and **Fruithunter**. After the ridge the road descends steeply into a valley, before joining a network of logging roads on the flanks of Gunung Murud. In this area there have been reports of **Bulwer's Pheasant** calling in the forest, although sightings tend to be fleeting and sporadic.

Access & Accommodation

Bakelalan can be reached either by MAS-Wings operated flights from Lawas (currently scheduled on three days per week), or by dirt road, which requires the hire of a four-wheel drive vehicle. The drive from Lawas takes 5–9 hours depending on the condition of the road, or 4–6 hours if driving from Payeh Maga. Vehicles are often available for hire in Bakelalan and are necessary for visiting some of the nearby sites for the Dulit Frogmouth and the Payeh Lam Road.

There are currently two places that offer visitor accommodation in Bakelalan. Apple Lodge is a simple, family-run business directly opposite the airstrip that offers meals and small, clean rooms. A homestay operated by Mr Sang Sigar and his wife is situated a short walk further on and accommodates groups of up to a dozen people. Sang is also the primary guide for trips to observe the Dulit Frogmouth.

Conservation

Most mammals are still heavily hunted in this region, and encounters with wildlife other than birds are fleeting at best. Land clearing remains a threat to some forests, particularly those that are not on steep terrain. The new access provided by the Payeh Lam Road has allowed people to penetrate into the montane forest for indiscriminate woodcutting, and this is likely to continue as the road is extended. Donations are collected by Sang Sigar and are directed towards conservation efforts and environmental education in the community.

Buntal Bay

Yeo Siew Teck

Abdelhamid Bizid

*The mudflats of Buntal Bay support good numbers of the globally threatened **Far Eastern Curlew**, a shorebird that is otherwise very difficult to see in Southeast Asia.*

KEY FACTS

Nearest Major Town
Kuching

Habitats
Mudflats, mangrove forest, beach forest

Key Species
Lesser Adjutant, Malaysian Plover, Bridled Tern
Winter Chinese Egret, Asian Dowitcher, Eurasian and Far Eastern Curlews, Nordmann's Greenshank, Grey-tailed Tattler, Great and Red Knots

Other Specialities
Irrawaddy Dolphin, Estuarine Crocodile, Proboscis Monkey

Best Time to Visit
August–March; rainy season December–February

Located on the coast north of Kuching, Buntal Bay is one of the most important areas for migratory shorebirds in East Malaysia. Covering an extensive area of coastal mudflats, varied intertidal habitats and mangrove forests, Buntal Bay attracts thousands of migratory shorebirds annually. Many are drawn to the bay's nutrient-rich mudflats, and good numbers of shorebirds that are otherwise rare in Peninsular Malaysia spend the northern winter here annually, including the globally endangered **Far Eastern Curlew**. In all, more than 30 species of shorebird have been recorded, including many globally threatened species such as **Nordmann's Greenshank** and the **Great Knot**. Other sought-after East Asian shorebirds found here include the **Asian Dowitcher** and **Grey-tailed Tattler**, while **Chinese Egrets** winter in good numbers. It is possible to see a good selection of the bay's shorebirds over a two-day trip while based in Kuching.

Birdwatching Sites

Mudflats on the Buntal River Mouth

This extensive area of coastal mudflats can be accessed by a small *sampan* (boat) from across the Buntal River. The mudflats are best accessed during low tide. This is one of the few places in the region where you can see good numbers of **Eurasian** and **Far Eastern Curlews**, both increasingly uncommon shorebirds in Southeast Asia. Additionally, the **Asian Dowitcher**, **Grey-tailed Tattler**, both knots and godwits, and small numbers of the globally endangered **Nordmann's Greenshank** are encountered annually. The **Lesser Adjutant** is also sometimes seen alongside the large flocks of shorebirds.

Francis Yap

*Good numbers of the globally threatened **Chinese Egret** spend the northern winter at Buntal Bay annually.*

Coastal Beach

The long stretches of sandy beach here support a very different community of shorebirds. Small numbers of the resident **Malaysian Plover** can be seen all year round. In winter they are joined by flocks of **Kentish Plovers**, including individuals of the distinctive *dealbatus* subspecies, both **Lesser** and **Greater Sand Plovers**, **Ruddy Turnstones**, **Sanderlings** and **Red-necked Stints**.

Buntal Malay Fishing Village

There are a few seafood restaurants built alongside the waterfront along Buntal Bay. From here, one can see small numbers of **Chinese Egrets** feeding along the Buntal River.

Access & Accommodation

Buntal Malay fishing village is about 25km from Kuching and can be reached by either public transport or the regular shuttle services running between the main hotels in Kuching and the Damai area. A drop-off can be requested while passing near the village. By road, the village is 30 minutes from Kuching, and 15 minutes from Damai.

Conservation

Buntal Bay's mudflats are not officially protected even though they have been recognized for their importance to migratory shorebirds along the East Asian–Australasian Flyway. The mangrove forests fringing the bay are protected as forest reserves. In the past, poaching of shorebirds happened frequently, but education work led by the Malaysian Nature Society (MNS) has greatly improved the situation.

Francis Yap

*The **Asian Dowitcher** is a regular visitor to Buntal Bay in small numbers.*

Gim Cheong Tan

*The Endangered **Nordmann's Greenshank** can be seen in small numbers along the mudflats at Buntal Bay.*

Borneo Highlands

Yeo Siew Teck

Low Bing Wen

Submontane forest fringes the edge of the Borneo Highlands Resort.

KEY FACTS

Nearest Major Town
Kuching

Habitats
Hill and submontane rainforests

Key Species
Lowland Reddish Scops Owl, Banded Kingfisher, White-crowned Hornbill, Green Broadbill, Blue-banded Pitta*, White-necked and Grey-breasted Babblers, Eyebrowed Wren-Babbler, Grey-breasted Spiderhunter
Submontane Mountain Serpent Eagle*, Bornean Frogmouth*, Bornean Barbet*, Rail-babbler, Scaly-breasted Bulbul, Pygmy White-eye*
Winter Narcissus and Mugimaki Flycatchers

Other Specialities
Moonrat, Bornean Slow Loris*, Long-tailed Porcupine, Tufted Ground Squirrel*, Thomas's Flying Squirrel, Cream-coloured Giant Squirrel, Prevost's Squirrel, Black-eared Pygmy Squirrel*, Malay Weasel, Banded Palm Civet, Wallace's Flying Frog, Malayan Horned Frog, Bornean Rainbow Toad*

Best Time to Visit
February–July

Increasingly popular among birdwatchers due to its accessibility, Borneo Highlands is a resort complex located on a plateau in the Penrissen Range that straddles the rugged border of Sarawak and Kalimantan (Indonesian Borneo) some 60km south of Kuching. Located at an altitude of 1,000m asl, the site offers excellent access to virgin rainforest within walking distance of local accommodation. Past surveys have revealed a diverse array of flora and fauna, and it was in this region that the Bornean Rainbow Toad was rediscovered after an 87-year hiatus. More than 200 bird species have been recorded in the area, including rare and elusive species difficult to see elsewhere on Borneo like the **Blue-banded Pitta** and **Rail-babbler**, and (in recent years) the scarce **Mountain Serpent Eagle**. Other sought-after species that have been seen recently include a good mixture of lowland and submontane birds like the **Jambu Fruit Dove**, **Bornean Frogmouth**, **White-crowned Hornbill**,**Olive-backed Woodpecker**, **Scaly-breasted Bulbul** and **Grey-breasted Babbler**.

Birdwatching Sites

The Golf Course Area

The Borneo Highlands golf course and its main access road (Jalan Puncak Borneo) are fringed by extensive areas of good-quality submontane dipterocarp forest. Along the main road and the buggy trails criss-crossing the golf course are many excellent vantage points offering panoramic views of the surrounding forest. Fruiting trees should be checked for **Bornean**, **Golden-**

Bjorn Olesen

*Flocks of **Pygmy White-eyes** are a feature of the landscape around the Borneo Highlands Resort complex.*

whiskered, **Blue-eared** and **Brown Barbets**, as well as green pigeons and leafbirds. This area is also rich in bulbuls, and besides the usual species flocks of **Cinereous Bulbuls** and the localized **Scaly-breasted Bulbul** can be expected.

A number of specialities seldom encountered on standard birdwatching circuits in north Borneo have been recorded here and should be of interest to visiting birdwatchers, notably the **Mountain Serpent Eagle** and **Temminck's Babbler** (ssp. *erythrote*). Borneo Highlands is also arguably the best place in Borneo to see the **Pygmy White-eye**, and small groups of this diminutive species are easily seen in fruiting and flowering trees along the roadside. Dense forested gullies should be checked for the **Hill Blue Flycatcher**, here of the distinctive *montanus* subspecies.

Forest Trail on the Sarawak/ Kalimantan Border

Despite being an increasingly popular place for jungle trekking among tourists, this trail is also of great interest to birdwatchers. A number of highly sought-after species like the **Blue-banded Pitta**, **Rail-babbler**, **Grey-breasted Babbler** and **Eyebrowed Wren-Babbler** have been encountered here in recent years. A good strategy is to slowly walk the trail early in the morning and listen for these birds. Other lowland species of interest include **Diard's Trogon**, **White-crowned Hornbill**, **Olive-backed Woodpecker**, **Green Broadbill**, **Maroon-breasted Philentoma** and **Rufous-chested Flycatcher**.

Bjorn Olesen

*The handsome **Scaly-breasted Bulbul** is regularly seen in fruiting trees around the resort complex.*

Michelle & Peter Wong

*The exquisite **Narcissus Flycatcher** is an annual winter visitor to the Borneo Highlands region.*

Jungle Cabin Area

Extensive flowerbeds border the individual jungle cabins. Consequently, this area regularly attracts nectar feeders such as **Temminck's Sunbird** and **Grey-breasted**, **Spectacled**, **Thick-billed** and **Yellow-eared Spiderhunters**.

Access & Accommodation

The best way to access Borneo Highlands is to first take a taxi from Kuching to the foothills of the highlands. From here, a shuttle service runs regularly to the resort for a small fee. Accommodation is limited to the Borneo Highlands Resort, where you can choose to stay either in the clubhouse or in the jungle cabins.

Conservation

The forested areas around Borneo Highlands and surrounding parts of Gunung Penrissen are not protected but have been recognized as one of Sarawak's Important Bird Areas. However, much of the area is managed by Borneo Highlands Hornbill Golf and Jungle Club, which only guests are allowed to access.

Kubah National Park

Yeo Siew Teck

Kerangas forest along one of Kubah's trails.

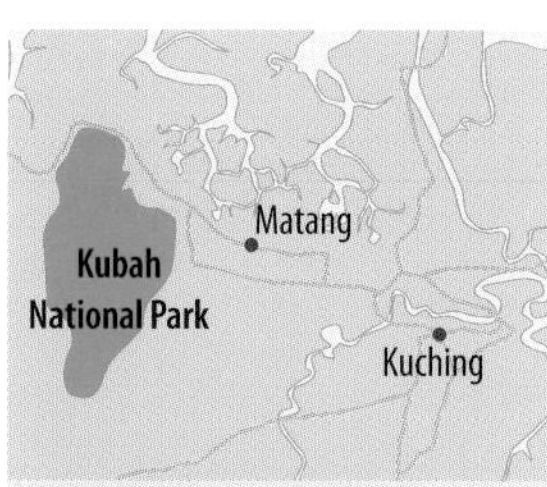

KEY FACTS

Nearest Major Town
Kuching

Habitats
Lowland dipterocarp forest, kerangas (heath) forest

Key Species
Great Argus, Reddish Scops Owl, Bonaparte's Nightjar, Short-toed Coucal, Rufous-collared, Banded and Blue-banded Kingfishers, Blue-banded Pitta*, Bornean Bristlehead*, Crested Jay, White-necked Babbler, Bornean Wren-Babbler*, Rufous-chested and Bornean Blue* Flycatchers
Winter Fairy Pitta

Other Specialities
Horsfield's Tarsier, Bornean Slow Loris*, Cream-coloured Giant Squirrel, Plain* and Black-eared* Pygmy Squirrels, Malay Weasel, Yellow-throated Marten, Wallace's Flying Frog, Malayan Horned Frog, Matang Narrow-mouthed Frog*

Best Time to Visit
February–July

Kubah National Park occupies an area of 22.3km² and is located about 20km to the west of Kuching. The park sits on a sandstone plateau and is part of the Matang Range, which comprises three major peaks, Gunung Selang, Gunung Senduk and Gunung Serapi. Gunung Serapi is the highest point in the park, rising to a height of 900m asl. This site is well known for its diversity of palms and amphibians. In recent years it has also become one of the most popular birdwatching destinations near Kuching. More than 250 bird species have been recorded here, including several difficult Bornean endemics such as the **Blue-banded Pitta**, **Bornean Wren-Babbler** and the enigmatic **Bonaparte's Nightjar**.

Birdwatching Sites

Main Road

The principal tarred road begins at the main entrance to the park and leads to the summit of Gunung Serapi at 900m asl. Along the way it passes through a variety of forest types ranging from lowland dipterocarp forest to kerangas and secondary forest. Birdwatching along the main road can be very productive, especially if the roadside *Macaranga* trees are in fruit. In particular, the stretch of road around the start of the Waterfall Trail is home to several **Blue-banded Pittas**, although seeing them takes some effort. Additionally, the roadside frog pond in the same area, apart from supporting a variety of amphibians including the extraordinary Wallace's Flying Frog after dark, is good for birdwatching in general. The shy **Blue-banded Kingfisher** has been seen here, and during the winter months the **Fairy Pitta** has been recorded in the area. Other notable species that have been recorded along the road include the **Rufous-collared Kingfisher**, **Crested Jay** and **Black-and-white Bulbul**. After dark, spotlighting may reveal the **Reddish Scops Owl** and **Blyth's Frogmouth**, as well as Horsfield's Tarsier.

Bjorn Olesen

*Kubah National Park is one of the most accessible sites in Borneo to see the stunning **Blue-banded Pitta**.*

Wong Tsu Shi

*The delightful **Crested Jay** is often seen along the main road through the park.*

The park is internationally renowned for its diversity of amphibians, including the amazing Wallace's Flying Frog.

Waterfall Trail

This 1.6km trail is popular among trekkers but also provides opportunities to view various understorey bird species, notably the **Blue-banded Pitta**. Other sought-after species that may be seen include the **Great Argus**, **Short-toed Coucal**, **Diard's Trogon**, **White-necked Babbler**, **Bornean Wren-Babbler** and **Bornean Blue Flycatcher**.

Rayu Trail

This 3.9km trail bisects the core of the park and leads to Matang Wildlife Centre. As it takes the better part of the day (6–8 hours return at birdwatching pace) to cover the trail, it is generally recommended for visitors who are staying within the park or have time to spare. The trail passes through some excellent tracts of dipterocarp forest, and notable species that have been recorded here include the sought-after **Bornean Bristlehead** and **Bornean Wren-Babbler**, as well as the **Great Argus**, **Green Broadbill** and **Rufous-tailed Shama**.

Access & Accommodation

Kubah National Park is a 45-minute drive from Kuching, and hiring taxis to reach the park from Kuching is easy. The park headquarters at the main entrance offer a range of accommodation, including chalets, a hostel and rest houses. Bookings can be made at the National Parks Booking office in Kuching or on its website (ebooking.com.my). A permit is required to enter the park and can be paid for at the main entrance. It allows access to both Kubah National Park and the nearby Matang Wildlife Centre.

Conservation

Kubah National Park was gazetted in 1989 and first opened its doors to visitors in 1995. The park's geology (a combination of sandstone and limestone) and rugged terrain have spared it from significant development pressure. However, pollution of its water sources resulting from irresponsible visitors and soil erosion is a management issue.

*The **White-crowned Hornbill** is regularly heard here, but is difficult to observe in the dense canopy.*

MYANMAR

The Union of Myanmar is a country of great geographical contrasts. From the snow-capped peaks of Hkakabo Razi in the far north of Kachin state to the dense lowland jungles of the Myeik (Mergui) Archipelago, Myanmar stretches 2,000km north–south and spans a wider latitude than any other Southeast Asian country. Much of Myanmar's forests, mountains and swamps were virtually unexplored for decades due to poor infrastructure and security concerns. Following the official dissolution of the military junta in 2011 and the subsequent outpouring of foreign investment, Myanmar is arguably more accessible now than it has ever been in recent memory. This bodes well for global birdwatchers with an interest in Southeast Asia's birds, since Myanmar has the second largest avifauna after Indonesia, and is home to no less than eight endemics as well as a host of near-endemic specialities, including the endangered Gurney's Pitta, here in its global stronghold.

Due to limited surveys in the past much remains to be learned about the birdlife of Myanmar. The mysterious Burmese subspecies (ssp. *altirostre*) of Jerdon's Babbler went missing for many years and was only rediscovered in swampy grasslands near Yangon in 2015. The elusive Pink-headed Duck has drawn ornithologists to the country's north, and while it has not been conclusively seen yet, the field trips have discovered vast swampy riverine plains that still support populations of Green Peafowl, Masked Finfoot and various vultures. Fortunately, many of the most sought-after birds in Myanmar can be seen in the vicinity of major tourist sites, and the experience of watching endemic Jerdon's Minivets against the backdrop of thousands of temples and pagodas at Bagan is likely to be etched in any birdwatcher's memory forever.

KEY FACTS

No. of Endemics
8

Country List
1,115 species

Top 10 Birds
1. White-browed Nuthatch
2. Gurney's Pitta
3. Ward's Trogon
4. Blyth's Tragopan
5. Masked Finfoot
6. Jerdon's Minivet
7. Spoon-billed Sandpiper
8. White-winged Duck
9. Beautiful Nuthatch
10. White-bellied Heron

Climate

Like much of continental Southeast Asia, Myanmar experiences a tropical monsoon climate with distinct wet and dry seasons. The period between December–February is the coolest and driest time of the year, following which temperatures gradually increase, peaking in April. Torrential rain occurs between May–September, and many roads are impassable during this period. Additionally, Myanmar's highest mountains in the Himalayas in Kachin state experience an alpine climate, characterized by low temperatures and snowfall in winter.

Access, Transportation & Logistics

The main international gateway into Myanmar is via its former capital Yangon (Rangoon). Most foreign visitors are required to have a visa in advance before entering Myanmar, although an e-Visa service is now available to citizens of many countries. Transport infrastructure between tourist destinations (and their associated birdwatching sites) like Bagan and Lake Inle is of a very high standard, and birdwatchers can utilize a wide variety of transport, such as domestic flights, buses and rental cars, to reach them. However, outside these sites, infrastructure may be limited and many regions in northern Myanmar are off limits to foreigners. Special permits (and potentially mandatory guides) need to be applied for when wanting to visit such areas. Some English is understood and spoken throughout the country, especially in the major cities and tourist sites.

Health & Safety

Vaccinations for Hepatitis A and B and typhoid are recommended for visiting birdwatchers. Malaria is prevalent in remote areas and prophylaxis should be taken, although this is less of a problem in the dry season when most birdwatchers visit. Other mosquito-borne viruses like dengue fever and Japanese encephalitis are present and appropriate precautions should be taken.

Various insurgent groups are still active in parts of Myanmar, particularly along its rugged border regions in Kachin and Shan State in the north-east, and parts of Kayin State. These areas should be avoided, at least for now – travelling to them generally requires a special permit. Birdwatchers keen on visiting the far north should check the latest security situation before planning their trips.

Birdwatching Highlights

A two-week birdwatching trip visiting the three main sites of Bagan, Kalaw and Nat Ma Taung (Mount Victoria) can be expected to yield around 350 species, including all the country endemics except for the enigmatic Naung Mung Scimitar Babbler. The dry scrub country in and around the renowned Bagan Archaeological Zone is home to four of the country's endemics, while several days exploring the wonderful montane forests of Mount Victoria offer opportunities for seeing the endemic White-browed Nuthatch and Burmese Bushtit, as well as a host of near-endemics shared with north-east India, including Striped and Brown-capped Laughingthrushes, the Mount Victoria Babax and the Grey Sibia. Finally, the cooler environs of the former colonial hill station of Kalaw and nearby Lake Inle support a rich birdlife ranging from Jerdon's Bush Chat and recently discovered populations of the Chinese Grassbird in lakeside vegetation to the localized Burmese Yuhina and Black-tailed Crake in the highlands.

Those with a sense of adventure and discovery may want to consider a visit to the Gulf of Mottama (for shorebirds) or the northern state of Kachin. Indawgyi Lake is where thousands of ducks winter, including the rare Baer's Pochard, and is also home to other ornithological highlights like Sarus Cranes and vultures. Myanmar's bird-rich Himalayan forests can be accessed from the Hponkan Razi Wildlife Sanctuary, and the wonderful forests here are home to the localized Sclater's Monal, Ward's Trogon, Ibisbill, fast-declining White-bellied Heron and Snowy-throated Babbler, the last a restricted-range species confined to hill evergreen forest in north-east India (for example in Namdapha National Park) and northern Myanmar.

James Eaton

*Endemic to Myanmar, the **White-browed Nuthatch** has the smallest distribution of any nuthatch.*

***White-throated Babblers** are the easiest of Myanmar's endemic birds to see.*

Hponkan Razi Wildlife Sanctuary

Thet Zaw Naing & Yong Ding Li

Most of the high ridges of Hponkan Razi are regularly covered in snow.

KEY FACTS

Nearest Major Town
Putao (Kachin)

Habitats
Mixed deciduous forest, submontane and montane evergreen forests, pine-rhododendron forest, alpine meadows

Key Species
Blyth's and Temminck's Tragopans, Sclater's Monal, Hill, Rufous-throated and White-cheeked Partridges, White-bellied Heron, Ibisbill, Speckled and Ashy Wood Pigeons, Blyth's Kingfisher, Ward's Trogon, Rufous-necked Hornbill, Green and Black-headed Shrike-babblers, Collared Treepie, Grey-bellied and Slaty-bellied Tesias, Grey-bellied Wren-Babbler, Cachar Wedge-billed and Snowy-throated Babblers, Yellow-throated and Manipur Fulvettas, Rufous-chinned, Chestnut-backed, Rufous-vented, Spot-breasted and Blue-winged Laughingthrushes, Himalayan Cutia, Beautiful Sibia, Fire-tailed Myzornis, Pale-billed and Rufous-headed Parrotbills, Spotted Elachura, Beautiful Nuthatch, Rusty-flanked Treecreeper, Long-tailed Thrush, Rusty-bellied Shortwing, White-browed, Rufous-breasted and Golden Bush Robins, Golden-naped and Scarlet Finches, Tibetan Serin, Collared Grosbeak, Black-headed Greenfinch

Other Specialities
Eastern Hoolock Gibbon, Shortridge's Langur, Asiatic Wild Dog, Asiatic Black Bear, Red Panda, Clouded Leopard, Black Musk Deer, Red Goral, Takin

Best Time to Visit
November–April; rainy season June–September

The eastern subranges of the Eastern Himalayas sweep into the northern tip of Myanmar near the confluence of eastern India and Chinese Tibet. Although not as lofty as the main Himalayan Range, the towering peaks here still soar above 5,600m asl. Kachin state's Hponkan Razi Wildlife Sanctuary sprawls across nearly 2,700km^2 of mountainous country across the border from Namdapha National Park in Arunachal Pradesh (India), reaching its highest point at the often snow-capped summit of Hponkan Razi (4,100m asl). Together with the adjacent Hkakabo Razi National Park to the north, Hponkan Razi forms one of the largest protected landscapes in Southeast Asia, and encompasses large tracts of pristine montane forests. The wilderness of Hponkan Razi is as remote as it gets anywhere in Southeast Asia, and getting to the key birdwatching sites requires many days of hiking in mountainous country. However, the fantastic habitat is home to many highly sought-after species like **Sclater's Monal**, the Critically Endangered **White-bellied Heron** and the localized **Ward's Trogon**. More than 300 bird species have been recorded in Hponkan Razi to date, and further exploration of the reserve should yield many interesting discoveries.

Yann Muzika

***Ward's Trogon** usually occurs in upper montane forests above 1,200m asl on the slopes of the eastern Himalayas.*

Birdwatching Sites

Trail from Shan Gaung to Hpatek Camp

The village of Shan Gaung can be reached via a short drive (c. 19km)

Yann Muzika

*The gorgeous **Fire-tailed Sunbird** regularly participates in mixed foraging flocks.*

*The **Hill Partridge** is one of three species of* Arborophila *partridge that occur at Hponkan Razi.*

from Putao. The forests immediately around Shan Gaung have been disturbed by clearance for cultivation, but a number of widespread montane species like the **Red-headed Trogon**, **Rufous-bellied Niltava** and various babblers still occur. Passing through extensive hill and submontane forests, the trail from Shan Gaung gradually climbs from 450m to 1,300m asl towards Hpatek Camp. The key highlight here is the restricted-range **Snowy-throated Babbler**, which occurs only in northern Myanmar and easternmost India. Small flocks of this skulking babbler forage in the dense undergrowth from 500m asl upwards, often in thickets of bamboo. Another sought-after species that should be looked for here is the **Chestnut-backed Laughingthrush**.

The vantage points on the hike to Hpatek allow for great vistas of the Burmese Himalayas towards the border with India, including the snow-capped peak of Hponkan Razi. Streams along the trail should be checked for **Blyth's Kingfisher**, the **Brown Dipper** and forktails. As the trail enters montane evergreen forests, large foraging flocks with a variety of species become regular. Clearings with bamboo should be checked for parrotbills, including the **Pale-billed Parrotbill**.

Benjamin Schweinhart

*The enchanting **Fire-tailed Myzornis** is a nomadic species that can be encountered in the high-elevation oak-rhododendron forest at Hponkan Razi.*

Hpatek Camp to Wasadam Village

From Hpatek campsite it is a 7-km hike to Wasadum village. The elevation eventually drops back to about 800m asl. Large arboreal mixed flocks contain various shrike-babblers, warblers, fulvettas and yuhinas, while flocks in the undergrowth usually contain scimitar babblers and laughingthrushes. This elevational band is also where the attractive **Beautiful Nuthatch** occurs, usually in family parties in association with mixed flocks of cutias, sibias and other babblers. Other species of interest include the **Silver-breasted Broadbill**, **Collared Treepie** and extremely skulking **Spot-breasted Laughingthrush**.

Wasadum to Hkarlon Village

The trail from Wasadum winds through 8km of mountainous country before entering the village of Hkarlon. Flocks containing the skulking **Yellow-throated Fulvetta**, **Rufous-chinned Laughingthrush** and bizarre **Slender-billed Scimitar Babbler** have been encountered along this stretch of forest. Other species of interest include the enigmatic **Naung Mung Scimitar Babbler** and **Spotted Elachura**.

Hkarlon to Ziyadam Village

From Hkarlon, the trail continues for about 9km before entering the village of Ziyadam at about 1,050m asl. The **White-cheeked Partridge** has been recorded in the forest along this stretch. There are a number of open clearings and banana plantations, from which **Rufous-necked Hornbills** can be observed regularly passing overhead. As the trail descends towards the Ziyadam River, the **White-bellied Heron** and monotypic **Ibisbill** are possible.

Ziyadam Village to Chaungsone Camp

From Ziyadam village the trail eventually climbs to 1,300m asl and dips towards the Monla River, where the **White-bellied Heron**, **Ibisbill**, **River Lapwing** and **Blyth's Kingfisher** can be seen. There are no more settlements beyond Ziyadam. The montane forest at this elevation is home to large mixed flocks comprising **Red-billed Scimitar Babblers**, **Assam Laughingthrushes**, **White-hooded Babblers** and a good representation of fulvettas and yuhinas. Check densely vegetated gullies for the **Cachar Wedge-billed Babbler**. Another highlight here is **Ward's Trogon**, a species that is best seen here in Southeast Asia.

Chaungsone to Thit Pin Gyi Camp

The trail steadily climbs for 8km to about 1,850m asl at Thit Pin Gyi, and passes through undisturbed upper montane forest and tracts of coniferous forest. **Hill** and **Rufous-throated Partridges** are regularly heard in the evergreen forests at these heights. Mixed flock participants include the **Black-headed Shrike-babbler**, **Rufous-winged** and **White-browed Fulvettas**, **Streak-throated Barwing**, **Rufous-backed Sibia** and various tits and warblers. A variety of scimitar babblers, tesias, wren-babblers and the localized **Rufous-breasted Bush Robin** skulk in the densely vegetated ravines.

Thit Pin Gyi to Kan Taut Myit Camp

Steadily climbing the upper slopes of Hponkan Razi, the trail ascends for 6km to more than 2,500m asl, passing through upper montane forests dominated by oaks, chestnuts and rhododendrons. While the mixed flocks remain a feature of the birdwatching at these heights, the main highlights are the pheasants. Both **Blyth's** and **Temminck's Tragopans** occur in the oak-chestnut forests here. **Sclater's Monal** dwells above 3,000m asl in spring, but descends to as low as 2,000m asl in winter and may be seen. The **Fire-tailed Myzornis** has been encountered along this stretch, and should be looked for in flowering trees.

Access & Accommodation

Putao, the main point of entry to Hponkan Razi, can be reached via a two-hour flight from Mandalay or Yangon. Hotels and guesthouses are available in Putao. Beyond Putao, the only accommodation involves staying at the villages on the hike up Hponkan Razi, arranged by bird-tour agencies based in Yangon. Past the last village of Ziyadam, camping is necessary. Temperatures regularly plunge to below 5° C and good sleeping bags are strongly recommended.

Conservation

Hponkan Razi Wildlife Sanctuary was established in 2003. The wider landscape around the Wildlife Sanctuary, including Hkakabo Razi National Park, is recognized for its outstanding representation of eastern Himalayan ecosystems, and has been nominated for UNESCO World Heritage Site status. As in many parts of northern Myanmar, illegal hunting for the wildlife trade threatens a number of large mammals and birds, and has resulted in the decline of large mammals such as bears, Black Musk Deer and Indochinese Tigers throughout the sanctuary. Illegal poaching of wild orchids has also decimated populations of a number of attractive species.

Benjamin Schweinhart

*Seeing the threatened **Blyth's Tragopan** is the highlight of a trip to Hponkan Razi.*

Michelle and Peter Wong

*Vocal parties of **Red-billed Scimitar Babblers** are regular participants of larger mixed flocks.*

Indawgyi Lake Wildlife Sanctuary

Thet Zaw Naing

Bjorn Olesen

Large flocks of Bar-headed and ***Greylag Geese*** *winter in Indawgyi Lake annually.*

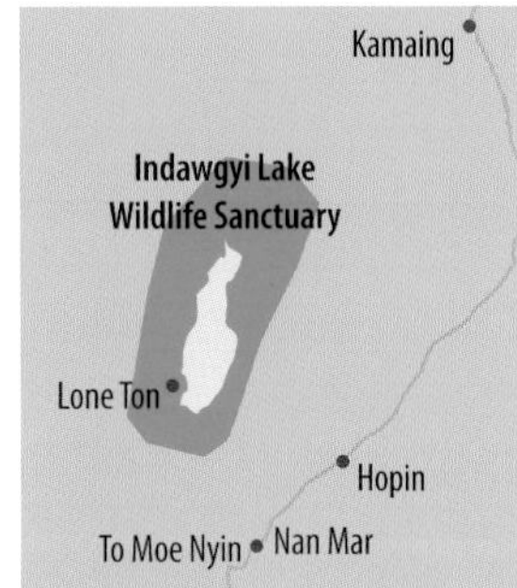

KEY FACTS

Nearest Major Towns
Moe Nyin, Myitkyina (Kachin)

Habitats
Freshwater wetlands, grasslands, dry deciduous and broadleaved evergreen forests, cultivation

Key Species
Indian Spot-billed Duck, Pink-headed Duck?, Woolly-necked and Black-necked Storks, Lesser Adjutant, Black-headed Ibis, Spot-billed Pelican, Oriental Darter, White-rumped and Slender-billed Vultures, Indian Spotted Eagle, Grey-headed Fish Eagle, Black-tailed Crake, Sarus Crane, River Lapwing, River and Black-bellied (rare) Terns, White-browed Piculet, Pale-headed and Great Slaty Woodpeckers, Collared Myna, Jerdon's Bush Chat
Winter Falcated Duck, Baikal Teal, Baer's Pochard, Ferruginous Duck, Himalayan Vulture, Greater Spotted Eagle, Pied Harrier, Yellow-breasted Bunting

Other Specialities
Chinese Pangolin, Bengal Slow Loris, Eastern Hoolock Gibbon, Shortridge's Langur, Assamese, Stump-tailed and Northern Pig-tailed Macaques, Asiatic Wild Dog, Sun Bear, Leopard, Asian Elephant, Sambar and Hog Deer, Red Muntjac, Gaur, Red Serow

Best Time to Visit
November–April; rainy season June–September

Indawgyi Lake Wildlife Sanctuary is located in Moe Nyin township, Kachin state, in the far north of Myanmar. Situated at an elevation of 175m asl, the sanctuary's main feature, Indawgyi Lake, lies within a basin that is orientated north–south and fringed with open plains. Covering 120km^2, Indawgyi Lake is Myanmar's largest natural lake and one of the largest freshwater lakes in Southeast Asia. It is fringed to the east and west by hills densely covered in mixed deciduous and moist evergreen forests. A number of small rivers drain into lake from these surrounding ridges, while the Indaw Chaung flows out from the north-eastern end of the lake, eventually draining into the Mogaung Chaung. From a birdwatching perspective, Indawgyi's main attraction is its massive flocks of wintering ducks, including the highly threatened **Baer's Pochard**. Other attractions include vultures and many large waterbirds that are now very rare in other parts of Southeast Asia, most notably the impressive **Sarus Crane**. The largest wintering congregations of **Common Cranes** in Southeast Asia occur in the wetlands around Indaw Chaung. There have also been anecdotal accounts of the mysterious **Pink-headed Duck** from this site in recent years.

Bjorn Olesen

Woolly-necked Storks *occur all year round at Indawgyi Lake.*

Birdwatching Sites

Indawgyi Lake

This lake hosts large congregations of wintering ducks from November onwards, including the Critically

Endangered **Baer's Pochard**, which has been reported here sporadically in recent years. Besides various geese (mostly **Greylag** and **Bar-headed Geese**) and widespread ducks like the **Gadwall**, **Eurasian Teal** and **Common Pochard**, the wintering flocks should be checked for groups of **Red-crested Pochard** and surprises like the **Falcated Duck** and **Baikal Teal**. On the fringes of the lake are extensive floating mats of vegetation that usually conceal rails (like the **Grey-headed Swamphen**) and two species of jacana. The best way to see waterbirds on the lake is from a canoe, which can be rented from the village of Lone Ton. Alternatively, one can be stationed at the Shwe Myin Tzu Pagoda overlooking the western edge of the lake, a suitable vantage point for viewing waterbirds.

Indaw Chaung Wetlands

The extensive, seasonally inundated grasslands situated between the mouth of the Indaw Chaung and the lake are among the best birdwatching sites within the sanctuary. Located close to Chaung Wa village, access is possible via a one-hour ride on motorized boats from Lone Ton. The most important population of the eastern subspecies (ssp. *sharpii*) of the **Sarus Crane** outside Cambodia is found here and can be seen all year round, while in winter small groups of **Common Cranes** are regularly encountered. Besides many common waterbirds and open-country species, the **Oriental Darter** and up to five stork species can be expected here in winter, including the declining **Black-necked Stork** and small numbers of wintering **Black Storks**.

While raptors like the **Greater Spotted Eagle** and various harriers are regular, another major attraction is the vultures. These have declined spectacularly all over Southeast Asia, and the largest populations in the region are restricted to northern Myanmar and Cambodia. Although usually seen in flight overhead, lucky visitors have seen groups of **White-rumped** and **Slender-billed Vultures**, alongside the occasional **Red-headed** and **Himalayan Vultures**, gathering at carcasses. Areas of tall grass and scrub should be checked for the localized **Collared Myna**, **Jerdon's Bush Chat** and wintering **Yellow-breasted Bunting**.

Naung Kwin Inn

This area fringes the northern end of Indawgyi Lake and is extensively covered with dense stands of tall elephant grass and scrub. **Sarus Cranes**, storks, vultures and many open-country species have been recorded here.

Bjorn Olesen

*Flocks of **Common Cranes** winter in the grasslands and wetlands around Indawgyi Lake.*

Bjorn Olesen

*The wetlands around Indawgyi are home to important populations of the **Sarus Crane**.*

Forested Hills at Nanyinkha

The extensive areas of dry deciduous and evergreen forests in the hills south-west of Lone Ton are worth exploring for a different suite of birds from the Indawgyi wetlands. Besides parakeets, barbets and the **Blue-bearded Bee-eater**, highlights include the **Great Hornbill** and more than 10 species of woodpecker, including the raucous **Great Slaty Woodpecker**.

Access & Accommodation

The best way to get to Indawgyi is to fly from Yangon or Mandalay to Myitkyina, the capital of Kachin state, and travel overland by car for about 170km to the sanctuary. Alternatively, take the train from Yangon or Mandalay to Moe Nyin, then head to Indawgyi by car (c. 62km). Accommodation is available at the various guesthouses in Lone Ton.

Conservation

Indawgyi Lake Wildlife Sanctuary was established in 1999 for the protection of waterbirds and their habitats, as well as the surrounding watershed forests. Logging activities in the hills surrounding Indawgyi Lake increasingly threatens the ecology of the lake system due to heavy sedimentation.

Kalaw & Lake Inle

Low Bing Wen & Thet Zaw Naing

Low Bing Wen

The scenic hills around the Yay-Aye-Kan reservoir comprises a variety of habitats in a small area that supports a surprising diversity of birdlife.

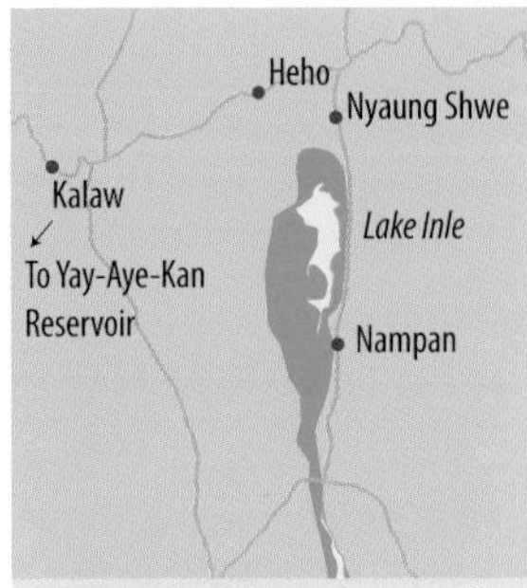

KEY FACTS

Nearest Major Towns
Kalaw, Nyaung Shwe (Shan State)

Habitats
Montane evergreen forest, coniferous forest, scrub and freshwater wetlands

Key Species
Kalaw Black-tailed Crake, Pin-tailed Green Pigeon, Blue-bearded Bee-eater, Blue-throated Barbet, Olive Bulbul, Spot-throated Babbler, Rusty-cheeked Scimitar Babbler, Spectacled Barwing, White-browed and Silver-eared Laughingthrushes, Dark-backed Sibia, Scarlet-faced Liocichla, Burmese Yuhina, Black-breasted Thrush, Black-headed Greenfinch
Kalaw (winter) Asian Stubtail, Martens's Warbler, Crested Bunting
Wetlands Eastern and Western Marsh Harriers, Pied Harrier, Wire-tailed Swallow, Chinese Grassbird, Collared Myna, Jerdon's Bush Chat
Wetlands (winter) Ferruginous Duck, Baer's Pochard (rare), Greater Spotted Eagle, Black-browed and Clamorous Reed Warblers, Baikal Bush Warbler, Yellow-breasted Bunting

Best Time to Visit
November–March; rainy season May–October

A relaxing boat ride around the banks of Lake Inle is likely to yield sightings of the patchily distributed ***Jerdon's Bush Chat****, here at one of its strongholds.*

A former colonial hill station used by the British to escape Myanmar's oppressive summer heat, Kalaw still retains its colonial vibes and laid-back atmosphere. Many examples of colonial architecture still remain which, together with the network of tidy roads lined with colourful shrubs and trees, make Kalaw more reminiscent of a British village than a town in Southeast Asia. For birdwatchers the main attractions lie in the montane evergreen and pine forests located close to the township, home to regional specialities such as the **Black-tailed Crake**, **Burmese Yuhina** and **Olive Bulbul**.

Located close to Kalaw is the famous tourist attraction of Inle, an oval-shaped lake with a maximum depth of 4m. It is part of the Shweyaung rift valley and features forested mountain ranges to the east and west, and the flat Nyaung Shwe plain to the north. The total drainage basin of Inle occupies an area of 3,700km^2 and is home to more than 120,000 people. Lake Inle and its surrounding forested mountains are protected as a wildlife sanctuary. The rich birdlife here includes 250 recorded species to date, including **Jerdon's Bush Chat** and **Chinese Grassbird** in the reedbeds, and the increasingly rare **Collared Myna** among its floating agricultural plots.

Birdwatching Sites

KALAW

Yay-Aye-Kan Reservoir

Arguably the most important birdwatching site around Kalaw, the reservoir is located 6km south-west of the hill station and is accessible via motorcycle taxis or hiking from town. The forested hills surrounding the reservoir support good numbers of the near-endemic **Burmese Yuhina**, which is usually seen among mixed foraging flocks. Other species that should be looked for in these flocks include the **Olive Bulbul**, **Spectacled Barwing** and **Silver-eared Laughingthrush**. At the northern end of the catchment area near Kalaw, regenerating stands of pine forest in the area hold the **Black-headed Greenfinch** and **Black-breasted Thrush**.

Cheng Heng Yee

*The **Black-breasted Thrush** is regularly seen in the pine forests around Kalaw.*

Myin Ka Village

The road from Kalaw to the village of Myin Ka, located about 1km east of Kalaw, still contains tracts of remnant montane evergreen forest and scrubby secondary growth that support a slightly different suite of birds from Yay-Aye-Kan. In particular, the **Spot-throated Babbler** and secretive **Scarlet-faced Liocichla** appear to be locally common at this locality. Other species that have been seen here include the **Asian Emerald Cuckoo**, **Dark-backed Sibia** (ssp. *castanoptera*), **Spectacled Barwing**, **Pin-tailed Green Pigeon** and skulking **Rusty-cheeked Scimitar Babbler**.

Con Foley

***Scarlet-faced Liocichlas** skulk in the dense undergrowth in the remnant forests near Myin Ka.*

LAKE INLE

Northern Shores of Lake Inle Near Nyaung Shwe

Lake Inle is usually visited on a day trip from Kalaw, although accommodation options are plentiful in the area should you desire a more thorough exploration of the wetlands and forest. There are two main birdwatching sites in the area. The first is around the Birdwatching Tower at the northern end of the lake and 20 minutes by boat from the town of Nyaung Shwe. The main ornithological interest here lies in the extensive reedbeds, which hold important populations of **Jerdon's Bush Chat** and **Chinese Grassbirds**, the latter discovered here only recently. In winter the area supports flotillas of wintering ducks, and careful scanning of these flocks may be rewarded with small numbers of **Ferruginous Ducks** and the rarely recorded **Baer's Pochard**.

Nampan

The second site popularly visited by birdwatchers in Lake Inle is around Nampan Village at the eastern end of the lake, reached via a boat ride of an hour and a half or a 25-minute drive from the town of Nyaung Shwe. The **Collared Myna**, a species difficult to see elsewhere in Southeast Asia, can be seen in

James Eaton

*The near-endemic **Burmese Yuhina**, found only in south-east Myanmar and adjacent parts of Thailand, is regularly seen in the forests around Kalaw.*

James Eaton

*The **Chinese Grassbird** was rediscovered in the wetlands around Lake Inle by James Eaton in 2012.*

Francis Yap

*The **Black-browed Reed Warbler** is a common winter visitor to the wetlands around Lake Inle.*

Thet Zaw Naing

Extensive wetlands fringe Lake Inle.

small numbers around the floating agricultural plots, and the area supports waterbird breeding colonies that are used by a variety of storks, ibises and herons. In winter it is possible to see several species of bunting, including the globally threatened **Yellow-breasted Bunting**.

Access to both of these sites is facilitated by Lake Inle's famous leg rowers, boatmen who navigate their boats solely using their legs, and watching them in action is an experience in itself.

Access & Accommodation

Kalaw, and by extension Lake Inle, is most easily reached via regular domestic flights from Yangon (one hour 10 minutes each way) or Bagan (35 minutes each way) to Heho Airport, which is located an hour's drive away from the hill station. Birdwatchers can also opt to travel via lengthy bus or train rides from these starting locations. There is variety of accommodation options around Kalaw and at Lake Inle, which can also be visited as a day trip from Kalaw if desired. It takes about an hour and a half each way to drive from Kalaw to Lake Inle.

Conservation

Lake Inle is a wildlife sanctuary that was established in 1985, while Yay-Aye-Kan Reservoir is protected by district regulations which prohibit logging, cultivation and hunting. In recognition of its freshwater ecosystems and the rich heritage of the communities living around the lake, Inle was declared Myanmar's first Biosphere Reserve in 2015. However, water quality in Lake Inle is increasingly degraded by sedimentation due to farming activities around it, as well as pollution from pesticides, fertilizers and sewage run-off.

Bagan Archaeological Zone

Low Bing Wen

Low Bing Wen

The magnificent scenery of the Bagan Archaeological Zone is matched by few birdwatching sites in Southeast Asia.

Bagan offers a birdwatching experience few other sites can match in Southeast Asia. Once the site of the capital of the ancient Kingdom of Pagan, Bagan is one of Southeast Asia's most significant archaeological sites, with 2,200 pagodas and temples concentrated within a mere 104km². Against this magnificent backdrop birdwatchers search for no less than four Burmese endemics (**Jerdon's Minivet**, **Hooded Treepie**, **Burmese Bush Lark**, **White-throated Babbler**), along with distinctive subspecies of the **Eurasian Collared Dove**, **Vinous-breasted Starling** and **Long-billed Pipit** in the country's dry scrub and xenophytic woodlands. If the exertion is too much to bear, a relaxing cruise along the Irrawaddy River is an option as well, where notable species include the patchily distributed **River Lapwing**, **White-tailed Stonechat** and **Striated Babbler**.

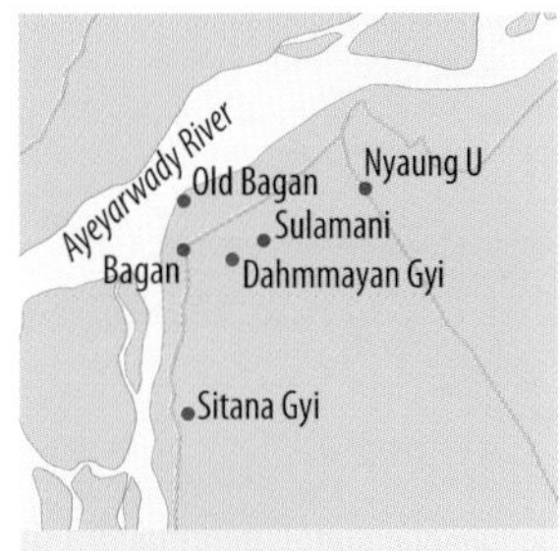

KEY FACTS

Nearest Major Town
Bagan (Mandalay Region)

Habitats
Open woodland, cultivation, scrub, reedbeds and sandbanks (along Irrawaddy River)

Key Species
Rain Quail, White-eyed Buzzard, River Lapwing, Small Pratincole, Eurasian Collared Dove, Laggar Falcon, Jerdon's Minivet*, Rufous and Hooded* Treepies, Burmese Bush Lark*, Sand Lark, Wire-tailed Swallow, Grey-throated Martin, Striated and White-throated* Babblers, Vinous-breasted Starling, White-tailed Stonechat, Long-billed Pipit
Winter Eurasian Wryneck, Tickell's Leaf Warbler, Booted Warbler Bluethroat, Black-headed Bunting

Best Time to Visit
November–March; rainy season May–October

Birdwatching Sites

Sulamani & Dahmmayan Gyi Temples

The area's specialities may be seen anywhere in the Archaeological Zone, but the area between the two temples seems particularly productive. **Burmese Bush Larks** and **White-throated Babblers** are common, conspicuous and easy to observe, but **Jerdon's Minivet** and **Hooded Treepie** pose a greater challenge. The marshy areas around the temples support a variety of waterfowl, while **Rain Quails** and **Siberian Rubythroats** lurk in dense thickets. A pair of resident **Laggar Falcons** is frequently seen perched on top of the Sulamani Temple.

Yong Ding Li

A boat trip along the Irrawaddy River is often enlivened by the sight of the patchily distributed White-tailed Stonechats.

*The unpredictable **Hooded Treepie** is arguably the toughest of Myanmar's dry-zone endemics to see.*

*Unlike most other minivets, the endemic **Jerdon's Minivet** forages in low scrub, and regularly descends to the ground to feed.*

*One of the few long-distance migratory woodpeckers, the **Eurasian Wryneck** winters in the scrubby country around Bagan.*

*The **Burmese Bush Lark** is easy to see in open, scrubby areas in Bagan.*

Sitana Gyi Temple

This temple, a short drive south of New Bagan, is a good back-up site for the area's specialities. The acacia woodland is denser here and there is a sprinkling of cacti throughout the landscape, conducive conditions for foraging **Hooded Treepies**, which are often seen unobtrusively foraging for invertebrates among the cacti and thorn scrub. **Jerdon's Minivet** is also present here, along with the distinctive subspecies of the **Eurasian Collared Dove** (ssp. *xanthocycla*), **Vinous-breasted Starling** (ssp. *burmannicus*) and **Long-billed Pipit** (ssp. *yamethini*). There is also a resident pair of **Laggar Falcons**.

Irrawaddy River Cruise

These cruises begin from the jetty in Old Bagan. From here, it is a one-hour boat ride downstream to a series of sandy islands in the middle of the river featuring tracts of tall grass and *Phragmites*. Disembarking and exploring the islands on foot is recommended. Waterfowl occur in low density, with **Indian Spot-billed Ducks** and **Ruddy Shelducks** possibilities. The declining **River Lapwing** and **Small Pratincole**, two species sensitive to increasing disturbance of their riparian habitats around Southeast Asia, may also be seen here. The sandy banks of the islands are the foraging ground for small groups of **Sand Larks**, and the riparian grasslands hold good numbers of **White-tailed Stonechats**. Other birds possible in this area include the **Striated Babbler**, **Red Avadavat** and wintering **Bluethroat**, as well as the **Wire-tailed Swallow** and **Grey-throated Martin**.

Access & Accommodation

As the focal point of Myanmar's growing tourism industry, tourist infrastructure around Bagan is very well developed. Bagan can be reached by air (via Nyaung U Airport), land and even river from Yangon, and a variety of accommodation options catering to a range of budgets is available. Visiting birdwatchers are free to explore the area without the need for a local guide or permits. Do note, however, that an entrance fee of 25,000 Myanmar Kyats, or 20 USD, is payable on arrival at the city.

Conservation

In recognition of its rich cultural and archaeological heritage, Bagan has been nominated for UNESCO World Heritage Site status, although this is still pending. The dry-zone habitats in Bagan are well protected on account of their location, being situated in an area of cultural and historical significance.

Nat Ma Taung National Park

Low Bing Wen

The pristine oak-rhododendron forest on the slopes of Mount Victoria is home to a variety of regional endemics and Eastern Himalayan specialities.

Nat Ma Taung, also known as Mount Victoria, is the highest peak in the Chin Hills, a mountain range that extends northwards from eastern Myanmar into north-east India. The intrepid explorer-botanist Frank Kingdon-Ward was among the first to document the flora of these lush mountains. For birdwatchers, Mount Victoria is the focal point of the Nat Ma Taung National Park, a large protected area that covers 720km^2 of montane and lowland forests. At 3,053m asl, the mountain supports a variety of habitats that are home to an impressive array of montane species, ranging from eastern Himalayan specialities to a number of country endemics. A visit to Mount Victoria is the centrepiece of any birdwatching tour to Myanmar, and birdwatchers generally need the better part of a week to come to grips with the birdlife in each of the different habitats on the mountain and its foothills. The coniferous forests on Mount Victoria's upper slopes support the endemic **White-browed Nuthatch** and **Burmese Bushtit**, while lower down a number of restricted-range species like the **Brown-capped Laughingthrush**, **Grey Sibia** and **Chin Hills Wren-Babbler** lurk in the understorey of the montane evergreen forest. Even the scrubland and forest patches on the lower slopes should not be neglected, for they harbour the near-endemic **Striped Laughingthrush** and **Spot-breasted Scimitar Babbler** alongside a variety of bush warblers, flycatchers and laughingthrushes.

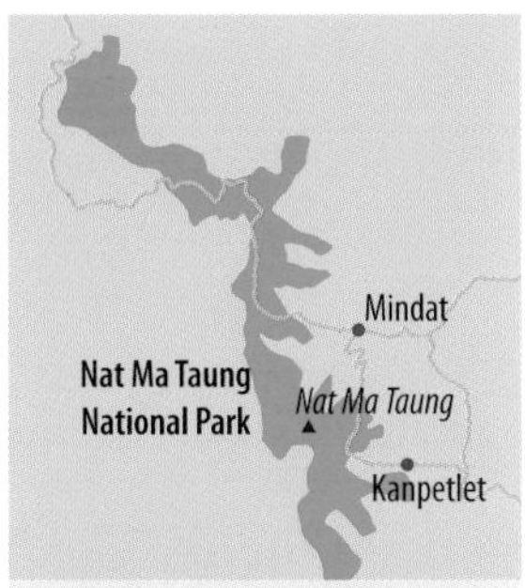

KEY FACTS

Nearest Major Towns
Kanpetlet (Chin State), Bagan

Habitats
Submontane and montane forests (including oak-rhododendron), coniferous forest, scrub

Key Species
Montane Mrs Hume's Pheasant, Blyth's Tragopan (rare), Hill Partridge, Hodgson's Frogmouth, Yellow-billed Blue Magpie, Green and Black-headed Shrike-babblers, Black-bibbed Tit, Burmese Bushtit*, Spot-breasted and Black-throated Parrotbills, Black-throated Prinia, Spotted Elachura, White-tailed, Chestnut-vented and White-browed* Nuthatches, Himalayan Cutia, Red-faced Liocichla, Assam, Blue-winged, Striped and Brown-capped Laughingthrushes, Mount Victoria Babax, Grey Sibia, Streak-throated Barwing, Chin Hills Wren-Babbler, Slender-billed and Spot-breasted Scimitar Babblers, Fire-tailed Sunbird, Yellow-bellied Flowerpecker
Winter Grey-sided Thrush, Slaty-backed Flycatcher
Lowland Grey-headed Parakeet, Pale-headed Woodpecker, Collared Falconet, White-rumped Falcon, Burmese Nuthatch

Other Specialities
Bengal Slow Loris, Western Hoolock Gibbon (rare), Red Goral (rare)

Best Time to Visit
November–March; rainy season May–October

***Mrs Hume's Pheasant** is relatively thin on the ground, but may sometimes be encountered near the edge of the track that ascends Mount Victoria.*

Birdwatching Sites

Road from Bagan to Kanpetlet

The nine-hour journey to Mount Victoria from Bagan passes through some excellent mixed deciduous forests and extensive stands of bamboo in the foothills of the Chin Hills, which are home to several sought-after species. Suggested stops include the Nagabwet Forest Reserve and the forests near the village of Saw before the ascent to Kanpetlet. Birdwatching around these areas can be very fruitful, and desirable species include the difficult to find **Grey-headed Parakeet**, here at one of its global strongholds, as well as the **White-rumped Falcon**, **Himalayan Flameback**, **Pale-headed Woodpecker** and **Burmese Nuthatch**. Streams should be checked for the **Black-backed Forktail**.

Mixed Coniferous Forest at 2,500m

From the Pine Wood Villas it is an hour's drive up the slopes of Mount Victoria through a band of moist evergreen forest to reach the mixed coniferous forest on its upper slopes.

James Eaton

The globally threatened ***White-browed Nuthatch*** *is endemic to Mount Victoria and its adjacent peaks.*

Wang Bin

Birdwatchers with the stamina for spotlighting may be rewarded with sightings of the unobtrusive ***Hodgson's Frogmouth****.*

This is the best place to encounter the endemic **White-browed Nuthatch**, which often follows mixed flocks and can be seen creeping along the trunks of moss- and epiphyte-clad oaks (*Quercus semecarpifolia*), rhododendrons and conifers. The same mixed flocks may also contain the endemic **Burmese Bushtit** and **Black-bibbed Tit**. Flowering rhododendrons attract beautiful **Fire-tailed Sunbirds**, while noisy parties of the near-endemic **Mount Victoria Babax** and **Brown-capped Laughingthrush** lurk in the scrubby understorey. **Mrs Hume's Pheasant** is sometimes seen along the road or grassy clearings in the early morning at this elevation.

Montane Evergreen Forest at 2,000m

Lower down the vegetation type consists of montane evergreen forest, dominated by towering, moss-covered oaks and a dense understorey of bamboo and vine tangles. These forests are home to many birds characteristic of the Eastern Himalayas. Mixed species foraging flocks are dominated by the restricted-range **Grey Sibia** and a variety of woodpeckers, minlas, fulvettas, warblers and yuhinas. These flocks should also be checked for parties of the uncommon **Black-headed Shrike-babbler**, **Himalayan Cutia** and localized **Streak-throated Barwing**. The understorey is home to the distinctive subspecies (ssp. *ripponi*) of the **Black-throated Parrotbill**, **Broad-billed Warbler** and skulking **Chin Hills Wren-Babbler**. Nocturnal spotlighting may yield **Hodgson's Frogmouth**. **Blyth's Tragopan** is still present despite heavy hunting pressure, and a group of five, including a splendid male, was seen on a visit in 2012.

James Eaton

Streak-throated Barwing *forage amongst the roadside vegetation in noisy parties.*

James Eaton

*The **Brown-capped Laughingthrush** (ssp. victoriae) is a restricted-range species endemic to the mountains on the Indo-Burmese border.*

Thet Zaw Naing

*Parties of **Burmese Bushtit** are regular in the pine forests.*

Scrub & Forest Remnants at 1,850m

Due to the practice of shifting cultivation among local people, a large portion of the mountain's lower slopes comprises dense secondary scrub that has regenerated on abandoned farmland. These mosaics of scrub, grass and remnant montane forests are surprisingly rich in birdlife and can be explored via several narrow trails leading towards distant villages from the access road up the mountain. Prime targets here are the near-endemic **Striped Laughingthrush** and **Spot-breasted Scimitar Babbler**. Other species include the monotypic **Spotted Elachura**, **Blue-winged Laughingthrush**, **Red-faced Liocichla** and secretive **Brown Bush Warbler**.

Access & Accommodation

Mount Victoria can be reached via a bumpy nine-hour drive from Bagan on unpaved roads, although this could change in the future with extensive roadworks planned for the route. A permit and local ranger are required to enter the park, and although it might be possible to arrange for this at the park headquarters in the township of Kanpetlet on the eastern boundary of the park, it is best done in advance via one of the local tour agencies in Bagan or Yangon. Accommodation on the mountain is limited, with a choice between Pine Wood Villas (where most birdwatchers stay) and the newer Mountain Oasis Resort, the latter located closer to Kanpetlet.

Conservation

Habitat clearance for agriculture is rampant on the lower slopes of Nat Ma Taung National Park, with some evidence of disturbance by slash-and-burn farming at upper elevations as well. Fires set by shifting cultivators have damaged vegetation in various areas throughout the park. Tea plantations have also encroached beyond the boundaries of the park. The very low density of large mammals, including primates, in the forests of Mount Victoria indicates high hunting pressure suffered by the wildlife here. Nat Ma Taung National Park has been nominated for UNESCO World Heritage Site status.

James Eaton

*The **Chin Hills Wren-Babbler** is a shy skulker that may be seen in the roadside forest with a bit of effort.*

Yann Muzika

*The **Blue-winged Laughinghthrush** is one of at least eight possible laughingthrushes on Mount Victoria.*

Gulf of Mottama

Pyae Phyo Aung

BANCA

Coastal mudflats along the Gulf of Mottama.

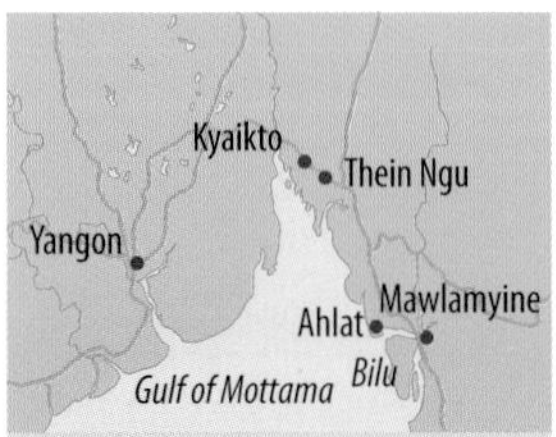

KEY FACTS

Nearest Major Towns
Kyaikhto, Mawlamyine (Mon State)

Habitats
Coastal mudflats, beaches, mangroves, scrub, coastal marshes and aquaculture ponds

Key Species
Painted Stork, Lesser Adjutant, Black-headed Ibis
Winter Ruddy Shelduck, Pied Harrier, Eurasian Curlew, Bar-tailed and Black-tailed Godwits, Great Knot, Broad-billed Sandpiper, Curlew Sandpiper, Spoon-billed Sandpiper, Red-necked Stint, Spotted Redshank, Brown-headed Gull, Pallas's Gull

Other Specialities
Finless Porpoise (rare)

Best Time to Visit
October–April (for migratory shorebirds)

*Significant populations of the Endangered **Great Knot** winter on the Gulf's mudflats.*

Extensive areas of intertidal mudflats and estuarine mudflats span Pyapon (Ayeyarwady Region) and Mudon townships (Mon State), along the shallow Gulf of Mottama. This relatively undisturbed and funnel-shaped stretch of tidal flats receives a constant inflow of sediments from the mighty Ayeyarwady, as well as the smaller Sittaung and Thanlwin (Salween) rivers. At the same time, it acts as nutrient sink, providing spawning and nursery grounds for aquatic species beyond the Gulf and far into the Bay of Bengal. Its tidal cycle is extremely pronounced in speed and amplitude, thus driving a very powerful tidal bore. The mudflats support large numbers of waterbirds, of which 12 species are globally threatened. Approximately half of the global population of the Critically Endangered **Spoon-billed Sandpiper** is known to winter in the Gulf. Key wintering populations of other shorebirds include both **Black-tailed** and **Bar-tailed Godwits**, **Eurasian Curlews**, and **Great** and **Red Knots**. Additionally, the coastal wetlands are used by large waterbirds such as the **Painted Stork** and **Black-headed Ibis**, along with a variety of raptors including the striking **Pied Harrier**. Visitors to the Gulf will discover some of the least known wetlands in Southeast Asia, and bear witness to dedicated conservation work addressing the needs of both coastal communities in Myanmar and migratory shorebirds.

Birdwatching Sites

Thein Ngu

A multiple-day boat trip from Thein Ngu village (Bilin Township) in the northern Gulf of Mottama is an excellent and adventurous means to observe flocks of wintering **Spoon-billed Sandpipers** among other waterbirds. These include a significant proportion of the East Asian Flyway populations of the threatened **Great Knot** and **Nordmann's Greenshank**, as well as large numbers of the distinctive **Broad-billed Sandpiper**, and many terns and gulls. These extensive intertidal flats are estimated to support in excess of 150,000 shorebirds in mid-winter, based on recent surveys.

Björn Olesen

*Parties of **Green Bee-eaters** are a familiar sight in Myanmar's countryside.*

Peter and Michelle Wong

*The largest congregations of the globally threatened **Spoon-billed Sandpiper** in Southeast Asia occur on the mudflats of the Gulf.*

Over a few days, birdwatchers can stay on the boat and access areas of mudflats under suitable tidal conditions to observe flocks of wintering shorebirds alongside smaller numbers of the resident **Black-headed Ibis**, a species that is fast declining across Southeast Asia.

Ahlat

The village of Ahlat sits in the southern Gulf of Mottama, not far from Mawlamyine (formerly Moulmein) Town. There are two nearby high-tide roosts that are used by **Spoon-billed Sandpipers** and other smaller shorebirds, as well as smaller numbers of larger shorebirds like **Eurasian Curlews**, together with many herons and egrets. To access this site, you can be based at Mawlamyine and make daily excursions to Ahlat for birdwatching. It is possible to see most of the key species here over two days. From Mawlamyine, nearby Bilu island supports thriving cottage industries for handicrafts and may be of interest to some visitors.

Accommodation & Logistics

Kyaikhto town can be accessed by car or public bus from Yangon; the travel time is about three and a half hours. Currently, visitors may stay at the Biodiversity and Nature Conservation Association (BANCA) Environmental Education and Sustainability Centre (in Bilin township), and arrange local bird guides, boats and field logistics there. Boats are crewed by former bird hunters now working with BANCA to promote local conservation activities. Mawlamyine town is about 300km away from Yangon and can be reached by public bus or car on a 5–6-hour road trip. Many hotels are available, especially along the Thanlwin River, where some opportunistic birding is possible. An hour's drive from Mawlamyine are the limestone caves of Hpa-An (Kayin state) where **Limestone Wren-Babblers** (ssp. *crispifrons*) are regularly seen.

Conservation

Hunting and mist-netting are the major immediate threats to the waterbirds, although conservation work in recent years has significantly reduced this threat in many areas. More than 70 per cent of local communities living along the Gulf of Mottama coastline depend on fishery resources for their livelihoods. Due to unsustainable fishing and shellfish harvesting practices, local communities were keen to participate in designating the wider Gulf as a Ramsar Site. Through the involvement and strong commitment of the Mon State Government, working closely with local people and the central government, the north-east coastline along the Gulf of Mottama was finally designated as a Ramsar Site in May 2017. Currently administered by the Forest Department, this site is expected to be managed by the provincial government and local communities in the near future.

Yann Muzika

*Perhaps the best looking of the Asian harriers, the **Pied Harrier** may be seen in the mosaic of coastal scrub and agriculture.*

Lenya National Park

Thet Zaw Naing

Thet Zaw Naing

Lenya National Park protects the largest expanse of lowland rainforest in peninsular Myanmar.

KEY FACTS

Nearest Major Towns
Myeik, Bokpyin (Tanintharyi Division)

Habitats
Lowland and hill rainforests, monsoon forest, scrub

Key Species
Long-billed, Ferruginous and Crested Partridges, Crested Fireback, Great Argus, Storm's Stork (rare), Wallace's Hawk-Eagle, Large Green Pigeon, Red-billed, Chestnut-bellied and Black-bellied Malkohas, Moustached Hawk-Cuckoo, Large and Blyth's Frogmouths, Blue-banded Kingfisher, White-crowned, Helmeted, Tickell's Brown and Plain-pouched Hornbills, Red-throated Barbet, Bamboo and Great Slaty Woodpeckers, Green Broadbill, Gurney's Pitta, Maroon-breasted Philentoma, Crested Jay, Black Magpie, Chestnut-capped Thrush, Red-throated Sunbird, Pin-tailed Parrotfinch

Other Specialities
Sunda Pangolin, Sunda Colugo, Lar Gibbon, Dusky Langur, Tenasserim Lutung, Indochinese Tiger, Clouded Leopard, Asiatic Black and Sun Bears, Asian Elephant, Malayan Tapir, Gaur, Sambar, Lesser Mousedeer, Southeast Asian Box Turtle

Best Time to Visit
November–June; May–June for Gurney's Pitta

Straddling 1,766km^2 of the Tenasserim Range that runs along the long and rugged Thai-Burmese frontier, Lenya National Park lies in the remote and densely forested Tanintharyi (Tenasserim) Region in southernmost Myanmar. Lenya sprung into the ornithological limelight with the discovery of a significant population of the then Critically Endangered **Gurney's Pitta** in 2003. The park's conservation importance lies in the large tracts of lowland dipterocarp forest that support a number of Sundaic species whose ranges do not extend further north beyond the Thai-Malay Peninsula. Surveys here have found charismatic species more typical of Malaysia and western Indonesia, like the **Great Argus**, **Large Green Pigeon**, **Large Frogmouth** and **Helmeted Hornbill**, as well as significant populations of the migratory **Plain-pouched Hornbill**. Although 267 bird species have been recorded to date, this vast park remains relatively unexplored and there is significant potential for further discoveries. Getting to Lenya can be a challenge as infrastructure remains poor, but for the intrepid birdwatcher the park offers an excellent alternative to other lowland rainforest sites along the Thai-Malay Peninsula.

Birdwatching Sites

Khae Chaung to 'English Chaung'

The lowland rainforests of Lenya are best accessed from the village of Khae Chaung, which can be reached from the road that connects Myeik to Kawthaung. The unpaved track from Khae Chaung passes the foothills of Kane (*c.* 10km) through mostly cultivated land. Beyond Kane, the next 6km of track towards English Chaung Camp passes through hilly forest and is excellent for birdwatching. Hornbills like **Great** and **Tickell's Brown Hornbills** regularly pass overhead. Of particular

Con Foley

*The **Chestnut-breasted Malkoha** is commonly sighted in the park's forests.*

interest is the presence of both **Long-billed** and **Ferruginous Partridges**, although they tend to be more easily heard than seen. The forests should also be checked for a number of other Sundaic specialities that do not extend further north into continental Myanmar, such as the **Maroon-breasted Philentoma**, as well as noisy parties of broadbills including both **Long-tailed** and **Silver-breasted Broadbills**.

'English Chaung' Camp to Wettaung Chaung Camp

From English Chaung Camp it is another 5km on the rough track to Wettaung Chaung, which offers excellent access to lowland rainforest. Four long forest trails radiate from Wettang Chaung and are the focus of any birdwatching trip here. A good mixture of lowland and hill rainforest species like the **Collared Babbler** can be found here. However, the main highlight in the forests around Wettang Chaung is **Gurney's Pitta**, which should be searched for in dark gullies that contain extensive clumps of rattan palms. While checking the forest floor for pittas and other ground birds, also look out for the **Plain-pouched Hornbills** that regularly pass overhead, mixed foraging flocks, four species of lowland broadbill and more than 10 woodpecker species that occur here, including the sought-after **Bamboo** and **Great Slaty Woodpeckers**.

Myeik (Mergui) Islands

The remote and relatively undisturbed Mergui Islands can be accessed from Myeik by chartered ferries. Of greatest interest to birdwatchers are the larger islands of Bada and Lambi, of which the latter is protected under the Lambi Island Marine National Park. Recent biodiversity surveys have found flocks of **Plain-pouched Hornbills** on some of these islands, as well as important populations of the **Great-billed Heron** and **Beach Stone-curlew**. The islands also support key populations of the **Nicobar Pigeon**.

Con Foley

*The endearing **Black-and-yellow Broadbill** is more easily heard than seen in the dense canopy of the Burmese lowland rainforest.*

Access & Accommodation

The entry point to the park is the southern city of Myeik, which is regularly serviced by flights from Yangon. Khae Chaung can be reached from Myeik via the Myeik-Kawthaung road and is about 130km away. A cheaper option is to travel overland by bus from Yangon to Myeik, then travel onwards to Khae Chaung by rented car. A good variety of hotels is available in Myeik, but at the key birdwatching sites it is necessary to camp. Given the currently poor infrastructure and political instability in some parts of the region, the best way to access Lenya is to make arrangements with bird-tour companies based in Yangon.

Conservation

The lowland forests of Lenya were first proposed as a national park in 2002. Although parts of the originally proposed area have been gazetted, Lenya National Park receives very little protection and currently lacks a proper management plan. Due to poor enforcement, encroachment by oil-palm plantations, logging and hunting remain rampant.

*The most important population of the endangered **Gurney's Pitta** dwells in the lowland rainforest of Lenya National Park and its environs.*

Thet Zaw Naing

*Lenya is probably the best place to see the threatened **Plain-pouched Hornbill** outside northern Peninsular Malaysia, with breeding reported here as well.*

THE PHILIPPINES

Spread over 1,500km of the western Pacific, the 7,100 islands of the Philippine archipelago are one of the world's great biodiversity hotspots. Much of the country is rugged and mountainous, with peaks like Pulag, Apo and Dulang-dulang soaring to nearly 3,000m asl. Due to its proximity to continental Asia and millennia of isolation, the Philippines supports a highly distinct biota, with more than half of its mammal and resident bird species endemic to the archipelago. The Philippines boasts more than 230 avian endemics, from the mighty Philippine Eagle to the diminutive Pygmy Flowerpecker. Given that the country continues to lose 2% of its forest cover per year, it is no wonder that birdwatchers cite it as a place to visit sooner rather than later.

Over half of the Philippine endemics can be seen on a trip focusing on the three main islands of Luzon, Mindanao and Palawan, and one or two of the Visayan islands like Bohol and Negros. However, many endemics occur well off the beaten track, especially on the eastern Visayan islands of Samar and Leyte, posing a logistical challenge for birdwatchers with time at hand and a sense of adventure.

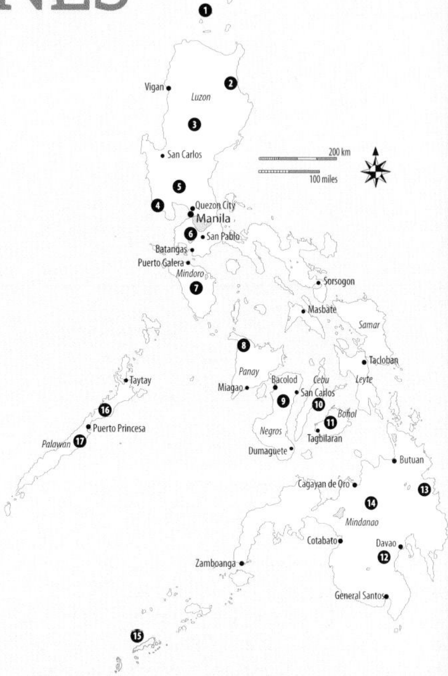

1. Calayan Island
2. Sierra Madre Mountains
3. Mount Polis
4. Subic Bay
5. Candaba Wetlands
6. Mount Makiling
7. Sablayan
8. Panay
9. Mount Kanla-on
10. Tabunan
11. Rajah Sikatuna
12. Mount Apo
13. PICOP
14. Mount Kitanglad
15. Tawi-Tawi
16. Puerto Princesa
17. Narra Forests

Climate

The climate is tropical, being characterized by high rainfall, temperature and humidity throughout the year. However, this can vary significantly across the archipelago. February–April is usually the driest and coolest time of the year, while the rainy season between June–November is associated with frequent thunderstorms and the possibility of destructive typhoons and tropical storms.

Access, Transportation & Logistics

The vast majority of international travellers first arrive in the sprawling capital city of Manila on the island of Luzon. Manila serves as the domestic gateway to numerous destinations on other Philippine islands, although Cebu is also very well connected to a number of major cities in the region due to tourism. Most foreign visitors can visit the country for up to a month without a visa. Overland journeys are likely to involve the iconic jeepney – the modified jeep that, while not exactly comfortable, is able to traverse a wide variety of terrain including unpaved mountain roads. Visiting birdwatchers should also be aware of the need for lengthy and potentially strenuous hikes, particularly when searching for montane endemics, as much of the remaining forest is restricted to the higher elevations that are often inaccessible to vehicles. English is widely spoken throughout the Philippines.

KEY FACTS

No. of Endemics
c. 240

Country List
684 (Wild Bird Club of the Philippines, 2016)

Top 10 Birds
1. Philippine Eagle (Luzon, Mindanao)
2. Palawan Peacock-Pheasant (Palawan)
3. Flame-templed Babbler (Negros)
4. Wattled Broadbill (Mindanao)
5. Whiskered Pitta (Luzon)
6. Giant Scops Owl (Mindanao)
7. Rufous Hornbill (Luzon, Mindanao)
8. Black Shama (Cebu)
9. Walden's Hornbill (Panay)
10. Sierra Madre Ground Warbler (Luzon)

Health & Safety

Various mosquito-borne diseases, including dengue, malaria and Japanese encephalitis, occur and appropriate precautions should be taken. There are few natural annoyances aside from mosquitoes, but leeches are present in remote forested areas.

The rainforests of the Philippines are packed with endemic sunbirds including the little known ***Lina's Sunbird.***

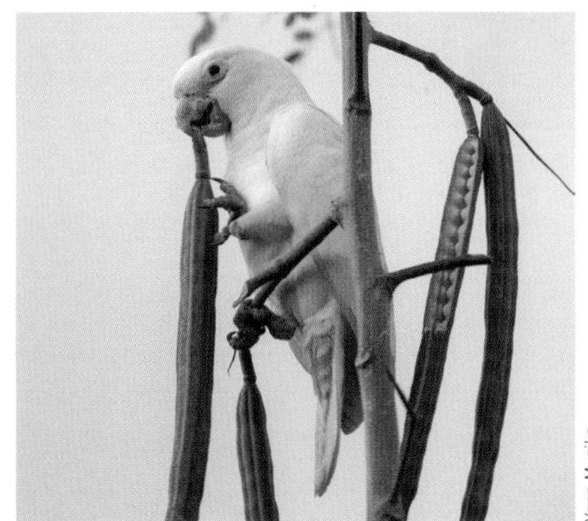

Yann Muzika

The attractive ***Red-vented Cockatoo*** *is now best seen in Palawan's forests.*

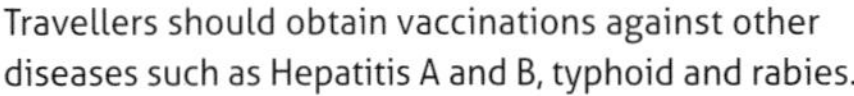

Travellers should obtain vaccinations against other diseases such as Hepatitis A and B, typhoid and rabies.

The Philippine archipelago is generally safe, but travellers are advised to avoid the Sulu Archipelago due to the prevalence of insurgent activity and associated risk of kidnapping incidents. There have also been sporadic clashes between insurgents and security forces on the southern island of Mindanao. Travellers should seek advice from their local embassies for updates on the dynamic security situation.

Birdwatching Highlights

A three-week trip taking in Luzon, Mindanao and Palawan generally yields more than 150 endemic birds, and over 300 species in total. This includes a host of highly sought-after species, such as the incomparable Philippine Eagle in Mindanao, the dazzling Palawan Peacock-Pheasant on Palawan, and the Flame-breasted Fruit Dove, Ashy Thrush and Philippine Eagle-Owl on Luzon, with the last two now showing regularly in forest patches near Manila.

If time permits, an extension into the Sierra Madre Mountains of north-eastern Luzon yields rich rewards. The star attraction in these remote mountains is the elusive Whiskered Pitta, complemented by a host of endemics including the Cream-breasted Fruit Dove, Luzon Bleeding-heart, Sierra Madre Ground Warbler and Grand Rhabdornis. In addition, a visit to the endemic-filled Visayan islands of Negros, Bohol and Cebu yields charismatic specialities like the Visayan Broadbill, Flame-templed Babbler and Black Shama.

Birdwatchers keen to venture off the beaten track have no shortage of options. Surveys on the eastern Visayan islands of Samar and Leyte have identified extensive forests with hard to find species like the Mindanao Bleeding-heart and little-known Visayan Miniature Babbler. Further south on Mindanao, many of the forested mountains outside of Kitanglad and Apo (including Malindang and Matutum), and outlying islands like Pollilo, Camiguin Sur and Siargao, remain poorly explored. Last but certainly not least, right on the border with the Malaysian state of Sabah are the Sulu Islands, notably Tawi-tawi, which is home to the Critically Endangered Sulu Hornbill and Blue-winged Racket-tail.

Francis Yap

The ***Philippine Eagle*** *is the national bird of the Philippines.*

Calayan Island

Carmela P. Española

Paul Quiambao

The coasts of Calayan Island

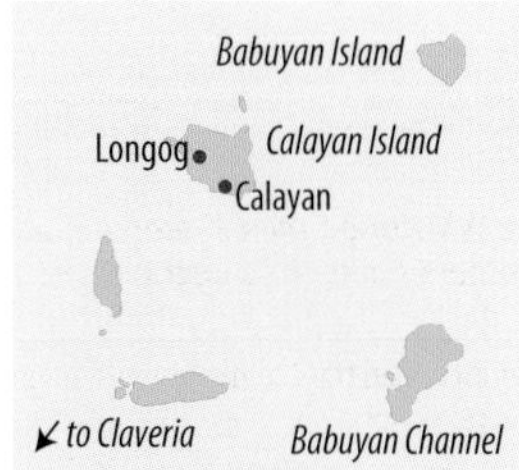

KEY FACTS

Nearest Major Towns
Aparri, Claveria (Cagayan)

Habitats
Lowland evergreen forest, scrubland, sandy beaches, rocky shores, coastal cliffs

Key Species
Calayan Rail*, Plain Bush-hen*, Malaysian Plover, Philippine Cuckoo-Dove, Whistling Green Pigeon, Ryukyu Scops Owl, Northern Boobook, Japanese Paradise Flycatcher, Brown-eared Bulbul, Lowland White-eye
Passage Grey-faced Buzzard
Sea crossing Wedge-tailed Shearwater, Bulwer's Petrel

Other Specialities
Mcgregor's Flapped-legged Gecko*, Ross' Calayan Gecko*, Philippine Warty Pig*, Humpback Whale, Ryukyu Flying Fox

Best Time to Visit
January–May (rainy season starts in June with typhoons passing often the rest of the year)

The recently discovered ***Calayan Rail*** *inhabits the forested interior of the island.*

At 196km², Calayan is the largest and most populous island in the Babuyan group in the far-northern Philippines. Never previously connected to mainland Asia or Luzon, the island is volcanic in origin and its highest peak, Mount Calayan (499m asl) is an extinct volcano. Bordered by rocky and white sand beaches, Calayan is peppered throughout with coralline limestone outcrops and drained by four rivers. Heavily forested hills run through the centre of the island, where the island-endemic **Calayan Rail** was first described for science in 2004. More than 100 bird species have been recorded, including a number that are of interest to birdwatchers, such as the **Whistling Green Pigeon** and **Ryukyu Scops Owl**, species otherwise typical of the Ryukyu Islands, and occurring here at the southern end of their distribution. During autumn passage and the winter months, landbirds on migration southwards pass the Babuyan Islands, and the threatened **Yellow Bunting**, as well as various passerines that are rare in the Philippines, have been recorded here. The Pacific Ocean off the Babuyan Islands is a known cetacean feeding and breeding ground, notably for the majestic Humpback Whales from February to April. The boat crossing from mainland Luzon to Calayan offers a chance to see whales, dolphins and a variety of seabirds including the **Wedge-tailed Shearwater**.

Birdwatching Sites

Longog, Magsidel

Situated in the centre of Calayan Island and reached after a three-hour trek through dense forest, Longog is where the sought-after **Calayan Rail** was first discovered. The rail can be reliably seen here in the dense forest understorey, preferring areas near streams in dense primary or secondary forest. Longog also sports a large clearing with rice paddies as well as banana and coconut groves at the periphery, entirely surrounded by forest. From here,

The ***Whistling Green Pigeon*** *occurs on Calayan at the southern limits of its distribution.*

Pygmy Swiftlets are regular and the **Pacific Swift** can be seen during the migratory period. The banana groves are regularly visited by the **Plain Bush-hen**, which feeds on fallen fruit. The deeper forests of central Calayan offer a chance to see the **Whistling Green Pigeon** (ssp. *filipinus*), **Philippine Cuckoo-Dove** (ssp. *phaea*), **Black-chinned Fruit Dove** and possibly **Metallic Pigeon**, which was historically recorded. From Calayan's forests to its scrubland, the conspicuous **Brown-eared Bulbul** (ssp. *fugensis*) never fails to make its presence known by its loud calls. After dark, the numerous snags or standing dead trees by the clearing attract the **Ryukyu Scops Owl** (ssp. *calayensis*) and **Northern Boobook**.

Dadao, Dilam & Dibay

The coastline along these barangays is interspersed with steep limestone cliffs and secluded sandy or rocky beaches that offer spectacular lookouts. Here, foraging **Malaysian Plovers**, **Pacific Reef Herons**, and many other migratory shorebirds (including the **Grey-tailed Tattler**) can be encountered, as well as foraging **Western Ospreys** and **White-bellied Sea Eagles**. A little further inland at Dilam is scrubland where the **Island Collared Dove** and **Common Emerald Dove**, among other open country species, are frequently seen. Birdwatchers visiting in the migratory passage months in September–October should look out for rare eastern Palearctic migrants, including the **Yellow Bunting** and **Brambling**.

Access & Accommodation

The jump-off points to Calayan are the towns of Aparri and Claveria in Cagayan province. Buses depart daily from Metro Manila to Cagayan, with a travel time of approximately 12 hours. Weather permitting, the island may be reached via ferry from Aparri or small outrigger boats off Claveria. The 5–8-hour boat ride, depending on where the jump-off point is, can get rough so waterproofing of bags is recommended. You must prepare for the possibility of being stranded on the island for days due to inclement weather. Local guides can be hired at the municipal tourism office where you may also find recommendations for a homestay or inn at the town proper. Power supply is limited to 12 hours a day, from noon to midnight. Local transport is mainly by motorbike or on a hand tractor converted into a mass transport vehicle, called 'kuliglig'.

Conservation

Conservation work in the Babuyan Islands began with the Calayan Rail as the flagship species. Municipal Ordinance No. 84 was passed by the Calayan Municipal Council prohibiting the capture, sale, possession and collection of the species. Since then, the Calayan Wildlife Sanctuary (29km^2) was established in Sitio Longog, Barangay Magsidel, in May 2011 with a group of forest wardens hired to patrol the sanctuary. Nevertheless, habitat loss from agriculture ('kainguin') and development continues, resulting in declining forest cover.

Calayan Island is home to a resident population of the ***Japanese Paradise Flycatcher****.*

Sierra Madre Mountains

Robert Hutchinson & Abdelhamid Bizid

Ding Li Yong

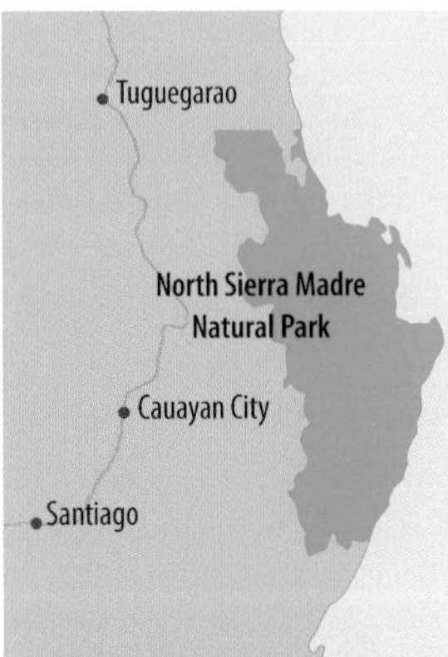

The Northern Sierra Madre Natural Park is the largest protected area in the Philippines, and still supports good populations of many Philippine endemics that are difficult to see elsewhere.

The Sierra Madre is the longest mountain range in the Philippines, forming the backbone of the island of Luzon, where it runs from Cagayan Province in the far north to Quezon Province in the south. Besides containing the largest expanse of rainforest left in the Philippines, the Sierra Madre is also an important source of many of Luzon's major rivers, including the Cagayan, the longest river in the country. Parts of the northern Sierra Madre fall within the extensive Northern Sierra Madre Natural Park, which at 3,594km² is the largest protected area in the Philippines and home to many avian endemics that are difficult to see elsewhere. Chief among these sought-after birds is the fabled **Whiskered Pitta**, endemic to the mountains of Luzon. The gaudy **Flame-breasted Fruit Dove**, a species on the must-see list of many visiting birdwatchers, is also regularly encountered in the forests of the Sierra Madre, as is the **Sierra Madre Ground Warbler**, a species once thought to be a wren-babbler. While some areas in the Northern Sierra Madre are accessible to visitors, careful planning is required to reach the key sites as long treks and several days of camping are needed. An exciting new alternative is the Central Sierra Madre mountains, which supports a very similar birdlife and is accessible from the township of Baler in Aurora province.

KEY FACTS

Nearest Major Towns
Tuguegarao (Cagayan Province), Baler, San Luis (Aurora Province)

Habitats
Lowland, hill and montane rainforests, grassland, cultivation

Key Species
Philippine Eagle* (rare), Philippine Hawk-Eagle*, Island Collared Dove, Luzon Bleeding-heart*, Flame-breasted* and Cream-breasted Fruit Doves*, Spotted Imperial Pigeon*, Rufous Coucal*, Scale-feathered Malkoha*, Philippine Dwarf Kingfisher*, Rufous Hornbill*, Whiskered Pitta*, Blackish Cuckooshrike*, White-lored Oriole*, Slender-billed Crow, White-fronted Tit*, Sierra Madre Ground Warbler*, Luzon Striped Babbler*, Golden-crowned Babbler*, Grand Rhabdornis*, Ashy-breasted Flycatcher*, Blue-breasted Blue Flycatcher*, Furtive Flycatcher*, Green-faced Parrotfinch*
Winter Pechora Pipit

Other Specialities
Golden-crowned Flying Fox*, Philippine Deer*, Philippine Warty Pig*

Best Time to Visit
January–April

Birdwatching Sites

Hamut Camp & Sawa Camp

These two camps are the two most frequently visited sites in the northern Sierra Madre Mountains. Located on the forested slopes of Mount Dos Cuernos, both sites are within a few kilometres of each other and the birdlife at both locales is generally similar. Access to both sites first involves a long walk across denuded hills covered with scrub and cogon grass,

*The **Flame-breasted Fruit Dove** is widely considered to be the most attractive of all the Philippine fruit doves.*

Yann Muzika

*The Sierra Madre Mountains are by far the most reliable place in the Philippines for seeing the fabled **Whiskered Pitta**.*

a feat best completed in the early morning or late afternoon when temperatures are most comfortable. Among the common open-country species to look out for are the rare **Island Collared Dove** (ssp. *dusumieri*), which still survives here in small numbers, **Pied Harrier** and flocks of **Philippine Green Pigeon** and the scarce **Spotted Imperial Pigeon** flying overhead.

The forest at lower elevations is dominated by large stands of bamboo, which is the preferred habitat of the skulking **Furtive Flycatcher**, a species best located by following its sweet song. When the bamboo is in flower, large flocks of **Green-faced Parrotfinches** may be seen, but the species is highly nomadic and such an event may occur only once every few years. Roving feeding flocks at this elevation should be scrutinized for both the **Luzon Striped Babbler** and **Golden-crowned Babbler**, while in the canopy the **Blackish Cuckooshrike**, **White-lored Oriole** and **White-fronted Tit** are all possible.

With increasing elevation the quality of the forest improves and a new suite of birds may be observed. The **Cream-breasted Fruit Dove** (ssp. *faustinoi*) is frequently heard calling from the canopy, while the spectacular **Flame-breasted Fruit Dove** may be encountered in the vicinity of fruiting figs. Diligent scanning of the trails and forest floor may be rewarded with views of the **Luzon Bleeding-heart**, which is still locally common here. The northern subspecies

Abdelhamid Bizid

*The recently described **Sierra Madre Ground Warbler** is very difficult to observe in the dense understorey.*

(ssp. *hydrocorax*) of **Rufous Hornbill** is also present, and although it is devilishly hard to see through the dense canopy, its loud, booming calls are one of the characteristic sounds of the forest. Another distinctive sound to listen out for is that of the very scarce **Slender-billed Crow**, here of the distinct subspecies (ssp. *sierramadrensis*) endemic to Luzon. If heard, it is well worth tracking down as its characteristic calling display with wings raised high above its back is very entertaining to observe. However, the most desirable species in these forests are also the most elusive. Both the **Sierra Madre Ground Warbler** and **Whiskered Pitta** are likely to be heard during a stint here, but seeing them can be an entirely different matter in the dense understorey.

Baler Camp, Aurora

The Aurora Mountains, near the small fishing town of Baler, in the central Sierra Madre, offer a good alternative to the northern Sierra Madre. Overall, the forests here support fewer species, but many of the key targets here are easier to find than at Sawa or Hamut camps. The forests can be accessed on a steep, muddy trail. The trail follows the main stream on that side of the mountain, quickly entering lowland rainforest, home to sought-after species such as the **Rough-crested Malkoha**, **Philippine Pitta**, **White-fronted Tit**, **White-browed Shama** and the striking **Naked-faced Spiderhunter**. Camping in the forests at higher elevations (700m and above) on the north-eastern slopes of the mountains is needed to see the main targets. Mixed flocks at this elevation contain a similar selection of species as at Sawa, including **Golden-crowned Babblers**. The **Short-crested Monarch**, **Rufous Paradise Flycatcher** and **Blue-breasted Blue Flycatcher** are more easily seen here than in Sawa. The **Philippine Dwarf Kingfisher** (ssp. *melanurus*) is regularly encountered along the forest streams.

Abdelhamid Bizid

Flocks of ***Rufous Hornbills*** *still survive on the forested slopes of the Sierra Madre.*

Cream-breasted Fruit Doves and other pigeons, as well as **Luzon** and **Rufous Hornbills**, can be seen in the fruiting trees. A quiet walk in the early morning along the smaller forest trails near the camp provides a good chance to see the **Luzon Bleeding-heart** and the skulking **Sierra Madre Ground Warbler**. **Whiskered Pittas** are often seen in gullies here, and most regularly encountered in March–May.

Access & Accommodation

Both Camp Sawa and Hamut are accessible from Tuguegarao, the capital of Cagayan Province, which has at least three daily flights from Manila taking just under an hour, or numerous buses covering the same route, although most do the journey overnight and take around 10 hours. For overnight stays Tuguegarao has a range of hotels and guesthouses to suit all budgets. Aurora is best accessed by road from and to Manila. However, you should be mindful of the traffic jams, which increase when getting closer to the capital. Baler offers accommodation but an early start from Manila should give enough time to reach the main camp before dusk on the same day. One important feature of Aurora is that a large stream needs to be crossed at the start of the hike, and torrential rain may thus lead to lengthy delays. Organizing the camps and long hikes with a team of porters is not easy, and while those with time can do it independently with local guides, advanced organization with a local bird-tour operator is worth considering.

Conservation

Both Hamut and Sawa camps fall within the Northern Sierra Madre Natural Park, which at nearly 3,600km^2 forms the largest protected area in the whole of the Philippine archipelago. Sadly this protection does not include physical patrols and deforestation is rampant, with most lowland forest at the fringes already lost, and hunting of birds and other wildlife for food is rife, resulting in much reduced numbers of larger species. The forests in Aurora have benefited from ecotourism initiatives, but local villagers still commonly hunt mammals and large birds such as pigeons for food. Deforestation may be less pronounced than further north, but the Sierra Madre here is narrower, more isolated and vulnerable to logging; a newly built road through the Sierra Madre in the north threatens to introduce further deforestation.

Cordillera Central, Mount Polis

Mark Jason Villa

Yann Muzika

The forested slopes of the Cordillera Central near Banaue offer easy roadside birdwatching and a chance to see the spectacular rice terraces carved onto the mountain slopes.

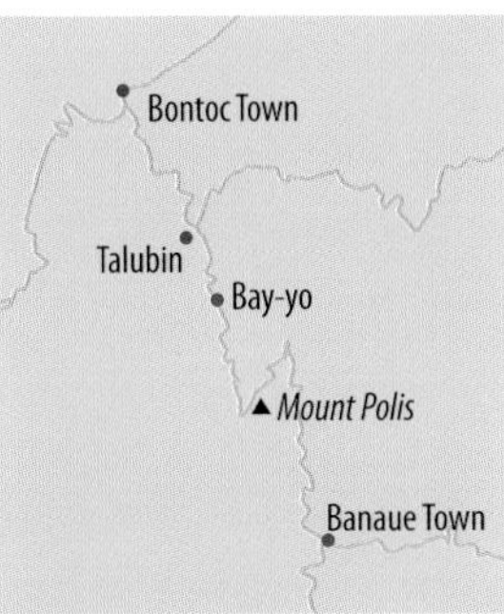

KEY FACTS

Nearest Major Town
Banaue (Ifugao Province)

Habitats
Pine forest, montane rainforest, cultivation, scrub

Key Species
Flame-breasted Fruit Dove*, Luzon Scops Owl*, Philippine Swiftlet*, Montane Racket-tail*, Green-backed Whistler*, Mountain Shrike*, Negros Leaf Warbler*, Long-tailed*, Philippine* and Benguet* Bush Warblers, Chestnut-faced Babbler*, Island Thrush, Bundok Flycatcher*, White-browed Jungle Flycatcher*, Luzon Water Redstart*, Flame-crowned Flowerpecker*, Luzon Sunbird*, Red Crossbill, White-cheeked Bullfinch*
Winter Eyebrowed and Brown-headed Thrushes, Grey Wagtail, Olive-backed Pipit, Little Bunting (rare)

Other Specialities
Northern Luzon Giant Cloud Rat*, Giant Bushy-tailed Cloud Rat*, Luzon Striped Rat*

Best Time to Visit
All year round

Con Foley

*The localized **Luzon Water Redstart** is regularly seen along the river near Bay-Yo town.*

Luzon's Central Cordillera is the second longest mountain range on the island after the Sierra Madre, encompassing a total area of 18,300km². The rugged Central Cordillera is home to the highest peak on Luzon and the second highest in the Philippines, Mount Pulag, which rises to a maximum height of 2,922m asl. The mountains of the Central Cordillera still retain much of their forest cover, with extensive tracts of pine forests dominated by Benguet

Pine (*Pinus kesiya*) and montane rainforests at lower elevations. For birdwatchers, the Central Cordillera is of great interest as it is home to many sought-after endemics such as the **Montane Racket-tail**, **Whiskered Pitta**, **Philippine** and **Benguet Bush Warblers**, and highly localized **Luzon Water Redstart**. The iconic **Philippine Eagle**, previously thought to only occur on the Sierra Madre mountain range in Luzon, was recently discovered in the Central Cordillera.

Birdwatching Sites

Mount Polis

The most frequently visited site in the Central Cordillera Region is known to many birdwatchers as Mount Polis. At more than 1,900m asl, it is the highest point on a ridge and is located near the boundary of Mountain and Ifugao Provinces. The closest town is Banaue, a famous tourist site renowned for its impressive rice terraces dating back as far as 2,000 years ago.

Much of the birdwatching on Mount Polis is done along the road, and with the vegetation at eye-level this provides numerous opportunities for close observations of mixed species foraging flocks. The flocks often comprise the **Elegant Tit**, **Negros Leaf Warbler** (ssp. *benguetensis*), **Chestnut-faced Babbler**, **Sulphur-billed Nuthatch**, **Luzon Sunbird** and occasionally the uncommon **Flame-crowned Flowerpecker** (ssp. *anthonyi*). When roadside trees are fruiting, the stunning **Flame-breasted Fruit Dove** may also be seen perched in the canopy. Spending some time in areas of scrub by the roadside may yield the **Mountain Shrike** and **White-cheeked Bullfinch**.

To see some of the skulking species, one must enter a forest trail to access the mossy montane forest. One such trail can be found behind the large Virgin Mary Statue near the shops at the start of Mount Polis. Spending some time on this trail may be rewarded with sightings of the **Philippine Bush Warbler** and the skulking **Long-tailed Bush Warbler**, as well as the local subspecies (ssp. *poliogyna*) of **White-browed Shortwing**. The **Bundok Flycatcher** (formerly Snowy-browed Flycatcher) may also be encountered. The legendary **Whiskered Pitta** used to be recorded in the forests here but due to habitat loss is now either very rare or locally extirpated.

In order to observe the localized **Luzon Water Redstart**, the short journey to the nearby town of Bay-Yo is required. The river here supports several individuals that can be viewed from the road with the aid of a telescope; closer views may be obtained by descending into the valley. Occasionally, a pair may also be seen around the tiny roadside waterfall. En route, stands of pines near the road should be checked for mixed flocks and the local subspecies (ssp. *luzoniensis*) of the **Red Crossbill**.

Access & Accommodation

Banaue, the gateway to Mount Polis, is an 8–9-hour drive north of Manila. There are also regular public buses departing Manila to

Con Foley

*The **Green-backed Whistler** is a regular participant of the mixed flocks in the Central Cordilleras.*

Bim Quemado

*The **Philippine Bush Warbler** skulks in dense herbage on the trail sides.*

Bim Quemado

The ***Chestnut-faced Babbler****, an aberrant white-eye, regularly joins many mixed flocks.*

Banaue daily. From Banaue it is a further 20km on the road heading to Bontoc and Sagada to reach the birdwatching area. The best accommodation in Banaue is the state-owned Banaue Hotel, but many other cheaper inns and hostel options are available. For transport, having a personal vehicle is ideal, but hotels also rent out jeepneys to guests going to many destinations, including Sagada and the nearby rice terraces.

Conservation

Much of the Central Cordillera does not receive any formal protection, the exceptions being Mount Pulag, Mount Balbalasang-Balbalan and Mount Data National Park. Habitat clearance for agriculture and quarrying activities are conspicuous and a major cause of landslides in the region. Additionally, the area is infamous for the age-old practice of hunting birds for consumption during the migratory season.

Dubi Shapiro

The ***Scale-feathered Malkoha*** *is arguably the best-looking member of the family.*

Yann Muzika

The uncommon ***Flame-crowned Flowerpecker*** *is sometimes seen in foraging flocks moving through the canopy.*

Subic Bay Watershed Reserve

Mark Jason Villa

Yong Ding Li

The Subic Bay Watershed Reserve currently encompasses one of the largest remaining areas of lowland rainforest left on the island of Luzon.

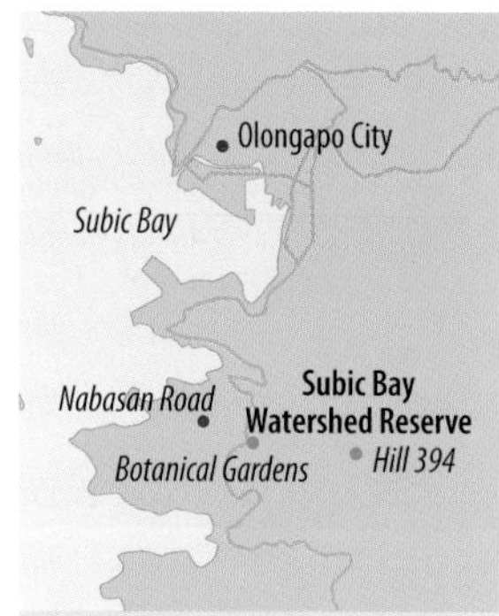

KEY FACTS

Nearest Major Town
Olongapo (Zambales Province)

Habitats
Lowland rainforest (primary and secondary), forest edge, scrub

Key Species
Philippine Hawk-Eagle *, Spotted Imperial Pigeon*, Philippine Green Pigeon*, Amethyst Brown Dove*, Rufous Coucal*, Purple Needletail, Rufous (rare)* and Luzon* Hornbills, Luzon Flameback*, Sooty Woodpecker*, Blue-naped Parrot, Green Racket-tail*, Bar-bellied and Blackish* Cuckooshrikes, White-lored Oriole*, White-fronted Tit*, Philippine Fairy-bluebird*
Winter Brown Shrike, Grey-streaked Flycatcher

Other Specialities
Long-tailed Macaque, Golden-crowned*, Large and Island Flying Foxes

Best Time to Visit
All year round

Bim Quemado

The rainforests at Subic Bay are one of the few remaining strongholds of the ***Green Racket-tail****.*

Subic Bay is best known as a former US naval facility that has now been turned into a major free port and economic zone. Alongside the adjacent Bataan National Park, the Subic Bay area contains some of the last significant areas of lowland dipterocarp forest in western Luzon, and covers nearly 100km^2 of Zambales province. The presence of the US military previously afforded strict protection for the forests here, and now much of the area remains protected as a watershed reserve. Presently, Subic Bay is the only reliable site for the endangered **Green Racket-tail**, a Luzon endemic. It is also the only site on Luzon where the attractive **Blue-naped Parrot** occurs in good numbers. Besides parrots, other fruit-eating birds like the **Philippine Green Pigeon**, **Yellow-breasted Fruit Dove** and **Amethyst Brown Dove** remain fairly easy to see, and the rare **Spotted Imperial Pigeon** has occasionally been recorded. The **Luzon Hornbill** is quite numerous, but the **Rufous Hornbill** is now rarely seen. All of Luzon's four woodpecker species, including the spectacular **Sooty Woodpecker**, are quite common here. Interestingly, there are not many small passerine species in Subic Bay apart from bulbuls, the occasional **Elegant Tit** and the **Stripe-headed Rhabdornis**. A good selection of the forest species can be seen here during a visit of two days, including night rambles.

Birdwatching Sites

Nabasan Road & Trail

Many birdwatchers visit a site called the Nabasan Road and spend most of their time here. This site eventually leads to Nabasan Bay, but there is a network of roads that loop around

Robert O. Hutchinson

*The rare **White-fronted Tit** is infrequently seen in the rainforest around Subic Bay.*

Con Foley

Sooty Woodpeckers are locally common in the forests of Subic.

the forest here and the drive is easy along roads fringed by rainforest. A trail provides access into the forest as well. Birdwatching here can be especially productive in the early mornings and late afternoons. Some of the important Subic specialities to look out for are the **Philippine Green Pigeon**, **Sooty Woodpecker** (ssp. *funebris*), **Green Racket-tail**, **Blackish Cuckooshrike** and **Trilling Tailorbird**. The unusual **Rufous Coucal**, a species that forages mainly in the canopy, unlike many other coucals, is quite easy to hear and see in the forests. By midday the forest can become very quiet as bird activity drops sharply. Spotlighting can produce a good selection of nocturnal species, including the **Chocolate Boobook**, **Luzon Hawk-Owl**, **Philippine Scops Owl** and, with luck, **Philippine Eagle-Owl**. **Great Eared** and **Philippine Nightjars** are also known to occur.

Con Foley

*The striking **Luzon Hornbill** is commonly seen in flocks around the forest at Subic Bay.*

Botanical Garden

Despite being referred to as such by many visiting birdwatchers, this site is not really a botanical garden. There are, however, extensive areas of open roadside forests that are conducive for birdwatching. The road here passes many abandoned bunkers and Quonset huts that are now used by nesting **Grey-rumped** and **Pygmy Swiftlets**. It eventually leads to Hill 394, a popular birdwatching site in the past but access is now prohibited. The **Rufous Hornbill**, **White-lored Oriole** and **White-fronted Tit** are among the sought-after birds that have been recorded here.

Access & Accommodation

Subic is about 80km or a drive of two and a half hours from Manila. From Manila, get on the North Luzon Expressway, then on to SCTEX (Subic Clark Tarlac Expressway) to Rizal Highway in Subic Bay Freeport Zone. There are many accommodation options in the Subic Bay area. The key birdwatching sites are only about 30 minutes from Subic town.

Conservation

In the past, the US Navy has afforded strict protection to the Subic Bay Watershed Reserve. Subsequently, although Subic was converted into a free-port zone, it has managed to retain its former protection. The presence of police and forest wardens also provides continued protection of the forest here. The planned development of tourist infrastructure and leasing of some of the bunkers for businesses may increase disturbance to the forest in the future. Some hunting occurs but this is currently quite limited.

Candaba Wetlands

Irene Dy

Irene Dy

A major pond at the Candaba Wetlands with Mount Arayat in the backdrop.

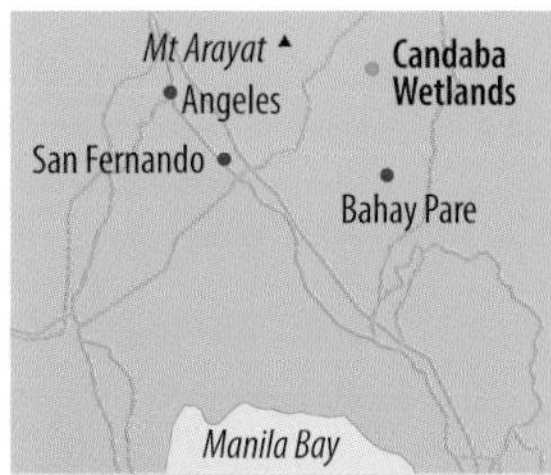

KEY FACTS

Nearest Major Town
Baliuag (Bulacan Province)

Habitats
Freshwater marshes, scrub, orchards, ricefields, fish ponds

Key Species
Wandering Whistling Duck, Philippine Duck*, Cinnamon Bittern, Barred and Buff-banded Rails, Philippine Swamphen*, Pheasant-tailed Jacana, Philippine Coucal, Eastern Grass Owl, Striated Grassbird
Winter Baikal Teal, Garganey, Gadwall, Falcated Duck, Common and Baer's Pochards (rare), Ferruginous and Tufted Ducks, Von Schrenck's Bittern (rare), Black-faced Spoonbill, Oriental Pratincole, Japanese Leaf Warbler, Middendorff's Grasshopper Warbler, Siberian Rubythroat

Best Time to Visit
October–April.

Straddling the provinces of Pampanga and Bulacan, the wetlands of Candaba encompass an extensive mosaic of freshwater ponds, marshes, seasonally inundated grassland, ricefields and scrub over the alluvial plains of the Pampanga River about 60km north of Manila. The entire area is estimated to cover well over 25,000ha, although much of this landscape is increasingly drained, disturbed or cleared for agricultural activities, threatening its populations of waterbirds. The diverse Candaba wetland landscape supports important populations of many waterbirds, including the endemic **Philippine Duck** and **Philippine Swamphen**, as well as more widespread species such as the **Wandering Whistling Duck** and **Oriental Pratincole**. In winter, large flocks of migratory ducks, including the **Garganey**, **Tufted Duck** and other species, congregate here, often joined by rarities such as the **Falcated Duck** and **Baer's Pochard**. The Endangered **Speckled Reed Warbler** was once recorded here, but there are no recent records. The main sites for birdwatchers, often referred to as Candaba Wetlands, are located north-east of the Pampanga River mouth, and an adjoining portion of privately owned fish ponds in northern Manila Bay is a magnet for waders and other migrants during high tide.

*Candaba supports some of the largest concentrations of the endemic **Philippine Duck** in the country.*

Birdwatching Sites

Candaba Wetlands

The popularly visited birdwatching site often referred to as the 'Candaba Wetlands' forms only a small fraction of this vast wetland landscape and encompasses about 70ha under private ownership. Within this area, there are three major ponds important to waterfowl, and these are used by large numbers of Philippine Ducks, and joined by big flocks of wintering ducks of multiple species in December–January. The adjoining areas are mostly ricefields and when flooded attract a number of migratory waders including the **Black-tailed Godwit**, **Ruff**, and

Irene Dy

*The **Philippine Swamphen** was formerly considered a subspecies of the widespread Purple Swamphen.*

Broad-billed and **Sharp-tailed Sandpipers**, and occasionally rare vagrants like the **Long-billed Dowitcher**. More typically, these habitat are used by large numbers of **Black-winged Stilt**, **Marsh Sandpiper**, **Whiskered Tern** and various plovers. A good diversity of passerines can be seen, including both **Striated Grassbirds**, which are ubiquitous, and in winter, the migratory **Oriental Reed Warbler** and skulking **Middendorff's Grasshopper Warbler**.

In recent years, agricultural encroachment has extensively disturbed two of these freshwater ponds, leaving only one where good numbers of waterfowl still occur. The surrounding areas of scrub in the property support the **Eastern Grass Owl** (ssp. *amauronota*), which is best seen in June–July, but accessibility on the dirt road can be difficult during this period due to rain.

Access & Accommodation

The best way to visit the marshes is by rented vehicle. Access to the Candaba Wetlands is usually on the North Luzon Express Way (NLEX) through the Balintawak toll entry, then exiting at the Sta. Rita Exit or Pulilan Exit. From either exits, the main point of entry to the site is from Barangay Bahay Pare. The 55km drive from Balintawak usually takes one and a half hours. The main birdwatching areas are located at 5–8km from the roadside to the ponds and need to be accessed by a high-clearance vehicle due to the poor road conditions. Since Candaba is reasonably near Metro Manila, it is feasible to stay at the variety of hotels and lodges at EDSA, Quezon City and the Balintawak area. Another option is to stay at Chocolate N Berries Inn along the national highway in Baliuag to be nearer to the site.

Conservation

The extensive wetlands of Candaba support significant populations of both resident and migratory waterbirds, and have been recognized as an Important Bird Area (IBA). However, much of these wetlands remain under threat from agricultural clearance, grass burning and the draining of main ponds that are used by waterbirds. Hunting of waterbirds for food is another issue that threatens various species. The wetlands of Candaba currently do not receive any formal protection, and surveys have revealed declining populations of most waterbirds.

Con Foley

*The widespread **Barred Rail** is commonly seen along the roads around the wetlands of Candaba.*

Mount Makiling Forest Reserve

Mark Jason Villa

Yann Muzika

*The **Ashy Thrush** occurs in the forests on Makiling, and can be seen at first light on the trail.*

Mount Makiling (1,090m asl) is an isolated dormant volcano in southern Luzon just over an hour's drive from Manila. The forest reserve, managed by the University of the Philippines Los Banos (UPLB), encompasses 42.4km² of mostly lowland and hill dipterocarp forests, but also contains some montane forest on its upper slopes. Being only 65km south of Manila, Mount Makiling is one of the easiest sites to explore, with a good number of widespread Philippine endemics. Local specialities include the stream-loving **Indigo-banded Kingfisher** (ssp. *cyanopectus*) and beautiful **Spotted Wood Kingfisher**, which is common here. Two endemic malkohas, the bizarre **Rough-crested Malkoha** and exquisite **Scale-feathered Malkoha,** occur only on Luzon and are best encountered at Mount Makiling. It is also a very good site for many sunbirds and flowerpeckers including **Red-keeled** and **Buzzing Flowerpeckers**, and **Flaming** and **Handsome Sunbirds**.

Birdwatching Sites

Mount Makiling Forest Trail

Excellent birdwatching starts immediately at the foot of Mount Makiling around the TREES hostel and office. There is a dead tree across the hostel where a family of **Philippine Falconets** regularly perches. A section of the trail up the mountain has recently been paved and vehicles are now able to drive up comfortably for the first kilometre or so – after that the trail becomes rocky and uneven. On the trail, look out for **Black-chinned Fruit Dove**, **Luzon Hornbill** and **Philippine Trogon**. This is also the best area to look for **Rough-crested** and **Scale-feathered Malkohas**. The melodious song of the skulking **White-browed Shama** (ssp.

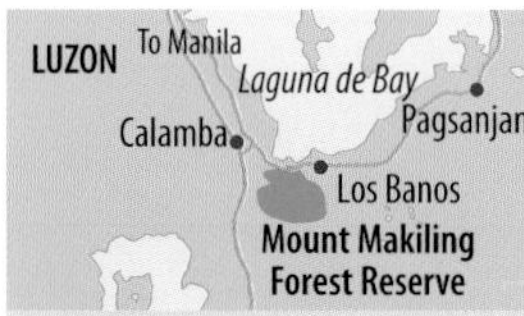

KEY FACTS

Nearest Major Town
Los Banos (Laguna Province)

Habitats
Lowland and hill rainforests, scrub and cultivation, grasslands

Key Species
Spotted Buttonquail*, Black-chinned Fruit Dove, Luzon Bleeding-heart*, Rough-crested* and Scale-feathered* Malkohas, Philippine Trogon*, Indigo-banded Kingfisher^A^, Spotted Wood Kingfisher*, Philippine Falconet*, Philippine Pitta, Elegant Tit*, Grey-backed Tailorbird*, Stripe-headed Rhabdornis*, White-browed Shama*, Ashy Thrush*, Flaming Sunbird*, Naked-faced Spiderhunter*
Winter White's and Brown-headed Thrushes, Pechora Pipit

Other Specialities
Southern Luzon Giant Cloud Rat, Rafflesia (*Rafflesia manillana*)

Best Time to Visit
February–June

Spotted Wood Kingfishers are not uncommon in Makiling's forest, but are most active just after dawn.

*The bizarre **Rough-crested Malkoha** is regularly encountered on Mount Makiling.*

luzoniensis) is often heard, but seeing it in the dense understorey can be tricky. The **Luzon Bleeding-heart** may be heard or chanced upon walking along the sides of the trail, while the **Philippine Pitta** and **Ashy Thrush** (now more easily seen at La Mesa Eco Park in Manila) are also possible. Dawn and dusk are the best times to listen for **Spotted Wood Kingfishers**, when they are vocal and most conspicuous. Nocturnal birdwatching walks may yield both **Luzon Hawk-Owls** and **Philippine Scops Owls**.

Botanic Gardens

A quick visit to the Botanic Gardens can be worthwhile, particularly when the trees are in flower. During such periods a multitude of sunbirds and flowerpeckers may be easily seen, including **Bicolored**, **Red-keeled**, **Buzzing**, **Striped** and **Pygmy Flowerpeckers** alongside **Flaming** and **Handsome Sunbirds**. The stream (Molawin Creek) within the gardens is also a regular haunt of the attractive **Indigo-banded Kingfisher**.

UPLB Dairy Husbandry Area

The grasslands and scrub near the dairy husbandry section of the campus are regular sites for **Spotted** and **Barred Buttonquails**, which may walk across or sand bathe on the path. The **Barred Rail** and **Plain Bush-hen** may also make an appearance. Grassland species including the **Tawny Grassbird** and **Golden-headed Cisticola** are regular in the area.

Los Banos Campus

Walking around the campus grounds offers a good chance of encountering flocks of **Lowland White-eyes** and other open country species.

Access & Accommodation

Most birdwatchers and recreational hikers access Mount Makiling through the University of the Philippines – Los Banos Campus, although access is also possible from the side of the mountain near Batangas. To get to the university from Manila, take the South Luzon Expressway, then the Calamba exit, before proceeding into the town of Los Banos. A permit is needed if you intend to bring a vehicle up the forest trail, but is not required if going on foot. There are several accommodation options within the university campus.

Conservation

Mount Makiling Forest Reserve is a protected area and is administered by the University of the Philippines. Strict guards and permit requirements help provide protection, but hunting does occur.

*The **Indigo-banded Kingfisher** is regular in the streams at the Mount Makiling Botanical Gardens.*

Abdelhamid Bizid

***Yellow-wattled Bulbuls** are best encountered around flowering and fruiting trees at the forest edge.*

Sablayan Penal Colony

Mark Jason Villa

Tan Ju Lin

The edges of Lake Libuao remained cloaked in rainforest.

The rugged island of Mindoro lies south-west of Luzon. For the birdwatcher intending to do a comprehensive tour of the Philippines, visiting Mindoro is a must as it contains no fewer than eight endemic birds found nowhere else in the country. Two (the Mindoro Imperial Pigeon and Mindoro Scops Owl) are restricted to the island's montane forest, but the remaining six can be seen in the shrinking lowland forest. The largest remaining tract of lowland dipterocarp forest in Mindoro lies within the Sablayan Prison and Penal Farm. While the **Mindoro Bleeding-heart** is unlikely to be seen by all but the luckiest birdwatchers, all the remaining island endemics are possible with a few days of effort. There are also several Philippine endemics that are easier to see here compared to elsewhere in the country, notably **Pink-bellied** and **Spotted Imperial Pigeons**, and the **Black-bibbed Cicadabird** (ssp. *elusa*).

Birdwatching Sites

Sablayan Penal Colony (Mount Siburan)

Extensive tracts of lowland rainforest remains at this important conservation site. Minimum security prisoners act as guides to the different trails inside the penal colony. It is possible to encounter the rare **Mindoro Bleeding-heart**, but a great deal of luck is needed. With a bit of effort, you can see the site's main attraction – the Critically Endangered **Black-hooded Coucal**. Despite its skulking habits, its loud, booming calls often give away its position. Avoid confusion with the more widespread **Philippine Coucal**, here of a distinctive black subspecies (ssp. *mindorensis*). Other birds likely to be encountered in the forest

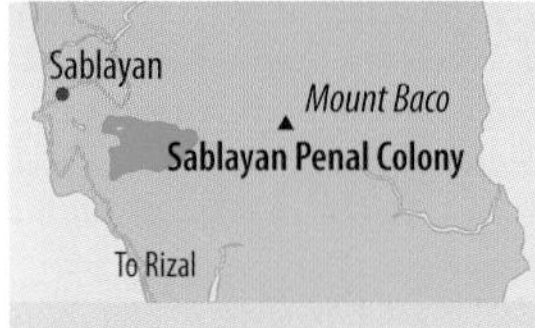

KEY FACTS

Nearest Major Town
Sablayan (Mindoro Occidental)

Habitats
Lowland and hill rainforests (primary and secondary), freshwater wetlands

Key Species
Philippine Duck*, Mindoro Bleeding-heart*, Pink-bellied* and Spotted Imperial* Pigeons, Black-hooded Coucal*, Mindoro Scops Owl*, Mindoro Hawk-Owl*, Mindoro Hornbill*, Mindoro Racket-tail*, Blue-naped Parrot, Black-bibbed Cicadabird*, Black-and-white Triller*, Scarlet-collared Flowerpecker*

Other Specialities
Mindoro Stripe-faced Fruit-bat*, Mindoro Pallid Flying Fox*, Oliver's Warty Pig*, Mindoro Dwarf Buffalo*, Bangon Monitor Lizard*

Best Time to Visit
All year round

while searching for the coucal are the **Mindoro Hornbill**, **Mindoro Racket-tail**, **Black-bibbed Cicadabird** and attractive **Scarlet-collared Flowerpecker**. After dark, spotlighting may reveal the recently split **Mindoro Hawk-Owl**.

Lake Libuao

A short drive from the forest but still within the penal colony is Lake Libuao. This large lake is surrounded by patches of secondary and primary forest. The **Wandering Whistling Duck**, **Philippine Duck**, **Stork-billed Kingfisher** and other waterfowl are readily found here. Scanning the surrounding forest can be rewarded with views of the **Pink-bellied Imperial Pigeon**, **Mindoro Hornbill**, **Blue-naped Parrot** and **Slender-billed Crow**.

Access & Accommodation

It is easiest to fly to San Jose on Occidental Mindoro, then drive about 80km north to the town of Sablayan. The penal colony is about 30 minutes away from the town proper. If driving, take the road that leads to Abra de Ilog and at Km 315 turn right and drive another 3km to the penal colony's gate. The penal colony has a basic guesthouse for visitors, although food might be an issue. A suitable alternative is to stay at one of the various small hotels and beach resorts in Sablayan town. A permit is required from the Department of Environment (DENR) and also the penal farm headquarters for entry.

Conservation

Although recognized by BirdLife International as an Important Bird and Biodiversity Area, Mount Siburan is currently not under any official protection. Restricted access to the forest due to its proximity to a penal colony does provide some degree of informal protection. Hunting and habitat disturbance due to prisoner activity such as collection of firewood, rattan and bamboo are threats to the remaining habitat.

Robert O. Hutchinson

*The recently split **Mindoro Hawk-Owl** is conspicuous in the forest at the Mount Siburan Penal Colony.*

Robert O. Hutchinson

*The globally endangered **Mindoro Hornbill** is one of eight endemic birds found only on the island of Mindoro.*

Con Foley

*Flocks of the elegant **Philippine Duck** may be seen at Lake Libuao.*

Sibaliw Forest & Central Panay Mountains

Robert Hutchinson

Yong Ding Li

The lowland rainforest around Sibaliw Research Station is one of the best-protected rainforests left in the Philippines.

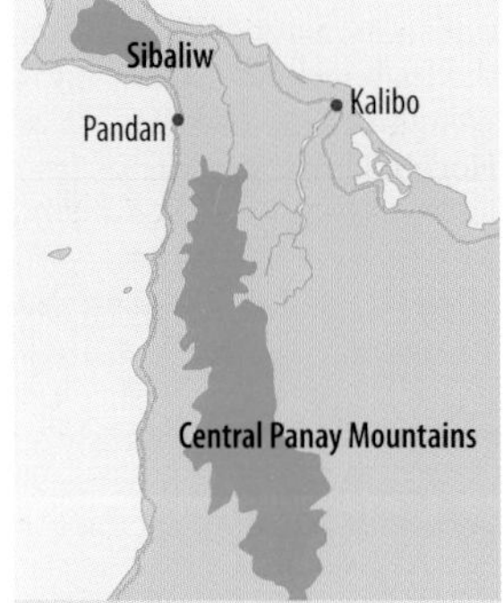

KEY FACTS

Nearest Major Towns
Kalibo, Caticlan (Aklan Province)

Habitats
Hill and montane rainforests

Key Species
Philippine Frogmouth*, Negros Bleeding-heart*, Pink-bellied Imperial Pigeon*, Amethyst Brown Dove*, Negros Scops Owl*, Walden's* and Visayan* Hornbills, Yellow-faced Flameback*, White-winged Cuckooshrike*, Panay Striped Babbler*, White-browed Shama*, White-throated Jungle Flycatcher*, Yellowish White-eye*

Other Specialities
Panay Cloudrunner*, Visayan Warty Pig*, Panay Monitor Lizard*, Philippine Bent-toed Gecko*

Best Time to Visit
February–April; wettest months October–December

Although Panay shares all but one of its endemic birds with the adjacent island of Negros, some of them are much easier to see on Panay. The wonderfully pristine rainforests surrounding the Sibaliw Research Station in Panay's Northwest Peninsula is the prime location to see two such species, the Critically Endangered **Negros Bleeding-heart** and **White-throated Jungle Flycatcher**. In addition, several pairs of Walden's Hornbill formerly occurred at this site, and despite the lack of sightings in recent years some locals believe the species still persists here.

Further to the south, Panay's Central Mountain Range, which includes the peaks of Mount Madja-as and Mount Baloy, still hold large tracts of submontane and montane rainforests and is home to two other Panay specialities. One is the Critically Endangered **Walden's Hornbill**, with some pairs being accessible on day hikes in the area. Panay's only true endemic, the **Panay Striped Babbler**, occurs only in these mountains on the upper elevations of Mounts Madja-as and Baloy.

Birdwatching Sites

Sibaliw Research Station

The forest surrounding the research station must hold one of the highest densities of **Negros Bleeding-heart** anywhere in the species' limited range. However, it is elusive so even here it may take some days to track down. The **White-throated Jungle Flycatcher** is unobtrusive but locally common, particularly around the dry Bulanao River bed. Other endemics like the central Visayan subspecies (ssp. *maculipectus*) of **Amethyst Brown Dove** and the **Yellow-faced Flameback**, possibly the most beautiful woodpecker in the Philippines, are also present.

*The threatened **White-throated Jungle Flycatcher** is not uncommon in the forests around Sibaliw.*

Mount Madja-as

The popular mountaineering route that leads up the slopes of Mount Madja-as is one of the most accessible sites to see the endemic **Panay Striped Babbler**. To see it necessitates accessing montane forests above 1,000m asl. While it is possible to do this in a day, it is best to plan for a night of camping in the event of bad weather. Once above 1,000m asl, the babbler is not uncommon and it becomes more conspicuous the higher up the mountain you travel.

To find the Critically Endangered **Walden's Hornbill**, local knowledge is very helpful – get in touch with local NGOs (panaycon.org) in advance and organize to be taken to suitable areas by those involved in the 'Nest Protection Scheme'. The hornbills are always present, but a visit during March–May increases the chance of being able to visit a nest tree and see the male delivering food to the incubating female locked inside a nest cavity.

Access & Accommodation

Both Panay sites are accessible from Kalibo and Caticlan Airports, which have several flights a day from Manila since they are both gateways to the popular beach resort island of Boracay. For Sibaliw, overland transport needs to be chartered since the starting point of the trek, Barangay Bulanao in Libertad, is off the tourist circuit. It is essential to organize permission in advance to visit the research station and ensure that there is space there to stay, as it can be fully booked by researchers. The two trails to the research centre both take several hours. One is longer but avoids rivers; it is recommended if you lack rock-climbing experience.

Paul French

*The enigmatic **Negros Bleeding-heart** occurs at particularly high densities around Sibaliw Research Station.*

In order to visit the mountains of central Panay, it is very important to gain permission from local police and local authorities before visiting either of these areas, as well as to contact local conservation NGOs to organize access to the best hornbill sites.

Conservation

Sibaliw is one of the few truly well-protected and guarded areas in the Philippines, with regular patrols reducing illegal logging and hunting to a minimum. The same cannot be said of Panay's Central Mountain Range, where due to rampant deforestation the forest is now mostly restricted to the higher altitudes. However, the 'Nest Protection Scheme' for Walden's Hornbills has been very successful in locating and protecting nests, as well as raising awareness of the bird's plight among local communities.

Lorenzo Vinciguerra

*A chance to see the Critically Endangered **Walden's Hornbill** is often the main reason why birdwatchers opt to include Panay in their itinerary.*

Mount Kanla-On Natural Park

Mark Jason Villa

Yong Ding Li

The rugged Mount Kanla-on Natural Park protects some of the last remaining tracts of rainforest on the island of Negros, and some of the largest roosting colonies of flying foxes in the region.

Much of the forest on the island of Negros has been cleared for sugar-cane plantations. Only about 4% of Negros remains forested and much of it comprises montane rainforest above 1,000m asl. Mount Kanla-on Natural Park, covering an area of roughly 246km^2 in the provinces of Negros Occidental and Negros Oriental, protects some of the island's last major tracts of montane forest. The park includes several peaks and volcanic craters, and incorporates Mount Kanla-on, which at 2,435m asl is the highest peak in Negros, as well as the Visayan islands. From an ornithological perspective, Mount Kanla-on is notable as the only site where the mythical **Negros Fruit Dove**, known from a single type specimen, has ever been recorded. For birdwatchers, the park offers easy access to montane rainforest, where a host of Visayan endemics can be found, in particular the **White-winged Cuckooshrike**, **Black-belted Flowerpecker** and charismatic **Flame-templed Babbler**. Fortunate visitors have also seen rarer species, including the **Negros Bleeding-heart** and **Visayan Hornbill**.

Birdwatching Sites

Mount Kanla-on

The most popular trail among birdwatchers starts from Barangay Minoyan, where the forest descends to about 400m asl. Except for a few steep sections, the ascent is easy and gradual. The submontane rainforests and mahogany plantations on the lower sections of the trail are good for observing the distinctive white

Bacolod
NEGROS
Guimaras
San Carlos City
Mount Kanlaon Natural Park

KEY FACTS

Nearest Major Towns
Murcia, Bacolod (Negros Occidental)

Habitats
Lowland and montane rainforests, mahogany plantations, cultivation and scrub

Key Species
Negros Bleeding-heart*, Negros Scops Owl*, Philippine Spine-tailed Swift*, Spotted Wood Kingfisher*, Visayan Hornbill*, Blue-crowned Racket-tail*, White-winged Cuckooshrike*, White-vented Whistler, Philippine Oriole*, Balicassiao*, Visayan Fantail*, Lemon-throated Leaf Warbler*, Philippine Tailorbird*, Flame-templed Babbler*, Yellowish White-eye*, Stripe-breasted Rhabdornis*, White-browed Shama*, Black-belted Flowerpecker*
Winter Japanese Night Heron (rare)

Other Specialities
Golden-crowned*, Island and Large Flying Foxes, Visayan Warty Pig*

Best Time to Visit
February–April

With a bit of luck, the elusive ***Visayan Hornbill*** *can be seen on the slopes of Mount Kanla-on.*

bellied race (ssp. *mirabilis*) of the **Balicassiao**, the **Visayan Bulbul** and the **White-browed Shama** (ssp. *superciliaris*). At 600–900m asl, look for the highly sought-after **Flame-templed Babbler** – it is usually first detected by its distinct song. It often associates with mixed foraging flocks comprising the **White-vented Whistler**, **Visayan Fantail**, **Elegant Tit**, **Lemon-throated Leaf Warbler** (ssp. *cebuensis*), **Philippine Tailorbird** and **Yellowish White-eye**. Some large flocks also contain the **Yellow-faced Flameback** and **Blue-crowned Racket-tail**. Also focus on the forest canopy, where the **White-winged Cuckooshrike** and **Philippine Oriole** may be seen. The cuckooshrike seems to be most common at above 1,000m asl.

Mambukal Mountain Resort Area

Mambukal resort is popular with local tourists looking to bathe in the hot springs and swimming pools. It can be very busy, particularly on weekends, but the large trees in the resort are good places to look for the locally common **Black-belted Flowerpecker**. A short trail skirts a forested valley and leads up the hills behind the resort. The **Spotted Wood Kingfisher** (ssp. *moseleyi*), **Mangrove Blue Flycatcher** (ssp. *philippinensis*) and **Magnificent Sunbird** may be encountered here. Another attraction is the large roosting colonies of three species of flying fox, namely Island, Large and Golden-crowned Flying Foxes, the last of which is recognized as the largest bat in the world.

Access & Accommodation

There are daily flights from Manila to Bacolod. Mambukal Mountain Resort is located 20km or an easy 40-minute drive south of Bacolod, where many birdwatchers are based when exploring Kanla-on. There are numerous hotels catering to a range of budgets in Bacolod. Mambukal Resort also has some rooms and cottages for rent. The road leading to Mount Kanla-on is very rocky, so a sturdy jeepney or a four-wheel drive vehicle is recommended for negotiating the trail.

Conservation

Although gazetted as a natural park, deforestation for agriculture, timber and charcoal burning continues to threaten the remaining forest habitats on the slopes of Mount Kanla-on. Much of the fringes of the park have already been converted to cultivation by settlers.

Robert O. Hutchinson

*The flamboyant **Negros Scops Owl** may be seen during spotlighting sessions around Mount Kanla-on.*

Abdelhamid Bizid

*The highly sought-after **Flame-templed Babbler** may be seen in foraging flocks moving through the montane forest on Mount Kanla-on.*

Tabunan Forest

Yong Ding Li

Yong Ding Li

The imposing limestone karsts of central Cebu protect the island's last remaining rainforests and the endemic birds associated with them.

The island of Cebu is often cited by conservationists as a worst-case scenario of tropical deforestation, with much of the landscape dominated by farmland and coconut plantations on the journey to Tabunan. Less than 1% of Cebu remains forested, and the remnant patches of rainforest persist mostly on steep limestone terrain or hill slopes inaccessible to all but the most persistent woodcutter and farmer. As a result of the drastic loss of habitat, Cebu's handful of endemic birds are all at very high risk of extinction. This includes the recently described **Cebu Hawk-Owl**. The most accessible site to see the majority of Cebu's endemic birds is Tabunan, a small patch of forest 4km² in area nestled within the larger Central Cebu Protected Landscape (283km²). With some effort and luck, the hawk-owl, **Black Shama** and **Streak-breasted Bulbul** can all be seen on a two-day trip. The same cannot be said for the **Cebu Flowerpecker**. There have been very few records of this scarcest of the Cebu endemics at Tabunan, and the fact that it forages high in the canopy adds to the challenge of finding this mega rarity.

Robert O. Hutchinson

The recently described **Cebu Hawk-Owl** *may be seen around the edges of the remaining forest at Tabunan.*

Birdwatching Sites

Tabunan Forest

Most birdwatchers visit Tabunan forest due to its proximity to Cebu City, although the Alcoy forest further south is becoming increasingly popular. Situated on steep, rocky limestone hills,

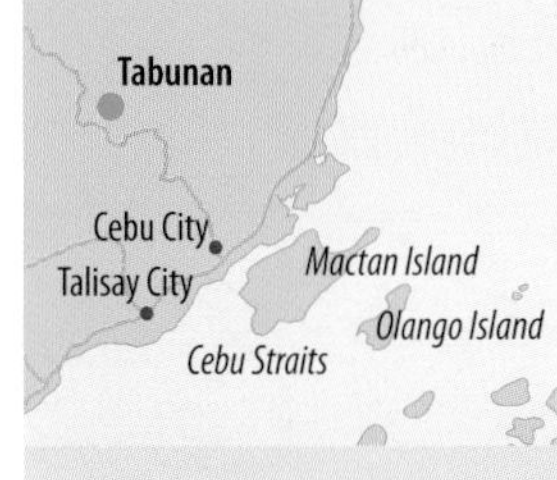

KEY FACTS

Nearest Major Town
Cebu City

Habitats
Hill and submontane rainforests (mostly secondary), scrub and cultivation

Key Species
Philippine Serpent Eagle*, Black-chinned Fruit Dove, Philippine Hawk-Cuckoo*, Cebu Hawk-Owl*, White-vented Whistler, Balicassiao*, Streak-breasted Bulbul*, Black Shama*, Cebu Flowerpecker*, Magnificent Sunbird*, Cebu Flowerpecker* (rare)

Other Specialities
Large Flying Fox

Best Time to Visit
January–June; rainy season June–December

Tabunan's forest is mainly secondary in nature, with numerous vines and few large trees. The many limestone fragments can make some parts of the trail physically challenging. The main trail starts from the roadside, cutting through cultivation for almost 1km before entering the forest, and passes two look-out platforms. The **Black Shama** is easy to hear in the forest, but seeing it takes patience and effort as these shy birds skulk in the dense understorey. The endemic subspecies (ssp. *monticola*) of the **Streak-breasted Bulbul** is uncommon, but may be seen with some effort. Before long one will encounter a **Mangrove Blue Flycatcher** or a **White-vented Whistler**, the latter the source of one of the dominant sounds in the forest.

Con Foley

*The distinctive subspecies (ssp. cebuensis) of the **Coppersmith Barbet** is common along the forest edge at Tabunan.*

Con Foley

*The endangered **Black Shama** is a vocal resident of Tabunan Forest.*

From the second platform, which sits on a low cliff, it is possible to see the **Philippine Serpent Eagle**, the distinctive subspecies (ssp. *cebuensis*) of the **Coppersmith Barbet**, **Magnificent Sunbird** (split from Crimson Sunbird), **Streak-breasted Bulbul**, **Balicassiao** (ssp. *mirabilis*) and common **Red-keeled Flowerpecker**. The **Cebu Hawk-Owl** is best encountered in the early morning near the forest edge, and seeing one requires an overnight stay at Tabunan. The **Philippine Megapode** is sometimes heard as well.

Olango Island

Depending on tidal conditions, it may be worth visiting Olango Island, a low-lying island just across the straits from Mactan. The main reason to visit are the large groups of wintering shorebirds and egrets between December–February, notably the **Asian Dowitcher**, **Far Eastern Curlew** and **Chinese Egret**.

Access & Accommodation

Accommodation is easily available in Cebu City, from where it is a 45-minute to one-hour drive to Tabunan, depending on traffic conditions. Birdwatchers planning to stay at Tabunan to see the Cebu Hawk-Owl need to contact the local guide, Lucresio Oking, who is based there.

Conservation

The near-complete deforestation of Cebu, hastened during the Spanish colonial period, left behind a number of remnant forest patches of extremely high conservation value. Most of these are under some form of formal or local protection. Tabunan falls under the Central Cebu Protected Landscape and receives formal protection from the Department of Environment and Natural Resources (DENR). Small-scale woodcutting and hunting still occur along the remote fringes of the forest, but some local enforcement appears to have kept this under control.

Rajah Sikatuna Protected Landscape

Yong Ding Li

Tan Ju Lin

The forest clearing at Camp Magsaysay is good for observing nightbirds and hornbills.

KEY FACTS

Nearest Major Towns
Bilar, Carmen (Bohol)

Habitats
Lowland rainforest (on limestone), cultivation, tree plantations

Key Species
Mindanao Bleeding-heart*, Black-faced Coucal*, Everett's Scops Owl*, Philippine Frogmouth*, Philippine Trogon*, Winchell's* and Northern Silvery* Kingfishers, Samar Hornbill*, Visayan Broadbill*, Azure-breasted Pitta*, Black-bibbed Cicadabird* (rare), Yellow-breasted Tailorbird*, Black-crowned Babbler*, Striated Wren-Babbler*, Philippine Fairy-bluebird*, Rufous-tailed Jungle Flycatcher, Handsome* and Bohol* (split from Metallic-winged) Sunbirds

Other Specialities
Philippine Tarsier*, Philippine Colugo*, Philippine Warty Pig* (rare)

Best Time to Visit
January–September; rainy season October–December

Named after the local 16th-century chieftain, Datu Sikatuna, the Rajah Sikatuna Protected Landscape is probably one of the easiest places to be acquainted with the forest avifauna of the eastern Visayas. Most of Bohol has been heavily deforested in the past four centuries for agriculture. Rajah Sikatuna is an exception and retains nearly 60km^2 of lowland forests – the largest remaining patch of forest cover on Bohol. Due to its proximity to Tagbilaran, Rajah Sikatuna is popular with birdwatchers. A well-marked network of forest trails provides excellent access to the hilly forest, where key species like the **Visayan Broadbill** and **Azure-breasted Pitta** can be seen. The clearings around the park provide good lookout points for viewing hornbills, pigeons and nightjars flying over, and in the past even the Critically Endangered **Red-vented Cockatoo** was reported here. It is possible to see more than 70 species on a trip of 3–4 days.

Birdwatching Sites

Camp Magsaysay Clearing

Located about 2km from the town of Bilar and not far from the famed Chocolate Hills, this part of Rajah Sikatuna is where most birdwatchers eventually arrive before entering the forest trails. The forest edge provides an excellent vantage point for seeing swifts, raptors and the **Samar Hornbill**, besides some of the common forest species. In the evening **Great Eared Nightjars** hawk over the clearing for insects, and a short ramble along the forest edge at night may yield views of **Philippine Frogmouths**, **Everett's Scops Owls** and Philippine Colugos.

Con Foley

The eye-catching ***Azure-breasted Pitta*** *is regularly seen along the trail network at Rajah Sikatuna.*

*The unobtrusive **Visayan Broadbill** is one of the main targets for birdwatchers visiting Rajah Sikatuna.*

Camp Magsaysay Trails

To see the charismatic species that draw birdwatchers to Rajah Sikatuna, one must spend at least a full day exploring the trail network that starts from behind Camp Magsaysay. For visitors with limited time, the shorter Tarsier, Oriole and Pitta Trails can be walked, while those with a few days to explore the park can explore the longer Tarictic and Trogon Trails. **Black-faced Coucals**, **Philippine Trogons** and **Azure-breasted Pittas** are regularly seen and heard along the trails, and mixed flocks usually contain bulbuls, **Visayan Blue Fantails**, **Black-crowned Babblers** and white-eyes. The sought-after **Visayan Broadbill** may join some of the mixed flocks, but due to its habit of keeping to the canopy diligence and knowledge of its call are needed to find it. Small, noisy parties of **Brown Tit-Babblers** forage in the undergrowth and are easier to see than the **Striated Wren-Babbler**, which tends to be shyer and less easily seen.

Fruit doves and green pigeons are more often heard than seen. The best chance to see them is to look on and around a fruiting tree and check the species attracted to it. Besides **Yellow-breasted** and **Black-chinned Fruit Doves**, **White-eared** and **Amethyst Brown Doves**, parties of **Samar Hornbills** and **Philippine Fairy-bluebirds** (ssp. *ellae*) can be seen. Imperial pigeons are thin on the ground, and the near-mythical **Mindanao Bleeding-heart** can only be found with a great deal of luck.

Loboc River

The resorts along the Loboc River are popular with general tourists and are a good base for visiting Rajah Sikatuna. Besides most of the widespread Philippine endemics like the **Philippine Hawk-Cuckoo, Red-keeled Flowerpecker** and many open-country species, the **Northern Silvery Kingfisher** has been seen here regularly.

Access & Accommodation

While camping is possible at the Camp Magsaysay clearing, many visitors base themselves in the resorts at Bilar or Loboc River for a more comfortable stay. If transport can be arranged, you can also opt to stay in Tagbilaran and make daily drives to the birdwatching sites.

Conservation

The Rajah Sikatuna Protected Landscape is the most important area for conservation on Bohol and protects the largest patch of forest left on the island. While some sections along the flatter fringes of the park have been converted to cultivation, the rugged and hilly limestone landscape has made logging and farming difficult, and thus helped save the forests from clearance.

Con Foley

*The handsome **Yellow-breasted Tailorbird** is endemic to the Visayan Islands and regularly seen at Rajah Sikatuna.*

Con Foley

*Seeing a roosting **Philippine Frogmouth** can make any birdwatcher's day.*

Mount Apo Natural Park

Pete Simpson

Pete Simpson

The montane forest of Mount Talomo above Eden, part of the Mount Apo Natural Park.

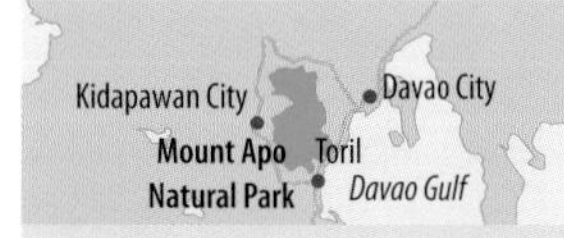

KEY FACTS

Nearest Major Towns
Davao City, Toril (Davao del Sur)

Habitats
Submontane and montane rainforests, scrub and cultivation

Key Species
Philippine Eagle*, Pinsker's Hawk-Eagle *, Bukidnon Woodcock*, Amethyst Brown* and Mindanao Brown* Doves, Giant*, Everett's* and Mindanao* Scops Owls, Southern Silvery Kingfisher*, Writhed* and Mindanao* Hornbills, Mindanao Racket-tail*, Mindanao Lorikeet* (rare), Black-and-cinnamon Fantail*, Mindanao White-eye*, Apo Myna*, Island Thrush, Bagobo Babbler*, Slaty-backed Jungle Flycatcher*, Bundok* and Cryptic* Flycatchers, Whiskered* and Olive-capped* Flowerpeckers, Grey-hooded* and Apo* Sunbirds, Cinnamon Ibon*, Red-eared Parrotfinch*
Winter Japanese Night Heron, Chestnut-cheeked Starling, Middendorff's Grasshopper Warbler

Other Specialities
Philippine Colugo*, Mindanao Flying Squirrel*, Philippine Pygmy Squirrel*, Philippine Deer*, Philippine Warty Pig*

Best Time to Visit
January–June; wettest months May–June

Mount Apo, an active stratovolcano on the Apo-Talomo range in south-central Mindanao, rises to 2,964m asl and is the highest peak in the Philippine archipelago. Much of Mount Apo and the surrounding mountains fall within the Mount Apo Natural Park, a large protected area that is known to support the majority of Mindanao's endemic montane birds. Consequently the park, particularly the densely forested slopes of Mount Talomo (2,674m asl) above the Eden Nature Resort, is increasingly popular as a destination among birdwatchers keen to see a good selection of Mindanao's endemics. Given its accessibility from Davao City and a variety of accommodation options, the park offers a viable alternative to visiting Mount Kitanglad. Besides a good selection of submontane and montane species, the **Cryptic Flycatcher** and **Whiskered Flowerpecker** are two birds that are more easily seen here than anywhere else on Mindanao. However, the greatest draw to this rugged part of Mindanao in recent years is the ease with which the nesting behaviour of the spectacular **Philippine Eagle** can be observed.

Birdwatching Sites

Eden Nature Resort

Birdwatching opportunities in this area start from within the comforts of the resort at an altitude of 900m asl. A short walk from the start of the Mountain Trail is where the **Cryptic Flycatcher** can be encountered. Noisy parties of **Brown Tit-Babblers** and **Orange-tufted Spiderhunters** regularly occur with mixed feeding flocks. Further downstream the **Southern Silvery Kingfisher** is regularly seen. Those staying overnight in the resort may see **Everett's** and **Giant Scops Owls** during an owling excursion within its premises. Recently, a stakeout for wintering **Japanese Night Heron** has been found in a nearby private property.

Pete Simpson

*The **Cryptic Flycatcher** is a localized, mid-altitude Mindanao endemic that can be found within Eden Resort.*

Pete Simpson

A newly fledged ***Philippine Eagle*** *on Mount Talomo.*

Mount Talomo Trail

About 2km past the Eden Nature Resort is the start of the trail that creeps through groves, cultivation and a couple of villages before entering the montane forests on the slopes of Mount Talomo. Starting at 1,000m asl, this trail eventually enters the forest at about 1,200m asl. The sought-after **Whiskered Flowerpecker** can usually be found after a short walk uphill past the last village. The same trees are also visited by **Buzzing**, **Fire-breasted** (ssp. *apo*) and **Olive-capped Flowerpeckers**. **Naked-faced Spiderhunters** are sometimes seen. **Long-tailed Bush Warblers** (ssp. *unicolor*) call regularly from the long grass adjacent to the trail, but are difficult to see well. From a lookout point before the forest edge, **Pinsker's Hawk-Eagle**, and very occasionally the **Philippine Eagle**, can be seen soaring over the forest. As you enter the forest the **Red-eared Parrotfinch** can sometimes be found feeding on the ground.

After the trail enters the forest, it is usually referred to as the 'Waterpipe Trail' by birdwatchers. It is easy to hike it up to 1,400m asl, from where it becomes steep and difficult to follow. In the forests up to about 1,500m asl, check out the many mixed foraging flocks that contain a range of specialities, including the **Black-and-cinnamon Fantail** and **Mindanao White-eye**. Pigeons like the **Yellow-breasted Fruit Dove** and **Amethyst Brown Dove** are more easily heard than seen. Three highly sought-after birds that have been recorded above 1,400m asl on this trail but are extremely difficult to find include the **Mindanao Brown Dove**, **Bagobo Babbler** and **Slaty-backed Jungle Flycatcher**.

Above 1,500m asl, **Apo Mynas** are easy to locate by their distinctive calls, as are **Mindanao Racket-tails**, but the latter are usually only briefly seen in flight over the forest canopy. There is a campsite by the trail at 1,500m asl where the **Mindanao Scops Owl** is present and the **Bukidnon Woodcock** can be heard above the canopy at dawn and dusk. For a day trip, the campsite makes a good lunch stop before the descent. The **Island Thrush** (ssp. *kelleri*) is often seen just above the campsite and the **Apo Sunbird** may be found visiting some of the flowers.

Philippine Eagle Centre

Not far out from Davao on the Calinan-Baguio road is the Philippine Eagle Centre. Although it is best known for its captive Philippine Eagles, some forest birdwatching is possible here on the forest edge, including sightings of bulbuls and flowerpeckers. The Malagos Watershed Forest behind the centre needs an access permit, and is where **Southern Silvery** and **Winchell's Kingfishers** have been recorded.

Access & Accommodation

From a hotel in downtown Davao City or the international airport, the town of Toril is a 45-minute drive on the MacArthur Highway. Eden Nature Resort is a 20-minute drive uphill from Toril. While the resort is a convenient place in which to stay, other options are available in Toril. A four-wheel drive vehicle is required to negotiate the road above Eden Resort to the drop-off point for the 'Waterpipe Trail'. The security situation in this part of Mindanao remains volatile and apart from the Eden Resort area, the areas described should not be accessed without a local guide, and coordination with local Barangay officials and the military.

Conservation

Mount Apo was gazetted as a natural park in 1936 and currently protects an area of just under 550km^2. More than 60% of the park's area is no longer under original forest cover, having being converted to farmland by settlers. Illegal encroachment at the forest edge, logging and poaching remain continued threats to the wildlife in the park.

Everett's Scops Owl *may be encountered along in the forest edge at Eden Resort.*

Mount Talomo is the most reliable site to see the highly localized ***Whiskered Flowerpecker****.*

Pete Simpson

The ***Mindanao Scops Owl*** *occurs in the upper montane forests on the slopes of Mount Talomo.*

Rufous-headed Tailorbirds *are commonly heard at higher elevations.*

PICOP Forest

Pete Simpson

Low Bing Wen

PICOP's remaining lowland forests are being converted into land for cultivation at an alarming rate.

The Paper Industries Corporation of the Philippines (PICOP) was awarded a logging concession encompassing 1,867km^2 of forested land (mostly lowland rainforest) in eastern Mindanao in 1952. In the late 1980s, after nearly three decades of logging activities, the corporation failed and settlers rapidly moved in. Since then the remaining forest has continued to diminish steadily from illegal logging and slash-and-burn agriculture (*kaingin* in Tagalog). Although the forests of PICOP are still a reliable site for a number of lowland specialities, like the stunning **Wattled Broadbill** and **Celestial Monarch**, many birds, especially the larger pigeons and hornbills, are becoming increasingly difficult to find or have disappeared from the area altogether. For the time being the forests of PICOP offer an easy introduction to the lowland forest avifauna of Mindanao and it may still be possible to see nearly 100 species on a visit of between 4–5 days.

Birdwatching Sites

The 'Cemetery'

The disturbed forests in this area are good for seeing mixed foraging flocks, which may comprise the **Mindanao Blue Fantail**, **Rufous Paradise Flycatcher**, **Stripe-headed Rhabdornis** and – of great interest to many birdwatchers – the bizarre **Wattled Broadbill**. Also look out for large fruiting trees as these often attract the **Mindanao Hornbill**, **Writhed Hornbill** and the distinctive southern subspecies (ssp. *mindanensis*) of **Rufous Hornbill**. From the roadside it is possible to see the **Little Slaty Flycatcher** and **Black-headed Tailorbird** skulking in the dense understorey at the forest edge. A 200m-long trail that branches off from the road here and is known to local bird guides is good for

KEY FACTS

Nearest Major Town
Bislig (Surigao del Sur)

Habitats
Lowland rainforest (mostly logged; also forest on limestone), lower montane forest, cultivation, wetlands

Key Species
Crested and Philippine* Honey Buzzards, Pinsker's Hawk-Eagle *, Mindanao Bleeding-heart*, Pink-bellied Imperial Pigeon*, Giant* and Everett's* Scops Owls, Chocolate Boobook*, Mindanao Hawk-Owl*, Philippine Trogon*, Hombron's*, Winchell's* and Southern* Silvery Kingfishers, Philippine Dwarf Kingfisher*, Rufous*, Writhed* and Mindanao* Hornbills, Buff-spotted Flameback*, Blue-crowned Racket-tail*, Wattled Broadbill*, Azure-breasted Pitta*, Black-and-white Triller*, Black-bibbed Cicadabird*, Mindanao Blue Fantail*, Short-crested* and Celestial* Monarchs, Rufous-fronted* and Black-headed* Tailorbirds, Striated Wren-Babbler*, Mindanao Pygmy Babbler*, Rusty-crowned Babbler*, Philippine Fairy-bluebird*, Stripe-headed Rhabdornis*, Philippine Leafbird*, Metallic-winged Sunbird*

Other Specialities
Philippine Colugo*, Philippine Tarsier*

Best Time to Visit
January–June; rain can occur throughout the year

Miguel David De Leon

*The **Philippine Dwarf Kingfisher** is the most difficult of PICOP's kingfishers to see.*

the **Azure-breasted Pitta** and **Striated Wren-Babbler**.

The 'Quarry'

A 1km-long narrow trail passes through open and recently cut areas of forest, but there is still fairly intact, closed-canopy forest towards the end of this trail. Night rambles can be productive around the 'quarry' area, with **Giant** and **Everett's Scops Owls**, **Chocolate Boobook** and **Mindanao Hawk-Owl** all possible. **Winchell's Kingfishers** call from the tops of high trees at dawn and are usually difficult to see well. The **Pink-bellied Imperial Pigeon** is often heard here but difficult to see. Numbers of the attractive **Blue-crowned Racket-tail** have declined significantly, but these parrots are noisy and thus easy to locate if present.

Pete Simpson

*The **Buff-spotted Flameback** is one of four woodpecker species that may be encountered at PICOP.*

Metallic-winged Sunbirds can be seen at the forest edge, usually associating with mixed flocks towards the end of the trail. These flocks may contain the **Mindanao Blue Fantail**, **Rufous Paradise Flycatcher**, **Rufous-fronted Tailorbird**, **Rusty-crowned Babbler** and **Mindanao Pygmy Babbler**, and occasionally also the **Wattled Broadbill** and **Short-crested Monarch**. A well-known roadside pond along the road leading out of the PICOP area from the Quarry often holds the **Southern Silvery Kingfisher**.

Road 42

Parking at the junction of roads 42 and 42A, a walk of 2km takes you through some of the most accessible good-quality lowland forests left at PICOP. Mixed flocks should be checked for the **Black-and-white Triller**, **Philippine Leafbird** and **Black-bibbed Cicadabird**, and both **Short-crested** and **Celestial Monarchs**. The shy **Hombron's Kingfisher** sometimes emerges from the steep valley adjacent to the roadside. Both the **Wattled Broadbill** and **Azure-breasted Pitta** have been known to nest in this area in the past. The **Philippine Fairy-bluebird** is still present and is occasionally recorded, but has become scarcer in recent years.

*The stunning **Wattled Broadbill** is one of the most sought-after endemic birds in the Philippines.*

*The **Celestial Monarch** is best located by its distinctive three-note call.*

Bislig Airfield

This accessible area of scrub and wetlands is a 15-minute drive from Bislig City. From the disused airport runway you can scan the marshes for the resident **Philippine Duck**, **Philippine Swamphen** and **Watercock**. During the winter months both **Gray's** and **Middendorff's Grasshopper Warblers** have been recorded here.

*The **Southern Silvery Kingfisher** was only recently recognised as a distinct species.*

Access & Accomodation

Bislig City is accessible by largely paved roads from the airports at Butuan to the north or Davao City to the south. Most birdwatchers travel via Davao as there are more connecting flights to the airport there. Drive time from either airport is 4–5 hours by private transfer or 6–7 hours by public bus. The comfortable Paper Country Inn in Bislig City has hosted birdwatchers for more than 20 years, and the well-known local birdwatching guide Felizardo Goring is contactable through the hotel. From Bislig it is another 1–2 hours over rough roads – a four-wheel drive vehicle is required for the trip.

Conservation

Despite a nationwide logging ban, the destruction of PICOP's remaining rainforests continues unabated due to illegal logging. It has been estimated that all of PICOP's forest could be lost within the next 10 years. Remaining areas of rainforest remain on the most rugged of the PICOP landscape, especially limestone areas and mountains further inland.

*One of the best-looking coucals, the handsome **Black-faced Coucal** is often seen clambering up branches in the lower canopy.*

*As forest coverage declines steadily, the **Rufous Hornbill** is becoming increasingly difficult to see at PICOP.*

Pete Simpson

***Writhed Hornbills** remain relatively common in the forests of PICOP.*

Mount Kitanglad Range Natural Park

Pete Simpson

Montane forests on the slopes of the Kitanglad Range (erroneously spelled 'Katanglad' in many older reports).

KEY FACTS

Nearest Major Town
Malaybalay (Bukidnon Province)

Habitats
Lower and upper montane evergreen forests, cultivation and scrub, grasslands

Key Species
Philippine Eagle*, Pinsker's Hawk-Eagle *, Bukidnon Woodcock*, Giant Scops Owl*, Philippine Frogmouth*, Philippine Spine-tailed Swift*, Purple Needletail, Hombron's Kingfisher*, Mindanao Racket-tail*, Mindanao Lorikeet* (rare), McGregor's Cuckooshrike*, Black-and-cinnamon Fantail*, Rufous-headed Tailorbird*, Mindanao White-eye*, Apo Myna*, Bagobo Babbler*, White-browed Shortwing, Slaty-backed Jungle Flycatcher, Olive-capped Flowerpecker*, Grey-hooded and Apo* Sunbirds, Cinnamon Ibon*, Red-eared Parrotfinch*, White-cheeked Bullfinch*

Other Specialities
Golden-Crowned Flying Fox*, Philippine Colugo*, Philippine Tarsier*, Mindanao Moonrat*, Philippine Deer*, Philippine Warty Pig*

Best Time to Visit
February–May; no pronounced rainy season

Covering an area of more than 310km², the Mount Kitanglad Range Natural Park was established in 1990 to conserve the mid-montane evergreen forests and mossy forests on the Kitanglad Range in Bukidnon Province, home to two of the highest peaks in the Philippines (Mounts Kitanglad and Dulang-dulang). The park sprawls over Malaybalay City and the municipalities of Lantapan, Impasugong, Sumilao and Libona. Mount Kitanglad, an inactive volcano in the Kitanglad Range, is the fourth highest mountain in the Philippines at 2,899m asl.

Mount Kitanglad has been a popular birdwatching site since the 1970s, particularly the area around the 'Del Monte Lodge'. Officially called the Philippine Eagle ecotourism lodge, the lodge at Sitio Lalawan was constructed and donated by Del Monte Philippines Inc. in 1993. The original building still stands, although in a dilapidated state. All of Mindanao's endemic montane birds have been recorded around the lodge and along the trail up Mount Kitanglad, with the exception of the localized Lina's Sunbird.

Birdwatching Sites

Trail from Damitan to Del Monte Lodge

The two-hour walk from the village of Damitan at 1,100m asl to Del Monte Lodge passes through mostly open farmland and grassy areas where **Striated** and **Tawny Grassbirds** are common, while the **Purple Needletail** and **Philippine Spine-tailed Swift** may be seen overhead.

Remnant montane forest around the lodge at 1,360m asl is good for mixed

*The **Giant Scops Owl** is frequently heard around the grounds of the Del Monte Lodge in the dead of night.*

foraging flocks containing the **Yellow-bellied Whistler**, **Black-and-cinnamon Fantail**, **Negros Leaf Warbler**, **Sulphur-billed Nuthatch**, **Turquoise Flycatcher** and intriguing **Cinnamon Ibon**, an aberrant sparrow of Mindanao's montane forests. In addition, the **Olive-capped Flowerpecker**, **Flame-crowned Flowerpecker** and distinctive subspecies (ssp. *apo*) of the **Fire-breasted Flowerpecker**, as well as the **Grey-hooded Sunbird**, have been recorded here. Both **White-eared** and **Amethyst Brown Doves** have been recorded on fruiting trees around the lodge alongside the **Yellow-breasted Fruit Dove**. The shy **Hombron's Kingfisher** is often heard calling at dawn in the dense forested gullies behind the lodge and may require some effort to see.

For safety and logistical reasons, spotlighting is usually carried out only around the lodge, where the **Bukidnon Woodcock**, **Giant Scops Owl**, **Philippine Frogmouth** and **Philippine Nightjar** have all been seen. The **Mindanao Scops Owl** is also present but rarely recorded here.

*The **Mindanao Hornbill** is the most regularly encountered of Mindanao's three hornbill species.*

Trail from Del Monte Lodge to 'Eagle Watch Point'

The two-hour hike from the lodge to the 'Eagle Watch Point' at 1,550m asl continues through mostly farmland, open scrub and a mosaic of small forest patches. However, it is possible to add a number of new birds to the list, including **McGregor's Cuckooshrike** and the **Mindanao White-eye**. The **Red-eared Parrotfinch** sometimes feeds among the extensive clumps of wild sunflowers along the trail. At this higher elevation, the **Apo Myna** and **White-cheeked Bullfinch** can also occur. The **Long-tailed Bush Warbler** (ssp. *unicolor*) calls frequently from the long grass and dense thickets adjacent to the trail, but is generally difficult to see.

'Eagle Watch Point'

The main target here is the **Philippine Eagle**, but other birds of prey such as the **Crested** and **Philippine Honey Buzzard**, **Pinsker's Hawk-Eagle** and **Rufous-bellied Eagle**, are also recorded. Noisy parties of **Mindanao Racket-tails** are often seen flying over the area. The wacky **Apo Myna** and **Stripe-breasted Rhabdornis** are commonly seen perched on the tops of exposed branches.

*The legendary **Bagobo Babbler**, one of the toughest Philippine endemic to see, is sometimes recorded on the slopes of Mount Kitanglad.*

Miguel David De Leon

*The **Grey-hooded Sunbird** replaces the Apo Sunbird in the lower montane forest.*

The poorly known **Bukidnon Woodcock** *is most regularly observed in the montane forests of the Kitanglad Range.*

A male **Hombron's Kingfisher** *poses with a freshly captured beetle in Mindanao's forests.*

Trail above 'Eagle Watch Point'

The trail from the 'Eagle Watch Point' heading up the slopes of Mount Kitanglad soon enters unbroken montane forest, and after another two-hour walk forest at 1,850m asl is reached, where the **Apo Sunbird** may be observed. Other Mindanao montane endemics, such as the **Mindanao Lorikeet**, **Bagobo Babbler** and **Slaty-backed Jungle Flycatcher**, are all present but rarely seen.

Access & Accomodation

To access Mount Kitanglad and stay at the Del Monte Lodge, an entry permit should be obtained in advance from the Mount Kitanglad Range Natural Park office in Malaybalay. Its website (mkrnp.org) provides costs and contact information, and park officials can also check the availability of the Del Monte Lodge and contact the local bird guide, Carlito Gayramara, who can help organize birdwatching tours.

Access to Malaybalay (Bukidnon) is usually done through the city of Cagayan de Oro, which is serviced by flights to Laguindingan Airport some 35km to the west of the city, or by ferry from the Visayan Islands. From Cagayan de Oro, it is a further 35km drive east to the Sayre Highway that runs to Malaybalay city.The first drop point for the journey up to the Del Monte Lodge is about 15km before Malaybalay city on the Sayre Highway at the little enclave of Dalwangan. From Dalwangan, a 6km drive on rugged roads in a four-wheel drive vehicle takes you to the final drop point at Damitan. From here, luggage and supplies can be carried up on horseback.

The Del Monte Lodge is a large but very basic wooden structure with no electricity supply. It is a two-storey structure, with the dormitory on the second level above the dining area. A sleeping mat, sleeping bag, blankets, pillow and towel are provided. All meals are usually prepared by Carlito's family. There are flushing toilets and the shower utilizes buckets of cold water, although hot water can be provided after a long walk up the mountain.

Conservation

The boundaries of the park are not well defined and encroachment by farmers continues unabated in the area. A large part of the park has already been cleared in the past and is covered by grassy scrub, dense sunflower patches, secondary woodland and farms.

An **Amethyst Brown Dove** *attends to a fruiting tree.*

Tawi-Tawi Island

Bim Quemado & Lisa J. Paguntalan

The steep limestone hills in the interior of Tawi-Tawi are still covered in rainforest.

KEY FACTS

Nearest Major Town
Bongao (Tawi-Tawi)

Habitats
Lowland and hill rainforests, coastal forest, cultivation, scrub, mangroves and tidal flats

Key Species
Philippine Duck*, Christmas Frigatebird, Metallic Pigeon, Sulu Bleeding-heart* (no recent records), Tawitawi Brown Dove*, Sulu Hawk-Owl*, Winchell's Kingfisher*, Sulu Hornbill*, Sulu Pygmy Woodpecker*, Red-vented Cockatoo*, Blue-winged Racket-tail*, Blue-naped and Blue-backed Parrots, Yellowish Bulbul, Brown Tit-Babbler
Winter Chinese Egret, Far Eastern Curlew, Great Knot

Other Specialities
Philippine Slow Loris, Estuarine Crocodile, 'Sulu Warty Pig'*, Large and Island Flying Foxes

Best Time to Visit
All year round

Located in the far south of the Philippines, the rainforests of the Sulu Archipelago, especially that on Tawi-Tawi and its satellite islands, harbour a number of globally threatened species found nowhere else in the world. For visiting birdwatchers, the lowland forest around Panglima Sugala is of great interest as it is home to many sought-after Sulu endemics such as the **Blue-winged Racket-tail**, **Sulu Hornbill**, **Sulu Pygmy Woodpecker**, **Sulu Hawk-Owl** and **Tawitawi Brown Dove**. As not many birdwatchers make it this far, most of these species remain poorly known, and are represented by very few sightings in recent times. In addition, migratory waterbirds such as the **Great Knot**, **Far Eastern Curlew** and **Chinese Egret** can also be seen along the coastal mudflats of Tawi-Tawi during the winter months.

Endemic to the Sulu Archipelago, the ***Sulu Pygmy Woodpecker*** *persists in a range of habitats on Tawi-Tawi.*

Bim Quemado

Bim Quemado

*The **Sulu Hawk-Owl** may be seen around the fringes of Panglima Sugala town and other settlements.*

Birdwatching Sites

Upper Malum, Panglima Sugala

The most recently explored site on Tawi-Tawi is the Upper Malum watershed in Magsaggaw, Panglima Sugala. At a little over 600m asl in elevation, this area constitutes the largest remaining forest cover in the Sulu archipelago and overlaps with the municipalities of Languyan and Tandubas. More than 70 per cent of the forest falls within the jurisdiction of Panglima Sugala, and the best way to access these forests is through Upper Malum in Magsaggaw.

It takes an hour of road travel to reach the forest edge of the Upper Malum Watershed before birdwatchers commence a 3km hike that involves some stream crossings. Commonly encountered forest species here include the **Yellowish Bulbul** (ssp. *haynaldi*), **Brown Tit-Babbler** (ssp. *kettlewelli*), **Philippine Hanging Parrot** (ssp. *bonapartei*) and the dapper **Winchell's Kingfisher** (ssp. *alfredi*). At the forest edge, it is also possible to see screeching **Red-vented Cockatoos**, **Blue-naped Parrots** and – with luck – the rarer **Blue-backed Parrot** on bare trees.

Deeper into the forests of Tawi-Tawi, the real specialty is the Critically Endangered **Sulu Hornbill**, perhaps the rarest of the world's hornbills, estimated to number less than 100 individuals. The impressive supporting cast here includes the striking **Blue-winged Racket-tail**, **Tawitawi Brown Dove**, **Sulu Pygmy Woodpecker** and **Philippine Oriole**, as well as more widespread species like the **Black-naped Fruit Dove**, **Greater Coucal** and **White-bellied Woodpecker** (ssp. *suluensis*). Close to streams and in the dark forest understorey, there is a good chance to see **Winchell's Kingfisher**, the recently split **Dimorphic Dwarf Kingfisher** and the **Philippine Pitta** (ssp. *yairocho*). The legendary **Sulu Bleeding-heart**, first described based on specimens collected in 1891, used to be recorded in these forests but has not been seen in decades and is now either very rare or extirpated.

Bonggao

The coastal mudflats around Bonggao offer an excellent opportunity to see migratory shorebirds and frigatebirds. With effort, an excursion to the trails at Bud Bonggao and around Mindanao State University may reveal small numbers of the globally threatened **Far Eastern Curlew**, **Great Knot** and **Chinese Egret**. The woods around the Tawi-Tawi College of Technology and Oceanography campus

*The **Sulu Hornbill** is the most threatened of the world's hornbills.*

are a regular place to find the elusive **Sulu Hawk-Owl**, locally known as 'lukluk' due to its vocalizations.

Spending some time on the Bud Bonggao trails in the early morning may be rewarded with sightings of both **Grey Imperial** and **Pied Imperial Pigeons**, as well other common forest species. Visiting birdwatchers have in the past reported an unidentified flycatcher taxa here, which definitely deserves further investigation. Recent developments in the area have led to increased visitor traffic in the form of local tourists, and remaining natural vegetation is limited to very steep, inaccessible slopes.

Access & Accommodation

Bonggao may be reached by plane (one hour) from Zamboanga City. There are daily direct flights from Manila, Davao and Cebu to Zamboanga City. There are also regular boats departing Zamboanga for Bonggao and from Bonggao to Zamboanga. From Bonggao, you may arrange private transportation for the 30-minute journey to Panglima Sugala. It is another 20km to travel from Panglima Sugala town to the Upper Malum watershed. There are also jeepneys carrying farm products that may pick up visitors. Accommodation in Bonggao is available at the Sand Bar Resort, but there are also cheaper inns and hostel options. Birdwatchers to Tawi-Tawi are advised to obtain the latest security updates and coordinate their visits carefully with the local government tourism officer or the mayor's office in Panglima Sugala. Visitors should consider contacting Nicky Icarangal, who is one of the most experienced bird guides for the area.

Conservation

The Upper Malum Watershed in Panglima Sugala has been proposed as a protected area by the regional Department of Environment and Natural Resources (DENR) of the Autonomous Region of Muslim Mindanao (ARMM). Much of the remaining forests in Tawi-Tawi have no formal protection, the exception being Bonggao Peak, which is recognized as a 'Local Conservation Area'. Habitat clearance for agriculture and illegal logging are major causes of forest loss in the region.

*The **Tawitawi Brown Dove** is the most localized of the* Phapitreron *species.*

Bim Quemado

*The **Blue-winged Racket-tail** may sometimes be seen perched on bare branches by forest clearings.*

Bim Quemado

***Winchell's Kingfishers** can be remarkably unobtrusive when they call from the forest canopy.*

Puerto Princesa Subterranean River National Park

Yong Ding Li

Francis Yap

Much of the national park consists of rainforest situated on a rugged karst landscape.

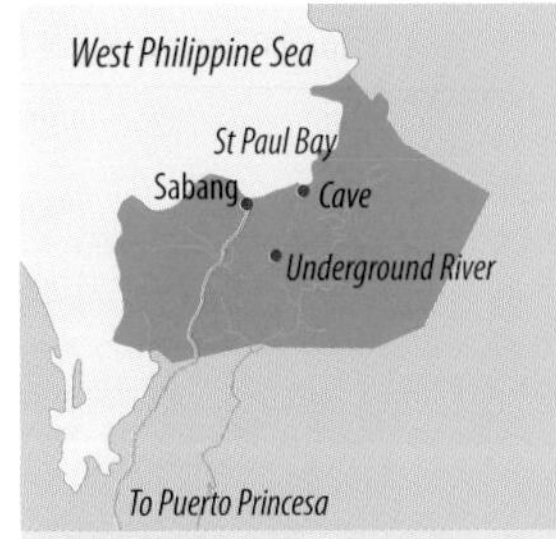

KEY FACTS

Nearest Major Towns
Sabang, Puerto Princesa (Palawan)

Habitats
Lowland and hill rainforests (including limestone forest), swamp forests, coastal beach and mangrove forests

Key Species
Philippine Megapode, Palawan Peacock-Pheasant*, Black-chinned Fruit Dove, Palawan Scops Owl*, Palawan Frogmouth*, Palawan Hornbill*, Great Slaty Woodpecker, Spot-throated* and Red-headed* Flamebacks, Blue-headed Racket-tail*, Blue-naped Parrot, Red-vented Cockatoo*, Philippine Pitta, Blue Paradise Flycatcher*, Palawan Tit*, Palawan* and Sulphur-bellied* Bulbuls, Ashy-headed Babbler*, Falcated Wren-Babbler*, White-vented Shama*, Palawan Blue Flycatcher*, Striped*, Pygmy* and Palawan* Flowerpeckers
Winter Grey-tailed Tattler, Grey-streaked Flycatcher, Pechora Pipit

Other Specialities
Palawan Pangolin*, Palawan Treeshrew*, Palawan Stink Badger*, Binturong, Leopard Cat, Palawan Bearded Pig*, Northern Palawan Tree Squirrel*, Palawan Horned Frog*, Palawan Forest Turtle*

Best Time to Visit
December–June; Rainy season from June-October

Famous for the underground river that flows beneath limestone caverns and a myriad of subterranean formations, the Puerto Princesa Subterranean River (formerly St Paul's National Park) is Palawan's flagship wildlife destination, and while compacted into an area of no more than 222km^2, it presents an unbeatable combination of Palawan's terrestrial and coastal ecosystems, including pristine mangroves, limestone, lowland and hill dipterocarp forests, with forests on its remote eastern boundary cloaking the lower slopes of Cleopatra's Needle, one of Palawan's major peaks. Almost all of Palawan's endemic birds occur here, most importantly the magnificent **Palawan Peacock-Pheasant**. Many parrots and pigeons, in decline elsewhere on Palawan due to hunting pressure, remain locally common here. It is still possible to see good numbers of **Blue-naped Parrots** and **Blue-headed Racket-tails**, while the **Red-vented Cockatoo** has recently been seen along the access road leading to Sabang.

Birdwatching Sites

Central Ranger Station

The ranger station can be reached after a short hike and a creek crossing from the row of resorts and chalets that fringe the Sabang coastline. An early morning visit to the forest surrounding the station provides opportunities to encounter most of the common forest species, including the **Palawan Hornbill**, **Yellow-throated Leafbird**, and many bulbuls and flowerpeckers. The Jungle Trail starts from behind

Con Foley

*A male **Palawan Peacock-Pheasant** regularly visits the underground river ranger station.*

Francis Yap

The Underground River is a major tourist attraction.

Yann Muzika

*The handsome **Blue Paradise Flycatcher** is best detected by its ringing whistles.*

the Ranger Station and winds its way through hilly coastal forests for 1–1.5km before reaching a damaged section. Nevertheless, the trail remains excellent for seeing most of the forest birds, including the two resident pittas, Black-chinned Fruit Dove, Ashy-headed Babbler, White-vented Shama, Blue Paradise Flycatcher, Palawan Blue Flycatcher and the distinctive Palawan subspecies of Slender-billed Crow (ssp. *pusilla*) and Pin-striped Tit-Babbler (ssp. *woodi*). Green Imperial Pigeons are also regularly heard. The Stream Trail that heads inland past the creek has become overgrown, but is a regular site for seeing the elusive Falcated Wren-Babbler, parrots, kingfishers (like the Oriental Dwarf Kingfisher) and pigeons. Also keep an eye on the sandy beach for the resident Great-billed Heron and Malaysian Plover.

Subterranean River Ranger Station

Besides a number of common forest species, the star attraction here is a confiding male Palawan Peacock-Pheasant that forages in the surrounding forest, although finding him can be difficult once crowds of tourists arrives. Other highlights include groups of Philippine Megapodes, Ameline Swiftlets, Stork-billed and Ruddy Kingfishers, Palawan Hornbills, Philippine Pittas and Palawan Blue Flycatchers.

Access Road to Sabang

The last 3–5km of the Sabang road cuts through a mosaic of farmland, scrub and secondary forest and offers excellent opportunities for birdwatching given the relatively open nature of the forest. Many adaptable forest species, including the Pin-striped Tit-Babbler, Palawan Tit, Palawan Frogmouth, Spot-throated Flameback, Blue-headed Racket-tail, Hooded Pitta and the endemic bulbuls can be seen here. The undoubted highlight is the rare Red-vented Cockatoo, which can be seen in the early morning from a lookout point about 12km before Sabang. Spotlighting around the entrance of the all-terrain vehicle (ATV) track may yield the Palawan Scops Owl, with other possibilities including the endemic subspecies (ssp. *wiepkeni*) of the Spotted Wood Owl and migratory Northern Boobook.

Access & Accommodation

Given the popularity of the park with general tourists, there is a variety of transport options to get there from Puerto Princesa city, including buses, vans and rental cars. A permit to enter the park can be obtained from the Parks Office in the Puerto Princesa City Coliseum. Most birdwatchers stay in the resorts along Sabang beach, which range from the basic to the comfortable.

Robert O. Hutchinson

*The melodious song of the skulking **Falcated Wren-Babbler** often betrays its presence.*

Conservation

The park is relatively well protected, and has received international recognition as a UNESCO World Heritage Site for its excellent representation of Palawan's biodiversity and unusual geological formations. Some encroachment by woodcutters and poachers may occur at the remote eastern fringes.

Narra Forests

Irene Dy

Irene Dy

A river crossing on the hike up Mount Victoria.

KEY FACTS

Nearest Major Town
Narra (Palawan)

Habitats
Hill and montane rainforests (including forests on ultramafic soil)

Key Species
Palawan Peacock-Pheasant*, Metallic Pigeon, Palawan Scops Owl*, Red-vented Cockatoo*, Blue-headed Racket-tail*, Palawan Tit*, Palawan Bulbul*, Falcated Wren-Babbler*, Palawan Blue Flycatcher*, Palawan Striped Babbler*, Palawan Flowerpecker*

Other Specialities
Endemic pitcher plants (like Attenborough's Pitcher Plant), Palawan Horned Frog, Palawan Toadlet

Best Time to Visit
April–June

The Municipality of Narra (abbreviated from 'National Rehabilitation Resettlement Administration') is located about 96km south of Puerto Princesa City in central Palawan. This area is so named because it was strategically chosen for the resettlement of farmers to establish areas for rice and corn cultivation in 1946. While much of Narra's coastal plain is now heavily cultivated, its rugged interior remains extensively cloaked in dense rainforests, including the imposing Victoria Peak, which rises to an elevation of well over 1,700m asl. These forests support most of Palawan's endemic birds and the mountain is perhaps the easiest of the mountains in which to see Palawan's only montane endemic, the **Palawan Striped Babbler**, an active participant of mixed species flocks. It is also one of very few sites (across the sea from Rasa Island) where the Critically Endangered **Red-vented Cockatoo** roosts and can still be seen in reasonable numbers. In Narra, local communities have actively participated in conservation activities led by the Katala Foundation for nearly 20 years, led by the notable conservationist Indira Lacerna Widmann.

Irene Dy

The ***Palawan Striped Babbler*** *is the only Palawan endemic restricted to the mountains*

Birdwatching Sites

Mount Victoria

Located about 17km north of Narra town, the well-forested Mount Victoria rises to an elevation of 1,707m and is widely regarded as the best site to see the **Palawan Striped Babbler** (the other being the more inaccessible Mount Mantalingahan to the

Con Foley

*The **Spot-throated Flameback** occurs in coastal and lowland forests around Narra.*

south), as well as most of Palawan's montane birds. Demanding a reasonable amount of fitness, the route to ascend Mount Victoria crosses a large number of streams (some at ankle depth, others to waist level) and was only opened in 2009 by a team of intrepid mountaineers led by Jehson Cervancia. Palawan Striped Babblers typically occur above 1,000m asl, but become regular only around 1,300m asl. Other high-elevation species to look out for include the skulking **Sunda Bush Warbler** (ssp. *palawanus*) and **White-browed Shortwing** (ssp. *sillimani*), while mixed flocks should be checked for parties of **Fiery Minivets**, **Negros Leaf-Warblers** (ssp. *peterseni*), **Yellow-breasted Warblers** (ssp. *xanthopygius*) and **Mountain White-eyes**. With luck, the **Palawan Scops Owl** can also be seen here, near the upper limits of its elevation range.

Pancan Dos Marcelo

This site is located not far from Narra town and contains an excellent lookout point where the threatened **Red-vented Cockatoos** can be reliably observed. Small groups of cockatoos can be seen flying in to feed on the Palawan mainland at sunrise, and returning to Rasa Island to roost at sundown.

Access & Accommodation

From Puerto Princesa Airport, Narra is a two-hour drive (about 95km) along the Puerto Princesa South Road. Lodging is easily available at Narra town. From the town, it is another 30 minutes' drive northwards to reach Sitio Mariwara, Barangay Princesa Urduja, where most hikes up Mount Victoria begin. The strenuous hike involves 15 river crossings in the first five hours. The distance to reach the main campsite is about 12.5km and requires a hike of 8–10 hours, depending on the condition of the streams. Past the campsite, it is another 3km before reaching the summit. A permit and local guides are required for the hike and can be arranged at the tourism office in Narra town. A stop at the barangay office to register is also required.

Conservation

Mount Victoria and the surrounding highlands contain extensive areas of montane forests, as well as significant tracts of forests on ultramafic soil, but receive little or no formal protection. Threats include illegal logging, encroachment by settlers, and in some areas, illegal hunting of various mammals. The Red-vented Cockatoo has been the focus of substantial conservation efforts in the Narra area, led by the Katala Foundation in collaboration with the local government. Because the cockatoos have a penchant for Malunggay (*Moringa oleifera*), the entire town has an ordinance where residents plant Malunggay to provide an ample food source for the birds. Thanks to the efforts of the Katala Foundation and local communities, the town received a Galing Pook Award in 2015, a national award that recognizes innovative practices by local governments in the Philippines.

Con Foley

*The **Palawan Flowerpecker** is one of the most easily seen endemics on the island.*

SINGAPORE

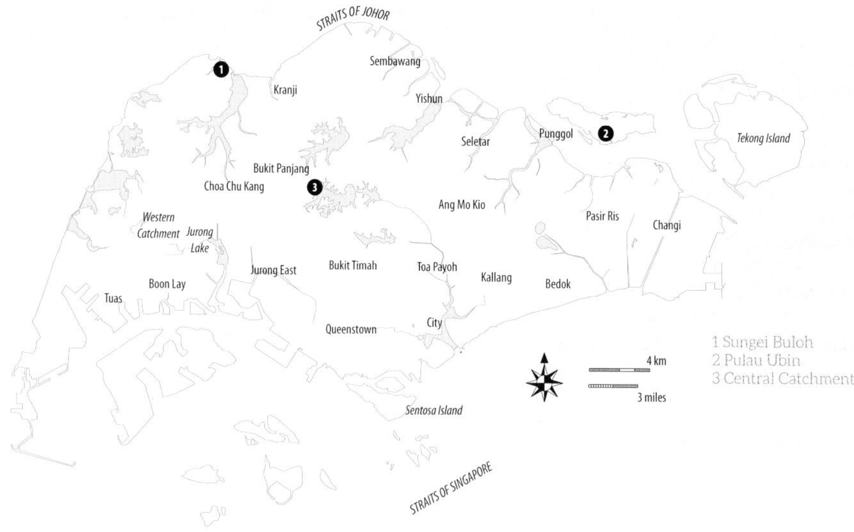

As Southeast Asia's smallest country, Singapore encompasses an area of just over 716km². The main island of Singapore (Pulau Ujong) is hilly near the centre, and is where the country's highest point, Bukit Timah (164m asl), stands. There are also extensive low hills on the west and south-west coast. A number of small rivers radiate from the centre of the island, the longest being Sungei Seletar (15km), but most of these have been dammed to form reservoirs. About 64 satellite islands surround Singapore, the largest being Pulau Tekong and Pulau Ubin.

Despite its status as a highly urbanized city-state, Singapore's protected areas like the Central Catchment Nature Reserve still support a wide variety of flora and fauna, and birds are no exception. More than 390 species have been recorded and this figure continues to grow annually with the help of an ever-increasing number of keen birdwatchers, photographers and nature enthusiasts.

Due to the country's aspiration to be a 'City in a Garden', many of the birds can be seen throughout an easily accessible network of public parks and nature reserves. Although Singapore lacks endemic birds, a number of regional specialities difficult to see elsewhere in Southeast Asia are locally common, notably the Grey-headed Fish Eagle, Red-legged Crake and Straw-headed Bulbul. The winter months herald the arrival of a diverse array of migratory species, including sought-after birds such as the Green-backed Flycatcher and globally threatened Brown-chested Jungle Flycatcher.

KEY FACTS

No. of Endemics
0

Country List
397 species

Top 5 Birds
1. Straw-headed Bulbul
2. Brown-chested Jungle Flycatcher
3. Von Schrenck's Bittern
4. Spotted Wood Owl
5. Jambu Fruit Dove

Francis Yap

*The dazzling **Oriental Dwarf Kingfisher** refuels in Singapore's forests during its annual migration.*

Climate

Singapore has an equatorial climate, being hot and humid throughout the year. Rain can occur at any time, but is most intense during the north-east monsoon between November–January. The inter-monsoonal period between February–May is generally the driest and hottest period of the year.

Access, Tranportation & Logistics

As a regional transit hub, Singapore is well connected by air to the rest of the world, with frequent flights arriving from every major continent. Visitors from many countries do not require a tourist visa, and can stay in the country for 14 days to three months, depending on their nationality. On the ground, the island's comprehensive public transport network, including buses, taxis and trains, allows you to reach birdwatching sites in comfort and with little hassle. Accommodation options are located throughout the island and range from budget hostels to international luxury hotel chains. English is widely spoken throughout the country.

Health & Safety

Dengue is the most prevalent mosquito-borne disease, and precautions should be taken in both urban areas and at birdwatching sites. There are few other natural irritations in the forests, although ticks are possible in parts of the nature reserves frequented by wild pigs (*Sus scrofa*).

Although crime rates in Singapore are of the lowest in the world, basic common-sense crime-prevention measures, such as not leaving valuables unattended, should be practised.

Crepuscular ***Red-legged Crakes*** *are more easily seen in Singapore than elsewhere in its range.*

Birdwatching Highlights

More than 120 species can be expected in two days of birdwatching, particularly during the northern winter in October–March. Resident species difficult to see elsewhere in the region include the Red-legged Crake, Spotted Wood Owl and Mangrove Pitta, and the globally threatened Straw-headed Bulbul and Copper-throated Sunbird. During the winter months the globally threatened Brown-chested Jungle Flycatcher passes through the island in good numbers between October–November en route to its wintering grounds in Sumatra, along with a variety of other passage migrants including the Asian Dowitcher, Bar-tailed Godwit, Blue-winged Pitta and Yellow-rumped Flycatcher. Winter visitors include sought-after species such as Von Schrenck's Bittern, Chinese Egret, Jerdon's Baza and Hooded Pitta.

Mohamad Zahidi

The nomadic ***Jambu Fruit Dove*** *can appear in urban parks when figs are fruiting.*

Sungei Buloh Wetland Reserve

Low Bing Wen

Wong Tuan Wah

An aerial panorama of Sungei Buloh Wetland Reserve.

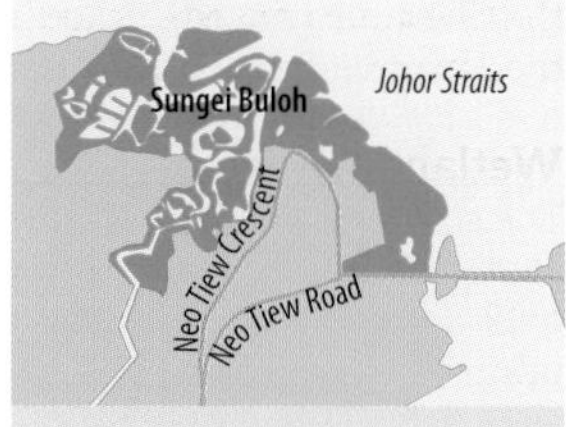

KEY FACTS

Nearest Major Towns
Woodlands, Choa Chu Kang

Habitats
Mangrove forests and coastal mudflats, secondary woodland, freshwater wetlands

Key Species
Red Junglefowl, Great-billed Heron, Grey-headed Fish Eagle, Buffy Fish Owl, Oriental Pied Hornbill, Straw-headed Bulbul, Copper-throated Sunbird
Passage Asian Dowitcher, Black-tailed and Bar-tailed Godwits, Great Knot, Broad-billed Sandpiper, Brown-chested Jungle Flycatcher, Yellow-rumped Flycatcher
Winter Von Schrenck's Bittern, Chinese Egret, Black Baza, Grey-tailed Tattler, Terek Sandpiper, Ruddy Turnstone, Chestnut-winged Cuckoo, Large, Malaysian and Hodgson's Hawk-Cuckoos, Black-capped Kingfisher

Other Specialities
Smooth-coated Otter, Estuarine Crocodile, Shore Pit Viper

Best Time to Visit
September–December

As a result of coastal reclamation and the damming of rivers to form reservoirs, Singapore's mangrove forests have all but disappeared, occupying only 0.5% of the country's land area. One area of extant mangrove forest on the main island is protected within the boundaries of the Sungei Buloh Wetland Reserve. Originally a complex of abandoned aquaculture farms, the site was designated a nature park in 1989 and subsequently gazetted a nature reserve in 2001. A further extension to the reserve was opened in 2014, which now protects 202ha of mangroves, freshwater marshes and secondary forest. Although much of its original mangrove avifauna has been extirpated, the **Copper-throated Sunbird** is still readily seen, along with sought-after resident species including the **Great-billed Heron**, **Grey-headed Fish Eagle** and **Straw-headed Bulbul**. During the northern winter the reserve is the most accessible wader-watching site on the island, hosting a range of shorebirds that use it as a high-tide roost. Diligent scanning may reveal small numbers of rarer waders, including the **Asian Dowitcher** and **Great Knot**, among the more common species.

Birdwatching Sites

Migratory Bird Trail

This 2km circular loop trail provides numerous opportunities to see roosting shorebirds from a variety of viewing hides, screens and shelters. During October–December, large concentrations of shorebirds regularly

Greater Crested Terns are easily encountered on Coastal Trail along the Straits of Johor.

roost close to the Main Hide, Hide 1C and Hide 1D. These flocks regularly contain uncommon species such as the **Black-tailed Godwit**, **Ruddy Turnstone** and **Broad-billed Sandpiper**, and free-flying Milky and Painted Storks from private collections! The mangrove trees lining the trail are often visited by **Copper-throated Sunbirds** when in flower, and support uncommon migrants during the northern winter, including the **Chestnut-winged Cuckoo** and occasionally the **Japanese Paradise Flycatcher**.

Wetland Centre

The secondary forest and freshwater ponds around the centre support a good variety of birdlife. A pair of **Oriental Pied Hornbills** utilizes a nearby nest box and sometimes allows close approach. Other species that may be seen throughout the year include the **Red Junglefowl**, **Straw-headed Bulbul** and introduced **Javan Munia**. In winter, **Black** and **Von Schrenck's Bitterns** have been recorded around the ponds, while the wooded areas support wintering passerines such as the **Blue-winged Pitta**, and **Yellow-rumped** and **Mugimaki Flycatchers**.

Coastal Trail

This 1.3km trail traverses the length of the reserve's new extension and offers good birdwatching opportunities, including several vantage points for scanning the Straits of Johor for seabirds. A high-tide vigil at Eagle Point and Kingfisher Pod may be rewarded with roosting **Black-headed Gulls** on the floating fish farms off the coast among the more numerous **Greater** and **Lesser Crested Terns**. Other terns recorded here include the **Black-naped Tern** and rare **Caspian Tern**. Resident birds include the **Copper-throated Sunbird**, **Buffy Fish Owl** and **Plaintive** and **Rusty-breasted Cuckoos**, while in winter a variety of warblers, cuckoos and flycatchers can be seen as well. Raptor watching here can also be rewarding, with regular sightings of the **Black Baza**, **Grey-headed Fish Eagle** and **Changeable Hawk-Eagle**.

Access & Accommodation

The reserve can be visited as a day trip from anywhere in Singapore. If utilizing public transport, alight at Kranji Mass Rapid Transit (MRT) Station and take the regular bus service 925, which stops opposite the reserve's visitor centre. On weekends the bus route is extended to incorporate the wetland centre. The reserve is also easily reached by taxi. It is open from 7 a.m. to 7 p.m. daily, and no entrance fees or permits are required.

Francis Yap

__Great-billed Herons__ have been recorded in the reserve in recent years.

Francis Yap

The striking __Copper-throated Sunbird__, an important pollinator of mangrove trees, is common within the reserve.

Conservation

Sungei Buloh Wetland Reserve is one of Singapore's four nature reserves, and also became an ASEAN Heritage Park in 2003. It is well protected and rangers patrol it frequently. Problems faced by the reserve include transboundary pollution from coastal developments along southern Johor, as well as intensification of commercial farm development along its fringes.

Pulau Ubin

Low Bing Wen

Yong Ding Li

Pulau Ubin, a small, 10.2km² island off northern Singapore, features a variety of terrestrial habitats that provide breeding and foraging areas for a wide variety of species.

Located off the north-eastern coast of Singapore, Pulau Ubin is an island covering 10.2km² with a maximum elevation of 75m at its highest point at Bukit Puaka. Commercial granite quarrying, once the lifeblood of the island, ceased in the 1960s, and today the island serves as a getaway from urban Singapore, providing locals with a glimpse into the lives of their forefathers and recreational space for various activities including cycling, camping and nature appreciation. Many initiatives have also been undertaken to conserve the island's rich biodiversity, and a mosaic of habitats including intertidal sandflats, mangrove forests, abandoned fruit orchards, grasslands and secondary rainforests supports more than 150 bird species, including many that are scarce on mainland Singapore, such as the **Green Imperial Pigeon**, **Mangrove Pitta**, **White-rumped Shama** and **Mangrove Blue Flycatcher**. During the winter months the island also serves as a wintering ground for a variety of migratory shorebirds, cuckoos, flycatchers and warblers, including the globally threatened **Chinese Egret**.

KEY FACTS

Nearest Major Town
Changi Village

Habitats
Mangrove and coastal beach forests, secondary woodland, orchards, intertidal flats

Key Species
Red Junglefowl, Oriental Darter, Great-billed Heron, Green Imperial Pigeon, Banded Bay, Plaintive and Rusty-breasted Cuckoos, all of Singapore's resident owls, Blue-eared Kingfisher, Oriental Pied Hornbill, Mangrove Pitta, Straw-headed Bulbul, White-rumped Shama, Mangrove Blue Flycatcher, Van Hasselt's Sunbird
Winter Chinese Egret, Grey Plover, Terek Sandpiper, Ruddy Turnstone, Chestnut-winged Cuckoo, Large, Malaysian and Hodgson's Hawk-Cuckoos, Oriental Scops Owl, Blue-winged Pitta, Streaked (rare) and Cinereous Bulbuls, Lanceolated Warbler

Other Specialities
Malayan Porcupine, Greater Mousedeer

Best Time to Visit
March–June; October–December for migrants

Birdwatching Sites

Chek Jawa Wetlands

Located at the south-eastern tip of Pulau Ubin, this 1km² site is unique for supporting several ecosystems within a small area, featuring sandy and rocky shores, coral rubble and seagrass lagoons bordered by coastal forests and mangroves. A network of boardwalks loops around the site and provides good opportunities for viewing avifauna. **Red Junglefowl** and **White-rumped Shamas** are often encountered in the coastal forest, and nearby mangrove forests host species such as **Abbott's Babbler**, **Mangrove Pitta** and **Mangrove Blue Flycatcher**. In the winter months good numbers of waterbirds can be seen foraging along

The ***Mangrove Pitta*** *is particularly vocal between April and June.*

the intertidal areas at low tide, including scarcer species such as the **Chinese Egret**, **Bar-tailed Godwit**, **Grey Plover** and **Ruddy Turnstone**.

Sensory Trail

Located within a five-minute walk of the main jetty on the island, this easy loop trail provides an excellent introduction to the island's birdlife. The trail meanders through abandoned fruit orchards, mangrove forests and freshwater ponds, which support a diverse array of birdlife. The globally threatened **Straw-headed Bulbul** is readily encountered, as is the **Oriental Pied Hornbill**. Other species like **Plaintive** and **Rusty-breasted Cuckoos**, **Long-tailed Parakeet** and **Van Hasselt's Sunbird** also occur here. The numerous fruiting trees along the trail occasionally attract avian dispersants from southern Malaysia, such as **Streaked** and **Cinereous Bulbuls**, and rarities like the **Cinnamon-headed Green Pigeon**.

Ketam Quarry

This former granite quarry, located on the western side of the island, has undergone habitat rehabilitation and is now popular as a mountain biking park. The well-wooded environs and large expanse of grassland on the southern section of the quarry also make it a popular birdwatching site. The mangrove forests leading to the quarry along Sungei Puaka are the best place to encounter the **Mangrove Pitta** during April–June, when it is particularly vocal. The grasslands around the quarry support a range of species, including the **Red-wattled Lapwing**, **Barred Buttonquail**, **Baya Weaver** and **Yellow-bellied Prinia**, supplemented by skulking **Pallas's Grasshoppers** and **Lanceolated Warblers** during the winter months. Good numbers of **Grey Herons** roost within the quarry, and the **Blue-eared Kingfisher** is sometimes seen. In recent years the **Oriental Darter** has been reported within the quarry.

Access & Accommodation

Pulau Ubin can be visited as a day trip from anywhere in Singapore, although those on a tight schedule can opt to stay at a range of hotels in Changi Village, the island's gateway. The island is accessed via regular 15-minute bumboat rides (S$3 per person, per trip) from Changi Point Ferry Terminal. Once on the island visitors can opt to walk, cycle or charter 12-seater van taxis to reach various areas throughout the island. There is also limited accommodation on the island for those who opt to spend the night here.

Francis Yap

All of Singapore's resident owls, including the stately ***Spotted Wood Owl****, inhabit the forests of Pulau Ubin.*

Conservation

Pulau Ubin has been designated as a Nature Area under the management of the National Parks Board. Due to its offshore location, threats to terrestrial ecosystems are comparatively minor, and include occasional forest fires during the dry season and small-scale poaching of wild pigs by the few remaining local villagers. Local marine ecosystems, on the other hand, have been impacted by frequent algal blooms in recent years, possibly as a result of coastal habitat degradation and development along the Straits of Johor.

Francis Yap

Pulau Ubin is a global stronghold of the threatened ***Straw-headed Bulbul****.*

Central Catchment & Bukit Timah Nature Reserves

Yong Ding Li & Lim Kim Seng

Yong Ding Li

Lowland dipterocarp forest along the edge of the MacRitchie Reservoir.

KEY FACTS

Nearest Major Towns
Bukit Timah, Ang Mo Kio

Habitats
Lowland rainforest, freshwater swamp forest, abandoned orchards

Key Species
Grey-headed Fish Eagle, Jambu Fruit Dove, Violet Cuckoo, Chestnut-bellied Malkoha, Buffy Fish Owl, Malaysian Eared Nightjar (rare), Red-crowned Barbet, Blue-rumped Parrot, Straw-headed Bulbul, Short-tailed Babbler, Greater and Lesser Green Leafbirds
Passage Oriental Dwarf Kingfisher, Brown-chested Jungle Flycatcher, Yellow-rumped Flycatcher
Winter Von Schrenck's Bittern, Ruddy Kingfisher, Blue-winged and Hooded Pittas, Sakhalin Leaf Warbler, Zappey's, Green-backed and Mugimaki Flycatchers, Siberian Blue Robin

Other Specialities
Sunda Colugo, Sunda Pangolin, Sunda Slow Loris, Banded Langur, Small-toothed Palm Civet, Malayan Porcupine, Lesser Mousedeer

Best Time to Visit
March–June, October–December; rainy season December–January

Heavily deforested during the colonial period, less than 1% of Singapore is now covered in the extensive lowland dipterocarp and freshwater swamp forests that used to cloak the island. However, all the remaining primary and tall secondary forests are protected in these two important nature reserves, which also form the main catchment area for the country's water supply. Bukit Timah, the island's highest point at 164m asl, has some of the best remaining hill dipterocarp forests, and was recently connected to the much larger Central Catchment Reserve via an ecological corridor completed in 2013. Alhough many forest species have been extirpated, the reserves boast more than 200 bird species and a good diversity of smaller forest mammals, including a number of birds difficult to see in other parts of Southeast Asia, such as the **Grey-headed Fish Eagle**, **Jambu Fruit Dove**, **Chestnut-bellied Malkoha** and **Red-crowned Barbet**, as well as a good representation of migratory songbirds like the poorly known **Sakhalin Leaf Warbler**.

Birdwatching Sites

MacRitchie Reservoir & Sime Road

An extensive network of trails starting from the entrance at Venus Drive provides access to some of the best lowland forest left in Singapore. Other important facilities are the 250m-long Treetop Walk, allowing observation of the

Wang Bin

*At least four pairs of **Grey-headed Fish Eagles** occur in the Central Catchment Nature Reserve.*

crowns of various forest trees, and the 30m-tall Jelutong Tower. From the tower the **Grey-headed Fish Eagle**, **Chestnut-bellied Malkoha**, **Blue-rumped Parrot**, flocks of **Long-tailed Parakeets**, **Asian Fairy-bluebirds**, and many sunbirds, bulbuls, leafbirds and flowerpeckers are regularly sighted. Alternatively, quietly walking along the forest trails might yield sightings of a number of babblers and cuckoos. In the winter months many shyer, ground-dwelling birds like thrushes, pittas and robins may be seen along the trails.

*The gaudy **Red-crowned Barbet** is the only forest barbet in Singapore's forests.*

Upper Peirce Forest

Accessed via Old Upper Thomson Road on the eastern boundary of the reserve, this area contains mainly secondary forest, with some patches of remnant freshwater swamp forest. It is less popular among joggers and casual hikers who are otherwise ubiquitous in the reserves, and the **Red-crowned Barbet**, **Long-tailed Parakeet** and two forest babblers, **Chestnut-winged** and **Short-tailed Babblers**, are more easily seen here. In winter various warblers, cuckoos, flycatchers and secretive **Malayan Night Herons** and **Von Schrenck's Bitterns** have been sighted.

Bukit Timah Nature Reserve

While most forest species are shared between this and the Central Catchment reserve, the area's speciality is a large fig tree at the summit of Bukit Timah hill, and a few other fig trees on the lower flanks of the reserve at the Dairy Farm Nature Park. When the trees are in fruit, the **Jambu Fruit Dove** is regularly seen, as well as good numbers of the **Thick-billed Green Pigeon**, **Asian Fairy-bluebird**, three leafbirds (**Blue-winged, Lesser** and **Greater Green**) and many bulbul species. In winter flocks of **Eyebrowed** and **Siberian Thrushes**, and the enigmatic **Sakhalin Leaf Warbler**, have been seen here.

Francis Yap

***Blue-rumped Parrots** are usually seen in flight, but may pose at fruiting trees.*

Access & Accommodation

Many public bus services stop near the main access points into the reserves, namely Venus Drive (for Central Catchment Reserve) and Hindhede Drive (for Bukit Timah Reserve). Bukit Timah Nature Reserve can also be accessed from the Beauty World Mass Rapid Transit (MRT) Station. Dairy Farm Nature Park can be accessed by taxi or a 15-minute walk from the nearby Hillview MRT Station.

Conservation

Conferred with nature reserve status since the 1980s, both reserves are well protected and regularly patrolled by rangers. Problems faced include development of infrastructure and high-rise housing at the reserve fringes, and heavy usage of the trail network by local and overseas visitors, including joggers. A proposed project to build an underground line through the Central Catchment forest may increase disturbance in the area.

THAILAND

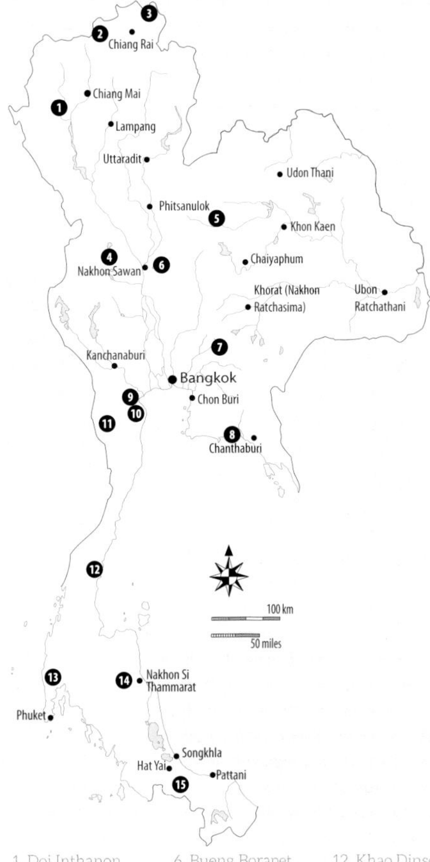

1. Doi Inthanon
2. North-western Highlands
3. Chiang Saen
4. Mae Wong & Huai Kha Khaeng
5. Nam Nao
6. Bueng Borapet
7. Dong Phayayen-Khao Yai
8. Ko Man Nai
9. Nong Pla Lai
10. Laem Phak Bia
11. Kaeng Krachan
12. Khao Dinsor
13. Sri Phang Nga Ao Phang Nga
14. Khao Luang
15. Khao Nam Khang & San Kala Khiri

Located in the centre of continental Southeast Asia, the Kingdom of Thailand is as geographically diverse as it is culturally rich. The rugged western frontier of Thailand is lined with mountains stretching from the Daen Lao Range, an outlier of the Shan Hills, in the far north to the Tanaosri (Tenasserim) Hills on the Thai-Malay peninsula, while the eastern frontier with Laos and Cambodia is equally mountainous in many areas. Much of central Thailand lies on the low-lying floodplain of the Chao Phraya, Thailand's main river system, which converges from its tributaries in Nakhon Sawan. The relatively flat Isan region on the north-east sits on the Khorat Plateau, where its many rivers drain into the Mekong River.

Boasting an excellent tourist infrastructure, great cuisine and friendly people, Thailand is often a global birdwatcher's first experience with birdwatching in Southeast Asia. The country supports a great diversity of habitats, ranging from lush tropical rainforest in the country's far south and expansive mudflats along the Gulf of Thailand, to high-altitude oak-rhododendron forests in the mountainous north. In addition, various sites in the Gulf of Thailand (like Ko Man Nai) and on the Thai Peninsula (like Chumphon) are where tens of thousands of migratory birds, including raptors, pass annually. This remarkable range of habitats and avian diversity, coupled with a burgeoning nationwide network of keen birdwatchers and bird photographers, has resulted in the avifauna in Thailand being among the best documented in all of Southeast Asia. A key attraction for birdwatchers is the opportunity to see birds associated with both the Himalayan and the Sundaic ecoregion, involving families such as cochoas, parrotbills and laughingthrushes in the north, and forest jewels such as trogons, broadbills and pittas in the south.

Climate

Thailand has a tropical monsoon climate with distinct wet and dry seasons. The Thai summer, the hottest time of the year, occurs between February–May. During the rainy season between May–October, torrential rain is a frequent occurrence and many national parks are closed. The mildest period of the year occurs between November–February, during which the mountainous regions of the far north may experience sub-zero temperatures overnight.

Access, Transportation & Logistics

The primary gateway into Thailand is the sprawling capital of Bangkok, although the popular tourist cities of Chiang Mai and Phuket receive international flights and offer a convenient launch pad to birdwatching sites in the northern and southern region respectively. Overseas visitors from many countries do not need a tourist visa to visit, and can stay in the country for 14 days to three months, depending on their nationality. The national road network is generally of good standard and most birdwatching sites can be reached by rented car, although the roads in some national parks may require a four-wheel drive or vehicle with high clearance to navigate. English is spoken in the main tourist areas. However, the situation is different in many rural and less visited areas, and some knowledge of basic Thai is very useful for visiting birdwatchers.

Health & Safety

Vaccinations for Hepatitis A and B and typhoid are recommended for birdwatchers visiting Thailand. Ticks are common in forested areas, especially during the dry season, and are replaced by leeches during the wet season. The risk of malaria is comparatively low, but precautions should be taken, particularly when staying in the rural lowlands. Biting flies, known locally as *khun*, are prevalent in some national parks and deliver itchy bites that heal slowly.

Thailand is generally a safe country but precautions against petty crime should be adhered to around towns and cities. Particular care should be exercised in the far south of the country, where insurgents have occasionally clashed with security forces, although tourists have not yet been targeted. Visitors should seek advice from local embassies.

Birdwatching Highlights

With a national list of more than 1,000 species, a comprehensive three-week trip across Thailand can yield over 450 species, particularly during the northern winter when many migratory species are present. In the country's far south, Sundaic species such as the Malayan Banded Pitta, Rail-babbler and a variety of broadbills, babblers and trogons take centre stage, while the mudflats around Bangkok support wintering populations of the globally threatened Chinese Egret, Nordmann's Greenshank and incomparable Spoon-billed Sandpiper.

Two of the country's best-known and largest national parks, Khao Yai and Kaeng Krachan, are not far from Bangkok and feature in nearly every birdwatcher's itinerary for Thailand. These spectacular parks provide an excellent introduction to Thailand's avifauna, besides featuring many sought-after species, including the Grey Peacock-Pheasant, Siamese Fireback, Coral-billed Ground Cuckoo, Blue and Eared Pittas plus more than seven species of hornbill, including the rare Plain-pouched Hornbill.

In the northern mountains of Thailand not far from Chiang Mai, the birdlife takes on a distinctly Himalayan flavour and features species such as Mrs Hume's Pheasant, Scarlet-faced Liocichla, and both Green and Purple Cochoas, together with babblers, laughingthrushes, parrotbills and finches. Those who make it to the wetlands in and around the Mekong River in the far north are also able to see a variety of migratory species during the winter months, including the stunning Firethroat, as well as highly localized residents such as Jerdon's Bush Chat.

KEY FACTS

No. of Endemics
3 (Siamese Partridge, Turquoise-throated Barbet, Deignan's Babbler)

Country List
1,036 species

Top 10 Birds
1. Spoon-billed Sandpiper
2. Coral-billed Ground Cuckoo
3. Mrs Hume's Pheasant
4. Giant Nuthatch
5. Blue Pitta
6. Grey Peacock-Pheasant
7. Siamese Fireback
8. White-fronted Scops Owl
9. Rail-babbler
10. Green Cochoa

Wich'yanan Limparungpatthanakij

Thailand has 37 species of woodpecker, including the splendid ***Black-headed Woodpecker****.*

Con Foley

Pheasants in Thailand can be quite approachable, such as this ***Grey Peacock-Pheasant****.*

Doi Inthanon National Park

Ayuwat Jearwattanakanok

Low Bing Wen

Doi Inthanon encompasses a wide range of habitats from lowland deciduous forests to montane bogs.

KEY FACTS

Nearest Major Town
Chom Thong (Chiang Mai)

Habitats
Lowland deciduous and mixed deciduous forests, submontane and montane evergreen forests, coniferous forest, freshwater swamps

Key Species
Lowland Black-headed Woodpecker, White-rumped Falcon, Black-backed Forktail
Montane Mountain Bamboo Partridge, Mrs Hume's Pheasant, Rufous-throated Partridge, Black-tailed Crake, Ashy and Speckled Wood Pigeons, Rusty-naped Pitta, Dark-sided Thrush, Green and Purple Cochoas, Black-throated Parrotbill, White-gorgeted Flycatcher, Yellow-bellied Flowerpecker, Green-tailed Sunbird
Winter Eurasian Woodcock, Fire-capped Tit, Grey-sided, Eyebrowed and Chestnut Thrushes, Mrs Gould's Sunbird

Other Specialities
Phayre's Leaf Monkey, Red Goral, Chinese Serow, Fea's Tree Frog, Siamese Big-headed Turtle

Best Time to Visit
December–March; rainy season June–October

Doi Inthanon, at 2,565m asl, is Thailand's highest peak and one of its most popular tourist destinations. Its diverse ecosystems, which include montane evergreen forest and dry deciduous forest at its foothills, make it one of Thailand's best birdwatching sites. Doi Inthanon holds many Himalayan species that are difficult to see elsewhere in Southeast Asia, including the **Black-tailed Crake**, **Dark-sided Thrush** and **Purple Cochoa**, as well as migratory species such as the **Eurasian Woodcock** and **Grey-sided Thrush**. The attractive subspecies (ssp. *angkanensis*) of the **Green-tailed Sunbird** is endemic to the montane forests on the mountain and is locally common. In the montane oak-chestnut forests, large mixed species foraging flocks containing many warblers, fulvettas and sibias are regularly encountered. The dry deciduous forests at lower elevations are home to several Indochinese specialities, including the **White-rumped Falcon** and **Black-headed Woodpecker**. To do justice to the wonderfully diverse birdlife on Doi Inthanon, birdwatchers need a stay of several days, visiting the different habitats at various heights.

Birdwatching Sites

The main road that leads from the foothills of Doi Inthanon near Chom Tong to the summit provides excellent access to the different habitats in the park. There are a number of birdwatching sites along this road, for which distance markers are provided here as reference.

Km 13 (Muang Ang)

The large parking area on the right is the start of a bumpy road to Muang Ang, a small village located deep in the lowland deciduous forest. A small bridge next to

*The unobtrusive **White-gorgetted Flycatcher** skulks in the forest understorey.*

Ayuwat Jearwattanakanok

*The highly sought-after **Green Cochoa** may be encountered at various sites throughout Doi Inthanon, particularly the Km 37 jeep track.*

Cheng Heng Yee

*A number of Himalayan specialities are more easily seen in Thailand, including this **Black-tailed Crake**.*

the parking area is a good place to look for the localized **Black-backed Forktail**. A high-clearance vehicle is needed for driving along the rugged road further into the forest. The **White-rumped Falcon** and **Collared Falconet** are most often seen during the dry season, when they begin to breed. The **Black-headed Woodpecker** is also regularly encountered here.

Khun Wang Campsite (Km 30)

A small patch of wetland, located about 200m inside the campsite next to an open lawn, is home to the rare and elusive **Black-tailed Crake**. With luck, the crake can be found venturing out of the wetlands to forage near the lawn, particularly at dusk. The main road after the campsite entrance leads to the Khun Wang Watershed Management Station. Fruit orchards and open forests along the road can be quite productive for birdwatching, with wintering **Fire-capped Tits** being one of the highlights.

Siri Bhum Waterfall

This waterfall is located not far from the campsite and can be seen from the road to Khun Wang. It is home to the **Slaty-backed Forktail**, **Plumbeous Water Redstart** and **White-capped Redstart** in winter. The gardens along the way up to the waterfall can also be quite productive, with highlights including the **Green Cochoa** and breeding **Brown-breasted Flycatcher**.

Mr Daeng's Shop (Km 31)

Narakorn Daengrassami, widely known as Mr Daeng, was one of the first birdwatchers to explore Doi Inthanon. His shop, which includes a restaurant and accommodation, is located on the left side of the road just after the park's headquarters. A damp gully located on the left of the shop often holds **Dark-sided Thrushes**, and occasionally wintering **Eurasian Woodcocks**. Mr Daeng also keeps a birdwatcher's log book where valuable records from visiting birdwatchers are kept and can be viewed.

Km 34.5 Track

The dirt track that leads to the Mae Uam Watershed Management Station is one of the most popular birdwatching sites on Doi Inthanon. Regular flocks of **Black-throated Parrotbills** (ssp. *feae*) can often be found among the bamboo groves, while the elusive **White-gorgeted Flycatcher** is often seen foraging in the forest undergrowth along the track. Many species of leaf warbler and babbler can be seen, particularly during winter, and rarities like the **Green Cochoa**, **Brown-breasted Flycatcher** and **Mrs Hume's Pheasant** are also present.

Ayuwat Jearwattanakanok

*The **Rufous-bellied Niltava** adds a splash of colour to the forests of Doi Inthanon.*

Second Checkpoint & Jeep Track (Km 37)

The second checkpoint is particularly good for birdwatching during the rainy season, when moths are abundant and are attracted by the night lights at the checkpoint. Many bird species come to pick up the moths in the morning. The **Green Cochoa** is regularly seen near the checkpoint. At night the **Brown Wood Owl** visits the checkpoint to hunt for large beetles, and the smaller **Mountain Scops Owl** is also present. The small track leading into the dense forest on the right used to be wide enough for a jeep to go through, hence the name 'Jeep Track', but has now been reduced to a narrow track accessible only by foot. The rare **Purple Cochoa** has occasionally been reported here. The dense understorey is home to many sought-after species such as the **Rusty-naped Pitta**, **Slaty-bellied Tesia**, **White-gorgeted Flycatcher** and **White-tailed Robin**.

Napamaytanidol & Napaphon Bhumisiri Chedi

The *chedis* (Buddhist stupas) are located near Doi Inthanon's summit. A small lawn next to the entrance often holds small groups of **Mountain Bamboo Partridges**, which normally come out at dusk or dawn. The gardens inside the chedis are good for smaller birds such as **Green-tailed** and **Mrs Gould's Sunbirds**.

Yann Muzika

__Black-throated Parrotbills__ should be looked for in dense groves of bamboo.

Kiew Mae Pan Nature Trail

Before reaching the summit, there is a large parking area on the left of the road where the Kiew Mae Pan Nature Trail starts. Many bird species can be seen just around the entrance, including the rare **Speckled Wood Pigeon**, **Yellow-bellied Flowerpecker** and wintering **Blue-fronted Redstart**. The trail passes through good-quality montane evergreen forest and scenic grasslands where the Chinese Goral, a goat-like ungulate, may sometimes be seen.

Doi Inthanon Summit Area

The montane forest at the summit area is the main highlight for birdwatchers. Many species are unusually tame due to the lack of hunting

Ayuwat Jearwattanakanok

The __Dark-sided Thrush__ skulks in the undergrowth in the montane bog at Ang Ka.

pressure. **Green-tailed Sunbirds** can be seen at close range feeding on flowering plants. Just behind the toilet, **Ashy Wood Pigeons** can sometimes be seen roosting and visiting the hidden salt lick down in the gully. Another highlight is the Ang Ka Nature Trail, which loops around a mountain bog surrounded by old rhododendrons and oaks. Many elusive species can be seen from the boardwalk, including the **Rufous-throated Partridge**, **Eurasian Woodcock**, **Dark-sided Thrush** and **White-browed Shortwing**. In winter flocks of thrushes are often seen visiting fruiting trees or flowering rhododendrons. These include the globally threatened **Grey-sided Thrush** and occasionally rarities such as the **Chestnut Thrush**, or even **Red-throated** and **Dusky Thrushes**.

Access & Accommodation

It generally takes about an hour by car from Chiang Mai to reach the park entrance near Chom Tong. The park offers a range of accommodation, including chalets that can be booked directly online (www.dnp.go.th/parkreserve/nationalpark.asp?lg=2), as well as campsites at several locations. Accommodation and food are available at Mr Daeng's shop located at the Km 31 marker. Alternatively, a variety of resorts is available along the main road before entering the national park.

Conservation

As a national park, Doi Inthanon is relatively well protected, although visitor numbers can be excessive, particularly during the New Year vacation period and at weekends in winter (December–February). Poaching and deforestation rates are currently low, but invasive alien species such as the Rainbow Trout and feral dogs and cats pose a problem to wildlife in the park.

North-Western Highlands

Ayuwat Jearwattanakanok

Low Bing Wen

The scenic peaks of the north-western highlands are the perfect backdrop for observing the region's avian specialities.

KEY FACTS

Nearest Major Towns
Chiang Mai (Chiang Mai)

Habitats
Hill and montane evergreen forests, mixed deciduous forest, coniferous forest, grasslands, cultivation

Key Species
Rufous-throated Partridge, Mountain Bamboo Partridge, Mrs Hume's Pheasant, Mountain Hawk-Eagle, Black-tailed Crake, Hodgson's Frogmouth, Crimson-breasted Woodpecker, Rusty-naped Pitta, Chestnut-headed and Slaty-bellied Tesias, Rusty-cheeked and Red-billed Scimitar Babblers, Collared and Spot-throated Babblers, Spot-breasted, Silver-eared and White-browed Laughingthrushes, Scarlet-faced Liocichla, Himalayan Cutia, Pale-billed and Spot-breasted Parrotbills, Spotted Elachura, Giant Nuthatch, Black-breasted Thrush, Green and Purple Cochoas, Jerdon's Bush Chat, Spot-winged Grosbeak
Winter Amur Falcon, Brown Bush Warbler, Grey-sided and Chestnut Thrushes, Grey-winged Blackbird, Golden Bush Robin, White-bellied Redstart, Sapphire and Ultramarine Flycatchers, Yellow-bellied Flowerpecker, Fire-tailed Sunbird, Scarlet Finch, Dark-breasted and Black- headed Greenfinch, Tristram's Bunting

Other Specialities
Chinese Goral, *Tylototriton uyenoi* (a salamander), Fea's Tree Frog

Best Time to Visit
December–March; rainy season June–October

Thailand's north-western highlands consist of several major mountains stretching from the border with Shan state, Myanmar, into the north-western tip of Chiang Mai. Of these mountains, those of major interest to the birdwatcher include Doi Lang and Doi San Ju, Doi Ang Khang and Doi Chiang Dao. They also represent the southernmost end of the distribution range for many Sino-Himalayan species, such as **Mrs Hume's Pheasant**, **Spot-breasted Parrotbill**, **Crested Finchbill** and **Giant Nuthatch**, all highly sought-after by birdwatchers. Unsurprisingly, the north-western highlands are one of Thailand's main birdwatching hotspots, attracting many birdwatchers every year. The best time to visit is during the northern winter months, when many migratory species occur and records of rare vagrants are not infrequent. However, note that some of the mountains are popular among local tourists and can get very crowded during this period.

Birdwatching Sites

Doi Lang & Doi San Ju

Located on the north-western side of Fang, Doi Lang (2,100m asl) can be accessed from two different routes. The southern route, alternatively known as Doi San Ju, is more accessible, with a well-paved road, and can be accessed via the Fang bypass road. The northern route starts from Mae Ai district, north of Fang. A four-wheel drive vehicle is best for this rather rough route.

Con Foley

*The attractive **Rufous-throated Partridge** occurs widely in northern Thailand's montane forests.*

The habitats along each route up the mountain are very different, which is reflected in a different set of birds. Doi San Ju comprises mainly pine- and oak-dominated montane forest with some patches of hill evergreen forest. **Mrs Hume's Pheasant, Hodgson's Frogmouth** and **Giant Nuthatch** are the key species to be found here.

The small and somewhat isolated patch of montane evergreen forest near the summit of Doi Lang can also be very productive in winter. Many scarce migrants including the **White-bellied Redstart**, **Sapphire Flycatcher** and **Scarlet Finch** can be found here, as well as scarce and localized residents like **Hodgson's Frogmouth**, **Spot-breasted Laughingthrush**, **Himalayan Cutia** and **Spot-breasted Parrotbill.**

The northern route (Mae Ai) passes through lush hill evergreen forest and secondary growth, as well as groves of bamboo where the **Pale-billed Parrotbill, Red-billed Scimitar Babbler** and attractive **Collared Babbler** can be found. A small hill-tribe village with ricefields and a stream is sited roughly halfway up this road. **Jerdon's Bush Chat** has been reported in this area. At the military checkpoint close to the summit, many montane specialists are attracted to food left by the soldiers. These include the handsome **Scarlet-faced Liocichla** and occasionally the **Rufous-throated Partridge**. The surrounding forest is home to many rarities such as the **Crimson-breasted Woodpecker**, **Chestnut-headed Tesia**, **Spotted Elachura**, and **Green** and **Purple Cochoas**. In winter, numerous species of migratory robins, thrushes and leaf warblers can also be found. The road beyond the checkpoint passes a very scenic viewpoint where the spectacular **Fire-tailed Sunbird** occurs in winter.

Ayuwat Jearwattanakanok

*The diagnostic and far-carrying call of the globally threatened **Giant Nuthatch** is a feature of the north-western highlands' pine forests.*

Doi Ang Khang

This is the most accessible and most popular tourist destination in Thailand's north-western highlands. It can be reached from either Chiang Dao or Fang district. For birdwatchers, the Chiang Dao route is recommended due to its relatively low traffic volume. It passes through several Chinese-Thai villages and a diversity of habitats, including cultivation, secondary growth, mixed pine and oak forests, and montane evergreen forest.

Birdwatchers can start their explorations from Ban Arunothai, a village located in the foothills of Ang Khang. These areas of cultivation, with scattered patches of dry deciduous forest, can be surprisingly productive. Birds found here include the **Rufous-winged Buzzard**, **Burmese Nuthatch** and even **Mandarin Duck** at the reservoir in winter. After Ban Arunothai the road goes uphill through a wide range of habitats. The **Mountain Bamboo Partridge**, **Mrs Hume's Pheasant, Grey-headed Parakeet**, **Rusty-cheeked Scimitar Babbler**, **Scarlet-faced Liocichla** and **Giant Nuthatch** can all be seen along the way.

At Ban Luang, a Chinese-Thai village near the summit, birds can be numerous when the Himalayan Cherry trees blossom in winter. Ban Luang Resort, a small resort nestled in a valley, is also a productive birdwatching site. Birds seen here include the **Black-breasted Thrush**, **Rufous-bellied Niltava**, **Black-headed Greenfinch** and occasional **Black-tailed Crake**.

The main tourist attraction at Doi Ang Khang is the Royal Project, which can be good for birdwatching too. In winter scarce migrants such as the **Grey-sided Thrush, Grey-winged Blackbird** and **Japanese Thrush**, and scarce residents like the **Black-breasted Thrush**, **Dark-sided Thrush**, **Scaly Thrush**, **White-tailed Robin** and **Rusty-naped Pitta** can be seen here.

Another birdwatching hotspot is the

Ayuwat Jearwattanakanok

*The elegant **Mrs Hume's Pheasant** is regularly seen on the road up to Doi San Ju in the early morning.*

*The robust **Spot-breasted Parrotbill** skulks in dense areas of bamboo and secondary growth on the north-western highlands.*

*The montane forests of the north-western highlands are one of the region's best places to see the secretive **Rusty-naped Pitta**.*

orchards near the Nor Lae military camp at the Thai-Myanmar border. This area usually contains flocks of **Black-headed Greenfinch**. Migrant rarities such as the **Brambling** and **Himalayan Vulture** have shown up here.

Doi Chiang Dao

At 2,175m asl, this is Thailand's third highest peak. Originally the place to look for **Mrs Hume's Pheasant** and **Giant Nuthatch** before Doi San Ju was discovered, it is the least accessible of the three mountains described here. A four-wheel drive vehicle is needed to reach the Den Ya Khat station near the upper slopes, where the road ends and the summit trail to Doi Chiang Dao begins. Mrs Hume's Pheasant and Giant Nuthatch can be seen around Den Ya Khat but not as easily as at Doi San Ju. Otherwise, bird species at this site are similar to those at the aforementioned sites.

Another interesting site at Doi Chiang Dao is the Wat Thum Pha Phlong Temple in the foothills of the mountains. The trail (with steps) that leads from the parking lot to the pagoda on the hill can be quite productive. Lower montane species such as the **Pin-tailed Green Pigeon**, **Blue Pitta** and **Streaked Wren-Babbler** can be found regularly. The Chiang Dao Wildlife Sanctuary headquarters located near the entrance of the temple is a good spot for birdwatching, especially at night for the **Oriental Bay Owl**, **Spot-bellied Eagle-Owl** and **Blyth's Frogmouth**.

Access & Accommodation

All three mountain sites can be accessed from Highway No. 107. A range of hotels, motels and guesthouses is available at both Chiang Dao and Fang. At Doi Ang Khang, many resorts, guesthouses and campsites are available. For Doi Chiang Dao, only campsites are available if you intend to stay beyond the Den Ya Khat station. Alternatively, you can stay at the foothills near Wat Thum Pha Plong Temple and Chiang Dao Cave, where numerous

__Mountain Bamboo Partridges__ are easy to see at Doi Lang.

guesthouses and resorts are available. For Doi Lang and Doi San Ju, visitors can stay in Fang and drive up the mountain daily. Alternatively, a campsite is available on the northern (Mae Ai) route.

Conservation

Despite the important biodiversity found in this area, large areas of Thailand's north-western highlands remain unprotected. Doi Chiang Dao receives the highest level of formal protection as it is a wildlife sanctuary. Doi Lang and Doi San Ju remain unprotected, but are fairly undisturbed due to existing border tensions between Thailand and Myanmar. Even so, poaching of birds for food and the pet trade still occurs, especially among local people. Captive birds are readily seen in villages, and children holding slingshots are a common sight. Bush fires set by local people to clear vegetation for farming are regular at the end of the dry season, and may also negatively impact the birdlife.

Chiang Saen Wetlands

Ayuwat Jearwattanakanok

Ayuwat Jearwattanakanok

The harrier roost at Nong Lom is worth a visit as it allows close observation of a variety of wintering harriers, including the striking ***Pied Harrier.***

KEY FACTS

Nearest Major Towns
Chiang Saen, (Chiang Rai)

Habitats
Freshwater wetlands, grasslands, cultivation, mixed deciduous forest, secondary growth

Key Species
Rufous-winged Buzzard, River Lapwing, Small Pratincole, River Tern, Eastern Grass Owl, Freckle-breasted Woodpecker, Grey-throated Martin, Jerdon's Bush Chat
Winter Falcated Duck, Baer's and Common Pochards, Ferruginous Duck, Horned Grebe, Great Bittern, Brown-cheeked Rail, Long-billed Plover, Chestnut-crowned and Manchurian Bush Warblers, Blunt-winged and Paddyfield Warblers, Chinese, Baikal and Spotted Bush Warblers, Pallas's Grasshopper Warbler, Wallcreeper, Firethroat, Siberian and Chinese Rubythroats, Yellow-breasted Bunting

Best Time to Visit
December–March; rainy season June–October

Chiang Saen is located at the northernmost tip of Thailand on the tripoint with Myanmar and Laos. Chiang Saen district covers extensive areas of wetland, an increasingly rare habitat in north-west Thailand. Consequently, many species found here are not present or rare elsewhere in north-west Thailand. The area is also renowned for many national firsts and rare vagrants for both Thailand and Southeast Asia, including rarities such as the **Long-tailed Duck**, **Horned Grebe**, **Wallcreeper**, **Blackthroat** and **Firethroat**. In addition, Chiang Saen is the only known breeding site in Thailand for the **Eastern Grass Owl**. The wetlands of Chiang Saen are best visited during winter months when many migrants, including a number of sought-after species, can be found.

Birdwatching Sites

Chiang Saen Lake

Boats for birdwatching can be hired at the Nong Bong Kai Non-hunting Area headquarters. Each year more than 10,000 ducks winter at this large freshwater lake located south of Chiang Saen city. Highlights among the wintering ducks may include the **Long-tailed Duck**, **Baer's Pochard**, **Ferruginous Duck**, **Baikal Teal**, **Mandarin Duck** and **Falcated Duck.** All four grebe species found in Thailand have been reported here. Dense areas of vegetation near the entrance and around the headquarters are good for searching for skulkers like the **Brown-cheeked Rail**, **White-browed Crake**, **Thick-billed Warbler** and **Pallas's Grasshopper Warbler**. There is also a boardwalk that runs along the edge of the lake. A small park around the headquarters is where the **Freckle-breasted Woodpecker**, **Burmese Shrike**, **Chestnut-tailed Starling** and **Purple Sunbird** are often seen, and rarities like **Chinese Blackbird** have also been reported.

Woraphot Bunkhwamdi

During the dry season the wetlands around Chiang Saen attract normally elusive chats like this ***Chinese Rubythroat.***

In recent years a wintering male ***Firethroat****, a poorly documented winter visitor that breeds in Sichuan's highlands, has been regularly reported within the Nam Kham Nature Reserve.*

The ***Large-billed Reed Warbler****, a species only rediscovered in recent years after a hiatus of more than a century, has recently been mist-netted in Chiang Saen.*

Nam Kham Nature Reserve

The reserve covers an area of about 0.15km^2 of reedbeds and oxbow lakes. There are four main hides scattered within the reserve, two along the edge of a large pond (Avadavat Hide and Pied Kingfisher Hide), and the other two (Cettia Hide and Rubythroat Hide) among the reedbeds. Both Cettia and Rubythroat Hides have small waterholes to attract birds. Elusive species such as **Chestnut-crowned** and **Baikal Bush Warblers**, and **Siberian Rubythroats**, are regularly seen during the dry season (January–April), while uncommon species like the **Chinese Rubythroat**, **Firethroat**, **Chinese Bush Warbler** and **Jerdon's Bush Chat** have been reported. A loop trail through the reedbeds can be particularly productive in the early morning and is where the **Red Avadavat**, as well as wintering **Manchurian Bush Warblers** and **Blunt-winged Warblers**, can be expected. The large pond on the western end of the reserve is good for birdwatching before it dries up by March. Both the **Brown-cheeked Rail** and **Great Bittern**, the latter a rare winter visitor to Southeast Asia, have been reported here.

Mekong River at Chiang Saen

Formerly one of the best birdwatching sites in the area, the Mekong is now heavily impacted by the many dams upstream, with serious repercussions on the river's ecosystems. The river channels here periodically dry up during November–April, creating large, exposed areas of sand and gravel bars that are used by many birds as breeding and feeding grounds. **Small Pratincoles**, **Long-billed Plovers** and **River Lapwings** are found regularly, while regional rarities such as **Pallas's Gull**, **Mandarin Duck**, **Bar-headed Goose**, **Swan Goose**, **Northern Lapwing** and **Common Ringed Plover** have been reported. The **Grey-throated Martin** uses high bluffs along the riverbank as nesting sites. Riverside reedbeds are also good for **Jerdon's Bush Chat**. However, water levels during the dry season have recently become unnaturally high. Only small remnant areas of sandbars are left, so few birds remain.

Nong Lom

This large patch of scrub and wetlands south of Chiang Saen Lake is home to the highly localized **Eastern Grass Owl**, here at its only known breeding site in Thailand. During the winter it also serves as a roosting site for more than 400 harriers each year, mainly **Pied** and **Eastern Marsh Harriers**, although rarer species like **Hen** and **Western Marsh Harriers** also occur regularly in small numbers. Hides for observing the harriers at their roosting site have been built and are maintained by local residents. In winter, **Pallas's Grasshopper Warbler**, **Eurasian Wryneck** and **Siberian Rubythroat** are regularly seen.

Woraphot Bunkhwamdi

Eastern Grass Owls at a nest in Chiang Saen.

Access & Accommodation

The wetlands at Chiang Saen are easily accessed from Chiang Rai via Highway No. 1016 with connections to the Golden Triangle, Mae Sai and Chiang Khong. All the key birdwatching sites are in relatively close proximity and the access roads to the sites remain in good condition. A wide range of guesthouses, motels, hotels and resorts is available in Chiang Saen city, along the Mekong River and at Chiang Saen Lake.

Conservation

Chiang Saen Lake is protected as part of the Nong Bong Kai Non-hunting Area. Nam Kham Nature Reserve is a private reserve devoted to bird conservation and scientific research. Monthly bird-ringing sessions have been held since 2008, contributing significantly to knowledge of birds in Thailand and their movements. In contrast, the Mekong River is under serious threat from Chinese dams upriver that have altered its natural flow and severely damaged the Mekong's ecosystems. Similarly, Nong Lom is also not legally protected, and habitat clearance for cultivation and development are serious threats.

James Eaton

Chiang Saen's wetlands are the best place to observe ***Jerdon's Bush Chat*** *in Thailand.*

Peter and Michelle Wong

Brown-cheeked Rail *regularly winters in the wetlands at Chiang Saen.*

Mae Wong National Park & Huai Kha Khaeng Wildlife Sanctuary

Wich'yanan Limparungpatthanakij

The Western Forest Complex, of which Mae Wong and Huai Kha Kheng are part, is the largest remaining tract of contiguous forest in Thailand.

KEY FACTS

Nearest Major Towns
Lan Sak, Ban Rai, Mae Wong and Mae Poen (Uthai Thani Province), Pang Sila Thong (Kampaeng Phet Province), Umphang (Tak Province)

Habitats
Lowland to montane evergreen forests, dry deciduous forest, bamboo groves, grasslands

Key Species
Rufous-throated and Bar-backed Partridges, Kalij Pheasant, Grey Peacock-Pheasant, Green Peafowl, River Lapwing, Pin-tailed and Ashy-headed Green Pigeons, Blue-banded Kingfisher, Tickell's Brown Hornbill, Rufous-necked and Plain-pouched Hornbills, Black-headed and Bamboo Woodpeckers, White-rumped Falcon, Collared Falconet, Silver-breasted, Long-tailed and Dusky Broadbills, Blue and Rusty-naped Pittas, Olive and White-throated Bulbuls, Collared and Spot-necked Babblers, Large and Coral-billed Scimitar Babblers, Burmese Yuhina, Burmese Nuthatch
Winter Swinhoe's and Rosy Minivets, Silver Oriole, Sulphur-breasted, Martens's and Alström's Warblers, Spot-winged Starling, Grey-sided and Eyebrowed Thrushes

Other Specialities
Phayre's Leaf Monkey, Asiatic Wild Dog, Asiatic Black and Sun Bears, Banded Linsang, Binturong, Marbled Cat, Leopard, Indochinese Tiger, Asian Elephant, Malayan Tapir, Fea's Muntjac, Gaur, Banteng, Asiatic Water Buffalo, Chinese Serow

Best Time to Visit
All year round

Both the Huai Kha Khaeng Wildlife Sanctuary (2,780km²) and Mae Wong National Park (895km²) form part of the Western Forest Complex, a vast landscape of 19 protected areas that spans over 18,000km² of Thailand's rugged frontier with Myanmar. This landscape collectively protects the largest contiguous tracts of forest left in Thailand, and is one of the largest in Southeast Asia. Extensive areas of lowland riparian and dry forest still persist, especially in Huai Kha Khaeng and the adjacent Thung Yai Naresuan Wildlife Sanctuaries. These habitats support comparatively healthy populations of the endangered **Green Peafowl** and various globally threatened megafauna, including the Indonese Tiger, Banteng, and the last remaining population of Asiatic Water Buffalo in Thailand. Two of the best sites for the birdwatcher to visit are described here.

Birdwatching Sites

Huai Kha Khaeng Wildlife Sanctuary Headquarters

As Huai Kha Khaeng is a wildlife sanctuary, most of the reserve is not open to tourists and requires special permission to access. The headquarters and its immediate environs is an exception. This area has some excellent remnant lowland deciduous forest, which consists of both dry dipterocarp forest and secondary growth. Khao Hin Daeng nature trail is a

*The **Grey Peacock-Pheasant** is a highlight of any visit to Chong Yen Station.*

loop trail near the headquarters that features some of these mixed deciduous forests. Flocks of **Black-headed Woodpeckers** and **Red-billed Blue Magpies** are habituated to visitors and allow close approach. Other notable species here include the **Ashy-headed Green Pigeon**, **White-bellied Woodpecker**, **White-rumped Falcon**, **Collared Falconet**, **Indochinese Cuckooshrike** and **Tickell's Blue Flycatcher**. The **Brown Fish Owl** is regular along the streams at night.

Additionally, the access road passes through a few watchtowers that require permission at the headquarters to use. By sitting and waiting at a tower, you might encounter large mammals coming to drink at the waterholes and mineral licks, as well as the **Green Peafowl**.

Huai Mae Dee & Khao Bandai Ranger Stations, Huai Kha Khaeng

The habitats here are similar to those at the headquarters, with additional dry-forest specialities such as the localized **Burmese Nuthatch**. Huai Mae Dee has a campsite and is readily accessible. The road from Huai Mae Dee to Khao Bandai is best traversed using a four-wheel drive vehicle, while access to Khao Bandai itself is restricted. The open riparian woodlands and sandbars at Khao Bandai are good for the **Green Peafowl** and **River Lapwing**, as well as the Asiatic Water Buffalo. This is also a known breeding site of the **Plain-pouched Hornbill**.

Sai Bor Waterfall, Huai Kha Khaeng

The forest at this site is wetter than the deciduous forest around the sanctuary's headquarters. Consequently, a different suite of birds is present and species such as the **Great Hornbill** and the highly localized **Olive Bulbul** are possible here.

Francis Yap

*The **Rufous-necked Hornbill** persists in small numbers in the forests of western Thailand and can be encountered at both Huai Kha Khaeng and Mae Wong.*

Abdelhamid Bizid

*Huai Kha Khaeng is a stronghold of the **Green Peafowl** in Thailand.*

*The forests at Chong Yen are one of the best places to encounter the attractive **White-necked Laughingthrush**.*

Mae Wong National Park

Park Headquarters

The moist evergreen forest around the headquarters supports a good cast of forest species, including the **Great Slaty Woodpecker**, **Olive Bulbul** and **Blue-throated Blue Flycatcher**. The **Crested Kingfisher** is also regular along the stream next to the headquarters complex.

Chong Yen Ranger Station

This is the key birdwatching site within the park. The montane broadleaved evergreen forest around the ranger station is home to several regional specialities. It is one of the best places to see the sought-after **Rufous-necked Hornbill** in Thailand, which is occasionally heard and seen flying around the station. Mixed foraging flocks in and around Chong Yen may contain the **Coral-billed Scimitar Babbler**, **White-necked Laughingthrush** and **Burmese Yuhina** among the more common species. Other possibilities include the **Mountain Hawk-Eagle**, **Pin-tailed** and **Wedge-tailed Green Pigeons**, **Stripe-breasted** and **Bay Woodpeckers**, wintering **Swinhoe's** and **Rosy Minivets**, **White-throated Bulbul**, **Spot-necked Babbler**, **Rufous-backed Sibia** and **Black-throated Parrotbill**.

A walk down the 'Old Umphang Road' from the Chong Yen campsite offers excellent opportunities to see various ground dwellers such as the **Rufous-throated Partridge**, **Grey Peacock-Pheasant**, **Rusty-naped Pitta**, **Slaty-bellied Tesia**, **Eyebrowed Wren-Babbler** and **White-tailed Robin**.

The upper sections of the road from the park headquarters to Chong Yen are known wintering sites for globally threatened species such as the rare **Wood Snipe** and **Grey-sided Thrush**.

Access & Accommodation

The birdwatching sites in Huai Kha Khaeng and Mae Wong can only be reached by a rented vehicle. The entrance to the headquarters of Huai Kha Khaeng is located at Km 53 on Route 3438. To access the Khao Bandai Ranger Station, you need to drive through Route 3011 to the Huai Mae Dee Ranger Station. Khao Bandai is located further along the same road but requires a permit to access. The entrance to Sai Bor Waterfall is located along Route 3282.

Chong Yen Ranger Station is beyond the headquarters of Mae Wong National Park, and can be accessed by driving through Route 1117. The road is paved all the way through to Chong Yen. To reach Mae Rewa Ranger Station, follow Route 1072 and continue straight through the intersection with Route 3504 at Mae Le.

Con Foley

*The attractive **Bar-backed Partridge** is widespread across continental Southeast Asia, but is more easily heard than seen.*

Wich'yanan Limparungpatthanakij

*The Western Forest Complex is exceptionally rich in woodpecker species, including the localized **Bamboo Woodpecker**.*

Park lodges are available at Mae Wong National Park headquarters. However, almost all sites offer camp grounds. For more convenient accommodation, you may consider various private lodges located several kilometres outside the entrance.

Conservation

Mae Wong National Park and Huai Kha Khaeng Wildlife Sanctuary form the core area of Thailand's Western Forest Complex. In recognition of its outstanding value to biodiversity, Huai Kha Khaeng has been declared a UNESCO World Heritage Site. Due to its rich wildlife, particularly large mammals, poaching is a major threat and several forest guards have lost their lives in recent years during violent confrontations with heavily armed poachers.

Nam Nao National Park

Wich'yanan Limparungpatthanakij

Patrawut Sitifong

*The **Blossom-headed Parakeet** is often seen perched high in the canopy, or in flight overhead.*

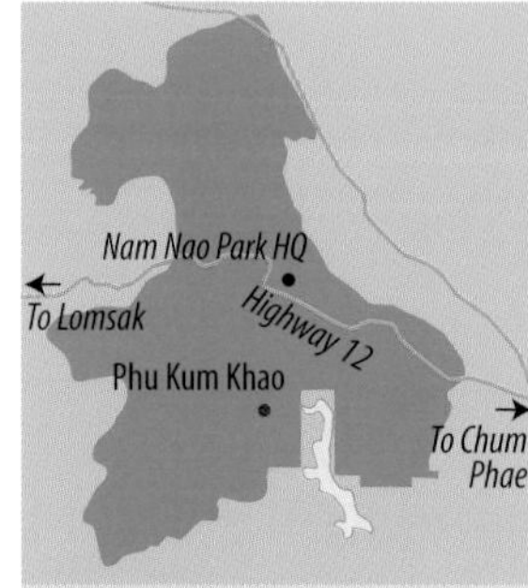

KEY FACTS

Nearest Major Towns
Lom Sak (Phetchabun Province), Chum Phae (Khon Kaen Province)

Habitats
Submontane mixed deciduous and broadleaf evergreen forests, pine forests

Key Species
Silver Pheasant, Siamese Fireback, Red-headed Trogon, Black-and-buff, Black-headed, Bamboo, White-bellied and Great Slaty Woodpeckers, Collared Falconet, Blossom-headed Parakeet, Silver-breasted and Long-tailed Broadbills, Blue and Rusty-naped Pittas, Indochinese Cuckooshrike, Red-billed Blue Magpie, Eurasian Jay, Brown Prinia, Large Scimitar Babbler, Red-billed Scimitar Babbler, Collared Babbler, Black-throated Laughingthrush, Burmese Nuthatch, Golden-crested Myna
Winter Swinhoe's and Rosy Minivets, Radde's, Two-barred and Claudia's Leaf Warbler, Sulphur-breasted Warbler, Eyebrowed Thrush

Other Specialities
Big-headed Turtle, Impressed Tortoise, Phayre's Langur, Marshall's Horseshoe Bat, Asiatic Jackal, Asiatic Wild Dog, Asiatic Black Bear, Sun Bear, Leopard, Indochinese Tiger, Red Muntjac, Sambar, Chinese Serow, Gaur, Asian Elephant

Best Time to Visit
November–April

At nearly 1,000km² in area, Nam Nao National Park sprawls across eastern Phetchabun province, with a smaller extent of the park lying within Chaiyaphum. The park encompasses an elevational range of 650–1,200m asl, with much of it located around 800m asl. It comprises extensive areas of dry deciduous forests, a habitat that has been heavily cleared or degraded across eastern Thailand. Nam Nao is one of few places where a number of species characteristic of such deciduous forests can still be found, particularly the **Indochinese Cuckooshrike**, **Brown Prinia** and localized **Burmese Nuthatch** (formerly a subspecies of the Chestnut-vented Nuthatch). The park also support a rich assemblage of woodpeckers, and the charismatic **White-bellied Woodpecker** is relatively easy to find here. It is possible to see nearly 100 species on a 2–3-day trip, targeting the different vegetation types in the park.

Birdwatching Sites

Access Road to Park Headquarters

The access road leading to the park headquarters from Highway 12, especially the stretch 2km before the main gate, can be very productive in the early morning, as are the grounds around the park's lodge and camping area. Many corvids, particularly the **Eurasian Jay** and **Red-billed Blue Magpie**, are very accustomed to large groups of visitors and can be easily seen here, while the loud call of the **White-bellied Woodpecker** is often heard. The **Great Slaty Woodpecker** is also a possibility.

The forests around the headquarters should also be checked for mixed species flocks. In particular, noisy flocks of laughingthrushes are a regular occurrence. Other species that join these flocks include various species of woodpecker, including both **Black-headed** and **Grey-headed Woodpeckers**. In addition, foraging flocks comprising smaller birds should be checked for **Burmese Nuthatch** and both **Indochinese** and **Black-winged**

Cuckooshrikes. The scrubby undergrowth of open breaks in the forest is where **Brown** and **Rufescent Prinias** can be found.

Nature & Loop Trails

The series of trails of varying lengths that start from around the park headquarters' visitor centre eventually enter some good-quality submontane evergreen forest and large stands of bamboo. A good strategy is to walk slowly, starting from the entrance, which follows a stream, and scanning the forest floor and undergrowth regularly. With a bit of luck, you should eventually cross paths with parties of **Bar-backed Partridges**, **Silver Pheasants**, **Long-tailed Broadbills**, **Blue Pittas** and **White-crowned Forktails**. When mixed laughingthrush flocks are located, one should also check for the **Red-headed Trogon** and **Common Green Magpie**, two species that frequently follow these flocks. Additionally, if **Red-billed Scimitar Babblers** are located, it is usually just a matter of time before you encounter parties of **Collared Babblers** (or vice versa).

Other Birdwatching Spots

Other spots in the park frequented by visiting birdwatchers include Ban Paek and the Phu Kum Khao pine forests. Both sites offer similar species as those in the forests around the vicinity of the park headquarters. The entrances to these two sites are along Highway 12 at the Km 49 and Km 53 markers respectively. It should be noted that accessing Phu Kum Khao area will require an SUV as it is some 14km from the main road on a rough track.

*The **Burmese Nuthatch** is one of the characteristic species of dry deciduous forests on mainland Southeast Asia.*

*Small parties of **Lesser Necklaced Laughingthrushes** are regularly seen at Nam Nao.*

Cheng Heng Yee

*The **Greater Yellownape** is one of more than 10 woodpecker species occurring in Nam Nao.*

Access & Accomodation

Highway 12 bisects Nam Nao National Park, making it fairly easy to access the park from either Lom Sak Subdistrict from the east or Chum Phae Subdistrict from the west. Public buses regularly shuttle between both subdistricts to the main gate near the park headquarters. Accommodation options are varied and you can consider staying at the lodges around the park headquarters or camp. Besides the visitor centre, other facilities include shops and restaurants near the headquarters that are all open daily.

Conservation

Nam Nao National Park is part of a complex of protected areas that border Tat Mok National Park and Phu Khiao Wildlife Sanctuary to the south. This complex of protected areas, known as the Western Isaan Forest Complex, spans nearly 5,000km^2 and forms one of the largest remaining wildlife refuges in this otherwise heavily cultivated area of Thailand.

Bueng Borapet

Wich'yanan Limparungpatthanakij

Ayuwat Jearwattanakanok

Bueng Borapet is the largest lake in central Thailand and supports a wide variety of waterfowl throughout the year.

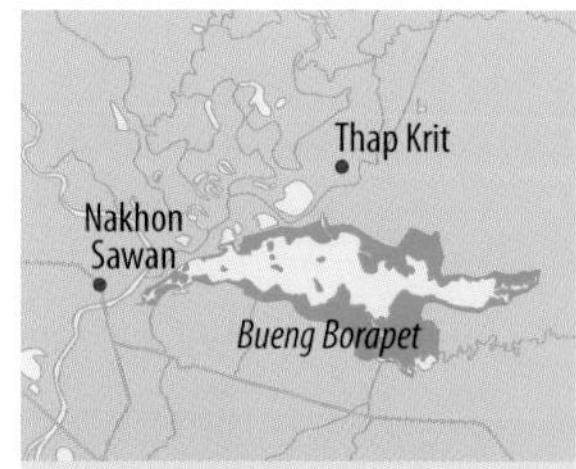

KEY FACTS

Nearest Major Towns
Mueang, Chum Saeng (Nakhon Sawan Province)

Habitats
Open waters, freshwater wetlands, fish ponds, paddy fields

Key Species
Indian Spot-billed Duck, Painted Stork, Black-headed and Glossy Ibises, Spot-billed Pelican, Indian Cormorant, Oriental Darter, White-browed Crake, Black-backed Swamphen, River Lapwing, Greater Painted-snipe, Pheasant-tailed and Bronze-winged Jacanas, Oriental Pratincole, Jacobin Cuckoo, Savanna Nightjar, Pied Kingfisher, Freckle-breasted Woodpecker, Indochinese Bush Lark, Golden-headed Cisticola, Asian Golden Weaver, Red Avadavat
Winter Knob-billed Duck, Ruddy Shelduck, Garganey, Ferruginous Duck, Baer's Pochard, Tufted Duck, Eurasian Bittern, Greater Spotted Eagle, Eastern Marsh and Pied Harriers, Grey-headed Lapwing, Black-tailed Godwit, Spotted Redshank, Small Pratincole, Blunt-winged Warbler, Manchurian and Large-billed Reed Warblers (rare), Paddyfield and Thick-billed Warblers, Lanceolated Warbler, Pallas's Grasshopper Warbler, White-shouldered Starling, Siberian Rubythroat, Yellow-breasted Bunting

Other Specialities
Siamese Crocodile

Best Time to Visit
December–February

Originally an extensive area of marshes and swampy floodplains, Bueng Borapet became the largest lake in central Thailand due to the construction of a dam in the 1930s, and now covers a much larger area of 224km^2. Water levels in the lake fluctuate greatly between seasons, being heavily influenced by the Ping and Nan Rivers flowing from northern Thailand. Frequently misspelled as 'Bueng Boraped' or 'Bung Boraphet', this large wetland is perhaps best known for being the only place in the world where the enigmatic and Critically Endangered **White-eyed River Martin** had been seen on several occasions throughout the 1960s and '70s. In more recent times the swamps and reedbeds of Bueng Borapet have garnered a reputation for being one of the best places to see waterbirds associated with freshwater wetlands in Thailand.

Birdwatching Sites

Headquarters & Waterbird Park

Here, along the southern side of the lake, birdwatching can be done around the parking lot in the small patch of open woodland and around the reedbeds along the lake margins. The reedbeds are good for a variety of skulking species, including the **Ruddy-breasted Crake**, **Thick-billed Warbler** and **Siberian Rubythroat**. On the floating vegetation around the lake margins, the **Black-backed Swamphen**, **Pheasant-tailed Jacana** and other waterfowl may be seen. The open woodland is home to the **Freckle-breasted Woodpecker** and winter visitors such as the **Eurasian Wryneck** and **Forest Wagtail**. Large flocks of starlings are a common sight in winter, and are dominated by **White-shouldered** and **Chestnut-tailed Starlings**.

Michelle & Peter Wong

*The Critically Endangered **Baer's Pochard** is occasionally seen on Bueng Borapet.*

A boat service at the pier near the parking lot takes visitors on sightseeing tours around the lake and can be hired to see waterfowl on the lake. A variety of wintering ducks is regularly seen, including the **Northern Pintail**, **Garganey**, **Ferruginous** and **Tufted Ducks**, and other waterfowl such as the **Indian Cormorant**, **Glossy Ibis** and **Spot-billed Pelican**. The Critically Endangered **Baer's Pochard** formerly wintered here regularly but is now only occasionally recorded.

Wich'yanan Limparungpatthanakij

*The **Cotton Pygmy Goose** is common at Bueng Borapet.*

Commercial Fisheries Area (Pramong Panich)

This extensive area is best explored with a car to locate specific areas where birds congregate. They include flooded fields, which are good for waders like the **Grey-headed Lapwing**, **Kentish Plover** and **Greater Painted-snipe**. The localized subspecies (ssp. *longicaudatus*) of **Long-tailed Shrike** frequents poles and scrub in the area with the regional endemic **Indochinese Bush Lark**. **Streaked** and **Asian Golden Weavers** are abundant. In winter it is worth scanning the weaver flocks for the rare and declining **Yellow-breasted Bunting**, once heavily harvested for religious release. The **Jacobin Cuckoo** is vocal and conspicuous during the rainy season.

Access & Accommodation

The wetlands of Bueng Borapet are best accessed by car, a three-hour drive heading north from Bangkok. To get to the lake take a right turn onto Route 3001 before entering the provincial town of Nakhon Sawan. The entrance to the headquarters of Bueng Borapet Non-hunting Area and the Waterbird Park is on the left. There is a gigantic figure of a White-eyed River Martin at the entrance. The Commercial Fisheries Area can be accessed by driving pass the recreation area along the northern margin of the lake on Route 225. Drive to the intersection with Route 3475 and turn right onto that road. The entrances to various laterite roads are along Route 3475.

Conservation

Research programmes such as the ongoing monitoring and ringing efforts have been carried out by Bueng Borapet Wildlife Research Station under the Department of National Parks, Wildlife and Plant Conservation. An area of private land near the lake containing waterbird colonies has recently been rented and is managed by the Bird Conservation Society of Thailand. Despite being declared a protected area, large-scale fishing still occurs and waterbirds and other wildlife frequently get tangled in fishing nets. Water levels in the lake have also been declining due to extraction for aquaculture and rice farming.

Wich'yanan Limparungpatthanakij

*The **Bronze-winged Jacana** occurs on the floating mats of vegetation on the lake.*

Dong Phayayen-Khao Yai Forest Complex

Wich'yanan Limparungpatthanakij

Bjorn Olesen

The Dong Phaya Yen-Khao Yai Forest Complex protects one of the largest contiguous tracts of forest remaining in eastern Thailand.

KEY FACTS

Nearest Major Towns
Wang Nam Khiao, Pak Chong (Nakhon Ratchasima Province), Amphoe Mueang (Nakhon Nayok Province), Amphoe Mueang (Prachin Buri Province),

Habitats
Hill and submontane evergreen forests, mixed and dry deciduous forests, grasslands, cultivation, secondary growth

Key Species
Silver Pheasant, Siamese Fireback, Ashy-headed Green Pigeon, Coral-billed Ground Cuckoo, Spot-bellied Eagle-Owl, Red-headed and Orange-breasted Trogons, Austen's Brown Hornbill, Wreathed and Great Hornbills, Green-eared and Moustached Barbets, Black-and-buff, Black-headed, Streak-throated, White-bellied and Great Slaty Woodpeckers, White-rumped Falcon, Collared Falconet, Eared and Blue Pittas, Banded, Long-tailed and Dusky Broadbills, Indochinese Cuckooshrike, White-browed Fantail, Red-billed Blue Magpie, Brown Prinia, Large Scimitar Babbler, Limestone Wren-Babbler, Burmese Nuthatch, Golden-crested Myna
Winter Masked Finfoot, Asian Emerald Cuckoo, Silver Oriole, Swinhoe's Minivet, Sulphur-breasted Warbler, Orange-headed and Eyebrowed Thrushes, Siberian Blue Robin, Mugimaki Flycatcher, White-throated Rock Thrush

Other Specialities
Sunda Pangolin, Bengal Slow Loris, Phayre's Leaf Monkey, Northern Pig-tailed and Stump-tailed Macaques, White-handed and Pileated Gibbons, Asiatic Wild Dog, Asiatic Black Bear, Yellow-throated Marten, Binturong, Marbled Cat, Clouded Leopard, Asian Elephant, Red Muntjac, Sambar, Chinese Serow, Banteng, Gaur, Siamese Crocodile

Best Time to Visit
November–April; rainy season June–October

The Dong Phayayen-Khao Yai Forest Complex encompasses 6,150km² of mountains and forests south of the Khorat Plateau in eastern and lower north-eastern Thailand, and consists of Khao Yai, Thap Lan, Pang Sida and Ta Phraya National Parks, and Dong Yai Wildlife Sanctuary. The reserves collectively protect the largest remaining tracts of contiguous forest in eastern Thailand. Of the reserves, Khao Yai National Park, Thailand's first national park, is the most popular among visiting birdwatchers. Here, Indochinese endemics such as the **Siamese Fireback**, **Coral-billed Ground Cuckoo** and **Moustached Barbet** dwell in the evergreen and mixed deciduous forests of the lower hills alongside endemic and distinctive subspecies of several forest birds such as the **Puff-throated Bulbul** (ssp. *isani*) and **Hill Blue Flycatcher** (ssp. *lekhakuni*), the latter named after the renowned Thai conservationist Dr Boonsong Lekagul. The limestone karsts along the western border of Dong Phayayen Mountains just outside Khao Yai (in Saraburi) are the only known locality of the highly distinctive subspecies (ssp. *calcicola*) of the **Limestone Wren-Babbler**. Although most of the ornithological highlights occur in Khao Yai, some of the less known reserves in the complex, such as Thap Lan National Park and Sakaerat Biosphere Reserve, also have their own charm and are worthy of further exploration.

Birdwatching Sites

Khao Yai National Park

As one of Thailand's most popular tourist attractions, birdwatchers should avoid visits to the park over the weekends and during public holidays. There are many birdwatching sites here.

Mike Kilburn

The forests of Khao Yai are a known wintering site for the globally endangered ***Silver Oriole****.*

Haew Narok Waterfall

This steep waterfall is located close to the southern gate of the park, although it is a long drive from here to other birdwatching spots that are closer to the northern gate. There are many flowering and fruiting trees around the parking lot, which attract various barbets, bulbuls, flowerpeckers and sunbirds. The refuse dump in the vicinity attracts a number of birds, including the sought-after **Coral-billed Ground Cuckoo**. The path to the waterfall is good for the **Orange-breasted Trogon**, **Heart-spotted Woodpecker**, **Eared Pitta** and **Large Scimitar Babbler**. Along forest clearings, the **Blue-bearded Bee-eater** is regular, while **Banded** and **Dusky Broadbills** often perch on dead trees.

Khao Khieo Viewpoint & Pha Dieo Dai Trail

Khao Khieo Viewpoint is the highest spot in Khao Yai accessible to tourists, and also the most accessible location for **Kloss's Leaf Warbler** in Thailand. The **Barred Cuckoo-Dove** and **Black-throated Laughingthrush** are often encountered. The slightly lower Pha Dieo Dai Trail is a looping boardwalk providing opportunities to see shy ground birds such as the **Silver Pheasant** and **Blue Pitta**. The Chinese Serow, a goat-like ungulate, is also seen here regularly. Pha Dieo Dai Viewpoint, inside the trail, is similar to Khao Khieo Viewpoint – they are both good for viewing raptors such as **Mountain Hawk-Eagles**, and hornbills including **Austen's Brown Hornbill**. The access road to Khao Khieo is particularly productive for roadside birdwatching in the morning, with regular sightings of **Siamese Firebacks** and **Coral-billed Ground Cuckoos**, as well as **Red-headed Trogons** and various barbets and bulbuls.

Pha Kluay Mai – Haew Suwat Trail

This 4km-long trail has become popular in recent years as a reliable site to see the Critically Endangered Siamese Crocodile, although the origin of this population is unclear. The globally endangered **Masked Finfoot** had been spotted occasionally along the flowing streams. The **Golden-crested Myna** is regularly found with flocks of **Common Hill Mynas** on tall emergent trees. The damp areas behind the toilet and restaurant can be good in the dry season for the **Asian Stubtail**, **Orange-headed Thrush** and **Siberian Blue Robin**, and the **Coral-billed Ground Cuckoo** and **Blue Pitta** may be seen.

Dave Sargeant

Khao Yai National Park is the most accessible site in the world for seeing the ***Coral-billed Ground Cuckoo****.*

Khao Yai Park Headquarters

The parking lots and cafeteria on the opposite side of the road are situated along flowing streams with many fruiting trees, offering great opportunities to see hornbills, barbets, bulbuls, flowerpeckers, leafbirds and the **Asian Fairy-bluebird**. Bamboo clumps adjacent to the parking lot are known to attract the **Pin-tailed Parrotfinch** when seeding. The loop trail behind the headquarters should be explored for a chance at the **Coral-billed Ground Cuckoo**, **Eared Pitta** and **White-throated Rock Thrush**. Foraging flocks are regularly encountered and feature a variety of birds ranging from laughingthrushes and scimitar babblers to **Swinhoe's Minivet** and the **Sulphur-breasted Warbler**. At the opposite site of the headquarters, one of the many entrances to the famous 'Trail B' can be found.

Trail B

This is the most popular trekking trail within Khao Yai. The northernmost entrance is located at Km 33 and can be very productive

Con Foley

The majestic ***Great Hornbill*** *is often seen around fruiting trees in Khao Yai National Park.*

Con Foley

*The **Blue Pitta** is not uncommon in the forests of Khao Yai, but seeing it takes patience.*

Con Foley

*The **Siamese Fireback** is regular on Khao Yai's Trail B.*

for birdwatching as well. There are a number of fig trees frequented by **Golden-crested Mynas**, hornbills, and gibbons including the Pileated Gibbon. Fallen fruits on the forest floor usually attract **Siamese Firebacks**. The endangered **Silver Oriole**, a localized winter visitor in mainland Southeast Asia, has been spotted feeding on flowering trees in this area. It is worth scanning the banks of the various forest streams for kingfishers and forktails. Note that some parts of the trail are densely vegetated and getting lost is a real possibility, especially if you do not use a guide. The nearest exit is a few kilometres away and leads to the Nong Phak Chi Watchtower, which provides a good vantage point for seeing raptors and hornbills, and grassland species such as the **Golden-headed Cisticola**. There are also exits at Wang Jampee, the park headquarters, Mo Singto Reservoir and opposite the TAT Pond. The latter is a regular spot to watch migratory **Silver-backed Needletails**, and the more common resident **Brown-backed Needletails** come in to drink in the late afternoon.

Sakaerat Biosphere Reserve

This is the most reliable site to see the **Siamese Fireback** – the national bird of Thailand and an endemic to eastern Thailand and Indochina. Sakaerat is dominated by dry evergreen and dipterocarp forests with a grassy understorey, which provides good habitat for the localized **Brown Prinia**. There are various trails close to the Sakaerat Environmental Research Station that are worth exploring. For mammal enthusiasts, the nearby Khao Phaeng Ma Non-hunting Area is the most easily accessible and reliable site to see large herds of Gaur in Thailand.

Sab Sadao Ranger Station, Thap Lan National Park

With the only remnants of undisturbed lowland dry dipterocarp forest left in eastern Thailand, Sab Sadao holds a number of species that are difficult to see elsewhere in Thailand. It is home to many hole-nesting species such as the **Black-headed Woodpecker**, **White-bellied Woodpecker**, **White-rumped Falcon**, **Collared Falconet** and **Burmese Nuthatch**. Roadside birdwatching and various forest tracks also provide good opportunities to see the **Rufous-winged Buzzard**, **Indochinese Cuckooshrike**, **White-browed Fantail** and **Brown Prinia**.

Access & Accommodation

At Khao Yai a range of accommodation and facilities is available both inside and outside the park, while paved roads throughout the park make it easy to drive around. The park is accessible from the northern and southern gates through Routes 2090 and 3077 in Nakhon Ratchasima and Prachin Buri respectively. Regardless of route choice, it takes about two and a half hours to reach Khao Yai from Bangkok. At Sakaerat and Sab Sadao, accommodation is not available within the reserve. However, there are hotels in the nearby Wang Nam Khiao District, a popular tourist site. Both sites are easily accessed via Route 304.

Conservation

Dong Phayayen-Khao Yai Forest Complex was established as a UNESCO World Heritage Site in 2005. The four national parks and a wildlife sanctuary were declared under the National Parks Act B.E. 2504 (1961) and the Wild Animal Reservation and Protection Act B.E. 2535 (1992) respectively. However, encroachment into the forest is a chronic problem. Harvesting of Eaglewood (*Aquilaria crassna*), timber, ivory and other wildlife products still occurs. Many resorts and cultivated areas are also developed illegally within the boundaries of these protected areas, further damaging what little forest remains.

Ko Man Nai and Pak Nam Prasae

Wich'yanan Limparungpatthanakij

Wich'yanan Limparungpatthanakij

A boat trip at Pak Nam Prasae allows close encounters with a variety of shorebirds during the northern winter.

Pak Nam Prasae, directly translated from Thai as Prasae River Mouth, is a long, sandy coastline dotted with coastal villages and interspersed by some areas of mangroves. It is a key site for migratory shorebirds and with the help of a boat, you can expect excellent photographic opportunities for a variety of shorebirds that roost on the tops of floating rafts during high tide, including globally threatened species such as **Nordmann's Greenshank**. Nearby, Mu Ko Man, or the Man Islands, are a cluster of three islands. Among birdwatchers and ornithologists, Ko Man Nai is well known as a stopover site for migratory birds, especially nocturnal land-bird migrants that fly across the Gulf of Thailand from the Thai-Malay Peninsula in spring (April–May). During this period numerous migrants from East Asia are present on the island, including globally threatened species such as the **Fairy Pitta** and **Brown-chested Jungle Flycatcher**.

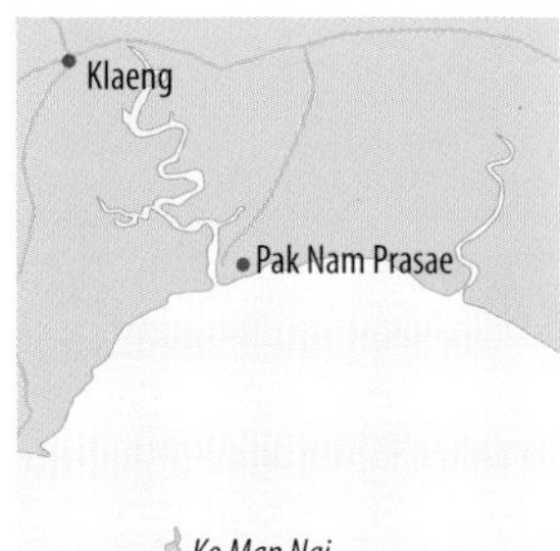

KEY FACTS

Nearest Major Town
Klaeng (Rayong Province)

Habitats
Sandy beaches, coastal forests, mangroves

Key Species
Winter and passage Chinese Egret, Slaty-legged Crake, Asian Dowitcher, Far Eastern Curlew, Nordmann's Greenshank, Grey-tailed Tattler, Terek Sandpiper, Great Knot, Spoon-billed Sandpiper, Chestnut-winged and Himalayan Cuckoos, Grey Nightjar, White-throated Needletail, Fairy Pitta, Tiger Shrike, Crow-billed Drongo, Japanese Paradise Flycatcher, Kamchatka, Pale-legged and Sakhalin Leaf Warblers, Black-browed Reed Warbler, Pallas's Grasshopper, Lanceolated Warbler, Eyebrowed and Siberian Thrushes, Brown-chested Jungle Flycatcher, Ferruginous, Zappey's Blue-and-white, Yellow-rumped, Green-backed, Narcissus and Mugimaki Flycatchers, Siberian Blue Robin, White-throated Rock Thrush
All year round Malaysian Plover, Bridled and Black-naped Tern, Mangrove Whistler

Other Specialities
Island Flying Fox, Indochinese Ground Squirrel

Best Time to Visit
September–May

Birdwatching Sites

Ko Man Nai

This is the most regularly visited island by birdwatchers and ornithologists as it is located closest to the mainland and contains trails through its forests. The sheer number of East Asian songbird migrants, such as the **Japanese Paradise Flycatcher**, **White-throated Rock Thrush**, and **Green-backed** and **Narcissus Flycatchers**, reported here alone

Francis Yap

Brown-chested Jungle Flycatchers are regular in the scrub on Ko Man Nai and best seen between April–May.

in a season is probably higher than the total figure from all other Thai birdwatching sites combined. Secretive ground dwellers like **Lanceolated** and **Pallas's Grasshopper Warblers**, **Eyebrowed Thrush** and **Siberian Blue Robin** are conspicuous here, perhaps exhausted from their long-distance journeys. Species of conservation concern, namely the **Fairy Pitta** and **Brown-chested Jungle Flycatcher**, are regular. Other birds include poorly known species such as **Kamchatka** and **Sakhalin Leaf Warblers**. Like in any migrant hotspot, vagrants enliven proceedings and past records include the **Christmas Frigatebird**, Thailand's first **Japanese Leaf Warbler**, **Alström's Warbler**, **Brown-breasted Flycatcher**, **Tristram's Bunting** and **Chestnut Bunting**. Shorebirds include the **Grey-tailed Tattler** on the rocky shore and resident **Malaysian Plover** on the beaches. The **Mangrove Whistler** is one of a few common resident species on the island.

Pak Nam Prasae

Thrust into the birdwatching spotlight due to the appearance of a vagrant **River Tern** in 2010, further visits to this site have revealed its importance for migratory shorebirds. During high tide most waders roost on floating bamboo rafts that locals use for hanging oysters.

Francis Yap

*In April and May, East Asian migrants are numerous at Ko Man Nai and include large numbers of paradise flycatchers, like this **Amur Paradise Flycatcher**.*

Bjorn Olesen

*Ko Man Nai is the most reliable site in Southeast Asia to see the globally threatened **Fairy Pitta** in spring.*

Globally threatened species including the **Chinese Egret**, **Far Eastern Curlew**, **Nordmann's Greenshank**, **Great Knot** and **Spoon-billed Sandpiper** have all been recorded, alongside regularly encountered birds such as the **Asian Dowitcher**, **Bar-tailed** and **Black-tailed Godwits**, **Eurasian Curlew**, **Red Knot**, **Red-necked Stint** and **Curlew Sandpiper**. The site currently boasts the highest count of **Nordmann's Greenshanks** at a single site in Thailand, with close to 100 birds recorded in some years.

Access & Accommodation

To reach the piers to Ko Man Nai, follow Route 3161 and turn left just before Route 3161 merges with the seaside Route 3145. From here you can hire a fisherman's boat to Ko Man Nai; make sure you arrange for transport on the way back beforehand. You cannot stay overnight at Ko Man Nai unless permitted to do so by the Department of Marine and Coastal Resources. On the other hand, Ko Man Klang and Ko Man Nok offer convenient lodges.

Pak Nam Prasae, located to the west, is accessible by driving to the end of Route 3152. On reaching the fishermen's village, visitors need to hire a long-tailed boat and state clearly that they would like to do a birdwatching trip. Resorts are available in both Klaeng town and near the beach.

Conservation

A ringing scheme led by leading Thai ornithologist Philip D. Round has been ongoing for several years. The islands of Mu Ko Man are protected and administered by the Department of Marine and Coastal Resources. At Pak Nam Prasae shorebirds are regularly monitored by volunteers from the Bird Conservation Society of Thailand during the annual Asian Waterbird Census, as well as by the Wildlife Research Division from the Department of National Parks, Wildlife and Plant Conservation.

Nong Pla Lai

Wich'yanan Limparungpatthanakij

Kin Yip Ho

The extensive ricefields at Nong Pla Lai provide enough prey items to sustain a wide variety of wintering raptors, including this young ***Greater Spotted Eagle****.*

Gulf of Thailand
Wat Khao Takhrao
Raptor observation point
To Petchaburi

KEY FACTS

Nearest Major Towns
Khao Yoi, Mueang, Ban Laem (Phetchaburi Province)

Habitats
Paddy fields, freshwater marshes and fish ponds

Key Species
Cotton Pygmy Goose, Painted Stork, Milky Stork, Black-headed Ibis, Spot-billed Pelican, Oriental Darter, Slaty-breasted Rail, Ruddy-breasted Crake, Watercock, White-browed Crake, Black-backed Swamphen, Greater Painted-snipe, Bronze-winged Jacana, Stork-billed Kingfisher, Freckle-breasted Woodpecker, Chestnut-capped Babbler, Streaked Weaver, Asian Golden Weaver
Winter Ruddy Shelduck, Garganey, Black-faced Spoonbill, Greater Spotted Eagle, Booted Eagle, Steppe Eagle, Eastern Imperial Eagle, Eastern Marsh and Pied Harriers, Great Stone-curlew, Grey-headed Lapwing, Pheasant-tailed Jacana, Spotted Redshank, Black-tailed Godwit, Asian Dowitcher, Yellow-breasted Bunting

Other Specialities
Smooth-coated Otter, Fishing Cat, Long-tailed Macaque

Best Time to Visit
August–April

Nong Pla Lai features extensive areas of semi-organic paddy fields, freshwater marshes and fish ponds that are influenced by seasonal floods. The area is well known as a wintering ground for a wide variety of migratory raptors, and the site not only features a raptor-watching viewpoint, but also formerly hosted an annual raptor-watching festival. Additionally, it is a regular wintering site for the Critically Endangered **Yellow-breasted Bunting**, here at one of its southernmost wintering sites. As the icing on the cake, the site is within two hours from downtown Bangkok by car, making it an excellent day-trip option for visitors on a tight schedule who wish to see a wide variety of open-country species.

Birdwatching Sites

Nong Pla Lai Ricefields

These ricefields support the largest concentration of wintering **Black Kites** in Thailand, with flocks numbering in the hundreds during the boreal winter. The best time of year to see migratory raptors, particularly eagles, is during the months of January and February, when the ricefields dry up and become more suitable as hunting areas for the birds. These fields should also be checked thoroughly for flocks of **Yellow-breasted Bunting**, which can number in excess of 50 individuals in some years. As

Ah Kei

*The **Eastern Imperial Eagle** is the least frequently encountered of the large 'Aquiline' eagles at this site.*

Koel Ko

*Small numbers of **Grey-headed Lapwings** occur in the flooded fields at Nong Pla Lai.*

an added distraction, various resident species can be observed engaging in courtship and nesting during this period, including **Baya**, **Streaked** and **Asian Golden Weavers**, which are all locally common and allow close approach. These weavers retain their attractive breeding plumage throughout much of the year to the month of August, by which time migratory species are already arriving and many wader species are already present in good numbers.

Wat Khao Takhrao

The marshes in the vicinity of this temple are one of the most reliable sites for the uncommon **Black-headed Ibis**. Fish and prawn ponds that have been drained attract concentrations of waterfowl such as **Painted Storks**, as well raptors including **Greater Spotted Eagles**. A large lake in the area supports rafts of wintering Palearctic ducks during the winter months, including the increasingly uncommon **Garganey**, as well as other waterfowl such as the occasional **Black-faced Spoonbill** and various waders such as the **Black-tailed Godwit**. In some years, more uncommon species such as the **Great Stone-curlew** and **Asian Dowitcher** have been recorded. Tracts of remnant mangroves near the temple are home to species such as the **Stork-billed Kingfisher** and **Freckle-breasted Woodpecker**.

Access & Accomodation

Visiting the key sites with a private car is recommended as Nong Pla Lai is an extensive area and usually requires thorough scanning along Routes 1023 and 1004 to locate sites where the birds are congregating. There are smaller, laterite roads that are worth exploring, but should be driven on with a high-clearance vehicle. Wat Khao Takhrao is located on Route 1004. At the temple, make a turn at the intersection following the sign Wat Thong Noppakhun, and proceed until you see a big lake on the right. A wide variety of hotels is available, especially in Petchaburi provincial town. However, the site can also be reached within two hours from Bangkok.

Conservation

Although the area is not protected by law, Nong Pla Lai is a well-known birdwatching site and local residents are familiar with the antics of visiting birdwatchers. Hunting does not seem to be a problem here. However, the legally permitted practice of putting nets over fish and prawn ponds to exclude birds from these areas results in the entanglement and mortality of birds of various species.

Cheng Heng Yee

*Unless the retiring **Greater Painted-snipe** is flushed, seeing it well can be a challenging endeavour.*

Laem Phak Bia & Pak Thale

Ayuwat Jearwattanakanok

The salt pans at Laem Phak Bia support important populations of wintering shorebirds.

KEY FACTS

Nearest Major Town
Phetchaburi town (Phetchaburi)

Habitats
Salt pans, mangroves, mudflats, wetlands, shrubland

Key Species
Painted and Milky Storks, Black-headed Ibis, Pacific Reef Heron, Spot-billed Pelican, Indian Cormorant, Slaty-breasted Rail, Malaysian Plover, Indian Nightjar, Mangrove Whistler, Plain-backed Sparrow
Winter Chinese Egret, Pied Avocet, Grey-headed Lapwing, Kentish Plover, Asian Dowitcher, Black-tailed Godwit, Eurasian and Far Eastern Curlews, Nordmann's Greenshank, Great and Red Knots, Ruff, Spoon-billed Sandpiper, Red-necked Phalarope, Pallas's and Lesser Black-backed Gulls, Rosy and White-shouldered Starlings

Other Specialities
Bryde's Whale, Irrawaddy Dolphin

Best Time to Visit
October–April; rainy season June–October

The inner Gulf of Thailand is one of the major wintering sites for migratory waders in Southeast Asia. Laem Phak Bia and Pak Thale are among the best places to see large numbers of these shorebirds. In October–April, thousands of waders can be seen gathering in the salt pans, where they rest and feed during high tide. The globally threatened **Nordmann's Greenshank** and **Spoon-billed Sandpiper** are the main attractions for birdwatchers as both are regularly found in this area. The Laem Phak Bia sandspit is also home to the resident **Malaysian Plover**, while the **Chinese Egret** and distinctive 'white-faced' subspecies (ssp. *dealbatus*) of **Kentish Plover** are regular winter visitors. From Laem Phak Bia, boat trips into the gulf to see the endangered Bryde's Whale can be arranged and afford an opportunity to see seabirds. The avian highlights and ease of access make this one of the most visited birdwatching sites in Thailand.

Birdwatching Sites

Salt Pans near Ban Laem

From the intersection between Highway 3177 and coastal road No. 4028, follow the coastal road northwards for 3.4km before reaching Ban Laem. After crossing the bridge, turn left onto a dirt track that passes through a landfill on the left. Migrant rarities have been reported around the landfill, including the **Great Stone-curlew** and **Rosy Starling**. Roadside shrubs can be quite productive, containing flocks of **White-shouldered Starlings** and the occasional **Eurasian Wryneck**. After passing the landfill a big abandoned building appears on the right. Turn right and follow the road through the salt pans where flocks of waders can be seen en route. **Black-tailed Godwits** are often seen in big flocks, sometimes with **Asian Dowitchers** or even a **Long-billed Dowitcher** within them. Scrutinizing flocks of small waders consisting of plovers and calidrid sandpipers may be rewarded with rarities such as the **Little Stint** and **Spoon-billed Sandpiper**.

Yann Muzika

The Critically Endangered ***Spoon-billed Sandpiper*** *is recorded annually at Pak Thale and is one of the key targets of any visiting birdwatcher.*

Laem Phak Bia King's Project

Follow the coastal road northwards to Ban Laem to the entrance to the King's Project on the right. The Environmental Research and Development Project, simply referred to as the 'King's Project', is a good place to be introduced to Thai avifauna. A series of ponds inside the project is well connected by small roads, making it easily accessible and convenient for bird photography from a vehicle. The ponds usually attract a number of waterbirds, notably the **Painted Stork**, **Spot-billed Pelican** and **Indian Cormorant**. In winter resident waterbirds are joined by flocks of waders including **Spotted Redshanks**, and **Red-necked**, **Temminck's** and **Long-toed Stints**. Scarcer species such as the **Ruff**, **Long-billed Dowitcher** and **Indian Stone-curlew** have also been reported. The mangroves surrounding the ponds are good for the **Mangrove Whistler**, a mangrove specialist. Other birds that can be found while driving around the project include the **Ruddy-breasted Crake**, **Greater Painted-snipe** and wintering **Black-capped Kingfisher**. Salt pans near the entrance of the project can offer some interesting species: look for the occasional **Milky Stork** where flocks of **Painted Storks** are found; even a threatened **Indian Skimmer** once showed up in these wetlands.

Laem Phak Bia Sandspit

Located north of the King's Project, the sandspit is home to the resident **Malaysian Plover** that breeds at this site. In winter it is regularly joined by flock of **Sanderlings**, **Greater** and **Lesser Sand Plovers** and distinctive 'white-faced' subspecies (ssp. *dealbatus*) of **Kentish Plover**. The sandspit is a regular wintering ground for the globally threatened **Chinese Egret**, which is often seen foraging close to the mangrove area alongside resident **Pacific Reef Herons**. Flocks of gulls and terns are usually seen gathering at the sandspit and sometimes contain **Pallas's Gull**. Although the sandspit can be reached on foot, it is best to access the area by boat.

Pak Thale

Driving northwards from Laem Phak Bia through the scenic salt pans for approximately 12km, you will eventually see a signpost for the Pak Thale Shorebird Conservation Area with an illustration of a Spoon-billed Sandpiper. Turn right and continue down the road leading through a village for about 600m before arriving at the BCST (Bird Conservation Society of Thailand) Conservation Center. On the right is a wide area of salt pans where thousands of

*A dark-morph **Pacific Reef Heron** hunts by the coast.*

waders regularly gather throughout the winter. Undoubtedly, the **Spoon-billed Sandpiper** is the most desired target here. Other rarities include the **Asian Dowitcher**, **Nordmann's Greenshank**, **Red-necked Phalarope** and **Slender-billed Gull**, while the **Eurasian Curlew**, **Terek Sandpiper**, **Great** and **Red Knots**, and **Broad-billed Sandpiper** are more readily encountered.

Access & Accommodation

The wetlands of Laem Phak Bia and Pak Thale are easily accessed by car via a 2–3-hour drive from Bangkok. Both sites are located on coastal road No. 4028. A range of guesthouses, resorts and hotels is available at both Ban Laem and Hat Chao Samran beach south of Laem Phak Bia. Alternatively, it is possible to stay in Phetchaburi, which is about 20km away.

Conservation

Most of the area is currently unprotected, except for the King's Project, where environmental research work is carried out on a regular basis. Conservation projects have been implemented by the Bird Conservation Society of Thailand (BCST), especially around Pak Thale and Laem Phak Bia Sandspit. By working with local people, BCST aims to create better awareness of the importance of these privately owned salt pans for wintering shorebirds, and advocate for their long-term protection.

*The sandspit at Laem Phak Bia is a breeding site of the patchily distributed **Malaysian Plover**.*

Kaeng Krachan National Park

Yong Ding Li & Wich'yanan Limparungpatthanakij

The waterhole at Ban Song Nook is a good place to connect with a number of ground birds.

KEY FACTS

Nearest Major Towns
Phetchaburi, Tha Yang (Phetchaburi)

Habitats
Lowland mixed forest, hill and submontane evergreen forests, cultivation (edge of park), wetlands

Key Species
Lowland, Hill Kalij Pheasant, Grey Peacock-Pheasant, Bar-backed, Green-legged and Ferruginous Partridges, Oriental Bay Owl, White-fronted Scops Owl, Spot-bellied Eagle-Owl, Orange-breasted Trogon, Rufous-collared and Blue-banded Kingfishers, Blue-bearded Bee-eater, Great, Wreathed, Plain-pouched (rare) and Tickell's Brown Hornbills, Bamboo, Streak-throated, Black-and-buff, Heart-spotted and Great Slaty Woodpeckers, Silver-breasted and Long-tailed Broadbills, Giant, Blue and Eared Pittas, Crested Jay, Large Scimitar Babbler, Spot-necked and Collared Babblers, White-tailed Flycatcher, Golden-crested Myna ***Submontane*** Mountain Hawk-Eagle, Yellow-vented Green Pigeon, Red-headed Trogon, Red-bearded Bee-eater, Rusty-naped Pitta, Ratchet-tailed Treepie, Pin-tailed Parrotfinch ***Winter*** Swinhoe's Minivet, Silver Oriole, Eastern Crowned, Sulphur-breasted and Radde's Warbler

Other Specialities
White-handed Gibbon, White-thighed Surili, Dusky Langur, Stump-tailed Macaque, Asiatic Black Bear, Leopard, Asian Golden Cat, Asiatic Wild Dog, Asiatic Jackal, Yellow-throated Marten, Asian Elephant, Banteng, Sambar

Best Time to Visit
November–July; rainy season August–October (park closed)

Kaeng Krachan National Park is one of very few large areas of wilderness left in Southeast Asia located in close proximity to a national capital – a mere 2–3 hour drive from Bangkok. This sprawling and rugged national park, the largest in Thailand, comprises 2,914km^2 of lowlands and hills of the Tanaosri (Tenasserim) Range in the provinces of Phetchaburi and Prachaup Khiri Khan, and borders Myanmar's Tanintharyi National Park in the west. With a highest point of over 1,200m asl, vast tracts of mixed evergreen and submontane evergreen forests, and the park's position on the Thai-Malay Peninsula, where the Sundaic avifauna transits into the continental Southeast Asian avifauna, there is exceptionally high bird diversity. More than 400 species have been reported, including numerous highly sought-after species like the elusive **Grey Peacock-Pheasant**, **Ferruginous Partridge**, **Giant Pitta** and globally threatened **White-fronted Scops Owl**, the latter a very localized and threatened night bird that is more regularly encountered here than anywhere else throughout its range. The submontane evergreen forests within the park are also the only place in Thailand where the highly localized and bizarre **Ratchet-tailed Treepie** can be found outside its mainly Indochinese distribution.

*The **Ferruginous Partridge** is regularly seen in the forests around Bang Krang.*

Con Foley

Tickell's Brown Hornbill has been recorded nesting around the Bang Krang area.

John & Jemi Holmes

The localized Ratchet-tailed Treepie is often observed associating with foraging flocks in the park's higher elevations.

Birdwatching Sites

Ban Song Nook

Although not technically in the park, this privately own garden and its adjoining woodland, just a few kilometres before the park's entrance, is particularly popular among visiting birdwatchers and photographers. A hide and well-maintained waterhole attract some of the more common forest species like the **Kalij Pheasant** (ssp. *crawfurdii*), **Green-legged Partridge**, both **Greater** and **Lesser Necklaced Laughingthrushes**, babblers and various crakes. The nearby Lung Sin waterhole offers a similar set of species, with the added bonus of the uncommon **Bar-backed Partridge** and occasional **Eared Pitta**.

Con Foley

Kalij Pheasants can sometimes be seen on the road as you head to Phanoen Tung.

Road to Bang Krang Camp

The forest along this road is generally low in stature and has been heavily disturbed at many points. Many of the common forest birds can be easily seen along the road, while bare trees may contain the scarce **Golden-crested Myna**. Hornbills may also be seen in flight overhead.

Stream Crossings after Bang Krang

This part of the park offers exceptionally exciting birdwatching opportunities. From Bang Krang onwards the forest increases in stature, and between the 16–18km marks it crosses three streams before gradually heading upslope. Besides the ubiquitous **Green-legged Partridge**, the **Grey Peacock-Pheasant**, and **Bar-backed** and **Ferruginous Partridges** are regularly heard, but seeing them requires patience and effort. This is also the case with various pittas, including **Blue** and **Eared Pittas**, while the **Giant Pitta** has only been recorded here very sporadically. Along the streams watch out for **Blue-banded** and **Oriental Dwarf Kingfishers**, as well as the **Chestnut-naped Forktail**, all of which are fairly regular here. This excellent stretch of forest harbours the localized **Tickell's Brown Hornbill** alongside a variety of woodpeckers, barbets and broadbills. **Silver-breasted Broadbills** are the most regularly encountered of the broadbills, but **Black-and-yellow**, **Banded** and **Dusky Broadbills** are all possible. After dark the forest comes alive with a host of night birds, including the **Oriental Bay Owl** and **White-fronted Scops Owl**, as well as the **Great Eared Nightjar**.

Road to Phanoen Tung

The section from the 24–30km marks on this road is where most birdwatchers spend some time on foot. Montane specialities like the **Rusty-naped Pitta**, **Long-tailed Broadbill**, **Red-headed Trogon** and **Red-bearded Bee-eater** can be encountered here, but the main attraction is the localized **Ratchet-tailed Treepie**, which sometimes associates with foraging flocks comprising babblers, scimitar babblers and laughingthrushes.

Phanoen Tung Viewpoint and Camp

The two viewpoints here are very popular among tourists for their magnificent and often misty vistas of the forested mountains looking west towards the Burmese border. Situated at about 1,000m asl, an excellent variety of submontane and montane species can be encountered at the viewpoints. Besides typical montane species like **Blyth's Shrike-babbler**, **Grey-chinned Minivet** and **Flavescent Bulbul**, check for fruiting trees around the viewpoints. When they are in fruit, a number of trees visible from Viewpoint One attract many barbets, bulbuls, leafbirds and – with luck – the scarce **Yellow-vented Green Pigeon**. Raptors like the **Mountain Hawk-Eagle** and groups of hornbills are often seen from the two viewpoints, and while the threatened **Plain-pouched Hornbill** has only been reported very occasionally, **Wreathed** and **Great Hornbills** can be seen. Beyond the Phanoen Thung clearing, the road continues on to the Tor Tip Waterfalls. A highlight here is the nomadic **Pin-tailed Parrotfinch**, which may be encountered when the bamboos along the trackside are in flower.

*The **White-fronted Scops Owl** is best detected by its unusual call, a low trilling sound that slowly rises in pitch, which is sometimes passed off by birdwatchers as sounds made by insects.*

Access & Accommodation

Most visitors drive into the park, but independent travellers can charter mini-buses, or 'Songthaew', from Phetchaburi to get to the park entrance, from where pick-ups can be arranged to access Bang Krang. The main jeep track that provides access to all the key sites within the park can be badly rutted at many points. It is thus best to enter the park using a vehicle with high clearance, such as a pick-up or four-wheel drive vehicle.

A number of homestays and lodges is available outside the park, with the most popular ones located around Ban Maka. For those intending to stay within the park, basic camping facilities are available at the campsites at Bang Krang and Phanoen Tung.

Conservation

Kaeng Krachan National Park is currently listed as an ASEAN heritage park. The larger Kaeng Krachan forest complex, which includes the national park and three other protected areas, was nominated for UNESCO World Heritage status in 2011. Besides the poaching of large mammals, there are a number of cultivated areas and plantations that encroach onto the boundaries of the national park.

Wang Bin

*The **Spot-necked Babbler** is regularly heard, but seeing one well can take some effort.*

Con Foley

*The **Collared Babbler** is not uncommon in Kaeng Krachan.*

Khao Dinsor

Wich'yanan Limparungpatthanakij

Ayuwat Jearwattanakanok

The relatively low hill of Khao Dinsor is the best place for viewing raptor migration in Southeast Asia during September–November.

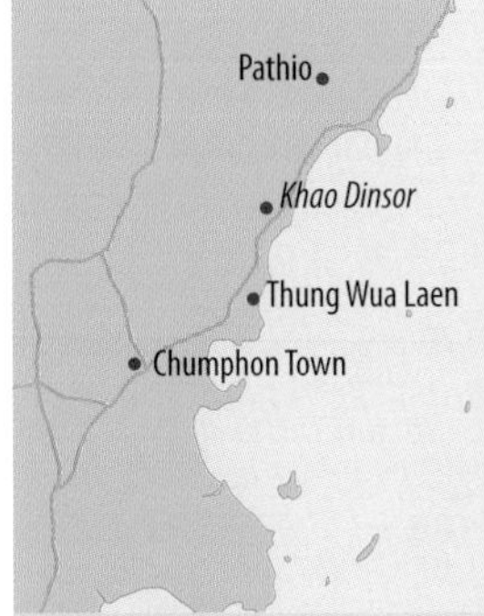

KEY FACTS

Nearest Major Towns
Pathio, Mueang (Chumphon Province)

Habitats
Secondary forest and scrub on rocky slopes

Key Species
Black-headed Ibis, Oriental Darter, Crested Honey Buzzard, Jerdon's and Black Bazas, Short-toed Snake Eagle, Greater Spotted and Booted Eagles, Shikra, Chinese, Japanese and Eurasian Sparrowhawks, Besra, Eastern Marsh and Pied Harriers, Black Kite, Rufous-winged, Grey-faced and Eastern Buzzards, Pale-capped Pigeon, White-throated, Silver-backed and Brown-backed Needletails, Common (rare) and Pacific Swifts, Blue-throated, Chestnut-headed and Blue-tailed Bee-eaters, Common Kestrel, Eurasian Hobby, Amur and Peregrine Falcons, Red-rumped Swallow, Asian House Martin

Other Specialities
Hog Badger, Lesser Mousedeer, King Cobra

Best Time to Visit
September–November

Khao Dinsor, a 350m asl hill flanked by the Isthmus of Kra to the west and Gulf of Thailand to the east, was placed firmly on the Southeast Asian birdwatching circuit in 2008 when information was circulated among birdwatchers and ornithologists that many bird of prey species could be viewed at close range as they passed over the hill en route to their wintering grounds further south in the Thai-Malay Peninsula. During September–November every year, up to half a million raptors pass over the hill under the watchful eye of a dedicated team of volunteers who have been monitoring their numbers annually for years. Some of the most numerous species on migration include the **Crested Honey Buzzard**, **Black Baza**, **Chinese Sparrowhawk**, **Japanese Sparrowhawk** and **Grey-faced Buzzard**. Non-raptor migrants include three species of needletail and several species of bee-eater.

Khao Dinsor

It takes about 30 minutes to make the uphill trek. Going up Khao Dinsor, shelters and platforms located along the 900m trek provide good stations for observing flyovers. Adaptable forest species such as the **Red Junglefowl**, **Black-headed** and **Stripe-throated Bulbuls**, **Abbott's Babbler**, **Puff-throated Babbler**, **Blue-winged Leafbird**, **Orange-bellied Flowerpecker**, **Crimson Sunbird** and **Little Spiderhunter** still occur on the scrubby hill slopes. Depending on the weather conditions on the day, more than 10,000 raptors may be

Ayuwat Jearwattanakanok

Tens of thousands of raptors fly over Khao Dinsor every day during the autumn migration period, including large flocks of ***Black Bazas****.*

Ayuwat Jearwattanakanok

*A juvenile **Crested Honey Buzzard** flies low over Khao Dinsor.*

Francis Yap

*Small numbers of **Jerdon's Baza** can be seen on migration at Khao Dinsor.*

seen during a single day's vigil, including many of the aforementioned species and some of the more uncommon ones, such as **Jerdon's Baza**, **Short-toed Snake Eagle**, **Crested Serpent Eagle** (ssp. *ricketti*), **Shikra**, **Besra**, **Eurasian Sparrowhawk**, **Rufous-winged Buzzard**, **Eastern Buzzard**, **Amur Falcon** and **Eurasian Hobby**.

Non-raptor migrants seen at Khao Dinsor include ever-increasing numbers of **Asian Openbill**, both subspecies of **White-throated Needletail** (ssp. *nudipes*, *caudacutus*), **Silver-backed Needletail** and **Asian House Martin**. Globally threatened species that regularly pass over the hill include the vulnerable **Greater Spotted Eagle** and **Pale-capped Pigeon**, which has a patchy distribution across Southeast Asia and eastern India. Thailand's first **Common Swift** was documented at this site in 2014, with three individuals photographed in the autumn of 2015.

Access & Accommodation

The entrance to Khao Dinsor is located along Route 3201, not far from the intersection with Route 3180, which leads to Thung Wua Laen Beach. A variety of accommodation options is available near this famous beach, where resident **Malaysian Plovers** are a regular sight.

Conservation

Annual monitoring of southbound passage migrants and ringing have been conducted by the Flyway Foundation, with cooperation from the Wildlife Research Division under the Department of National Parks, Wildlife and Plant Conservation. Numerous publications and articles have been published, and a summary of weekly counts is reported on a publicly accessible website. However, Khao Dinsor is not a protected area and some raptors seen passing over Khao Dinsor show distinct gunshot wounds before reaching the site. This is partially because some of them are known to prey regularly on birds kept in aviaries by local people living in the area.

Francis Yap

*An **Eastern Buzzard** on migration – the taxonomy of the buteonine buzzards of East Asia remains in flux.*

Sri Phang Nga & Ao Phang Nga National Parks

Ingkayut Sa-ar & Wichyanan Limparungpatthanakij

Yong Ding Li

Some of the most extensive areas of mangroves in peninsular Thailand can be found in Phang Nga province.

KEY FACTS

Nearest Major Town
Phuket city (Phuket Province)

Habitats
Lowland and hill rainforests, mangroves, scrub

Key Species
Sri Phang Nga Wallace's Hawk-Eagle, Red-billed Malkoha, Gould's Frogmouth, Red-bearded Bee-eater, Rufous-collared Kingfisher, Blue-banded Kingfisher, White-crowned Hornbill, Helmeted Hornbill, Green Broadbill, Malayan Banded Pitta, Crested Jay, Scaly-breasted and Grey-bellied Bulbuls, Greater Green Leafbird
Ao Phang Nga Brown-winged Kingfisher, Ruddy Kingfisher, Streak-breasted Woodpecker, Black-and-red Broadbill, Mangrove Pitta, Rufous-bellied Swallow
Winter Von Schrenck's Bittern, Pacific Swift, Orange-headed Thrush, Large Blue Flycatcher, White-throated Rock Thrush

Other Specialities
Lar Gibbon, Dusky Leaf Monkey, Malay Weasel

Best Time to Visit
November–May; rainy season June–August

The Khlong Saeng–Khao Sok Forest Complex forms one of the largest expanses of remaining broadleaved and mixed evergreen forests in peninsular Thailand. This vast forested landscape covers more than 5,450km^2 straddling the provinces of Surat Thani, Phang Nga, Chumphon and Ranong, and comprises 14 protected areas. Located on the southern end of the forest complex is Sri Phang Nga National Park, one of the most accessible of these protected areas. Besides the usual assemblages of woodpeckers, broadbills and babblers typical of peninsular Thailand, the park is probably the best place to see a number of localized species in Thailand, including **Gould's Frogmouth**. The Critically Endangered **Helmeted Hornbill**, one of Southeast Asia's most threatened species, continues to persist here. Sri Phang Nga is also one of very few places in the world where the enigmatic **Large Blue Flycatcher** can be found in the winter months. A visit to Sri Phang Nga could be complemented with a trip to the mangroves in nearby Ao Phang Nga to see a great selection of mangrove specialities, including the **Mangrove Pitta** and **Brown-winged Kingfisher**.

Birdwatching Sites

Sri Phang Nga National Park

Local people know and refer to Sri Phang Nga as 'Namtok Tam Nang', which is one of the better known waterfalls in Phang

Cheng Heng Yee

***Wallace's Hawk-Eagle** is the smallest of Asia's hawk eagles and is present in the lowland forests of Sri Phang Nga.*

*Breeding in the forests of the Eastern Himalayas, the poorly known **Large Blue Flycatcher** regularly winters in Sri Phang Nga.*

Nga province and located within the park. Easy birdwatching can be done near the park's entrance and camping ground, which offers good vistas of the surrounding forest-covered hills from an open clearing. In the morning, flocks of **Bushy-crested Hornbill** can be seen crossing the hills, as well as pairs of **Great Hornbills**, the **Chestnut-breasted Malkoha**, the **Spectacled Spiderhunter** and other common forest species.

The main road from the entrance to the park headquarters passes through excellent tracts of rainforest and bamboo groves, and offers good birdwatching. Regularly seen along this road are various Sundaic specialities, including **Wallace's Hawk-Eagles**, **Red-billed** and **Raffles's Malkohas**, **Banded** and **Maroon Woodpeckers**, **Green Broadbills** and the striking **Crested Jay**, which has now been determined by genetic studies to be an atypical shrike. After dusk, look out for **Blyth's** and **Gould's Frogmouths**, the latter best seen here in Thailand, and the **Buffy Fish Owl**. A nature trail enters the forest from the toilet blocks at the park headquarters, and passes through bamboo and a number of streams. This is perhaps the best areas in the park to look for the shy **Rufous-collared** and **Blue-banded Kingfishers**, **White-crowned Hornbill**, **Malayan Banded Pitta** and – from November to March – the **Large Blue Flycatcher**, a localized migrant from the Eastern Himalayas. With luck, the attractive **Chestnut-naped Forktail**, **Helmeted Hornbill** (far more often heard than seen) and **Lesser Fish Eagle** may all be seen.

Another site worth a visit, the Thung Chalee forest-protection station, is located about 30km to the north of the park headquarters. Although the forest cover here is mostly secondary, visitors may find a good selection of forest species, including the **Red-bearded Bee-eater**, **Black-and-yellow** and **Dusky Broadbills**, and various sunbirds and bulbuls. In winter, the skulking **Orange-headed Thrush** may be seen.

Cheng Heng Yee

*The **Green Broadbill** often perches quietly in the canopy, where its plumage provides excellent camouflage.*

Con Foley

Streak-breasted Woodpeckers are a regular sight in the mangroves of Ao Phang Nga.

Cheng Heng Yee

*Like other spiderhunters, the **Spectacled Spiderhunter** is best seen when it attends to flowering trees around the park headquarters.*

Ao Phang Nga National Park

This is undoubtedly the least disturbed and most extensive area of mangrove forests in Thailand. The key sites for birdwatching are all easily accessible without the need for a boat trip into the mangroves. Located about 8km from the town of Phang Nga, Ao Phang Nga sits on a dramatic interface between soaring limestone hills and coastal mangroves. Commonly encountered species include the **Streak-eared Bulbul**, **Orange-bellied Flowerpecker** and handsome **Rufous-bellied Swallow**, an endemic to the Thai-Malay Peninsula. A trail enters the mangroves near the park headquarters, and this is where the localized **Brown-winged Kingfisher** can be easily seen. Once in the mangroves, look out for the retiring **Ruddy Kingfisher** (ssp. *minor*), **Mangrove Pitta** and **White-chested Babbler** (often near the forest floor and mangrove stilt roots). In addition, **Von Schrenck's Bittern**, a local rarity, has been reported – check the lower mangrove stilt roots for this skulker. Other highlights include the localized **Chestnut-bellied Malkoha**, as well as **Streak-breasted** and **Rufous Woodpeckers**.

Travel & Accomodation

To access Sri Phang Nga National Park, one popular route is to first fly into Phuket, then rent a car (or taxi hire) and drive along Highway 4 in the direction of Takua Pa. Once on the highway, head to Kuraburi district, which is located about 30km to Ban Tum. At the road marker for Km 756, turn right on the junction, then continue for another 5km towards the park office. Accommodation options involve camping around the park office's campground area, or renting one of the park bungalows (reserve@dnp.go.th). Another alternative is to stay in the guest houses on the road along the Takua Pa-Kuraburi Highway. To get to Ao Phang Nga National Park, take Highway 4 from Phuket Airport towards Phang Nga province. After about 58km, turn right at the paved road, which will lead to the headquarters of the park after 2km.

Conservation

Although Sri Phang Nga National Park has been gazetted since 1988, there are still problems similar to those faced by protected areas in southern Thailand, including illegal clearance of forest for farms and vegetable gardens, and stands of palm-oil and rubber plantations. As such forms of encroachment are most prevalent in the lowland areas, lowland forest-dependent species are especially vulnerable. Poaching activities targeting birds to supply the wildlife trade are another prevalent problem.

Khao Luang National Park

Wich'yanan Limparungpatthanakij

Lowland and hill dipterocarp forests in the environs of the Krung Ching Waterfall.

KEY FACTS

Nearest Major Towns
Nopphitam, Tha Sala, Si Chon, Phrom Khiri and Mueang (Nakhon Si Thammarat Province), Don Sak (Surat Thani Province)

Habitats
Lowland and hill broadleaved evergreen forests, montane forest

Key Species
Great Argus, Bat Hawk, Wallace's Hawk-Eagle, Lesser Fish Eagle, Gould's and Blyth's Frogmouths, Cinnamon-rumped, Scarlet-rumped and Diard's Trogons, Rufous-collared Kingfisher, White-crowned, Helmeted and Black Hornbills, Turquoise-throated Barbet*, Malaysian Honeyguide, Green, Banded, Black-and-yellow and Dusky Broadbills, Malayan Banded Pitta, Maroon-breasted Philentoma, Fiery Minivet, Rail-babbler, Greater Green Leafbird, Scaly-breasted Bulbul, Rufous-bellied Swallow, Grey-headed and Black-throated Babblers, Fluffy-backed Tit-Babbler, Chestnut-capped Thrush, Fulvous-chested Jungle Flycatcher, White-crowned Forktail, Red-throated and Green-tailed Sunbirds
Passage and winter Plain-pouched Hornbill, Japanese Paradise Flycatcher, Zappey's and Green-backed Flycatchers

Other Specialities
Malayan Tapir, Clouded Leopard, Masked Palm Civet, Sunda Colugo, Sunda Slow Loris, Cream-coloured Giant Squirrel, Horse-tailed Squirrel, Three-striped Ground Squirrel

Best Time to Visit
March–October

Khao Luang National Park covers an extensive area of lowland, hill and montane evergreen forests, and is named for the highest peak in peninsular Thailand, Khao Luang (1,835m asl). In the past, some areas in Khao Luang, including Krung Ching, were controlled by political activists of the left-wing movement shortly after the national park was established in 1974. Today, the park is increasingly popular among tourists, especially birdwatchers, and is recognized as a good spot to see a variety of Sundaic lowland forest species such as the **Malayan Banded Pitta**, **White-crowned Hornbill** and the elusive **Malaysian Honeyguide**. More importantly, Khao Luang is possibly the best site to see the highly localized Thai endemic, the handsome **Turquoise-throated Barbet**, a species confined to the mountains of peninsular Thailand.

Birdwatching Sites

Park Headquarters & Krung Ching Waterfall

The forest lining the paved track leading to the Krung Ching Waterfall is excellent for a variety of lowland species. Among the many high-quality species recorded by numerous visitors along this track are the Critically Endangered **Helmeted Hornbill**, **Malaysian Honeyguide**, **Malayan Banded Pitta**, **Maroon-breasted Philentoma**, **Rail-babbler**, **White-crowned Forktail**, various trogons and a good diversity of babblers. The **Fulvous-chested Jungle Flycatcher** appears to be quite common in these forests. A number of narrow, unpaved forest trails splitting off from the main track are probably better for seeing the more elusive species, especially during holidays and weekends, when the main track is heavily utilized by tourists heading to the waterfall.

Con Foley

*The forest of Krung Ching is where the spectacular **Great Argus** has been observed at its dancing grounds in recent years.*

The abundant fruiting trees found in and around the park headquarters and camp grounds should be carefully checked for the **White-crowned Hornbill**, **Green Broadbill**, **Scaly-breasted Bulbul** and **Greater Green Leafbird**. The **Plain Flowerpecker**, a little-known species in southern Thailand, has also been photographed feeding in these trees. With luck, it may be possible to see the **Chestnut-capped Thrush**, a notably rare and retiring species in the Thai-Malay Peninsula.

The viewpoint on the slope just a few hundred metres before the park headquarters is a great spot for seeing raptors and hornbills in flight overhead. Flocks of the globally threatened **Plain-pouched Hornbill** are regularly spotted at eye level on their migrations to and from Peninsular Malaysia.

Khao Luang Summit

The highest point of Khao Luang (and effectively peninsular Thailand) is Khiriwong Peak, which rises to 1,835m asl. Accessing these higher elevations requires a local guide familiar with the area, as the trail is usually overgrown and difficult to trace. The upper slopes of the mountain are home to the endemic **Turquoise-throated Barbet** and **Malayan Laughingthrush**, the latter a Thai-Malay Peninsula endemic. Of interest to birdwatchers are a number of endemic subspecies of widely occurring montane birds, including the **Green-tailed Sunbird** (ssp. *australis*), **Blyth's Shrike-babbler** (ssp. *schauenseei*), **Chestnut-crowned Warbler** (ssp. *youngi*) and **Mountain Tailorbird** (ssp. *thais*). Getting to Khiriwong Peak usually takes

Wichyanan Limparungpatthanakij

*Flocks of **Plain-pouched Hornbills** make annual migrations from their breeding grounds further north in Myanmar to the forests of southernmost Thailand and Peninsular Malaysia*

more than a day's hike, and the popularly used camping area along the way is well within the elevation where most montane species, including the **Turquoise-throated Barbet**, may be found.

Other Birdwatching Spots

Khao Nan, Namtok Sikhit and Tai Rom Yen National Parks are all part of a contiguous complex of forests in Nakhon Si Thammarat and Surat Thani provinces. All three feature similar habitats and are accessible via paved roads, and their respective headquarters are excellent for birdwatching. The higher elevations of Khao Nong Peak (1,530m asl – accessible by hiking) of Tai Rom Yen National Park are among the very few other locations where the **Turquoise-throated Barbet** occurs.

Access & Accommodation

Car-rental services are available at Nakhon Si Thammarat Airport, a popular point of access to the park. There are multiple routes with fairly similar driving distances from the provincial town that leads north to Route 4140. From Route 4140, turn right onto Route 4186, followed by a left turn onto Route 4188, and another slight left following the main road immediately after crossing the Klai Canal. There is a campsite in the vicinity of the park headquarters. There are also a few accommodation options (small hotels) and restaurants several kilometres from the park gate.

Khiriwong Peak can be accessed after a strenuous hike, and can be facilitated by by local guides from Khiriwong Village as a starting point. The village is located on Route 4070. From the Provincial Town of Nakhon Si Thammarat, follow Route 4016, then turn left onto Route 4015. The right turn to Route 4070 will be seen upon approach to the bridge over the Tha Di Canal. It is highly recommended that visitors make arrangements with local guides and/or porters long before the trip.

Conservation

The forests of Khao Luang have received legal protection from the Thai government as a national park since 1974. It is arguably one of the most popular sites for birders visiting southern Thailand, while its summit remains hugely popular with hikers. To promote ecotourism and strengthen conservation activities, there are a number of cooperatives formed by local village communities around the national park.

Con Foley

__Cinnamon-rumped Trogons__ perch quietly in the understorey and are easily overlooked.

Wichyanan Limparungpatthanakij

The stunning __Malayan Banded Pitta__ is a regularly encountered denizen of the forests around the park headquarters

Khao Nam Khang & San Kala Khiri National Parks

Ingkayut Sa-ar & Wichyanan Limparungpatthanakij

Yong Ding Li

Hill dipterocarp forest at the Thai-Malaysian border, on the foothills of the Sankalakhiri mountains.

KEY FACTS

Nearest Major Town
Hat Yai (Songkhla Province)

Habitats
Lowland and hill rainforests, peat swamp forest

Key Species
Bat Hawk, Wallace's Hawk-Eagle, Large Green Pigeon, Jambu Fruit Dove, Mountain Imperial Pigeon, Malaysian Hawk-Cuckoo, Rufous-collared Kingfisher, Blue-banded Kingfisher, Red-bearded Bee-eater, Plain-pouched Hornbill, Great Slaty Woodpecker, Green Broadbill, Malayan Banded Pitta, Black Magpie, Spotted Fantail, Scaly-breasted and Puff-backed Bulbul, Fluffy-backed Tit-Babbler, Ferruginous Babbler, Chestnut-naped Forktail, Greater Green Leafbird, Scarlet-breasted Flowerpecker, Pin-tailed Parrotfinch

Other Specialities
Agile Gibbon, Siamang, Sunda Colugo, Dusky Leaf Monkey, Red Giant Flying Squirrel, Small-toothed Palm Civet, Sumatran Serow

Best Time to Visit
March–August; rainy season from October - January

Situated in the far south of peninsular Thailand near the border with Malaysia are two major protected areas, Khao Nam Khang and San Kala Khiri national parks. Part of the larger Hala-Bala Forest Complex, these two parks cover the northern end of the long Titiwangsa (or Sankalakhiri) Range, and contain significant areas of Sundaic lowland rainforests, a relatively limited forest assemblage in Thailand due to extensive historical clearance. These diverse and structurally complex rainforests support very high bird diversity, and are home to a number of globally threatened species such as **Wallace's Hawk-Eagle**, **Plain-pouched Hornbill**, **Great Slaty Woodpecker** and **Greater Green Leafbird**. The rainforests of Khao Nam Khang and San Kala Thiri are also among the few places where the localized **Large Green Pigeon**, **Black Magpie** and striking **Scarlet-breasted Flowerpecker** may still be found in Thailand.

Birdwatching Sites

Khao Nam Khang

This park is probably best known to local people for its historical underground tunnels excavated during the communist insurgency period that ended in the 1980s. The forests in the park offer an excellent introduction to Sundaic lowland forest species, including a number that are near the northern limits of their distribution in the Thai-Malay Peninsula. The forest edge

*The exquisite **Red-throated Barbet** spends much of its life in the forest canopy.*

*The **Scarlet-breasted Flowerpecker** is recorded here at the northern limits of its distribution in southernmost peninsular Thailand.*

by the park's office is a good place to see species like **Raffles's Malkoha** and **Blue-crowned Hanging Parrots**, while **Violet Cuckoos** are often heard in flight. The nearby Ton Lat Waterfall is a good spot to look for **Chestnut-naped Forktails**. Birdwatchers should check the many fig trees here as well. When in fruit, various bulbuls, leafbirds and pigeons congregate at the trees, including the handsome **Scaly-breasted Bulbul** and uncommon **Puff-backed Bulbul**. Other rarities reported here include the **Jambu Fruit Dove**, **Scarlet-breasted Flowerpecker** and **Thick-billed Spiderhunter**.

Heading out of the park's office along Highway 4243 towards Sadao district, a scenic viewpoint overlooking a forested valley is a good spot to observe **Bat Hawk** hunting bats at dusk. Also check the forests near the historical tunnels. Here, several species of babbler can be encountered, as well as trogons (like **Diard's Trogon**) and the locally rare **Spotted Fantail**.

San Kala Khiri National Park

Located between Songkhla and Yala, this park sits on the foothills at the northern terminus of the Sankalakhiri mountains, which extend into Malaysia (as the Titiwangsa Range). The area around the park's main office is mostly lowland rainforest below 200m in elevation, interspersed with extensive bamboo and secondary forest recovering from past logging. The forest edge by the park office is a good spot to check for a variety of birds, including flowerpeckers and bulbuls at flowering trees, the **Black-bellied Malkoha**, **Red-bearded Bee-eater**, **Sooty Barbet** and poorly known **Brown-streaked Flycatcher**. **Pin-tailed Parrotfinches** should be looked for in the bamboo stands. In the morning, small numbers of **Mountain Imperial Pigeons** may be sighted flying over the clearing, and if you are lucky, you may see the globally threatened **Large Green Pigeon**. Other Thai rarities, such as **Black Magpie**, **Scarlet-breasted Flowerpecker** and **Temminck's Sunbird**, have also been reported here.

*The **Black-capped Babbler** is one of the more terrestrial forest babblers found here.*

Con Foley

The ***Blue-banded Kingfisher*** *inhabits undisturbed forest streams in the hills of southernmost Thailand.*

Visitors should explore the 2.5km-long nature trail by the park office. This is an excellent area for birdwatching and you may encounter a variety of babblers, including **Ferruginous**, **Moustached**, **Black-capped** and **Short-tailed Babblers**, as well as woodpeckers ranging from the diminutive **Rufous Piculet** (check the bamboo groves) to the giant **Great Slaty Woodpecker**. Other interesting species include the **Malaysian Hawk-Cuckoo**, **Rufous-collared Kingfisher** and **Malayan Banded Pitta**. To see **Plain-pouched Hornbills**, it is best to visit the park during May–July, when small groups are known to migrate towards the extensive forests across the border in Malaysia's Royal Belum State Park.

Access & Accomodation

The city of Hat Yai is the main point of access to both parks. From Hat Yai airport, it is about an hour's drive (about 70km) to Na Thawi district before the turn-off to Highway 4113. From here, it is another 19km before arriving at the Khao Nam Khang park office. Basic accommodation is available, but there are no food stalls at the time of writing. San Kala Khiri can be accessed along Highway 43 which connects Hat Yai with Pattani. At Amphoe Thepha, turn into provincial highway 4085 and head towards Saba Yoi district, before driving along provincial road 4095 to the park office at Khao Daeng. From here, it is another 40km drive to the park. There are no facilities available and visitors need to camp.

Conservation

Both parks form part of the Hala-Bala Forest Complex, which constitutes the largest remaining area of rainforest in far-southern Thailand, and is thus very important for wildlife conservation. While Khao Nam Khang has existed since 1991, San Kala Khiri was only recently gazetted. Much of the surrounding landscape here has been cleared for oil palm plantations, and the remaining forest cover is being encroached upon by illegal clearing activities and wildlife poachers. Highly sought-after songbirds such as leafbirds and shamas are still being poached for illegal markets in southern Thailand.

Con Foley

Raffles's Malkohas *are usually seen clambering from branch to branch in search of insects.*

Con Foley

The ***Blue Nuthatch*** *occurs in Thailand only in the forests of Hala-Bala in the extreme south of the country.*

TIMOR-LESTE

Located at the eastern end of the Lesser Sundas, the elongated island of Timor is shared between Indonesia's West Timor and the Republic of Timor-Leste. The interior of Timor is very rugged and is dominated by a number of mountain ranges, which rise to their highest points on Gunung Mutis (2,427m asl) in the west, and Foho Tatamailau across the border in Timor-Leste, which at 2,986m asl is the island's tallest peak. Numerous short rivers drain the island's mountains, notably the Lois and Laro in the east. Timor-Leste occupies the eastern half of the island of Timor, and the enclave of Oecusse in the west.

As Southeast Asia's youngest country, Timor-Leste gained independence in 2002 after decades of Indonesian occupation following the departure of the Portuguese colonists in 1975. Since then it has received significant support from the international community, and it is now far safer than it was as a birdwatching destination. Although the country shares much of its birdlife with neighbouring West Timor in Indonesia, many birds occur at higher densities in Timor-Leste, particularly parrots, doves and pigeon, due to the lower hunting pressure and minimal trapping for the captive-bird trade. In general, forest cover in lowland areas and the mountains is also more extensive in Timor-Leste compared to West Timor. The country therefore offers an excellent alternative for the adventurous birdwatcher seeking Timor's many avian endemics, including several species that have become very rare on the Indonesian side of the island.

KEY FACTS

No. of Endemics
0

Country List
c. 264 species

Top 5 Birds
1. Yellow-crested Cockatoo
2. 'Timor' Coucal
3. Iris Lorikeet
4. Timor Imperial Pigeon
5. Timor Green Pigeon

Climate

Timor-Leste has a tropical monsoon climate, associated with high humidity, temperatures and seasonal rainfall. Rainfall is most intense between November–July during the north-east monsoon, making many roads impassable. The dry season in mountain areas (like Mundo Perdido) and the far east (as at Nino Konis Santana National Park) occurs mostly between July–October.

James Eaton

*The **Cinnamon-banded Kingfisher** occurs from Lombok eastwards to Timor and the Tanimbars.*

James Eaton

Timor is home to no fewer than six honeyeater species, including the endemic ***Streak-breasted Honeyeater****.*

Health & Safety

Vaccinations for Hepatitis A and B and typhoid are recommended for visiting birdwatchers. Malaria is prevalent and prophylaxis should be taken. Other mosquito-borne viruses like dengue fever and Japanese encephalitis are also present, although incidence is reduced if visiting during the dry season, and appropriate precautions should be taken.

The security situation in the country is uncertain and dynamic. Petty crime such as pick-pocketing is mainly restricted to the capital of Dili, and localized unrest in various districts is not uncommon. Prospective visitors should consult their local embassy for advice on the security situation before travelling, and maintain a high degree of vigilance within the country. Up-to-the-minute local advice can be found on Facebook sites.

Access, Transportation & Logistics

The main international airport in the capital Dili is serviced by regular international flights from Singapore, Darwin (Australia) and Bali (Indonesia). Most foreign visitors are able to obtain a 30-day visa on arrival at Dili's airport. There are local tour companies that can arrange flexible tours, vehicle, driver and translator for accessing birdwatching sites. Rented vehicles, primarily four-wheel drives, can be used to travel around the country, as many roads are in poor condition. While the coastal road in particular is scenic, potholes, lack of signposting and the presence of livestock such as water buffalo, cattle and pigs can increase travel times and the difficulty for drivers. English is not widely spoken in Timor-Leste. Travelling with a guide-translator conversant in either Portuguese or Tetum helps in aiding navigation in the countryside.

Birdwatching Highlights

Timor-Leste remains under-birded, and there are numerous opportunities for uncovering new sites and country records. Preliminary explorations suggest that a week around the country can yield in excess of 140 species. The principal avian attractions are frugivores such as pigeons and parrots, with the country still home to good numbers of the globally threatened Yellow-crested Cockatoo, Black Cuckoo-Dove, Timor Green Pigeon, Timor Imperial Pigeon and Iris Lorikeet, species that are very difficult to see in West Timor. Additionally, the country is home to a distinctive coucal with entirely white plumage (during the breeding season). Although presently lumped as a subspecies (ssp. *mui*) of the widespread Pheasant Coucal of Australasia, it is a probable future split that will provide the country with its first avian endemic.

Lim Kim Chuah

The ***Oriental Plover*** *is a regular boreal migrant to Timor-Leste.*

Mount Mundo Perdido

Colin Trainor

Pristine montane evergreen forest cloaks the slopes of Mount Mundo Perdido.

Mundo Perdido, or 'Lost World' in the Portuguese language, is a small (16.1km^2), isolated and densely forested mountain in central-east Timor-Leste. It is spread over two districts – Baucau and Viqueque. With a maximum height of only 1,760m asl, it is far from being the highest mountain on Timor, but appears to have some of the most extensive areas of montane evergreen forest. This starts at 900–1,000m asl, with small patches of *Eucalyptus urophylla* woodland at 1,000–1,300m asl, particularly in the north-east. This is the best site on Timor to see the **Timor Imperial Pigeon**, and most other montane birds present on Timor are common here. Among Timor's montane birds only the confusingly named Javan (formerly Timor) Bush Warbler is definitely absent. Many montane birds are represented by subspecies that are endemic to either Timor island or Timor-Leste (like the **Island Thrush**). Remarkably, an isolated population of **Pygmy Flycatchers**, far from its nearest distribution in Sumatra and Borneo, was discovered on the mountain in 2006.

*The **Timor Imperial Pigeon** is a common inhabitant on the mountain.*

KEY FACTS

Nearest Major Towns
Baucau, Ossu village

Habitats
Montane evergreen forest, *Eucalyptus* woodland, grassland

Key Species
Metallic Pigeon, Little, Timor and Black Cuckoo-Doves, Timor Imperial Pigeon, Cinnamon-banded Kingfisher, Iris Lorikeet, Flame-eared Honeyeater, Pygmy Wren-babbler, Spot-breasted Heleia, Sunda Thrush, Timor Blue and Pygmy Flycatchers

Best Time to Visit
August–October, rainy season November–July

James Eaton

*The **Pygmy Wren-babbler** is an abundant resident on Mundo Perdido, represented here by a distinctive subspecies (ssp.* timorensis*) endemic to Timor.*

James Eaton

*The uncommon **Black Cuckoo-Dove** is relatively easy to encounter in Timor-Leste.*

Birdwatching Sites

Ossu

Excellent montane evergreen forest can be accessed from the Ossu village situated at the highest point of the Baucau-Viqueque road – around 930m asl, with a trail leading up the gully between two limestone dome mountains. Most of the montane birds, such as the **Timor Imperial Pigeon**, occur down to about 1,200m asl. The **Metallic Pigeon**, **Little** (ssp. *orientalis*), **Timor** and **Black Cuckoo-Dove** as well as **Banded Fruit Dove** can be very common, calling regularly throughout the day in the dry season. Understorey skulkers including the **Pygmy Wren-babbler**, **Sunda Bush Warbler**, **Sunda Thrush**, **Chestnut-backed Thrush**, **Lesser Shortwing** and **Snowy-browed Flycatcher** (ssp. *clarae*) are common on the mountain.

Viewing birds at the forest edge can be one way to see species that prefer more disturbed habitats, such as the **Black-breasted Myzomela** and **Olive-brown Oriole**. The distinctive subspecies (ssp. *sterlingi*) of the highly variable **Island Thrush** is one of the few birds that occur outside the forest and can occasionally be seen on isolated rocks among the grasslands. These open areas are also home to **Timor Sparrows**.

Access & Accommodation

Mount Mundo Perdido is a 5–6 hour drive east from Dili along the main north-coast road to Baucau (three hours), then south along the Baucau-Viqueque road to the village of Ossu. The protected area is undeveloped, with little information available on road conditions and accommodation. Therefore, it is best to make enquires online or in Dili. There are few (if any) designated walking tracks, but with the help of local guides you can walk almost anywhere. A reasonable level of fitness is an advantage on moderately steep slopes. Accommodation options are limited. Camping is possible on the mountain, but there are no camping facilities (bring all gear and provisions), and it is essential to gain permission and assistance from local villagers. There is bungalow accommodation at Wailakurini (*c.* 10km south of Ossu) and Venilale (*c.* 20km north of Ossu) (Mazzarello Guesthouse), and many basic accommodation options exist in Baucau.

Conservation

Mount Mundo Perdido is listed as a 'protected wild area' under regulation No. 2000/19, and as an Important Bird Area by BirdLife International. There is, however, little on-ground management. The main threats to the site are intensive grazing by livestock leading to reduced forest regeneration, slump erosion and reduced water quality of springs and streams originating in the mountain. Birds, particularly large forest pigeons, are hunted but possibly not intensely enough to impact wild populations.

James Eaton

*The noisy, rattling call of the **Spot-breasted Heleia** is a characteristic background feature of Mundo Perdido.*

Nino Konis Santana National Park

Colin Trainor

Colin R. Trainor

A regionally rare unbroken reef to rainforest habitat view with the 1,000m-high Paitchau range as a backdrop, Vero River, Tutuala.

KEY FACTS

Nearest Major Town
Los Palos

Habitats
Lowland evergreen and mixed forests, dry deciduous forest, grasslands, freshwater lakes, floodplain sedgelands, beaches

Key Species
Orange-footed Scrubfowl, Great-billed Heron, Timor Cuckoo-Dove, Timor Green Pigeon, Pheasant Coucal, Southern Boobook, Large-tailed Nightjar, Yellow-crested Cockatoo, Green Figbird, Spot-breasted Heleia, Black-banded Flycatcher, Tricolored Parrotfinch

Other Specialities
Javan Rusa, Sunda Flying Fox, Estuarine Crocodile

Best Time to Visit
July–November; rainy season December–June

Named after the national resistance hero Nino Konis Santana, this sprawling national park (1,236km²) covers the entire eastern tip of Timor-Leste. Its highly diverse habitats range from 556km² of karst vegetation, including the largest area of evergreen and mixed forest left on Timor Island, to Lake Iralalaro, the largest freshwater lake in the Lesser Sundas, and extensive areas of coral reefs along the coast. Due to this variety of habitats a diverse range of birds has been recorded. Small flocks of the Critically Endangered **Yellow-crested Cockatoo** persist, and the park is probably the best place to see the endangered **Timor Green Pigeon**, although it continues to be hunted and can be difficult to locate. As noted by Alfred Russell Wallace, the **Orange-footed Scrubfowl** is oddly absent from Timor, but it is present here at Jaco Island, only 900m offshore. Other specialities include the endemic and highly distinctive subspecies (ssp. *mui*) of **Pheasant (Timor) Coucal**, the endemic subspecies (ssp. *fusca*) of **Southern Boobook**, and frugivores such as the **Timor Cuckoo-Dove**, **Banded Fruit Dove**, **Green Figbird** and **Olive-brown Oriole**.

Birdwatching Sites

Mahahara

This site supports the majority of Timor's lowland forest birds, with only the Wetar Ground Dove, Great-billed Parrot and Buff-banded Thicketbird absent. Specialities include the **Pheasant Coucal**, which occurs at the edges of gardens, **Black** and **Timor Cuckoo-Dove**, **Timor Green Pigeon**, **Yellow-crested Cockatoo**, **Jonquil Parrot**, **Southern Boobook**, **Cinnamon-banded Kingfisher**, **Black-necklaced Honeyeater**, **Timor Stubtail**, **Black-banded Flycatcher**, **Timor Blue Flycatcher** and **White-bellied Bush Chat**.

James Eaton

Extirpated throughout much of Timor, small numbers of ***Yellow-crested Cockatoos*** *continue to hold out in the national park.*

Colin R. Trainor

Lake Iralalaro is excellent for waterbirds, including a good representation of Australian species.

James Eaton

*The **Little Pied Cormorant** is regular at Lake Iralalalo.*

Seeing birds in the canopy or mid-storey of the 30-40m-tall evergreen forest is a challenge. Walking slowly along gardens and the forest edge is one way to increase the chance of seeing all but a few forest interior species.

Lake Iralalaro

More than 60 waterbird species have been recorded on the lake and associated Irasequiro River. The bird community and ease of bird observation is likely to vary greatly between low-water levels and periods of peak inundation. It is a good place to see a decent representation of species typical of Australia, such as the **Green Pygmy Goose**, **Hardhead**, **White-faced Heron**, **Pied Heron**, **Australian Pelican**, **Australasian Darter** and **Comb-crested Jacana**. Palearctic shorebirds are present in September–April, wintering **Oriental Plovers** in November–April. Rarities include the **Spotless Crake**, **Greater Painted-snipe** and **Little Curlew**.

Valu Sere Beach & Jaco Island

Jaco Island is a small limestone island covered mostly with dry forest. The **Orange-footed Scrubfowl**, **Barred Dove**, **Rose-crowned Fruit Dove**, **Fawn-breasted Whistler**, **Ashy-bellied White-eye**, **Broad-billed Flycatcher** and **Flame-breasted Sunbird** are common here. The **Black Cuckoo-Dove**, **Pink-headed Imperial Pigeon**, **Timor Green Pigeon** and – surprisingly – **Timor Sparrow** have also been seen. The **Brown Booby** and **Bridled Tern** might be seen on the short crossing to Jaso Island, but many other seabirds are likely to occur. Dry forest and woodland around Valu Sere Beach can be a good spot to search for the Black Cuckoo-Dove, Timor Green Pigeon and **Yellow-crested Cockatoo**.

James Eaton

*An endangered **Timor Green Pigeon** feeding in a fruiting fig tree near Tutuala.*

James Eaton

*The distinctive subspecies (ssp. mui) of the **Pheasant Coucal** in Timor-Leste is likely to be split in the future.*

James Eaton

***Tricolored Parrotfinches** often forage unobtrusively on the forest floor.*

Lore

This is one of the best places to see the **Timor Green Pigeon** (check fruiting figs), **Yellow-crested Cockatoo**, **Black-banded Flycatcher** and **Tricolored Parrotfinch**. The Namulutu River is a wild area – be on the lookout for large Estuarine Crocodiles in and around the river and ocean, and camps of *Pteropus* fruit bats roosting in the mangroves. Namulutu Estuary can be good for the **Great-billed Heron** and **Beach Stone-curlew**.

Access & Accommodation

Nino Konis Santana National Park is a 5–7-hour drive east of Dili, along the main north coast road. Valu Beach and Jaco Island are accessed from Tutuala village along a 7km-long track, and Jaco Island can be accessed by local canoe. Malahara is accessed via a 23km road from the main town of Los Palos, and Lore is also accessed from Los Palos via a *c.* 30km track. Conditions are often rough and four-wheel drive vehicles should be used. Cars can be hired in Dili, and several tour companies can arrange cars, drivers and guide interpreters if needed. Formal accommodation is limited in the national park, but simple bungalows are available at Valu Sere Beach, and there is budget hotel accommodation in Los Palos. With the permission of village chiefs or local villagers, it is possible to camp or stay in villages.

Conservation

Nino Konis Santana National Park was formally established in 2007 and covers 1,236km^2, including marine habitats. The park is managed by the Ministry of Agriculture and Fisheries, although limited budgets are a major constraint. Illegal logging and slash-and-burn agriculture are threats to tropical forest habitats, and hunting probably continues to limit populations of birds such as the Yellow-crested Cockatoo and Timor Green Pigeon.

James Eaton

*A **Southern Boobook** (ssp. fusca) with a Tokay Gecko near Tutuala.*

VIETNAM

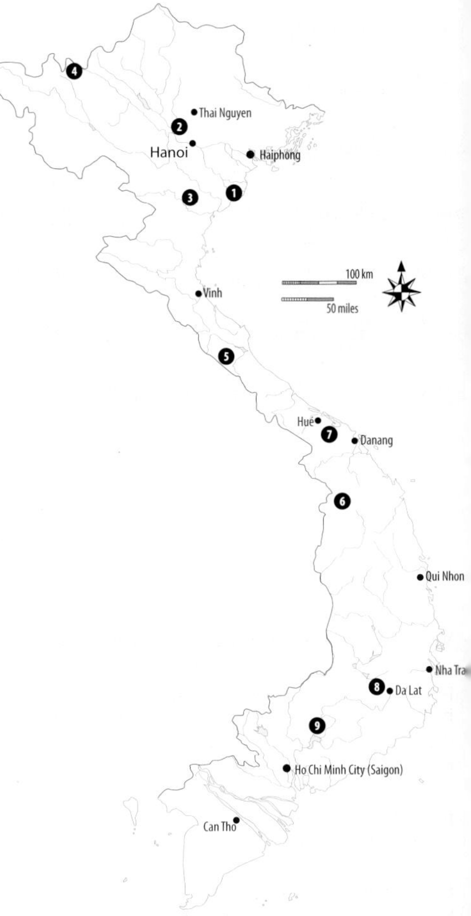

1 Xuân Thủy
2 Tam Đao
3 Cúc Phương
4 Hoàng Liên
5 Phong Nha-Kẻ Bàng
6 Ngọc Linh
7 Bạch Mã
8 Đà Lạt Plateau and Di Linh Highlands
9 Cát Tiên

Vietnam stretches over 1,500km in a compressed 'S' shape on the eastern fringe of the Indochinese peninsula. Much of the country is mountainous, with the wild and lengthy Trường Sơn (Annamite) mountains forming the main backbone on Vietnam's border with Laos. The far north of Vietnam is bisected by the Hoàng Liên Son, a subrange of the Himalayas, and rises to Vietnam's highest point on Fan Si Pan (3,140m asl). Towards the tail-end of the Trường Sơn is Tây Nguyên (Western Highlands), on which sit the Kon Tum and Đà Lạt Plateaus, before giving way to the flatlands of the vast Mekong Delta in southern Vietnam. Vietnam's montane and karst forests are rich in birds, and besides supporting 12 endemics, have been the scene of new discoveries, most recently the Limestone Leaf Warbler.

Despite being ravaged by two decades of war, Vietnam has emerged in the past three decades to become a regional economic powerhouse. However, this rapid economic growth has been a double-edged sword in relation to biodiversity conservation. Many of Vietnam's national parks have become more accessible than ever due to an ever-improving road network, but improving infrastructure and standards of living have also accelerated habitat destruction and widespread exploitation of biodiversity, particularly in the trade in bush meat and wildlife-derived products. Birdwatching in Vietnam is challenging in many areas, as due to generations of hunting the remaining wildlife is extremely wary and seldom allows close approach. However, there has been an increasing awareness (and appreciation) of nature and the need for conservation in recent years. With a good variety of country endemics and Indochinese specialities on offer, Vietnam is a necessary destination for any discerning birdwatcher visiting Southeast Asia.

Climate

Like its neighbours, Vietnam experiences a predominantly tropical monsoon climate, with distinct wet and dry seasons. The vast majority of annual rainfall falls during the south-west monsoon between May–September. The north-east monsoon between November–April is associated with the driest and coolest time of year, during which the northern mountains in the Hoàng Liên Son (like Sa Pa) may even experience sub-zero overnight temperatures.

Access, Transportation & Logistics

There are three main international airports in Vietnam – Hanoi in the north, Đà Nang in the centre, and Hồ Chí Minh (Saigon) in the south. Most foreign nationals require a tourist visa to enter Vietnam, and it is recommended that such a visa be obtained (via a local tour agency) before arrival in the country. Transport infrastructure is generally well developed, but note that international driving licenses are not accepted in Vietnam, and renting a car implies paying for the associated local driver as well. A cheaper way to commute, especially where short distances are involved (for instance within cities) is to rent a motorcycle. English is taught as part of the education system, so many young Vietnamese are able to communicate in English.

Laughingthrushes, including the endemic ***Orange-breasted Laughingthrush****, are well represented in Vietnam's avifauna.*

Health & Safety

Vaccinations for Hepatitis A and B and typhoid are recommended for visiting birdwatchers. Malaria is prevalent in remote areas and prophylaxis should be taken, although this is less of a problem in the dry season when most birdwatchers visit. Other mosquito-borne viruses like dengue fever and Japanese encephalitis are present and appropriate precautions should be taken.

Leeches and ticks occur in lowland and submontane forests. While ticks are more likely to be encountered during the dry season when most birdwatchers visit, leeches are in greater abundance during the wetter months of the year.

Birdwatching Highlights

On a comprehensive three-week tour covering sites across Vietnam, one can expect to see upwards of 300 species, including most of the country endemics and regional specialities. The majority of Vietnam's endemics can be found in the highlands in the south, and the city of Đà Lạt provides a comfortable base from which to search for them. Pride of place goes to the endangered Grey-crowned Crocias, which was rediscovered in the montane forests in nearby Chu Yang Sin in 1994. There is also a host of other charismatic endemics, including both Collared and Orange-breasted Laughingthrushes, and the Vietnamese Cutia and Vietnamese Greenfinch. The nearby Cát Tiên National Park in the lowlands north of Hồ Chí Minh City hosts a myriad of pittas, partridges and pheasants, including the range-restricted Orange-necked Partridge.

Moving north, central Vietnam's extensive mountains hold several recent discoveries, including the endemic Chestnut-eared Laughngthrush and near-endemic Black-crowned Barwing, both of which were described as recently as 1999. Seeing many of these highly localized species was once challenging, but better roads onto the Kon Tum Plateau have made things easier, and many birdwatchers nowadays visit the resort of Mằng Đen to see the endemics. In addition, the limestone forests of central Vietnam are home to highly localized species like the Sooty Babbler and Limestone Leaf Warbler. Once in northern Vietnam the birdlife becomes increasingly similar to that of southern China's subtropical evergreen forests, though regional specialities including the Short-tailed Parrotbill and White-eared Night Heron are easier to see here. A good variety of eastern Himalayan species can be seen in the Hoàng Liên Son Mountains of northern Vietnam, including sought-after species like the Slender-billed Scimitar Babbler and Red-winged Laughingthrush.

KEY FACTS

No. of Endemics
12

Country List
891 species

Top 5 Birds
1. Bar-bellied Pitta
2. White-eared Night Heron
3. Germain's Peacock-Pheasant
4. Grey-crowned Crocias
5. Chestnut-eared Laughingthrush

Xuân Thủy National Park

Le Manh Hung

Le Manh Hung

Aquaculture ponds that are fringed by natural stands of mangroves are the preferred roosting sites of Black-faced Spoonbills.

Xuân Thủy National Park is located in the coastal zone of the Red River Delta, at the mouth of the main channel of the Red River, also known as the Ba Lat River. This site comprises three islands interspaced with intertidal mudflats. Ngạn Island, the largest island, consists mainly of aquaculture ponds with remnant mangroves. Lu Island comprises open sandy and scrubby areas, as well as coastal marshes and a small area of aquaculture ponds. Xanh Island, the smallest, is a thin, sandy island that is still increasing in size as a result of deposition of sediment carried by the Red River. Xanh, together with parts of Lu Island, is submerged at high tide. Including artificial habitats created by fish farms, Xuân Thủy National Park supports a total of 14 distinct habitat types. As a result of its habitat diversity and relative lack of disturbance, Xuân Thủy National Park is an important staging and wintering area for large numbers of migratory waterbirds. A total of 325 bird species have been recorded. Ten species of globally threatened and near-threatened avian migrants regularly occur at the national park, namely the **Painted Stork**, **Black-faced Spoonbill**, **Chinese Egret**, **Spot-billed Pelican**, **Asian Dowitcher**, **Black-tailed Godwit**, **Nordmann's Greenshank**, **Spoon-billed Sandpiper**, **Saunders's Gull** and **Yellow-breasted Bunting**. In particular, Xuân Thủy supports the largest wintering population of **Black-faced Spoonbills** in Southeast Asia.

Abdelhamid Bizid

*The globally threatened **Saunders's Gull** is a sought-after waterbird on the mudflats of Xuân Thủy.*

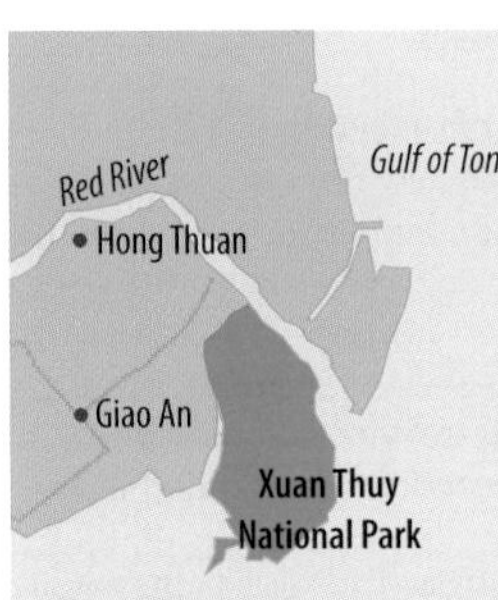

KEY FACTS

Nearest Major Towns
Quất Lâm town, Nam Định city

Habitats
Mangroves, aquaculture ponds, intertidal mudflats

Key Species
Summer Painted Stork, Spot-billed Pelican
Winter Baer's Pochard, Black-faced Spoonbill, Chinese Egret, Asian Dowitcher, Black-tailed Godwit, Far Eastern Curlew, Nordmann's Greenshank, Great Knot, Spoon-billed Sandpiper, Black-tailed, Lesser Black-backed and Saunders's Gulls
Passage Grey-faced Buzzard, Japanese Sparrowhawk, Grey-headed Lapwing, Fairy Pitta, Yellow-breasted Bunting

Best Time to Visit
October–April

Birdwatching Sites

Trails to the Ba Lat River Mouth & Garden Around Park Headquarters (or Ngạn Island)

The dense parkland and scrub around the Xuân Thủy National Park headquarters attracts a good diversity of migratory songbirds during the northern winter. Various thrushes including Japanese, Orange-headed, Grey-backed and Blue Whistling Thrushes have been regularly recorded, in addition to a variety of drongos, shrikes, warblers (like Eastern Crowned and Thick-billed Warblers) and the Siberian Rubythroat.

A 3.5km-long paved road extends from the headquarters to the Ba Lat River mouth, passing aquaculture ponds and mangroves en route. Various migratory songbirds and wintering waterbirds can be seen feeding or resting in this area, including the Garganey, Painted Stork (summer months only), Black-faced Spoonbill (October–April), Black-tailed Godwit, Red-necked Stint and Yellow-breasted Bunting.

Xanh Islet

Xanh islet can be accessed from the park headquarters by taking the boat starting from the Vop River in the direction of the Ba Lat River mouth. Along the riverbanks, look out for resident Pied Kingfishers and a host of migratory species, including the Spot-billed Pelican (non-breeding visitors in the summer months), Black-tailed Godwit, Eurasian Curlew, Spoon-billed Sandpiper (rare) and Black-capped Kingfisher.

Lu Islet (Con Lu), Giao Xuân Commune

From the park headquarters, you can access this site by taking a vehicle to the boat station at Giao Xuân commune (6km). A slow drive en route during the northern winter may yield species such as the Eurasian Wryneck, Amur Falcon, and both wintering Red-billed and White-shouldered Starlings.

From Giao Xuân, Lu Islet can be accessed by boat, which should be arranged and booked in advance through local birdwatching guides. It is important to plan the trip so that arrival on the Lu coincides with the highest tide. Otherwise, you need to be prepared to wade in water that can reach up to chest level to reach the islet. Species that can be seen on the mudflats and sandy areas include the Chinese Egret, Asian Dowitcher, Black-tailed and Bar-tailed Godwits, Eurasian Curlew, Nordmann's Greenshank, Great and Red Knots, Spoon-billed Sandpiper and Saunders's Gull. The vegetation (mostly scrub and Casuarina groves) in the middle of Lu Islet is also a resting area for many migrants. Species such as Band-bellied and Baillon's Crakes, Oriental Scops Owl, Fairy Pitta, White-spectacled Warbler, Radde's Warbler, Pallas's Grasshopper Warbler, Lanceolated Warbler and Blue-and-white Flycatcher have all been recorded during autumn passage (September–October).

Le Manh Hung

*The **Black-faced Spoonbill** is a locally common wintering waterbird at Xuân Thủy National Park.*

Yong Ding Li

***Red-billed Starlings** are regular winter visitors in northern Vietnam, although stragglers have reached other parts of Southeast Asia.*

Abdelhamid Bizid

*Lu Islet is excellent for observing migrating passerines in October, such as this **Siberian Thrush**.*

Access & Accommodation

Xuân Thủy National Park is 150km south-west of Hanoi in the coastal zone of the Red River Delta. There is no public transport that heads directly to the park from either Hanoi or Nam Định city. However, there are two main bus stations from Hanoi with services that run to the villages close to the park. Giáp Bát station has buses that run to Giao An village, Xuân Thủy district in Nam Định Province (twice daily). Mỹ Đình station has daily services that run to Giao An village in Xuân Thủy district (once daily). The buses stop about 3–5km from the respective villages, and motorcycles can be hired to ferry you to the site. Accommodation is available at the park headquarters, although advanced booking is strongly recommended if you are travelling with a large group or for weekend stays. More information about the park can be accessed from its website (vuonquocgiaxuanthuy.org.vn).

Conservation

Despite being protected as a national park and Vietnam's first Ramsar Site, Xuân Thủy's biodiversity is under threat from the continued expansion of aquaculture, and unsustainable levels of fishing and shellfish harvesting, within the park boundaries. Conservation actions are limited.

Le Manh Hung

Xuân Thủy is the only site in Vietnam where large flocks of wintering Palearctic ducks are regularly recorded.

Tam Đảo & Ba Vì National Parks

Le Manh Hung

Le Manh Hung

Montane evergreen forest at Tam Đảo National Park

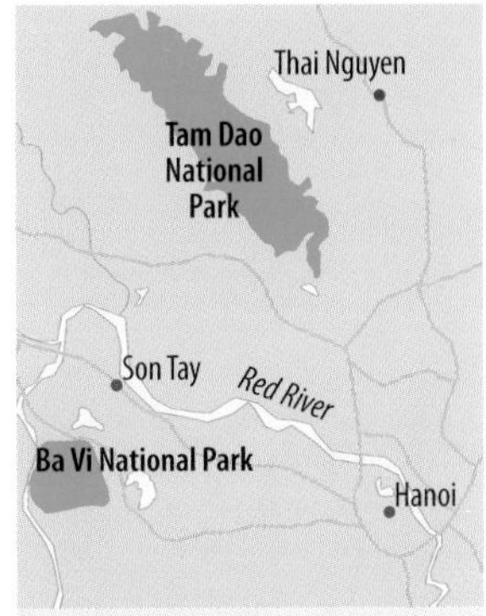

KEY FACTS

Nearest Major Towns
Tam Đảo, Sơn Tây (Hanoi Province)

Habitats
Hill, submontane and montane evergreen forests, bamboo forest, mixed coniferous forest

Key Species
Tam Đảo National Park Bar-backed Partridge, Silver Pheasant, Blue-naped Pitta, Ratchet-tailed Treepie, Indochinese Green Magpie, Chestnut Bulbul, Pale-footed, Russet and Brown Bush Warblers, Collared Babbler, Coral-billed and Streak-breasted Scimitar Babblers, Grey Laughingthrush, Short-tailed and Rufous-headed Parrotbills, Black-breasted Thrush, Purple and Green Cochoas, Fork-tailed Sunbird
Ba Vì National Park Red-vented Barbet, Rufous-cheeked Laughingthrush
Passage Black Baza, Grey-faced Buzzard, Greater Spotted Eagle, Amur Falcon
Winter Kloss's Leaf Warbler, Bianchi's Warbler, Orange-headed, Siberian, Scaly, Grey-backed and Japanese Thrushes, Chinese Blackbird, Red-flanked Bluetail, Fujian Niltava, Tristam's Bunting

Other Specialities
Vietnamese Salamander (Tam Đảo National Park)

Best Time to Visit
September–April ; rainy season May - September

Tam Đảo is a well-known hill resort sitting on top of a mountainous spur that rises from the flat lowlands of the Red River Delta some 80km from Hanoi. There are several forested peaks of more than 1,300m asl on the ridge, many of which lie within the Tam Đảo National Park (220km^2 of natural forest). The forests of Tam Đảo comprise mostly lush hill evergreen forest and lower montane evergreen forest. A total of 239 bird species has been recorded, such as the **Blue-naped Pitta**, **Pale-footed Bush Warbler**, **Rufous-headed Parrotbill** and **Purple Cochoa**.

About 50km south-west of Tam Đảo is Ba Vì National Park, which is dominated by Mount Ba Vì, an isolated massif east of the Da (Black) River. Mount Ba Vì has three peaks: the highest is Vua (1,296m asl), followed by Tản Viên (1,226m asl) and Ngọc Hoa (1,120m asl). At Ba Vì National Park, the remaining forests are mainly distributed at elevations above 600m asl, and are regularly shrouded in mist. The forest types at Ba Vì include mostly lower montane evergreen forest and some areas of lower montane mixed coniferous forest. A total of 191 bird species has been recorded here, including a number of sought-after specialities like the **Red-vented Barbet** and localized **Rufous-cheeked Laughingthrush**.

Abdelhamid Bizid

*The **Grey Laughingthrush** is a vocal and sought-after inhabitant of Tam Đảo National Park.*

*The local and patchily distributed **Rufous-cheeked Laughingthrush** is the main target for birdwatchers visiting Ba Vì National Park.*

*The gorgeous **Fork-tailed Sunbird** is regularly encountered in the mountains of Vietnam and Laos.*

Birdwatching Sites

The weather at both Tam Đảo and Ba Vì National Parks can be notoriously poor. Visiting birdwatchers are often frustrated by thick fog that covers the two sites for hours on end.

TAM ĐẢO NATIONAL PARK

Tam Đảo Town to Television Tower

From the town of Tam Đảo it is another 1,000 steps through submontane forest to reach the television tower, ascending from 900m to 1,290m asl. Along the steps several specialities can be seen, including the **Chestnut Bulbul**, **Grey Laughingthrush** and wintering **Siberian Thrush**. More rarely encountered species include **Green** and **Purple Cochoas**, as well as the **Short-tailed Parrotbill**, which may show up at the small bamboo patch near the top. During autumn and spring migration, large flocks of migrating raptors, including the **Grey-faced Buzzard** and **Amur Falcon**, can be viewed from the TV tower.

Tam Đảo Town to Tam Đảo 2 Trail (Formerly Water Tank Trail)

This 3km-long, broad trail passes through open scrubby forest along the first 2km before entering dense bamboo and broadleaved evergreen forest as a narrow trail. Key target species here include ground birds such as the **Bar-backed Partridge**, **Silver Pheasant** and elusive **Blue-naped Pitta**. Other species include parties of **Collared Babblers**, **Red-billed** and **Coral-billed Scimitar Babblers**, **Grey Laughingthrushes**, **Short-tailed** and **Rufous-headed Parrotbills**, and the occasional **Black-breasted** and **Japanese Thrushes**. Mixed species foraging flocks also contain **David's Fulvetta**, **Indochinese Yuhina**, **Fork-tailed Sunbird**, wintering **Fujian Niltava** and various leaf warblers.

National Park Headquarters

On days of dense mist around the hill station, it may be worth exploring the secondary growth and cultivation around the park headquarters. Possible highlights include a number of skulkers such as the migratory **Asian Stubtail**, **Pale-footed** and **Russet Bush Warblers**, and striking **Spot-necked Babbler**.

BA VÌ NATIONAL PARK

Summit Trail from 900m asl

Past the bus stop located at 900m asl on the access road is a steep trail that climbs 400 steps to the summit of Ba Vì through submontane forest. Birdwatching around the bus stop and the summit trail may yield the highly localized **Rufous-cheeked Laughingthrush**, a species restricted to parts of Indochina and Hainan Island in China. A host of other species, including the **Silver-breasted Broadbill**, **Eyebrowed Wren-Babbler**, **Pale Blue Flycatcher**

Le Manh Hung

*The **Green Cochoa** is an uncommon resident at Tam Đảo National Park.*

Yanm Muzika

*The **Short-tailed Parrotbill** is best searched for in dense stands of bamboo at Tam Đảo.*

and **Fork-tailed Sunbird** can be seen. In winter **Grey-backed**, **Orange-headed** and **Japanese Thrushes**, as well as the **Fujian Niltava**, occur here.

Guesthouse at 400m asl

Exploring the forests at this elevation may yield various lowland species such as the **Red-vented Barbet**. In winter a variety of warblers, including the **Buff-throated Warbler**, are also present. The area is also good for raptor watching with **Greater Spotted Eagle**, **Grey-faced Buzzard**, **Japanese Sparrowhawk** and **Amur Falcon** recorded during the autumn migration period in October.

Access & Accommodation

Tam Đảo National Park is about 80km from Hanoi city and can be reached by public transport, but there is only one direct bus per day. There are, however, plenty of buses travelling to Vĩnh Yên city (24km from Tam Đảo town), where taxis and motorbikes are available during the day. Accommodation is easy to find at Tam Đảo town (900m asl) and caters to a range of budgets.

Ba Vì National Park is about 50km from Hanoi city. There is no direct bus to the national park but buses call in at Sơn Tây town (10km from the park). Taxis and motorbikes are available for hire from Sơn Tây town. Accommodation is available both inside the national park (guesthouse at 400m asl) and at the bottom of the hill (numerous options available).

Michelle and Peter Wong

*The **Blue-naped Pitta** is a secretive inhabitant of bamboo thickets.*

Conservation

Around 150,000 people live in the buffer zone of Tam Đảo National Park, placing a great deal of strain on its ecosystems and natural resources. This problem is compounded by the weak enforcement of forest management regulations. Consequently, levels of timber and fuel-wood extraction, hunting and other illegal activities are high relative to other national parks in Vietnam. Illegal hunting of wildlife is a particular threat to many large bird and mammal species in Tam Đảo National Park, and wild game is freely sold in the restaurants at Tam Đảo town. Populations of some insect species are threatened by over-collection by locals, who sell the insects to collectors and tourists; certain insect groups are collected on a relatively large scale by organized groups. Medicinal plants and orchids are also over-exploited for sale to tourists.

Ba Vì National Park faces a slightly different set of problems. In recent years the biodiversity of the national park has undergone a dramatic decline. Logging activities, both by local people and forest enterprises, have cleared large areas of forest. Agricultural encroachment in the lowlands and shifting cultivation have also been responsible for the loss of large areas of forest. Widespread fuel-wood collection has resulted in forest degradation. Hunting pressure has been unsustainable, resulting in the loss of many mammal species. Forest fires are also a threat to remaining tracts of habitat.

Cúc Phương National Park & Vân Long Nature Reserve

Le Manh Hung

Le Manh Hung

Limestone forest at Cúc Phương National Park.

Cúc Phương, Vietnam's first national park established in 1962 by President Ho Chi Minh, is located in Ninh Bình Province, 135km from Hanoi. Cúc Phương's hilly karstic landscape is dominated by tall subtropical evergreen forest, lowland rainforest and extensive areas of secondary growth. In some parts of the park the forest is stratified into as many as five layers, including soaring emergent trees of *Parashorea stellata* and *Terminalia* species that reach up to 50m in height. However, due to the steep topography in many parts of the park, the forest canopy is often broken. To date, 313 bird species have been recorded in the park, including **Eared**, **Bar-bellied** and **Blue-rumped Pittas**, and the highly localized **Red-collared Woodpecker**.

About 30km north-east of Cúc Phương is Vân Long Nature Reserve in Gia Vien district. Vân Long is centred on a region of limestone karsts that rises abruptly from the flat coastal plain of northern Vietnam. The limestone landscape is surrounded by patches of wetlands, which comprise rivers and a shallow lake with extensive submerged vegetation. In recent years the area of this wetland has increased significantly following the construction of a dyke for irrigation purposes. While Vân Long's main attraction for naturalists is the significant population of Delacour's Langur, the reserve is also an important site for migratory waterbirds and resident specialities such as **Bonelli's Eagle** and the **Limestone Wren-Babbler** (ssp. *annamensis*).

Sam Thuong

*The **Limestone Wren-Babbler** is confined to forested limestone hills.*

KEY FACTS

Nearest Major Towns
Nho Quan town, Ninh Bình city

Habitats
Limestone and secondary forests

Key Species
Silver Pheasant, Austen's Brown Hornbill, Red-collared Woodpecker, Pied Falconet, Eared, Blue-rumped and Bar-bellied Pittas, White-winged Magpie, Indochinese Green Magpie, Ratchet-tailed Treepie, Limestone Leaf Warbler, Rufous-throated and Black-browed Fulvettas, Limestone Wren-Babbler, Hainan Blue and White-tailed Flycatchers
Winter White's, Black-breasted and Japanese Thrushes, Chinese Blackbird, Fujian Niltava, Rufous-tailed Robin

Other Specialities
Delacour's Langur, Owston's Civet

Best Time to Visit
October–May

Birdwatching Sites

Vân Long Nature Reserve

Located about two-thirds of the way from Hanoi city to Cúc Phương National Park, Vân Long Nature Reserve is a good stop for breaking the journey and to observe several interesting species, particularly during the northern winter. Migrants recorded in the wetlands and ponds in the reserve include the **Garganey** and **Eurasian Wryneck**, alongside a variety of raptors such as the **Black Baza**, **Grey-faced Buzzard** and **Japanese Sparrowhawk**. A boat ride is needed to access the steep limestone cliffs in the reserve, from where troops of Delacour's Langur can be seen.

CÚC PHƯƠNG NATIONAL PARK

Trails around Park Headquarters

A well-maintained network of trails around Cúc Phương's headquarters provides easy access into secondary and tall evergreen forests. One of the most popular is the botanic garden trail, where possibilities include soaring **Mountain Hawk-Eagles** and the localized **Pied Falconet** and **White-winged Magpie** – the latter two species are best seen on tall, open perches. During the northern winter a good variety of migratory songbirds is present, including **Swinhoe's Minivet**, **Japanese Thrush** and **Chinese Blackbird**.

Road from Headquarters to Bong Substation

This well-paved access road runs for 19km and provides good birdwatching opportunities in the national park. During the northern winter a variety of thrushes can be found feeding along the road, such as **Black-breasted**, **Grey-backed** and **Japanese Thrushes**. Shy residents including the **Silver Pheasant**, and **Blue-rumped** and **Bar-bellied Pittas**, have been recorded feeding along the road. At the Historical Man Cave (100m from the roadside), both the **Limestone Leaf Warbler** and the **Limestone Wren-Babbler** of the distinctive northern Vietnamese subspecies (ssp. *annamensis*) can sometimes be seen just in front of the cave.

Trails around Bong Substation

Located in the middle of the forest, there are several good trails for birdwatching, namely the Loop Trail, the Big Tree Trail and the open areas of forest around the guesthouse.

The Loop Trail covers a distance of about 7km and is home to specialities such as the unusual ground-frequenting **Red-collared Woodpecker**, **Long-tailed** and **Silver-breasted Broadbills**, **Blue-rumped**, **Bar-bellied** and **Eared Pittas**, and **Rufous-throated** (ssp. *stevensi*) and **Black-browed Fulvettas**. The **Grey Peacock-Pheasant** is sometimes heard but difficult to see. Some of the more readily encountered birds include the **Red-headed Trogon**, and **White-tailed** and **Hainan Blue Flycatchers**.

The Big Tree Trail is a good place to see a number of restricted-range corvids, notably the striking **Indochinese Green Magpie** and **Ratchet-tailed Treepie**. Other species of interest

Wang Bin

*The **Blue-rumped Pitta** is regularly encountered on Cúc Phương's forest trails.*

Sam Thuong

*The **Indochinese Green Magpie** is generally shy and easily overlooked.*

include parties of the skulking **Rufous-throated Fulvetta**, and wintering **Rufous-tailed Robin** and **Asian Stubtail**. The small botanic gardens and forested clearing around the bungalows are a good place to see the **Pied Falconet** and **White-winged Magpie**, and **Green** and **Purple Cochoas** have also been recorded. **Yellow-rumped Flycatchers** (on passage) and wintering **Japanese Robins** and **Tristram's Buntings** (rare) may also occur.

Access & Accommodation

Cúc Phương is a three-hour drive from Hanoi city (135km), or an hour and a half from Ninh Bình city (about 50km). There is no direct public transport to the national park from either city, but public buses run to Nho Quan town (about 12km from the park), from where taxis and motorbikes can be hired. Accommodation is available at both Cúc Phương park headquarters and the Bong substation, but advance booking for the latter is recommended due to the limited number of rooms.

Conservation

The buffer zone of the national park is home to around 50,000 people, many of whom depend on the natural resources of the park. The most widely exploited forest products are timber and firewood. The collection of snails, mushrooms, tubers and bamboo shoots for food is common, as is the collection of banana stems for animal fodder. Hunting, both for subsistence and commercial purposes, takes place at unsustainable levels, and threatens to eradicate numerous species from the park. The large number of tourists who visit Cúc Phương each year pose problems. Waste disposal, plant collection and excessive noise created by large tour groups are all problems that the national park staff have yet to effectively control. More significantly, the management agenda of the park is heavily focused on tourism development, at the expense of biodiversity conservation. This has resulted in the development of tourism infrastructure with negative environmental impacts.

Abdelhamid Bizid

*The **Fujian Niltava** is a winter visitor to northern Vietnam's forests.*

Le Manh Hung

__Delacour's Langur__ is Vân Long Nature Reserve's main attraction for birdwatchers and naturalists alike.

Hoàng Liên National Park & Sa Pa

Le Manh Hung

Le Manh Hung

Montane forest at 2,300m asl in Hoàng Liên National Park.

KEY FACTS

Nearest Major Town
Sa Pa (Lao Cai Province)

Habitats
Lower and upper montane evergreen forests, subalpine forest, dwarf bamboo stands, grassland, scrub and cultivation

Key Species
Hill Partridge, Yellow-billed Blue Magpie, Crested and Collared Finchbills, Black-streaked, Streak-breasted and Slender-billed Scimitar Babblers, Golden-breasted Fulvetta, Pale-throated* and White-throated* Wren-Babblers, White-browed, White-throated, Scaly, Black-faced, Silver-eared, Red-winged and Red-tailed Laughingthrushes, Scarlet-faced Liocichla, Spectacled Barwing, Spot-breasted, Vinous-throated, Ashy-throated and Golden Parrotbills, White-collared Yuhina, Beautiful and White-tailed Nuthatches, Gould's Shortwing, White-browed Bush Robin, Little Forktail, Yellow-bellied Flowerpecker, Black-headed Greenfinch

Other Specialities
Black-crested Gibbon; over 80 species of amphibians including the Giant Fire-bellied Toad, Sterling's Toothed Frog and Tam Dao Salamander

Best Time to Visit
February–June; rainy season July–September

Situated near the northern tip of Vietnam not far from China's Yunnan province, Hoàng Liên National Park and the township of Sa Pa are located in the Hoàng Liên Sơn Mountains, an eastern outlier of the Himalayas where Vietnam's loftiest peak, Mount Fan Si Pan or Phan Xi Păng (3,143m asl) is situated. The lowest point on the mountains is 380m asl, but much of the national park lies above 1,000m asl. The park supports a wide variety of habitat types – elevations below 1,800m contain extensive lower montane evergreen forest; elevations at 1,800–2,500m support upper montane evergreen forest; elevations at 2,500–2,800m support sub-alpine forest, and the vegetation above 2,800m is dominated by stands of dwarf bamboo with scattered, stunted trees. Below 1,000m natural forest has been almost entirely cleared and replaced with anthropogenic habitats such as grassland, scrub and cultivation. Secondary forest habitats are also found at higher elevations. In recognition of its importance to a number of restricted-range and globally threatened bird species, Hoàng Liên National Park qualifies as an Important Bird Area. For birdwatchers, this site offers a chance to see a good representation of eastern Himalayan specialities not far from Hanoi, including more than eight laughingthrushes.

Benjamin Schweinhart

The skulking ***Slender-billed Scimitar Babbler*** *is a perennial favourite among birdwatchers.*

Le Manh Hung

*The **Red-tailed Laughingthrush** is fairly common in Hoàng Liên National Park.*

Con Foley

*The **Crested Finchbill** is best seen in scrubby areas near the forest edge.*

Birdwatching Sites

Ham Rong Mountain

Le Manh Hung

*The **Black-faced Laughingthrush** is only found at the highest elevations of the park.*

Located close to Sa Pa town at a height of 1,800m asl, this mountain is covered mainly by agricultural gardens and plantations alongside remnant patches of dwarf bamboo and secondary scrub. Along the trails to the top, some of the area's specialities and species typical of disturbed areas, such as the **Crested Finchbill**, **White-browed Laughingthrush**, **Vinous-throated Parrotbill**, **Ashy-throated Parrotbill** and **Black-headed Greenfinch**, can be encountered.

Golden Stream (Via Bac Waterfall)

Located at the top of Ô Quy Hồ Pass, this site is about 15km from Sa Pa town in the direction of Lai Châu province. A trail that heads down to the waterfall passes through secondary growth, dwarf bamboo and remnant lower montane evergreen forest. Species of interest here include Himalayan specialities such as **Black-streaked** and **Streak-breasted Scimitar Babblers**, **Spectacled Barwing**, **White-collared Yuhina** and **Yellow-bellied Flowerpecker**. The Bac Waterfall is home to water redstarts and the **Little Forktail**.

Trail to Fan Si Pan Summit

This fairly steep (and wet) trail starts from the top of Ô Quy Hồ pass (1,900m asl) and climbs to the summit of Fan Si Pan (3,143m asl), Vietnam's tallest point. To do justice to the habitat, the hike normally takes 2–3 birdwatching days. There are two camping sites, located at 2,200m and 2,800m asl. The habitats are diverse, with scrub at the beginning of the trail, transitioning into secondary forest, grassland, dwarf bamboo, lower montane evergreen forest, upper montane evergreen forest and finally sub-alpine forest dominated by rhododendrons.

It does not take long to come across the large mixed foraging flocks that characterize these montane forests, which should be checked for yuhinas, fulvettas and the fiery **Mrs Gould's Sunbird**. Unsurprisingly, many of the best birds in the park should be searched for along this trek, including **Pale-throated** and **White-throated Wren-Babblers**, **Red-winged Laughingthrush**, **Scarlet-faced Liocichla**, **Himalayan Cutia**, **Golden Parrotbill**, **Beautiful Nuthatch**, **Gould's Shortwing** and **White-browed Bush Robin**, to name a few. In particular, pay attention to the dense bamboo groves around the second (summit) camping site. With a bit of effort, more than three species of laughingthrush, parrotbills and the bizarre **Slender-billed Scimitar Babbler** (the babbler equivalent of a curlew) may be found. The **Bar-winged Wren-Babbler** was recently recorded just below the Fan Si Pan summit.

Access & Accommodation

Sa Pa town is a long 380km drive from Hanoi and is best reached by overnight train to Lào Cai city, followed by a drive to the town proper in rented cars or taxis (40km). As a former French hill resort and popular tourist destination, Sa Pa provides a range of accommodation for all budgets. Recently, a cable car began operating from Sa Pa to the summit of Fan Si Pan, with a travel time of 25 minutes each way.

Firoz Hussein

Seeing the secretive ***Gould's Shortwing*** *is a highlight for any birdwatcher visiting the park.*

Conservation

Hoàng Liên National Park is recognized as an ASEAN Heritage Park for its biological diversity. Over-exploitation of natural resources, together with deforestation for cultivation of cardamom and by accidental fires, is rapidly destroying the natural vegetation of the park. Natural forest presently covers only around half the area of the park, and continued human disturbance threatens the remaining areas. Unsustainable tourism development represents a potential future threat to biodiversity at Hoàng Liên. Sa Pa town is already a major tourist destination, and there are plans to develop it into a resort city. These activities would facilitate continued exploitation of natural resources, increased risk of accidental forest fires from trekkers and increased demand for forest products such as orchids.

Sam Thuong

Vietnam's first record of the ***Bar-winged Wren-Babbler*** *was from the subalpine forests of Fan Si Pan.*

Phong Nha-Kẻ Bàng Karst Area

Le Manh Hung

Le Manh Hung

Limestone forest at Phong Nha-Kẻ Bàng National Park.

Phong Nha-Kẻ Bàng National Park (858km²) is located on the western end of Quảng Bình province in the central Trường Sơn Range, close to the international border with Laos. The park is situated in one of the largest areas of contiguous limestone karst landscapes in Indochina, which also include the Hin Namno National Biodiversity Conservation Area across the border in Laos. Phong Nha-Kẻ Bàng National Park is home to several globally threatened primates, including the charismatic Red-shanked Douc, one of Southeast Asia's most attractive primates. It is also a major tourist attraction well known for its limestone karst formations and extensive cave systems. As such, the park has been designated a UNESCO World Heritage Site. However, due to its location on the border, access to the park proper can be difficult. Fortunately for birdwatchers, two of Phong Nha's main avian attractions, the recently described **Limestone Leaf Warbler** and **Sooty Babbler**, both limestone specialists, can be seen near the boundaries of the park. The limestone-dependent Bare-faced Bulbul, a Laotian endemic for the time being, may also occur in more remote parts of the park.

KEY FACTS

Nearest Major Town
Đồng Hới city (Quang Binh Province)

Habitats
Limestone forest, lowland and hill evergreen forests

Key Species
Green-legged Partridge, Crested Argus, Austen's Brown Hornbill, Red-collared Woodpecker, Limestone Leaf Warbler, Sooty Babbler, Short-tailed Scimitar Babbler

Other Specialities
Hatinh Langur, Red-shanked Douc, Owston's Civet, Chinese Serow, Annamite Striped Rabbit

Best Time to Visit
December–May; rainy season September–November

Birdwatching Sites

Trails Above Phong Nha Cave

Getting to the cave requires a one-hour boat ride from the village of Sơn Trạch. Above the cave's gate, a trail comprising several hundred steps leading up the hill is the first place to try for both the **Limestone Leaf Warbler** and the **Sooty Babbler**. Birdwatchers should aim to arrive as early as possible to avoid the hordes of noisy tourists later in the day.

West Ho Chi Minh Highway Road

About 15km from the national park headquarters to the west, birdwatchers can consider stopping along the road when passing good areas of forest habitat on the highway en route to Thiên Đường (Paradise) cave. The **Sooty Babbler** can be seen foraging on the karsts, and groups of the endangered Hatinh Langur and Red-shanked Douc may sometimes be seen in large trees.

Hang Chin Co Area (Nine Women's Cave)

About 25km from the park headquarters, the trail in this area enters some excellent stretches of broadleaved evergreen forest. This is a good place to try for a number of rare and localized species, including the distinctive and endemic subspecies (ssp. *tonkinensis*) of the **Green-legged Partridge**, **Austen's Brown Hornbill** and **Red-collared Woodpecker**. The **Limestone Leaf Warbler** and **Short-tailed Scimitar Babbler** have also been recorded here. For birdwatchers with an interest in mammals, the Red-shanked Douc can also be searched for here.

James Eaton

*The **Sooty Babbler** is regularly seen within the park.*

Access & Accommodation

Phong Nha-Kẻ Bàng National Park is about 50km from Đồng Hới city and a five-hour drive by road north of Bạch Mã National Park. There is no public transport going directly to the park, but as it is located along the Ho Chi Minh Highway, a number of buses pass through daily. Accommodation is available in Đồng Hới city (including at resorts and luxury hotels), and outside the park (more modest hotels and guesthouses).

Conservation

The biggest threat to biodiversity in Phong Nha-Kẻ Bàng National Park is illegal poaching. Illegal trapping takes place throughout the park, although it is most frequent within a day's walk from human habitation. Most hunting is commercially orientated, with a well-established wild animal trade in the area. There has been substantial commercial hunting of primates, which has resulted in major population declines. In addition, widespread snaring is of particular concern because of its effects on populations of terrestrial mammals and birds. Tourist developments are another threat to biodiversity at the site. Presently, the Quảng Bình Tourism Company is attempting to both promote Phong Nha-Kẻ Bàng's natural assets and manage the burgeoning numbers of visitors.

Le Manh Hung

*The **Limestone Leaf Warbler** is a newly described species that is fairly common around the park in limestone forests.*

Le Manh Hung

*The endangered **Hatinh Langur** is one of Vietnam's many endemic primates.*

Ngọc Linh Highlands & Lò Xo Pass

Le Manh Hung

Le Manh Hung

Forested slopes of the Ngọc Linh highlands as seen from Lò Xo Pass.

The southern Trường Sơn Range rises to its maximum elevation on Mount Ngọc Linh (2,598m asl). Some of the montane forests are protected within the Ngọc Linh (Kon Tum) and Ngọc Linh (Quang Nam) Nature Reserves, with the Lò Xo Pass nested between these two reserves. Lò Xo Pass (Đèo Lò Xo on AH16) is located along the newly built Hồ Chí Minh Highway, which traces the route of the legendary Hồ Chí Minh Trail along Vietnam's border with Laos, and can be reached via a three-hour drive from Đà Nẵng city. The main habitats around the pass are lower montane evergreen forest, secondary growth and scrub. The surrounding peaks, including the tantalizingly close but currently off-limits Mount Ngọc Linh, are mostly covered in primary evergreen forest, although the effects of wartime defoliation can clearly be seen at certain points along the highway. However, the **Black-crowned Barwing**, described in 1999 and known from only three sites in Vietnam and Laos, seems to favour the accessible scrub and secondary growth bordering the highway.

Birdwatching Sites

Road Along Lò Xo Pass

This road is readily accessible, but birdwatchers should be wary of vehicles speeding along the highway. From the top of the pass, walk down in a northerly direction. The **Russet Bush Warbler** (ssp. *mandelli*), **Spot-throated Babbler** and **Black-crowned Barwing** can all be encountered in the roadside scrub.

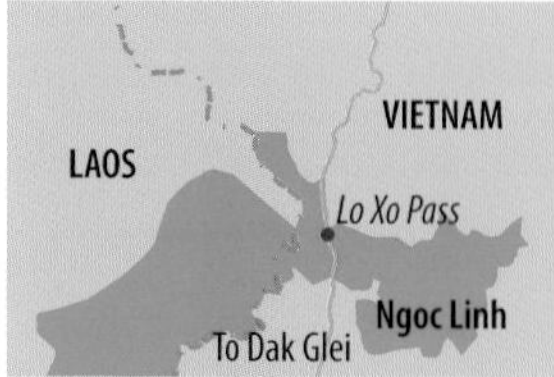

KEY FACTS

Nearest Major Towns
Khâm Đức (Quang Nam Province), Đắk Glei (Kon Tum Province)

Habitats
Montane evergreen forest, secondary forest and scrub

Key Species
Austen's Brown Hornbill, Indochinese and Red-vented Barbets, Blue and Rusty-naped Pittas, Indochinese Green Magpie, White-winged Magpie, Ratchet-tailed Treepie, Russet Bush Warbler, Red-billed and Short-tailed Scimitar Babblers, Spot-throated Babbler, White-cheeked, Black-hooded, Red-tailed, Chestnut-eared* and Golden-winged* Laughingthrushes, Black-crowned Barwing, Grey-crowned Crocias*, Yellow-billed Nuthatch, Pygmy Flycatcher
Winter Siberian Thrush, Narcissus and Blue-and-white Flycatchers

Other Specialities
Annamite Muntjac

Best Time to Visit
December–May; rainy season September–November

Thang Nguyen

*The **Black-crowned Barwing** is fairly common along the roadside at Lò Xo Pass.*

Con Foley

*The recently described **Chestnut-eared Laughingthrush** is best seen in the forests around Măng Đen.*

Road to the Laos Border

A 19km paved road runs to the border with Laos and features various checkpoints, requiring permission to enter. The first 3km consists of bare hillside with cultivation, but the next several kilometres feature excellent evergreen forest where various regional specialities can be seen, such as **Austen's Brown Hornbill**, **White-winged Magpie**, **Short-tailed Scimitar Babbler**, **Spot-throated Babbler**, **White-cheeked Laughingthrush**, **Black-hooded Laughingthrush**, **Red-tailed Laughingthrush** and **Yellow-billed Nuthatch**.

Ngọc Linh (Kon Tum) Nature Reserve

The nature reserve's headquarters is about 12km to the east of the highway, and is one of few places in its range where the endemic **Golden-winged Laughingthrush** can be seen. While it currently appears nearly impossible for the visiting birdwatcher to access the montane evergreen forests here, this may change in the future.

Mẵng Đen

About 60km south-east of Lò Xo is the mountain resort town of Mẵng Đen in Kon Tum province. There are still extensive areas of montane forest around the town and accessible by road. The main attraction is the localized **Grey-crowned Crocias**, which was only recently found here. Other species of interest include the **Black-hooded Laughinghthrush**, which is locally common, and a host of other birds shared with the forests around the Ngọc Linh area (like **Austen's Brown Hornbill**, **White-winged Magpie** and **Ratchet-tailed Treepie**). Further to the north (c. 17–19km) is the village of Mẵng Canh, from where birdwathers can access excellent areas of montane forest to find the recently described **Chestnut-eared Laughingthrush**.

Abdelhamid Bizid

*The **Indochinese Barbet** is commonly heard in the forests along Lò Xo Pass.*

Thang Nguyen

*The **Golden-winged Laughingthrush** was described as recently as 1999 by Jonathan Eames and colleagues.*

Access & Accommodation

There is no direct public transport to the area, but various buses ply the north–south route daily. It is easy to take any bus and ask to be dropped off at the top of Lò Xo Pass. The closest town is Đắk Glei in Kon Tum province (12km), where food and accommodation are available. The other town is Khâm Dức (40km in the direction of Đà Nẵng), where the standard of accommodation is higher than in Đắk Glei. Mẵng Đen can be easily accessed from Đà Nẵng and Pleiku by public bus or car.

Conservation

The main threats to biodiversity in the Ngọc Linh and Kon Tum highlands are hunting and over-exploitation of non-timber forest products. In addition, illegal clearance of forest for agriculture is rampant and hard to control.

Bạch Mã National Park

Le Manh Hung

Low Bing Wen

Montane evergreen forest in Bạch Mã National Park.

KEY FACTS

Nearest Major Towns
Lăng Cô town, Huế city (Thừa Thiên-Huế)

Habitats
Lowland and montane evergreen forests

Key Species
Rufous-throated and Green-legged (Annam) Partridges, Silver Pheasant, Crested Argus (rare), Coral-billed Ground Cuckoo, Red-headed Trogon, Blyth's Kingfisher, Austen's Brown Hornbill, Red-vented Barbet, Red-collared Woodpecker, Blue-rumped and Bar-bellied Pittas, White-winged Magpie, Indochinese Green Magpie, Ratchet-tailed Treepie, Sultan Tit, Spot-necked Babbler, Grey-faced Tit-Babbler, Short-tailed Scimitar Babbler, Black-browed Fulvetta, Masked Laughingthrush, Fork-tailed Sunbird
Winter Orange-headed, Siberian and Japanese Thrushes

Other Specialities
Red-shanked Douc

Best Time to Visit
December–May; rainy season September–November

Established in 1986, Bạch Mã National Park protects a diversity of habitats ranging from lowland evergreen forest to lush montane evergreen forest above 900m asl. Located near the cities of Huế, Đà Nẵng and Hội An, the park lies on a high mountain ridge that runs west–east from the Laotian border to the South China Sea at Hải Vân Pass on National Route 1A. This ridge interrupts the coastal plains of Vietnam and forms a biogeographical boundary between the flora and fauna of northern and southern Vietnam. The park has several peaks above 1,000m asl, the highest of which is Mount Bạch Mã at 1,448m asl. The lower slopes and hills are less steep, and are bordered by a narrow plain. To date, 249 bird species have been recorded in the park, and the number of species recorded in the wider Bạch Mã-Hải Vân area exceeds 330.

Birdwatching Sites

Trails around the Summit

An access road leads to the summit of Bạch Mã. There are several trails en route, and around the guesthouse where some sought-after species, such as **Austen's Brown Hornbill**, **Red-collared Woodpecker** and **Short-tailed Scimitar Babbler**, can be seen. The **Crested Argus** (ssp. *ocellata*) is sometimes distantly heard, but seeing it in the park is nigh impossible. Other more readily encountered species include the **Rufous-throated Partridge** (ssp. *guttata*), **Red-vented Barbet**, **White-winged Magpie**, **Indochinese Green Magpie**, **Ratchet-tailed Treepie**, **Grey-faced Tit-Babbler**, **Black-browed Fulvetta** (ssp. *grotei*) and **Fork-tailed Sunbird**. Look out also for the distinctive subspecies of the **Sultan Tit** (ssp. *gayeti*), especially in mixed feeding flocks.

Yann Muzika

*The park is one of the best places to see the unusual **Short-tailed Scimitar Babbler**.*

Crested Argus (Trĩ Sao) Trail

This trail, spanning about 5km from the top to the bottom, is a good site to try for ground dwellers including the endemic subspecies (ssp. *merlini*) of the **Green-legged Partridge**, **Blue-rumped Pitta** and **Bar-bellied Pitta**, and can be reached via the access road to the summit. The **Coral-billed Ground Cuckoo** and **Silver Pheasant** have also been encountered along the trail. While named after the **Crested Argus**, this legendary pheasant has not been seen here for many years – one may at best hear its distant calls from surrounding mountains. The Trĩ Sao Trail is also a back-up site for key species such as **Austen's Brown Hornbill**, **White-winged Magpie** and **Short-tailed Scimitar Babbler** if they were not seen around the summit.

Ah Kei

Blyth's Kingfisher occurs in the streams of Ḅach Mã.

Access Road

The well-paved access road that runs from the park headquarters to the summit provides a good vantage point to observe raptors and canopy birds flying over the forest, such as the **Black Eagle**, **Austen's Brown Hornbill** and **White-winged Magpie**. Mixed foraging flocks along the road can contain the **Ratchet-tailed Treepie**, **Sultan Tit**, **Grey-faced Tit-Babbler**, **Black-browed Fulvetta** and **Fork-tailed Sunbird**.

National Park Headquarters

This area is dominated by secondary growth and plantations, accessible via several trails. A different suite of birds occurs here, including a good variety of skulkers such as the **Masked Laughingthrush** and **Spot-necked Babbler**. Remnant degraded secondary forest in the area offers a final opportunity to see the **Green-legged Partridge** if it has not been seen by then.

Access & Accommodation

This park is about 70km from Huế City and 40km from Lăng Cô town. There are no direct buses to it, but plenty of regular buses heading south (from Huế city) and north (from Lăng Cô town) stop at Highway No. 1A (about 6km from the park headquarters), where taxis and motorbikes are available. Only cars are allowed to drive to the mountain top (17km of paved road). There are villas and guesthouses at the top of the mountain that should be booked in advance if you intend to travel with a large group, or during weekends and major holidays.

Sam Thuong

The distinctive subspecies (ssp. gayeti) of the ***Sultan Tit*** *is fairly common at the upper elevations of the park.*

Conservation

Large areas of forest in the park were destroyed by defoliants, bombs and heavy machinery during the Second Indochina War, and subsequent commercial logging by state forest enterprises over a decade caused the removal of most commercially valuable trees. Since the cessation of official logging operations, large-scale illegal exploitation of timber and non-timber forest products has continued. Forest fires, particularly in the regenerating forest, have contributed to further degradation and inhibited regeneration. Hunting pressure is high in the buffer zone.

Sam Thuong

An important population of the endangered ***Red-shanked Douc*** *occurs in Bach Mã and the nearby Sơn Trà Nature Reserve.*

Di Linh Highlands & Đà Lat Plateau

Le Manh Hung

Yong Ding Li

The montane forests along the Đà Lạt Plateau are home to the majority of Vietnam's endemic birds.

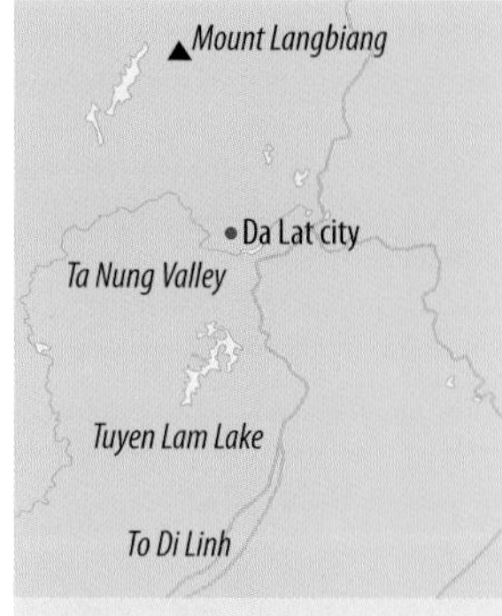

KEY FACTS

Nearest Major Towns
Di Linh town, Đà Lạt city (Lâm Đồng)

Habitats
Montane evergreen forest, secondary forest, coniferous forest, cultivation

Key Species
Bar-backed Partridge, Silver Pheasant, Pin-tailed and Wedge-tailed Green Pigeons, Red-vented and Indochinese Barbets, Bay Woodpecker, Blue and Rusty-naped Pittas, Dalat Shrike-babbler*, Indochinese Green Magpie, Grey-bellied Tesia, Dalat Bush Warbler*, Red-billed Scimitar Babbler, Black-crowned Fulvetta*, Black-hooded, White-cheeked, Orange-breasted* and Collared* Laughingthrushes, Vietnamese Cutia*, Grey-crowned Crocias*, Rufous-backed and Black-headed Sibias, Black-headed Parrotbill*, Yellow-billed Nuthatch, Green Cochoa, Vietnamese Greenfinch*, Red Crossbill
Winter Black Baza, Grey-faced Buzzard, White-throated Rock Thrush, Blue-and-white and Mugimaki Flycatchers

Other Specialities
Buff-cheeked Gibbon

Best Time to Visit
October–May; rainy season lasts June–September

Rising to its highest point on Chư Yang Sin (2,442m asl), the Đà Lạt and adjacent plateaus form an extensive area of highlands that contrast sharply with the lowlands that otherwise characterize much of southern Vietnam. There are extensive areas of coniferous forest (dominated by *Pinus kesiya*), montane evergreen forest and secondary forest left on the plateau despite high deforestation rates. Six bird species, namely **Collared** and **Orange-breasted Laughingthrushes**, **Black-crowned Fulvetta**, **Vietnamese Greenfinch** and the recently split **Dalat Shrike-babbler** and **Dalat Bush Warbler**, are only found within this upland region. The Đà Lạt Plateau also supports distinctive and endemic subspecies of the **Blue-winged Minla**, **Rufous-backed Sibia**, **Black-headed Sibia**, **Black-throated Sunbird** and **Red Crossbill**. There are excellent areas of remnant montane and pine forests within a short drive of the hill-resort city of Đà Lạt, where the majority of the endemics and specialities occur. The verdant montane forest on the Di Linh Pass, a two-hour drive from Đà Lạt, is also a great area to connect with some of these endemics.

Le Manh Hung

*The remaining montane forest on the plateau is home to the endemic **Grey-crowned Crocias**.*

Birdwatching Sites

Mount Liang Biang

Reaching an elevation of 2,167m asl and located about 12km from Đà Lạt, this isolated mountain forms part of the Bidoup-Núi Bà National Park. A well-paved road passes through coniferous forest on its lower reaches, where specialities such as the **Vietnamese Greenfinch** and **Red Crossbill** (ssp. *meridionalis*) can be seen. The montane evergreen forest accessed from the narrow forest trail leading to the summit is a good place to look for one of Đà Lạt's trickiest endemics, the splendid **Collared Laughingthrush**. Other specialities on the trail include the **Grey-bellied Tesia**, **Black-crowned Fulvetta**, **Vietnamese Cutia** and **Yellow-billed Nuthatch**.

Ta Nung Valley

A forested valley on the road from Đà Lạt to Ta Nung, usually simply referred to as Ta Nung, is located 14km from Đà Lạt city. This is the best place to find the localized **Grey-crowned Crocias**, along with a host of other local specialities such as the **Indochinese Barbet**, **Indochinese Green Magpie**, **Black-hooded**, **White-cheeked** and **Orange-breasted Laughingthrushes**, and a number of distinctive subspecies, such as the **Blue-winged Minla** (ssp. *orientalis*), **Rufous-backed Sibia** (ssp. *eximia*) and **Black-throated Sunbird** (ssp. *johnsi*), some of which are likely to be split in the future. Spotlighting here may yield **Hodgson's Frogmouth**, and there is also the possibility of the **Spot-bellied Eagle-Owl**.

Other species of interest include the **Mountain Hawk-Eagle**, **Pin-tailed Green**

Yann Muzika

*The beautiful **Collared Laughingthrush** is one of the trickiest endemics to see on the plateau.*

James Eaton

*The distinctive subspecies (ssp. annamensis) of the **Black-throated Bushtit** is frequently encountered in noisy flocks along the the roadside forest at Di Linh.*

Abdelhamid Bizid

*The endemic **Vietnamese Greenfinch** is readily seen in the pine forest on the plateau.*

Pigeon, **Black-headed Sibia** and **Black-headed Parrotbill**. During the northern winter various flycatchers, including **Blue-and-white** and **Mugimaki Flycatchers**, may be seen. However, due to extensive forest clearance and disturbance in the Ta Nung Valley, this site may not last for long.

Tuyến Lâm Lake

This large, man-made lake is surrounded by extensive hilly areas of pine forest, and is located a few kilometres from the centre of Đà Lạt. The pines around the lake shore are a good place to look for the **Oriental Hobby**, **Indochinese Cuckooshrike**, **Slender-billed Oriole**, **Vietnamese Greenfinch** and distinctive subspecies of the **Red Crossbill** (ssp. *meridionalis*) and **Black-throated Bushtit** (ssp. *annamensis*). The patches of remnant evergreen forest higher up are home to the **Bar-backed Partridge**, **Silver Pheasant**, **Red-vented Barbet**, **Indochinese Green Magpie**, and **Black-hooded**, **White-cheeked** and **Collared Laughingthrushes**. The sought-after **Grey-crowned Crocias** and **Yellow-billed Nuthatch** can also be found here.

Sam Thuong

*The pine forest on the slopes of Mount Liang Biang is home to the endemic **Vietnamese Cutia**.*

Sam Thuong

*The **Yellow-billed Nuthatch** has a disjunct distribution in south-central Vietnam and China.*

James Eaton

*The **White-cheeked Laughingthrush** is one of the more readily encountered laughingthrushes in Vietnam.*

Di Linh Highlands

About 80km from Đà Lạt city, the road passes through some excellent areas of montane evergreen forest. This mountain pass is known as Đeo Suoi Lanh (it is often mistakenly referred to in old birdwatching reports as 'Deo Nui San'), and lies on the main road from Di Linh to Phan Thiết. This site is one of the best places to look for the endemic **Orange-breasted Laughingthrush**. Other specialities include the **Red-vented Barbet**, **Blue Pitta**, **Dalat Shrike-babbler**, **Indochinese Green Magpie**, **Black-hooded Laughingthrush**, **Black-headed Parrotbill** and **Green Cochoa**. Large mixed flocks are regularly seen here and should be checked for various sibias, woodpeckers, babblers (like the **Red-billed Scimitar Babbler**) and the **Long-tailed Broadbill**.

Access & Accommodation

Đà Lạt is about 300km from Ho Chi Minh City and can be easily reached by road via Highway QL 20, by rented car or public bus services. There are also regular flights connecting Đà Lạt with Hồ Chí Minh City. There are plenty of hotels and guesthouses in Đà Lạt city, while Di Linh offers a limited number of guesthouses.

Conservation

One of the greatest threats to the forests of the Đà Lạt highlands is shifting cultivation. In addition to forest loss, the associated fires promote the spread of coniferous forest dominated by *Pinus kesiya*, which have lower biodiversity value. Another major threat is charcoal production and fuel-wood collection, which has already led to the destruction of large tracts of evergreen forest on Mount Langbiang. Other threats to biodiversity on the Đà Lạt Plateau include the over-exploitation of non-timber forest organisms such as orchids and birds, which are sold freely in Đà Lạt city.

Cát Tiên National Park

Yong Ding Li

Nguyen Manh Ha

Bau Sau features an extensive network of lakes and marshland that are regularly visited by the Green Peafowl.

KEY FACTS

Nearest Major Town
Tân Phú (Đông Nai)

Habitats
Lowland evergreen forests (including forest on limestone), bamboo groves, wetlands

Key Species
Germain's Peacock-Pheasant, Green Peafowl, Siamese Fireback, Orange-necked and Green-legged Partridges, Woolly-necked and Painted Storks, Lesser Adjutant, Ashy-headed Green Pigeon, Great Hornbill, Red-vented Barbet, Great Slaty, Pale-headed, Heart-spotted and Black-and-buff Woodpeckers, Collared Falconet, Bar-bellied and Blue-rumped Pittas, Grey-faced Tit-Babbler, Golden-crested Myna
Winter Black Baza, Grey-faced Buzzard, Grey-headed Lapwing, Pale-legged Leaf Warbler, Blunt-winged Warbler

Other Specialities
Pygmy Slow Loris, Black-shanked Douc, Northern Pig-tailed and Stump-tailed Macaques, Indochinese Lutung, Buff-cheeked Gibbon, Large-toothed Ferret Badger, Yellow-throated Marten, Asiatic Black Bear, Asian Elephant, Gaur, Sambar, Lesser Mousedeer, Siamese Crocodile

Best Time to Visit
December–April; rainy season May–October.

Cát Tiên National Park protects some of the last major areas of lowland rainforest in southern Vietnam, and its proximity to Hồ Chí Minh City has made it particularly popular among visiting birdwatchers. Some parts of the park are covered in dense bamboo thickets and scrub, a legacy of logging and defoliation during the Vietnam War. As a result of this mosaic of habitats, bird diversity is particularly high, with nearly 350 species recorded although some, like the **White-winged Duck** and **White-shouldered Ibis**, are very rare and localized in sectors of the park seldom visited by birdwatchers (like Cát Loc). However, many of Vietnam's most sought-after species, including **Germain's Peacock-Pheasant** and **Bar-bellied Pitta** can be easily seen around the park's headquarters. With some luck and effort, it is possible to see the localized **Orange-necked Partridge**, and Cát Tiên is one of the most accessible places in mainland Southeast Asia to find the **Green Peafowl**, undoubtedly Asia's most spectacular pheasant.

Birdwatching Sites

Trails around Park Headquarters

A well-maintained network of trails around the park headquarters provides easy access to lowland and limestone forests. One of the most popular trails is the

Wong Lee Hong

*The **Bar-bellied Pitta** is locally common at Cát Tiên.*

Sam Thuong

***Germain's Peacock-Pheasant** is often heard on the trail leading to Bau Sau. Lake.*

Zhang Ming

*The **Pale-headed Woodpecker** is best seen in the dense bamboo thickets in the park.*

Lagerstroemia Trail, which leads into excellent forest. Most birdwatchers encounter their first **Blue-rumped** and **Bar-bellied Pittas** while quietly walking along this trail. Also keep an eye on the many semi-bare trees visible from the trail, which often play host to perching **Collared Falconets**, **Golden-crested Myna** or flocks of noisy **Red-breasted Parakeets**. The shy Buff-cheeked Gibbons are frequently heard along this trail, but seeing them well is tough.

Forest Track to Bau Sau

This wide, unpaved track running from the park headquarters past Heaven's Rapids to the Bau Sau area follows the Dong Nai River to the east for some distance, before entering large stands of secondary riverine forest. Groups of **Red Junglefowl**, **Siamese Firebacks** and occasionally **Blue-winged Pittas** can be seen along the track, even from a vehicle. Alternatively, birdwatchers can hike for a few kilometres and slowly bird along the

Yong Ding Li

Lowland evergreen forest on limestone at the Lagerstroemia trail in Cat Tien National Park.

track, from where **Blue-bearded Bee-eaters**, **Black-and-red Broadbills**, starlings, various woodpeckers, bulbuls and babblers may be seen.

Bau Sau Lake & Trail

A 30-minute drive from the park headquarters takes birdwatchers to the start of this 5km-long trail, which snakes its way through limestone forest, before entering a clearing near Bầu Sấu (Crocodile Lake). The forest along the trail is excellent for **Germain's Peacock-Pheasant**, **Green-legged Partridge**, pittas and a wide range of forest birds. Keep an eye on the numerous mixed flocks that occur, which contain anything from **Black-and-buff Woodpeckers** to **Grey-faced Tit-Babblers**, minivets, woodshrikes, **Great Iora** and, in winter, various *Phylloscopus* warblers. The Bầu Sấu observation hide at the lakeside is excellent for watching waterbirds, including three species of stork (**Painted** and **Woolly-necked Storks**, and the **Lesser Adjutant**), the **Bronze-winged Jacana**, **Asian Golden Weaver** and, in the evening, small parties of **Green Peafowl** that come to feed and drink at the lake.

Đac Lua

An area of bamboo-covered hills, Đắc Lua is best known for the **Orange-necked Partridge**, a southern Indochinese speciality that is best seen in Cát Tiên, or across the border in eastern Cambodia's Mondulkiri Province (for example at Seima). While the partridge is readily heard, getting a good view is an entirely different matter. Other specialities include the bamboo-loving **Pale-headed Woodpecker** and **Large Scimitar Babbler**. A small pond here may draw waterbirds such as wintering **Grey-headed Lapwings**.

Patrawut Sitifong

*The diminutive **Collared Falconet** regularly perches on tall, bare trees.*

James Eaton

*Cát Tiên is a stronghold for the endangered **Buff-cheeked Gibbon**.*

Wang Bin

*The **Blue-rumped Pitta** is often seen in the forests around the park headquarters.*

Access & Accommodation

Cát Tiên National Park is about 174km from Hồ Chí Minh City, and can be reached by public transport. Regular buses heading north to Đà Lạt pass by the post office in Tân Phú town, from where motorcycles or taxis for the 24km journey to the park can be hired. As the Đồng Nai River separates the park from the surrounding countryside, a barge service regularly shuttles visitors across the river. Accommodation in the form of chalets is available at the park headquarters, from where permits to enter the park can be obtained.

Conservation

Despite the bad publicity suffered in recent years from the poaching of the last Javan Rhinoceros in Vietnam and effectively mainland Southeast Asia, Cát Tiên is relatively well protected as a national park. The park is divided into multiple management zones patrolled by rangers, and is partly surrounded by a buffer zone. Some encroachment and poaching occur in the more remote sections of the park (like Cát Loc Sector).

Recommended Reading

Southeast Asia Region

Arlott, N. (2016) *Collins Field Guide: Birds of South-east Asia*. William Collins. 448 pp.

Robson, C. (2014) *Birds of Southeast Asia*. 2nd edn. Christopher Helm. 544 pp.

Lee, W-S., Choi, C-Y. & Kim, H-K. (2018). *Field Guide to the Waterbirds of ASEAN*. ASEAN-Korea Environmental Cooperation Unit.

Brunei Darussalam

Myers, S. (2016) *Birds of Borneo*. Christopher Helm. 336 pp.

Phillipps Q. & Phillipps K. (2014) *Phillipps' Field Guide to the Birds of Borneo: Sabah, Sarawak, Brunei and Kalimantan*. John Beaufoy Publishing. 372 pp.

Cambodia

Goes F. (2014) *The Birds of Cambodia: An Annotated Checklist*. Fauna & Flora International Cambodia Programme. 504 pp.

Indonesia

Coates, B. J. & Bishop, K. D. (1997) *A Guide to the Birds of Wallacea: Sulawesi, the Moluccas and Lesser Sunda Islands, Indonesia*. Dove Publications. 535 pp.

Eaton, J. A., van Balen, B., Brickle, N. W. & Rheindt, F. E. (2016) *Birds of the Indonesian Archipelago: Greater Sundas and Wallacea*. Lynx Edicions. 496 pp.

MacKinnon, J. & Phillipps, K. (1993) *A Field Guide to the Birds of Borneo, Sumatra, Java and Bali*. Oxford University Press. 496 pp.

Strange, M. (2012) *A Photographic Guide to the Birds of Indonesia*. Periplus Editions. 544 pp.

New Guinea

Pratt, T.K. & Beehler, B.M. (2014) *Birds of New Guinea*. 2nd edn. Princeton University Press. 528 pp.

Gregory, P. (2017) *Birds of New Guinea Including Bismarck Archipelago and Bougainville*. Lynx Edicions. 464 pp.

Malaysia

Davison, G. & Yeap, C. A. (2013) *A Naturalist's Guide to the Birds of Malaysia*. John Beaufoy Publishing. 176pp.

Jeyarajasingham, A. & Pearson, A. (2012) *A Field Guide to the Birds of Peninsular Malaysia and Singapore*. Oxford University Press. 628 pp.

MNS Conservation Council (2015) *A Checklist of the Birds of Malaysia. Second Edition*. Malaysian Nature Society. 60 pp.

Myers, S. (2016) *Birds of Borneo*. Christopher Helm. 336 pp.

Phillipps Q. & Phillipps K. (2014) *Phillipps' Field Guide to the Birds of Borneo: Sabah, Sarawak, Brunei and Kalimantan*. John Beaufoy Publishing. 372 pp.

Wong, T. S. (2018) *A Naturalist's Guide to the Birds of Borneo*. Third Edition. John Beaufoy Publishing. 176 pp.

Myanmar

Rasmussen, P. C. & Anderton J. C. (2012) *Birds of South Asia: The Ripley Guide*. 2nd edn. Lynx Edicions. 1072 pp.

Smythies, B. E. (2001) *The Birds of Burma*. Natural History Publications (Borneo). 601 pp.

The Philippines

Kennedy, R. S., Gonzales, P. C., Dickinson, E., Miranda, H. & Fisher, T. (2000) *A Guide To the Birds of the Philippines*. Oxford University Press. 368 pp.

Tañedo, M., Hutchinson, R. O., Constantino A. & Constantino, T. (2015) *A Naturalist's Guide to the Birds of the Philippines*. John Beaufoy Publishing. 176 pp.

Singapore

Lim, K. S. (2009) *The Avifauna of Singapore*. Nature Society (Singapore). 500 pp.

Yong, D. L., Lim, K. C. & Lee, T. K. (2017) *A Naturalist's Guide to the Birds of Singapore*. Third Edition. John Beaufoy Publishing. 176 pp.

Thailand

Lekagul, B. & Round, P. D. (1991) *A Guide to the Birds of Thailand*. White Lotus. 457 pp.

Robson C. (2016) *A Field Guide to the Birds of Thailand*. Bloomsbury Publishing. 272 pp.

Round, P. D. (2000) *Field Check-List of Thai Birds*. Bird Conservation Society of Thailand, Bangkok.

Round, P. D. & Limparungpatthanakij, W. (2018) *A Naturalist's Guide to the Birds of Thailand*. John Beaufoy Publishing. 176pp.

Timor-Leste

Crosby M. (2007) *Important Bird Areas in Timor-Leste: Key Sites for Conservation*. BirdLife International. 86 pp.

Trainor, C. T., Coates, B. J. & Bishop, K. D. (2007) *The Birds of Timor-Leste/As Aves de Timor-Leste*. Dove Publications. 113 pp.

Other useful references on Southeast Asian Biodiversity

Francis, C.M. (2008) *A Guide to the Mammals of Southeast Asia*. Princeton University Press. 392 pp.

Das, I. (2010) *A Field Guide to the Reptiles of South-East Asia*. New Holland. 376 pp.

Acknowledgements

A book of this nature and extent would not be possible without the dedicated effort of many individuals. First and foremost, we are indebted to the team of authors who agreed to contribute to the book without hesitation despite their busy schedules. Many of our friends and colleagues put their best images at our disposal and we are thankful for their support. They include Ah Kei, Arman A, Shahril Ahmad, John Ashish, Woraphot Bunkhwamdi, Cheng Heng Yee, Jimmy Chew, Miguel David de Leon, James Eaton, Gerard Francis, Paul French, Sreedharan Gopalsamy, Firoz Hussain, Marcel Holyoak, Markus Handschuh, Ho Kin Yip, John and Jemi Holmes, Mike Kilburn, Koel Ko, Lee Tiah Khee, Jennifer Leung, Lim Kim Chuah, Lim Kim Keang, Alpian Maleso, John Mathai, Nguyen Manh Ha, Thang Nguyen, Bjorn Olesen, Mick Price, Bim Quemado, Quek Oon Hong, Greg Roberts, Benjamin Schweinhart, Dave Sergeant, Untung Sarmawi, Gloria Seow, Patrawut Sitifong, So Wai Ming David, Dubi Shapiro, Tan Ju Lin, Tan Gim Cheong, Myron Tay, Teo Nam Siang, Sam Thuong, Tong Menxiu, Uthai Treesucon, Daniel López Velasco, Philippe Verbelen, Vichit Viriyautsahakul, Lorenzo Vinciguerra, Wang Bin, Jason Wong, Michelle and Peter Wong, Wong Tuan Wah, Arne Wuensche, Francis Yap, Mohammad Zahidi, Zhang Ming and Robert Zhao. Abdelhamid Bizid, Yann Muzika, David Wilcove and Colin Trainor all gave useful input on the manuscript. Finally, we are thankful to our publisher, John Beaufoy, for putting yet another exciting idea into print and Rosemary Wilkinson for coordinating this project.

Conservation & Birdwatching Organizations

Southeast Asia Region
BirdLife International Asia (birdlife.org/asia)
Fauna & Flora International (fauna-flora.org)
Oriental Bird Club (orientalbirdclub.org)
The Nature Conservancy (nature.org)

Brunei Darussalam
Borneo Bird Club (borneobirdclub.blogspot.com)

Cambodia
Angkor Centre for the Conservation of Biodiversity (accb-cambodia.org)
NatureLife Cambodia (naturelifecambodia.org)

Indonesia
Burung Indonesia (burung.org)
Cikanaga Wildlife Centre (cikanangawildlifecenter.com)
Friends of the National Parks Foundation (fnpf.org)

Malaysia
Malaysian Nature Society (mns.my)
Borneo Bird Club (borneobirdclub.blogspot.com)
Wild Bird Club Malaysia (facebook.com/Wild-Bird-Club-Malaysia 910076699049453/)

Myanmar
Biodiversity and Nature Conservation Association (banca-env.org)
Myanmar Bird and Nature Society (myanmarbirdnaturesociety.com)

The Philippines
Haribon Foundation (haribon.org.ph)
Katala Foundation Inc (philippinecockatoo.org)
Wild Bird Club of the Philippines (birdwatch.ph)
Philippine Biodiversity Conservation Foundation (pbcfi.org.ph)

Singapore
Nature Society (Singapore) (nss.org.sg)

Thailand
Bird Conservation Society of Thailand (bcst.or.th)
anna Bird and Nature Conservation Club (lannabird@gmail.com)

Vietnam
Vietnam Birdwatching Club (hanoibirdclub2006@yahoo.com)
Viet Nature Conservation Centre (thiennhienviet.org.vn)

Facebook Groups, Blogs & Birdwatching Reports

Global Birding Trip Reports
Birdtours (birdtours.co.uk)
Cloudbirders (cloudbirders.com)
Fat Birder (fatbirder.com)
Surfbirds (surfbirds.com)

Cambodia
Birds of Cambodia Education & Conservation (facebook.com/groups/birdsofcambodia/?fref=ts)

Indonesia
Birding Bali and Indonesia (facebook.com/birdingbaliandindonesia/?fref=ts)
Birding Sumatera (facebook.com/groups/432398480272850/)
Pengamat Burung Indonesia (facebook.com/groups/PengamatBurungIndonesia/)
Burung Nusantara (burung-nusantara.org)

Malaysia
Borneo Bird Club (facebook.com/groups/40520417236/)
Johor Birders (facebook.com/groups/320474148146791/)
MNS Selangor Branch Bird Group (facebook.com/groups/sbbgbirdgroup/?fref=ts)
Malaysia Birding (malaysiabirding.org)

Myanmar
Myanmar Bird & Nature Society (facebook.com/groups/myanmarbirdnature/)

Philippines
Wild Bird Club of the Philippines (facebook.com/BirdwatchPhilippines/?fref=ts)

Singapore
Bird Sightings (facebook.com/groups/birdsightings/)
Singapore Birders (facebook.com/groups/singaporebirdgroup/)

Thailand
Thai Bird Report (facebook.com/groups/thaibirdreport/)

Vietnam
Vietnam Bird News (vietnambirdnews.blogspot.com)

Other Birdwatching Resources

Bird Sounds
Xeno-canto (xeno-canto.org)
AVoCet (avocet.zoology.msu.edu)
Macaulay Library of Natural Sounds (macaulaylibrary.org)

Bird Image Repositories
Borneo Bird Images (borneobirdimages.com)
Oriental Bird Images (orientalbirdimages.org)
China Wild Bird Images (cnbird.org.cn)

Bird Records Submission
Ebird (ebird.org)

Birdwatching Tours & Related Services

Regional
Birdtour Asia (birdtourasia.com)
Birdquest (birdquest-tours.com)
Rockjumper Birding Tours (rockjumperbirding.com)
Tropical Birding (tropicalbirding.com)
Victor Emmanuel Nature Tours (ventbird.com)

Cambodia
Sam Veasna Centre (samveasna.org)
Cambodia Bird Guide Association (birdguideasso.org)

Indonesia
Aceh Birder (acehbirder.com)
Bali Bird Walk (balibirdwalk.com)
Birding Kerinci (facebook.com/birdingkerinci)
Birdpacker (birdpacker.com)
Ecosafari Indonesia (ecosafariindonesia.com)
Jakarta Birder (facebook.com/JakartaBirder)
Flores Birding (floresbirdwatching.com)
Malang Birding Tour (malangbirdingtour.wix.com)
Malia Tours (malia-tours.com)
Magnificus Expeditions (magnificusexpeditions.com)
Sultan Birding Tours (sultan-birding.com)

Laos
Nam Et-Phou Louey Eco-tours (namet.org/wp/namnern/)

Malaysia
Birding in Borneo (birdingborneo.com)
Birding in Kinabatangan (birdinginkinabatangan.com)
Bird Malaysia (bird-malaysia.com)
Borneo Birds (borneobirds.com)
Endemic Guides (birdtourmalaysia.com)
Jungle Walla (junglewalla.com)
Penang Birder (penangbirder.blogspot.com)
Nature2Pixel (nature2pixel.my)
Wild Borneo Photography (wildborneo.com.my)

Myanmar
Myanmar Birding (myanmarbirding.com)
River Mekong Travel and Tours (birdingmyanmar.com)
SST Tourism (ssttourism.com)
Wildbird Adventures Travel and Tours (wildbirdtt.com)

The Philippines
Birding Adventure Philippines (birdingphilippines.com)
Birdguiding Philippines (birdguidingphilippines.com)
Birding in Mindanao (birdingmindanao.com)

Singapore
Birding Singapore (birdingsingapore.com)

Thailand
Nature Trails (naturetrails-thailand.com)
Nature & Bird Site Exploration (thailandbirdwatching.com)
South Thailand Birding (souththailandbirding.com)
Thai Birding (thaibirding.com)
Wild Bird Eco Tour (wildbirdeco.net)

Vietnam
Vietnam Birding (vietnambirding.com)
Vietnam Wildlife Tours and Research (vietnamwildtour.com)

Contributing Authors

Agus Nurza Zulkarnain

Agus Nurza is a veterinarian by profession. He started off as a birdwatcher in his hometown of Aceh in northern Sumatra, Indonesia. Since 2007 he has been involved in many studies on the ecology and distribution of Sumatra's bird species, and has published numerous reports and papers. More recently, Agus and his colleagues founded Aceh Birder to develop and promote birdwatching tours in the north Sumatra region. He is an accomplished bird photographer.
Email: agus.nurza@gmail.com

Chairunas Adha Putra

Chairunas Adha Putra is currently pursuing his master's degree at Bogor Agriculture University, where he conducts research on waterbirds. Since 2008 his interest in photography, birds and herpetofauna has taken him to many remote regions in Sumatra. He hopes to promote the east coast of the North Sumatra region as a major site for waterbird research, while raising the conservation profiles of waterbirds to the wider public.
Email: chairunasadha@ymail.com

Hanom Bashari

A highly experienced field ornithologist, Hanom worked for the Wildlife Conservation Society – Indonesia Program (WCS-IP) in Sulawesi from 2000 for five years, during which he carried out extensive biodiversity surveys across northern, central and southern Sulawesi. Since then he has worked for Burung Indonesia (Indonesia's national partner for BirdLife International), and has travelled widely across Indonesia to conduct surveys of birds and biodiversity assessments.
Email: h.bashari@burung.org

David Blair

A forest ecologist working in Australia with the Australian National University, David Blair is studying birds, possums and gliders, and post-fire vegetation recovery. David has worked as a nature photographer and writer for several years, covering threatened species and habitats in Indonesia and Australia. He has co-authored two books and more than 20 scientific papers on the ecology of Victoria's Mountain Ash forests, where he works.
Email: david.blair@anu.edu.au

Heru Cahyono

Heru Cahyono has more than 20 years of field experience on the birdlife and biodiversity of eastern Java. Heru graduated from the State University of Malang in 2011 with a degree in biology. Besides bird conservation, Heru also runs Malang Birding Tours to promote ecotourism in east Java. He has written two books on the birdlife of the region, including the *Field Guide to Raptors in East Java* (in Indonesian), published by BKSDA JATIM in 2013, and is currently working on *The Birds of Malang*.
Email: hc_garuda@yahoo.com

Con Foley

Con Foley has lived in Singapore for nearly 30 years. He has been interested in nature and photography since his childhood. After a long career in the corporate world, Con is now finding the time to pursue his passions again. He is the past president of the Nature Photographic Society (Singapore), and continues to teach bird and nature photography actively, on top of conducting birdwatching and photography tours. Besides a number of articles, Con's massive collection of images across the region has been widely published.
Email: con@confoley.com

Frederic Goes

An independent conservationist and ornithologist, Frederic Goes is currently based in Corsica, France. In 1994–2006, he lived and travelled extensively in Cambodia, conducting pioneering surveys and conservation work, as well as guiding and training a new generation of local ornithologists. During this period he worked for the Wildlife Conservation Society (WCS). Frederic is also the founding editor of *Cambodia Birding News*. Aside from the numerous reports, papers and books he has written, Frederic's passion for Cambodia's avifauna led him to write the country's annotated checklist.
Email: fredbaksey@yahoo.com

Ayuwat Jearwattanakanok

Ayuwat is a wildlife artist and birdwatcher based in Chiang Mai, Thailand. Ayuwat is a committee member of the Lanna Bird and Nature Conservation Club (LBNC) and Bird Conservation Society of Thailand (BCST), where he leads a number of conservation projects. Ayuwat is now working as an independent artist and is the sole illustrator of the upcoming *Raptors of Thailand (& South-East Asia)* and *A Handbook of Shorebirds of the World* books.
Email: ayuwje@gmail.com

Le Manh Hung (Dr)

Professional ornithologist and conservationist Le Manh Hung is based in Hanoi, Vietnam. He completed his doctorate at the Institute of Ecology and Biological Resources in Hanoi, where he studied raptor migration in Vietnam and the wider Asian region. He currently manages the Zoological Museum at the Vietnam Academy of Science and Technology. Hung also founded the Vietnam Birdwatching Club (VBC). In addition to being an experienced birdwatching guide, Hung is an accomplished bird photographer. Besides numerous papers, Hung wrote and illustrated the *Photographic Guide to the Birds of Vietnam* (in Vietnamese), the first such guide to the birdlife in the country.
Email: hungniltava@gmail.com

Ch'ien C. Lee
Originally from California, Ch'ien C. Lee has lived in Borneo since 1996, where he is an active member of the Malaysian Nature Society and has served on its bird records committee. A highly experienced field naturalist, Ch'ien regularly carries out bird and plant surveys in remote parts of Sarawak. Since 2005, Ch'ien has worked as a full-time freelance wildlife photographer, and travels widely across Malaysia and Indonesia to document the region's many endemic birds, on top of leading occasional birdwatching tours to remote parts of Borneo and elsewhere in Indonesia.
Email: mail@wildborneo.org.my

Cheong Weng Chun
Weng Chun's professional background is in parks management and conservation, having majored in Parks and Recreational Management for his degree at Universiti Putra Malaysia (UPM). Shortly after graduation he went on to manage the Paya Indah Wetland Sanctuary and Taman Wetland Putrajaya, two of the most important wetlands around Kuala Lumpur. An avid nature lover, Weng Chun has been birdwatching for nearly 20 years throughout Peninsular Malaysia and surrounding countries. He is currently a freelance bird guide based in Kuala Lumpur.
Email: wengchun@gmail.com

Lim Kim Seng
Lim Kim Seng has been birdwatching since the age of 10 and has more than 40 years experience across Southeast Asia. He received his master's degree in environmental management from the University of Adelaide and is currently an associate lecturer at Singapore's Republic Polytechnic. He is also a professional bird guide, having guided birdwatchers in Singapore and Malaysia since 1992. Kim Seng has authored more than 10 books on birds and has been an active member of Nature Society (Singapore) since 1975. Kim Seng is currently Singapore's top lister, with 362 species.
Email: ibisbill@yahoo.com

Low Bing Wen
A terrestrial ecologist based in Singapore, Low Bing Wen has a particular interest in birds. He started birdwatching at the age of 10 and has since amassed more than 20 years of field experience travelling around Asia and Australasia to document the biodiversity of these regions. He majored in zoology and conservation biology at the University of Western Australia, and has also been involved in various conservation initiatives, including assisting with the eradication of Black Rats and feral cats on Christmas Island (Australia), and conducting research on the pet-bird trade in Southeast Asia. He has published more than 30 scholarly and popular articles involving the biodiversity of Southeast Asia and Australia.
Email: halmaherastandardwing@gmail.com

Yann Muzika
Yann Muzika is a French birdwatcher and wildlife photographer currently residing in Hong Kong who has been based in Asia for almost 20 years. He regularly contributes his images to the Oriental Bird Image database, as well as a number of online media, books and magazines. Yann has travelled widely across Asia to document the continent's birdlife. Besides his photographic work, Yann contributes to a number of bird-conservation projects in Bangladesh, China and Indonesia.
Email: yannmuzika@mac.com

Pete Simpson
Pete Simpson has been birdwatching around the World since the 1980s, and started taking a serious interest in the birdlife of Mindanao in 2006. Now based in Davao City, he is an active member of the Wild Bird Club of the Philippines and regularly works with the regional Department of Environment and Natural Resources and with the Philippine Eagle Foundation to train forest rangers and birdwatching guides. Pete is an experienced bird guide and continues to explore new sites on Mindanao in search of its endemic birds.
Email:petesimpson@birdingmindanao.com

Thet Zaw Naing
Having obtained his master's degree in zoology at the University of Yangon in 1999, Thet Zaw Naing has since participated in and coordinated many monitoring and research programmes on Myanmar's birdlife. He is not only a highly qualified ornithologist but also a dedicated conservationist. He has been working for the Wildlife Conservation Society (WCS), Myanmar Program as a Bird Conservation technician since 2009. Thet is the current secretary of the Myanmar Bird and Nature Society.
Email: thetzawnaing@gmail.com

Colin R. Trainor (Dr)
Colin Trainor conducted his doctoral work through the Charles Darwin University on the conservation ecology of wildlife on Timor and its neighboring islands. Colin has been visiting Wallacea since 1997, initially as part of a conservation planning program with Burung Indonesia (BirdLife International, Indonesia Programme) on Flores, West Timor and Sumbawa. He has a keen interest in avian taxonomy, as well as in the conservation status and elevation and biogeographic patterns of Wallacean birds. To date, in addition to the *Field Guide to the Birds of Timor-Leste*, he has published more than 50 articles and reports based on his work in Wallacea.
Email: halmahera@hotmail.com

Mark J. Villa
Mark is a professional birdwatching guide based in Manila and has been guiding and organizing birdwatching tours since 2015. Mark has always had a deep interest for nature and wildlife since his childhood days. He started birdwatching seriously in 2003, and has actively explored many parts of the Philippines, including some of the remotest islands in the archipelago, as well as neighbouring parts of Southeast Asia, China and Europe. To support local conservation, Mark was involved in the Wild Bird Club of the Philippines as a volunteer, besides having served on its bird records committee.
Email: markjason@gmail.com

Yeo Siew Teck
Considered one of the most experienced bird guides in Borneo, Yeo Siew Teck's interest in birdwatching started in the 1990s. He has expanded this passion into an occupation – a travel agency that offers bird-tour packages in Borneo. Siew Teck is also active in bird conservation and has helped develop birdwatching tourism in Sarawak. A founding member of the Malaysian Nature Society's Kuching branch bird group, he has been its assistant coordinator since its formation in 2007, and also immediate past vice-chairman of the MNS Kuching branch. He is a regular speaker and trainer on birdwatching and birdwatching tourism.
Email: yeosiewteck@gmail.com

Yong Ding Li (Dr)
Ding Li carried out his doctoral research at the Australian National University, where he studied the effectiveness of biodiversity indicators. He previously carried out bird, mammal and insect surveys across Peninsular Malaysia and Borneo during his honours research at the National University of Singapore. Ding Li has had a deep interest in birds since his childhood, and has travelled widely across Asia to document the continent's wildlife. He also advises the IUCN Species Survival Commission's bird Red List authority on Southeast Asian birds. Since 2012 he has authored two books on the birdlife of the region, and published more than 40 papers on biodiversity conservation in Asia and Australia. Ding Li currently coordinates conservation and policy for the BirdLife International Regional Office in Singapore for the Asian region.
Email: zoothera@yahoo.com

Mikael Bauer (Dr)
Picking up birdwatching in his native Sweden, Mikael discovered the avian riches of Southeast Asia at the start of the new millennium and subsequently undertook several long trips traveling around the region. Taking up a special interest in Indonesia and its diverse avifauna, Mikael has travelled extensively throughout the country from Sumatra to West Papua. Mikael is a keen photographer and particularly enjoys the night birds of the archipelago. A biochemist by profession, Mikael works on developing enzymes as a green alternative to traditional chemicals.
Email: rhabdornis@yahoo.se

Abdelhamid Bizid (Dr)
Abdel is a French-Tunisian birdwatcher and photographer currently based in Hong Kong. His interests in nature photography developed during a professional stint in England 15 years ago. Abdel has travelled extensively across Asia during his free time, observing and photographing birds, wildlife, unusual plants, as well as looking for new cultural experiences. Abdel's wildlife images have been featured in numerous publications and books on the natural history and birdlife of Asia and North Africa.
Email: abizid@yahoo.com

Robert O. Hutchinson
Robert Hutchinson has been a birdwatcher since an early age, a passion that has taken him across the globe in search of birds, but with a particular passion for the Oriental region. In 2005 he co-founded Birdtour Asia for whom he still organises and guides tours throughout Asia on a full-time basis. Rob has published numerous articles and scientific papers on Asian birds, and co-authored *A Naturalist's Guide to the Birds of the Philippines*. Rob is also a keen sound-recordist and photographer. He currently resides in the Philippines.
Email: rob.birdtourasia@gmail.com

Wich'yanan Limparungpatthanakij
Wich'yanan "Jay" Limparungpatthanakij is a member of Thailand's Bird Records Committee. He is responsible for compiling records for the quarterly bulletin of the Bird Conservation Society of Thailand. He is also a reviewer on eBird. Wich'yanan has a master's degree in biology from Mahidol University. He is actively involved in various projects related to bird and nature conservation, and also regularly leads bird tours in Thailand.
Email: lim.wichyanan@gmail.com

Wong Tsu Shi
Wong Tsu Shi resides in Sabah, Malaysian Borneo, and is an accountant by profession. Tsu Shi is a keen birdwatcher and photographer, and started photographing birds 10 years ago. His bird images have been used in various natural history publications. He is the author and main photographer for *A Naturalist's Guide to the Birds of Borneo*.
Email: wongtshi@gmail.com

Khaleb Yordan
Based in Jakarta, Khaleb has been birdwatching for 10 years and currently works as a full-time bird guide. With over 5 years of guiding experience, Khaleb now runs Jakarta Birder, which organises bird tours in Jakarta and western Java. He is also a keen nature photographer and has documented many of Java's endemic birds during his field trips. Khaleb is also interested in bird conservation and has supported the Indonesian Seabird Survey (ISSUE) with his field expertise.
Email: jakartabirder@ymail.com

Sofian Zack
Formerly from Kuala Lumpur, Sofian is now based in Langkawi where he works as a freelance bird photography guide, on top of his full time job as a restaurant manager. He has pursued his interests in birds and birdwatching for more than 6 years and has a special interest in forest birding. He has also helped to organise and coordinate nature photography exhibitions on the wildlife of Langkawi in his free time.
Email: sofian.zack@yahoo.com

Irene Dy
Based in Manila, Irene Dy developed a keen interest in birdwatching through her experiences at the Wild Bird Club of the Philippines. Through numerous field trips, Irene eventually became interested in the use of photography to document her field observations. Over the years, Irene has explored the Philippine archipelago extensively, and contributed many local

and national records, including documentation of the recently described Bicol Ground Warbler. Combining her passion in birdwatching and photography, Irene hopes to contribute to wildlife and environment conservation in the Philippines.
Email: happyowl.images@gmail.com

Shita Prativi

Shita graduated from the Catholic University of Parahyangan in Bandung and began her professional career as a volunteer with NGOs affording support and education for homeless children and victims of domestic violence. After meeting Papua Bird Club founder Kris Tindige in 2005, Shita joined the Papua Bird Club in the role of legal advisor and began making use of her experiences to develop alternative education programmes for indigenous Papuan communities. Under the tutelage of Kris, Shita began leading wildlife tours across eastern Indonesia. After Kris's unfortunate death, Shita continued his work leading tours full-time and founded Magnificus Expeditions. She is now widely regarded as a premier bird tour leader in eastern Indonesia. Although much of her time is spent on tours, Shita continues to work to support local communities both through employment and income opportunities with her non-profit wing Papua Konservasi dan Komunitas.
Email: info@papuabirdclub.com

Carmela Espanola

Carmela started her career as a volunteer with the Philippine Endemic Species Conservation Project in Panay. After a year of volunteer work in Jolo, she took on an exploratory project on the island for Fauna and Flora International, which inspired the thesis she wrote on the birds of Capual Island for her MSc. While doing graduate work under the Russell E. Train Education for Nature Fellowship of the World Wide Fund for Nature, Carmela undertook several volunteer jobs, one of which resulted in the discovery of the Calayan Rail, a new species of bird endemic to Calayan Island. She also co-founded the Wild Bird Club of the Philippines, now the fastest growing and most active birdwatching organization in the country. Carmela then joined the Institute of Biology, University of the Philippines Diliman in 2006. Upon completing her PhD, she joined the board of Christians in Conservation. Carmela is also a country representative of the Oriental Bird Club.
Email: pespanola@up.edu.ph

Bim Quemado

Bim Quemado has travelled the Philippines extensively for birdwatching and bird photography since 1992, and is particularly interested in the birds of the southern Philippines. Through his field trips in the region, Bim has collected images of many of the rarest species in the Philippines, especially the Sulu Hornbill. Although Bim works full-time in the military, he is also a very keen conservationist and has worked with local communities in the southern Philippines to promote biodiversity conservation.
Email: grunt_92pmc@yahoo.com

Pyae Phyo Aung

As a program manager under the Biodiversity and Nature Conservation Association (BANCA), Pyae Phyo Aung is heavily involved in coordination work for various conservation projects in the East Asian-Australasian Flyway. Pyae Phyo also coordinates conservation activities of the Spoon-billed Sandpiper in Myanmar as the national representative to the Spoon-billed Sandpiper Task Force. He has been involved in the conservation of the Spoon-billed Sandpiper since 2010 and in 2013 took over management of BANCA's Spoon-billed Sandpiper conservation programme. He also played a major role in leading successful efforts to establish the Gulf of Mottama as a Ramsar Site. Pyae Phyo is currently on the Ramsar Convention Panel as National NGO Focal Point for Communication, Education and Public Awareness. He also has the role of Sub-Regional Representative for Asia (World Wetland Network).
Email: pyaephyo.banca.org@gmail.com

Anuar McAfee

Anuar McAfee has lived in Terengganu since 1991. Over the years he has developed an appreciation for the diverse flora and fauna found throughout the state. A pioneer of birdwatching in Terengganu, Anuar shares his passion with others by leading nature workshops for students and running ecotourism events for the state. In 2017, he published the first fully bilingual guide to the birds of Terengganu, 'Birds of Terengganu/Burung-burung di Negeri Terengganu', which highlights birding locations throughout the state, together with a complete checklist. Anuar currently teaches at Universiti Sultan Zainal Abidin.
Email: anuar.mcafee@gmail.com

Ingkayut Sa-ar

Ingkayut Sa-ar was born in Satun, Thailand. He has been a birdwatcher since an early age, and is especially interested in the ornithology of Peninsular Thailand. Ingkayut started his career as a conservation officer at the Bird Conservation Society of Thailand (BCST), where he coordinated many conservation projects in southern Thailand. Ingkayut currently works as a birding guide at Nature and Bird Site Exploration Co. Ltd., Thailand. Drawing from his diverse experiences, Ingkayut has written a number of articles on the birdlife of the Peninsula, and is responsible for compiling bird records in southern Thailand for the BCST bulletin.
Email: ingkayutsa@gmail.com

Janina Bikova

Janina works for the Wildlife Conservation Society's Laos Ecotourism for Nam Et-Phou Louey National Protected Area. She works closely with other stakeholders, especially local communities to promote sustainable ecotourism in the NEPL landscape.
Email: jbikova@wcs.org

Philippe Verbelen

Philippe Verbelen is a Belgium-based birder who has been birdwatching ever since he was capable of lifting a pair of binoculars. For over two decades, Philippe has been actively involved in rainforest conservation projects in the Congo Basin (Central Africa) on behalf of Greenpeace International. He started birding in Southeast Asia more than 25 years

ago and has travelled extensively throughout the region. Philippe has carried out numerous birding expeditions to many of Indonesia's least explored corners. He has a particular interest in recording bird vocalisations and nightbirds, and his work played a major role in the description of a number of new species for science, most recently the Rinjani Scops Owl and the Rote Myzomela.
Email: filip_verbelen@yahoo.fr

Marc Thibault

Marc Thibault is based in France as a project leader for the Tour du Valat Foundation where he works on wetland restoration projects. Outside of his work, Marc is a very keen birdwatcher with a strong fascination for the Oriental region in general, and the remote corners of the Indonesian archipelago in particular.
Email: marc.thibault@freesbee.fr

Carlos N. G. Bocos

Carlos Bocos has spent most of his life looking for birds. After a number of trips across the world and especially in Asia, he started to work as bird guide in 2012, joining BirdTour Asia in 2016. He is a very keen photographer, with hundreds of photos published in all sorts of publications. Currently he spends most of his time in Indonesia.
Email: CNBocos@gmail.com

Choy Wai Mun

Choy Wai Mun was born and raised in Penang. As a teenager, he started birdwatching back in 1990 and it immediately became one of his main passions. It was only after more than a decade that he decided to keep colourful and memorable birding experiencess alive through photography. Presently, Wai Mun considers himself an avid birder with an interest in bird photography. Wai Mun works as a freelance bird guide and regularly blogs about his birding adventures.
Email: peregrine@live.co.uk

Index